The Brushtail Possum

The Brushtail Possum

Biology, impact and management of an introduced marsupial

Edited by T.L. Montague

Manaaki Whenua
PRESS

Manaaki Whenua Press, Lincoln, New Zealand
2000

CATALOGUING IN PUBLICATION

The Brushtail Possum : Biology, impact and management of an introduced marsupial / T. L. Montague, editor. – Lincoln, Canterbury, New Zealand : Manaaki Whenua Press, 2000.

ISBN 0-478-09336-5

I. Montague, T. L. (Thomas L.), 1956-

UDC 599.223.1

Cover design by Kirsty Cullen, Landcare Research, Lincoln
Cover photograph by Keven Drew, Landcare Research, Lincoln
Copy-edited and proof-read by Christine Bezar
Typeset by Orca Publishing Services Ltd, Christchurch
Printed by The Caxton Press, Christchurch

Published by Manaaki Whenua Press, Landcare Research,
PO Box 40, Lincoln 8152, New Zealand, with the co-operation of Science and Research Unit, Department of Conservation, P.O. Box 10-420, Wellington.

Contents

Contributors

N. S. Allan
280 Ohiwa Harbour Road, RD 2, Opotiki,
New Zealand

N. D. Barlow
AgResearch, PO Box 60, Lincoln, New Zealand

K. B. Bicknell
Commerce Division, Lincoln University,
PO Box 84, Lincoln University, New Zealand

S. Butcher
Agriculture New Zealand, PO Box 10864,
Wellington, New Zealand

P. C. Caley
Landcare Research, Private Bag 11052,
Palmerston North, New Zealand

D. P. Choquenot
Landcare Research, PO Box 69, Lincoln,
New Zealand

J. M. Clark
Dudley Road, Inglewood, Taranaki, New Zealand

M. N. Clout
School of Environmental & Marine Sciences,
University of Auckland, PB 92019, Auckland,
New Zealand

J. D. Coleman
Landcare Research, PO Box 69, Lincoln,
New Zealand

P. E. Cowan
Landcare Research, Private Bag 11052,
Palmerston North, New Zealand

R. Cullen
Commerce Division, Lincoln University,
PO Box 84, Lincoln University, New Zealand

T. D. Day
AgResearch Ruakura Centre, Private Bag 3123,
Hamilton, New Zealand

C. T. Eason
CENTOX (Centre for Environmental Toxicology),
Landcare Research, PO Box 69, Lincoln,
New Zealand

M. G. Efford
Landcare Research, Private Bag 1930, Dunedin,
New Zealand

K. Ericksen
Department of Conservation, PO Box 10 420,
Wellington, New Zealand

G. P. Fitzgerald
Fitzgerald Applied Sociology, PO Box 8526,
Riccarton, Christchurch, New Zealand

T. P. Fletcher
Landcare Research, PO Box 69, Lincoln,
New Zealand
Present address: Perth Zoo, PO Box 489, South
Perth 6951, Western Australia

D. D. Heath
AgResearch, PO Box 40063, Upper Hutt,
New Zealand

R. J. Henderson
Landcare Research, PO Box 69, Lincoln,
New Zealand
Present address: Pest Solutions, PO Box 31191,
Ilam, Christchurch, New Zealand

G. J. Hickling
Dept of Entomology and Animal Ecology, Lincoln
University, PO Box 84, Lincoln University,
New Zealand

P. G. Livingstone
Technical Services Manager, Animal Health Board, PO Box 3412, Wellington, New Zealand

L. R. Matthews
AgResearch Ruakura Centre, Private Bag 3123, Hamilton, New Zealand

J. Meers
Institute of Veterinary, Animal and Biomedical Sciences, Massey University, Private Bag 11222, Palmerston North, New Zealand

T. L. Montague
Landcare Research, PO Box 69, Lincoln, New Zealand
Present address: Roe Koh Associates Ltd, PO Box 6254, Christchurch, New Zealand

D. R. Morgan
Landcare Research, PO Box 69, Lincoln, New Zealand

D. A. Norton
School of Forestry, University of Canterbury, Private Bag 4800, Christchurch, New Zealand

G. Nugent
Landcare Research, PO Box 69, Lincoln, New Zealand

C. E. O'Connor
Landcare Research, PO Box 69, Lincoln, New Zealand

J. P. Parkes
Landcare Research, PO Box 69, Lincoln, New Zealand

I. J. Payton
Landcare Research, PO Box 69, Lincoln, New Zealand

R. M. F. S. Sadleir
120 Sweetacres Drive, Belmont, Lower Hutt, New Zealand

L. S. Saunders
c/- 33b Middleton Rd, Christchurch, New Zealand

L. Selwood
Department of Zoology, La Trobe University, Plenty Rd, Bundoora 3083, Vic., Australia

E. B. Spurr
Landcare Research, PO Box 69, Lincoln, New Zealand

M. Stankiewicz
AgResearch, PO Box 40063, Upper Hutt, New Zealand
Present address: Animal and Food Sciences Division, Lincoln University, PO Box 84, Lincoln, New Zealand

P. A. Suisted
Landcare Research, PO Box 69, Lincoln, New Zealand
Present address: 28 Signal Hill Road, Opoho, Dunedin, New Zealand

P. J. Sweetapple
Landcare Research, PO Box 69, Lincoln, New Zealand

G. H. Tocher
PO Box 11 031, Musselburgh, Dunedin, New Zealand

C. J. Veltman
Conservation Sciences Centre, Department of Conservation, PO Box 10 420, Wellington, New Zealand

B. Warburton
Landcare Research, PO Box 69, Lincoln, New Zealand

R. L. Wilkinson
Landcare Research, PO Box 69, Lincoln, New Zealand

Foreword

The Hon. Simon Upton

"Know thy enemy." Sun Tzu's timeless injunction in the *Art of War* applies as well to New Zealanders facing the brushtail possum as it did to Chinese against invading hordes in the sixth century BC. The possum has been a spectacularly cunning and successful enemy.

We've been studying and plotting against *Trichosurus vulpecula* for years, but nowhere has there been a single, widely available published text which pulls all the latest information together into a convenient, authoritative and fully indexed volume.

The Brushtail Possum will fill that void admirably. It provides an up-to-date text for all those wanting a detailed and accurate reference point when seeking to answer questions about possums in New Zealand and the management of this introduced marsupial. The anatomy of the problem is viewed by New Zealand experts from every angle and it makes for fascinating and alarming reading.

That this book is so necessary is a sharp reminder – if we really needed it – of the special and fragile nature of our environment. Sixty-five million years of isolation have given New Zealand a unique ecology which, as we have learned the hard way, is peculiarly vulnerable to the introduction of new organisms.

With every page this book shows us the costs, environmental as well as financial, of an ill-considered introduction. These days, with the Hazardous Substance and New Organisms legislation, the Environmental Risk Management Authority, and an increased awareness of biosecurity imperatives, we are far better placed to avoid further such destructive incursions. We have to be.

Ever-expanding world trade and travel, an insatiable demand for new things to put in the garden, and the open horizons of genetic research mean that New Zealand will face more frequent and often more subtle questions about the introduction of new organisms into the future.

This book examines one of our most painful experiences with an introduced species. The story is, of course, far from over. It will take an enormous effort to control the possum. Having the latest information readily accessible is a small but important step.

Editor's Introduction

The cute and cuddly appearance of possums belies their status as a major wild animal pest in New Zealand. In Australia they are sometimes a pest in commercial forestry, but are usually a valued part of the native fauna. Transplanted to New Zealand, however, they prey on the eggs and chicks of native pigeons, kōkako, and other birds, and on native insects. Possums also defoliate favoured trees in huge areas of forest and, by spreading bovine tuberculosis, pose an immense threat to our major dairy and beef industries.

Introduced from Australia in the 1850s to establish an animal fur industry, they are now thought to number between 60 and 70 million in New Zealand. They have colonised 98% of the country and have gone from being a valued resource that was protected until 1921, to being harvested seasonally, and finally to being totally unprotected in 1946/47. New Zealand now spends approximately NZ$50 million each year to diminish their impacts, and the species has engendered more public debate, research, and management than any other introduced mammal. In 1994/95 the New Zealand Department of Conservation (DOC) controlled (trapped and poisoned) possums over 271 000ha or about 2% of conservation lands at a cost of NZ$6–8 million. In 1997/98, the organisation responsible for eradicating Tb from New Zealand's livestock, the Animal Health Board, funded Tb-vector-control operations (mostly for possums) that covered 10% of New Zealand (2.4 million ha) at a cost of $24 million. Possum research costs New Zealanders another NZ$20 million each year.

The magnitude and importance of the possum problem motivated Landcare Research to produce this comprehensive, fully indexed, reference book on possums in New Zealand. Most chapters are written by authors currently working on possums, many at Landcare Research and some with more than 25 years experience in their field. Each of the authors was given a chapter outline and asked to identify crucial gaps in our knowledge base. Indeed, one of the most valuable lessons in the process of producing this book has been the clarification of what we do not know about possums, especially when it comes to measuring the benefits of controlling them.

The range and diversity of the topics covered in this book means the chapters are essentially stand-alone reviews that direct readers to information. Much of the referenced information is unpublished and occurs only in contract reports, conference proceedings, and obscure annual reports of both extinct and extant organisations such as the Agricultural Pests Destruction Council. By allowing reference to such sources I hope that much of the useful information in the "grey press" is identified and made available to a wider audience. Not to use this unpublished information would be to ignore extensive work already done.

This book contains 26 chapters structured into five sections. The first section examines the introduction and spread of the possum throughout most of New Zealand and what we know about its biology, including its diet, home range and dispersal, social behaviour, reproduction, population dynamics, and predators, parasites and diseases. The second section looks at the impacts of possums and why they are considered a pest in New Zealand. It includes information on possums and Tb, as well as possum impacts on agriculture, native plants, and native animals. The third section details possum management techniques currently used to count, catch, kill, repel, or exclude possums; the models and economics of doing so; and some of the difficulties associated with controlling them (i.e., the "unwanted effects" such as the killing of non-target animals). The final two sections examine the benefits of controlling possums and give a glimpse of possum management in the future using immunocontraceptives and adaptive management techniques.

Most chapters have been written for readers with a general interest in possums as well as for scientific specialists and those involved in possum management. Some of the information presented in the chapter on possum reproduction and development is more difficult to read because it is hard to write about physiology and anatomy without using technical terms, and for this reason Chapter 6 contains its own glossary.

The publication of this book should make many more people aware of the immense breadth and complexities of the possum problem and assist their appreciation and understanding of a mammalian pest of prodigious significance and the means to resolve a most regrettable introduction.

Thomas L. Montague

Acknowledgements

Many people have contributed to this book. I am grateful to the 41 authors who wrote or co-authored chapters, and I am indebted to many others whose names do not appear in these pages. I would like to express my deepest gratitude to the Landcare Research in-house editors, Christine Bezar, Megan Ogle-Mannering, and Joanna Orwin who helped untangle the 160 000 words of text and checked references; Wendy Weller who patiently typed and retyped the book; Kirsty Cullen who did the graphics; Keven Drew who provided the possum photograph used on the cover of the book; Greg Comfort and Manaaki Whenua Press.

I also thank Jim Coleman, Graham Nugent, John Parkes, Bruce Warburton, and the numerous referees who commented on the manuscripts. Andy Pearce (Chief Executive Officer, Landcare Research), Phil Hart, and Kaye Green (Department of Conservation) gave their "behind the scenes" support throughout the project, which was funded by Landcare Research and the Department of Conservation, Wellington.

CHAPTER ONE

Anatomy of a Disastrous Success: The Brushtail Possum as an Invasive Species

Mick Clout and Kris Ericksen

The common brushtail possum (*Trichosurus vulpecula*) is a cat-sized Australian phalanger, weighing 2–4 kg. It is a solitary, nocturnal, arboreal marsupial that has a wide natural distribution in its native Australia and Tasmania, where it typically inhabits open eucalypt woodlands (Strahan 1983). In its native range the brushtail possum is only one of a suite of herbivorous arboreal marsupials. In New Zealand, where it has been introduced, it is the only possum present and has become a major invasive pest, causing severe damage to native ecosystems and acting as a vector of bovine tuberculosis. In this chapter, we explore how the possum became such a "disastrous success" in New Zealand.

History of introduction and control

The first successful introduction of brushtail possums to New Zealand was made in 1858 near Riverton, Southland (Pracy 1974), at a time when there was great enthusiasm among European settlers for the "artificial enrichment" of the fauna and flora (McDowall 1994). The prevailing view was that the native fauna, already depleted by recent extinctions, was largely doomed and that there was little value in what remained, other than as curiosities. The earliest deliberate introductions of animals by Europeans, starting with Captain Cook in the 1770s, were of domestic livestock. However, by the 1860s, settlers had organised themselves into "acclimatisation societies", dedicated to introducing new species to New Zealand (McDowall 1994). These societies institutionalised the previous informal process of introductions by individuals. They introduced animals such as deer, game birds, and trout for sport, a variety of European songbirds for nostalgic reasons, a range of other species for pure novelty value, and a few species for economic reasons. The introduction of the fur-bearing brushtail possum from Australia and Tasmania was in the latter category: it was motivated by economic profit.

By the middle of the nineteenth century, settlers in Australia had already discovered that brushtail possums had potentially valuable pelts, and were profiting from a fur trade. The American and Russian fur trades were very profitable at that time. It was therefore a logical step to attempt to establish a fur trade in New Zealand by introducing the fur-bearing brushtail possum. The very first possum importations, including an apparently unsuccessful one near Riverton in 1837 (Pracy 1974), were made by private individuals. The first successful introduction (also near Riverton) was evidently made in 1858. From the 1860s onwards most importations were made by acclimatisation societies, which imported many possums from Tasmania and Victoria, and encouraged and fostered their spread. There were at least 35 separate importations of possums, most of which took place during the 1890s (McDowall 1994). As possum populations became established, the acclimatisation societies not only continued to introduce them from Australia, but also spread New Zealand-bred animals throughout the country. From 1895 to 1906 the government also took an active role in the liberation of possums, so that in the 1890s, the acclimatisation societies and the New Zealand Government together made 86 separate liberations of possums (Pracy 1974). Between 1865 and 1926 a total of 127 recorded liberations were made by acclimatisation societies (108) and government agencies (19). Most of the liberations were of New Zealand-bred possums. A considerable number of other liberations made by private individuals should also be attributed to acclimatisation societies, since the individuals concerned were members of the societies and made releases of possums on their behalf (L.T. Pracy, retired, ex-Agriculture Pest Destruction Council, pers. comm.). The multiple liberations of possums throughout the country greatly supplemented and speeded up their natural spread. There was wide acceptance that having possums in New Zealand forests was beneficial.

By the early 1900s, the shooting and trapping of possums for furs had developed to the stage where the acclimatisation societies were becoming concerned at over-harvesting of this asset. They lobbied the government, and in 1911 unlicensed hunting was prohibited by declaring the possum as "imported game" under the Animals Protection Act 1908 (McDowall 1994). However, the government was also under pressure from farmers, orchardists, and horticulturalists to allow unrestricted hunting, because of increasing reports of damage by possums to crops. There followed a series of confusing changes in restrictions on the taking of possums. In 1912 all protection was removed, but in 1913 it was reinstituted in native forest areas only. Concern grew at the continuing liberation and spread of possums, and so in 1920 the government commissioned biologist Harold Kirk to report on the impacts of possums on forests and orchards. Kirk (1920) concluded that, while damage to orchards and gardens was indisputable, "the damage to New Zealand forests is negligible". He recommended that possums be released in all forest districts away from orchards and gardens and that the government retain a hunting season for possums, issue licences to trappers, and collect a levy on all skins traded. The Kirk report formed the basis of the ensuing 25 years of government policy on possums. In 1921 new regulations were introduced permitting licensed hunting during a restricted season and a levy was placed on the sale of skins. Acclimatisation societies were empowered to enforce the regulations and they shared the income from levies with the government.

Throughout the 1920s and 1930s there was continuing controversy over the effects of possums on native forests. The fledgling conservation movement pitched itself against the acclimatisation societies, which were gaining substantial revenues from the levy on skin sales (McDowall 1994). No more liberations were officially permitted after 1922, but applications continued to be made by several acclimatisation societies (L.T. Pracy pers. comm.) and possums continued to be aided by people in their spread throughout New Zealand. It was not until the 1940s that the first scientific evidence of the impacts of possums on native forests was collected (Kean & Pracy 1953). There was then a radical change in official policy. In 1947 the government cancelled all restrictions on the taking of possums, instituted penalties for harbouring or liberating them, and legalised the use of poisons for possum control. From 1951 to 1961 the government ran a bounty system as an interim measure to encourage possum control. The national responsibility for possum control initially lay with the Department of Internal Affairs, but was transferred in 1956 to the New Zealand Forest Service. Increased numbers of possums were killed as a result of the bounty. About 1 million per annum were taken for this purpose in the late 1950s (Parkes *et al.* 1996), in addition to those harvested for pelts as part of the continuing fur trade. Despite this, the bounty scheme failed to prevent the continued spread of possums and appeared ineffective in controlling numbers (McDowall 1994). There were even reports that some possum trappers were deliberately spreading possums into new areas to provide a source of income from bounties. Reports of similar illegal releases by commercial possum trappers in Northland and Coromandel followed the rising skin prices of the 1970s (L.T. Pracy pers. comm.). The annual number of possums harvested commercially for fur is directly related to the price of skins, which peaked in the early 1980s (Parkes *et al.* 1996). After the withdrawal of bounties, large-scale control of possums was conducted on Crown lands by the New Zealand Forest Service, initially under the Noxious Animals Act 1956 and then under the Wild Animal Control Act 1977. Since 1987, the Department of Conservation (successor to the now-defunct New Zealand Forest Service) has continued to control possums under the Wild Animal Control Act 1977. In farming areas control was conducted by Agricultural Pest Destruction Boards under the Agricultural Pests Destruction Act 1967. Under the Biosecurity Act 1993, local authorities and other interested parties are now empowered to control possums, provided that they have been declared pests under local Pest Management Strategies (Livingstone 1993).

Possum control is conducted not only to protect forests and native wildlife, but also to prevent the transmission of bovine tuberculosis (Tb), for which possums were first implicated as vectors in the late 1960s (Ekdahl *et al.* 1970). Possums later became recognised as most important vectors of Tb in cattle and farmed deer in New Zealand (Ministry of Agriculture and Fisheries 1986), which stimulated heavy expenditure on Tb control and research (estimated at $33 million in 1991/92 (Livingstone 1993)). Widespread possum control is now mainly by the aerial distribution of carrot or pellet baits containing 1080 (sodium monofluoroacetate), with bait stations containing either 1080-poisoned pellets or anticoagulant poisons being used for ground control in some smaller, more accessible areas. Recent confirmation that possums not only damage forest vegetation and carry Tb, but also prey on the nests of some threatened native birds (Brown *et al.* 1993) has further emphasised the importance of controlling possums on conservation lands. Possum control is now a major focus for the Department of Conservation, with Departmental expenditure for this purpose rising from $3 million in 1990/91 to $12 million in 1996/97 (Parkes *et al.* 1997).

Recent changes in distribution

The current distribution in New Zealand (Fig. 1.1) has changed little from that in 1984, as described by Cowan (1990a). The main changes in the past 15 years have been the complete colonisation of Northland, further spread in the south-west of the South Island, and the extermination of possums from some offshore islands (Table 1.1).

In Northland, possums were first reported at Te Kao in 1983, and at Te Paki, just south of Cape Reinga (Fig. 1.2) in late 1986 when possum droppings were found during a New Zealand Wildlife Service survey. In 1990, during a Department of Conservation survey of reserves at Te Paki, the only place at which a specially trained dog could not pick up the scent of possums was at North Cape Scientific Reserve (D. M. McKenzie, Department of Conservation, Whangarei, pers. comm.). The possum had effectively completed its century-long journey to the northern limit of New Zealand. On the Coromandel Peninsula, possums had

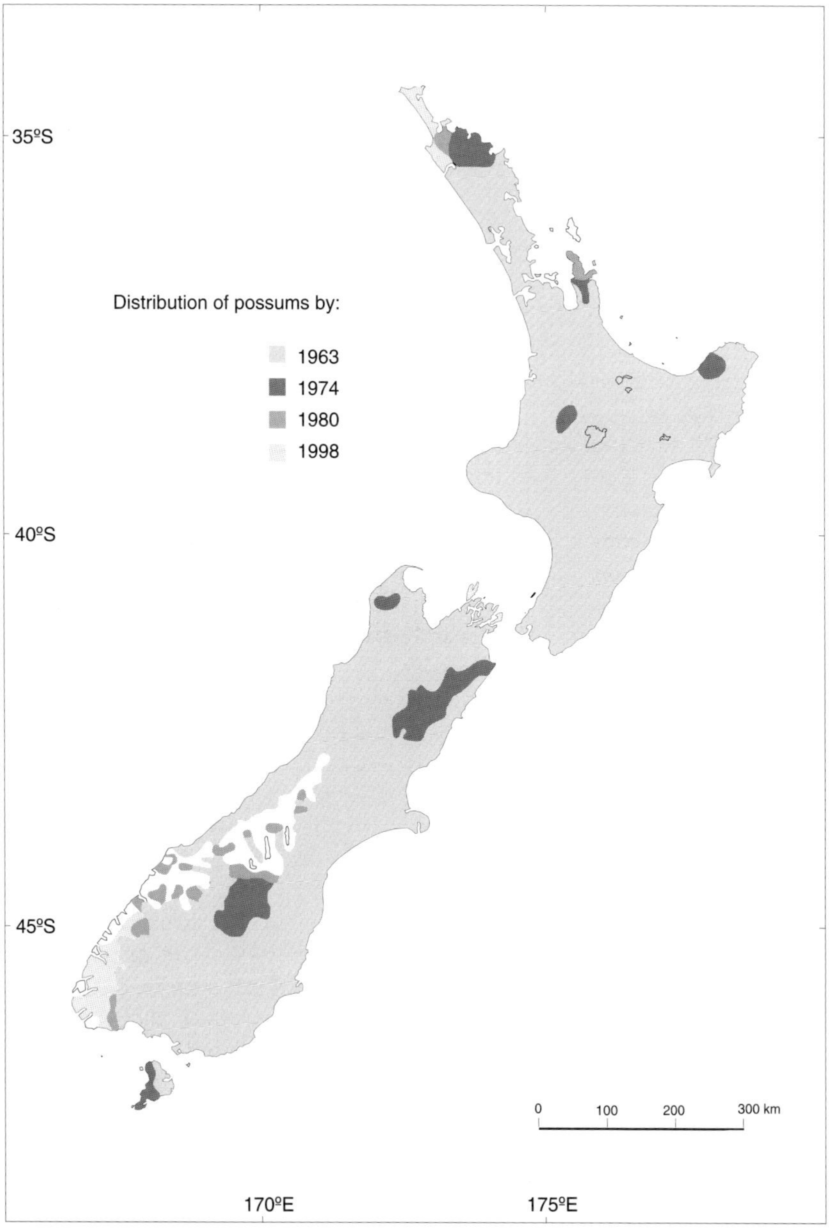

Fig. 1.1
Distribution of possums, at four stages of colonisation (after Cowan 1990a).

Table 1.1
New Zealand islands from which possums have been eradicated.

Island name	Area (ha)	Date possums introduced	Start year of eradication	Methods	Possum status in 1997	Sources
Northland						
Fortyseven	1	NA	1990	Talon in bait stations	Ongoing bait station operation	D. Taylor (Invercargill)
Harakeke	11	c. 1990	1992	Talon in bait station	Eradication confirmed 1994	R. Atkinson (Kerikeri)
Peach	11	NA	1990	Traps and Talon in bait stations	Unknown	R. Atkinson (Kerikeri)
Hauraki/Coromandel						
Rangitoto	2321	1868	1990	Poison, traps	Eradicated 1997	C. R. Veitch (ex Auckland)
Motutapu	1560	1868	1990	Poison, traps, and dog	Eradicated 1996	C. R. Veitch (ex Auckland)
Motutapere	50	c. 1902	1994	Aerial Talon 20P, then bait stations	Eradication confirmed 1997	R. Chappell (Coromandel)
Whanganui	220	c. 1902	1994	Cyanide, spotlighting, bait stations	Eradication confirmed 1997	R. Chappell (Coromandel)
Uretara (Bay of Plenty)	73.2	NA	1993	Pindone in bait stations	Ongoing bait station operation (close to mainland)	D. Williams (Rotorua)
Wellington/Marlborough						
Kapiti	1970	1892	1980	Poison, traps, and dogs	Eradicated 1986	Veitch & Bell 1990
Allports	16	<1980	1989	Poison	Eradicated 1990	D. Brown (ex Havelock Nth)
Tarakaipa	35	NA	1991	Poison, traps, and dogs	Eradication confirmed 1993	P. Brady (Picton)
Southern						
Pig (L. Wakatipu)	110	1975	1990	Poison and traps	Monitored annually for reinvasion	B. Barron (Wakatipu)
Pigeon (L. Wakatipu)	168	1975	1990	Poison and traps	Monitored annually for reinvasion	B. Barron (Wakatipu)
Stevensons (L. Wanaka)	40	<1993	1993	Bait stations	No sign since 1995. Monitored annually for reinvasion	J. Fleming (Wanaka)
Codfish	1396	<1925	1984	Poison, traps, and dogs	Eradicated 1987	Veitch & Bell 1990

NA = not available
Sources listed are/were Department of Conservation staff

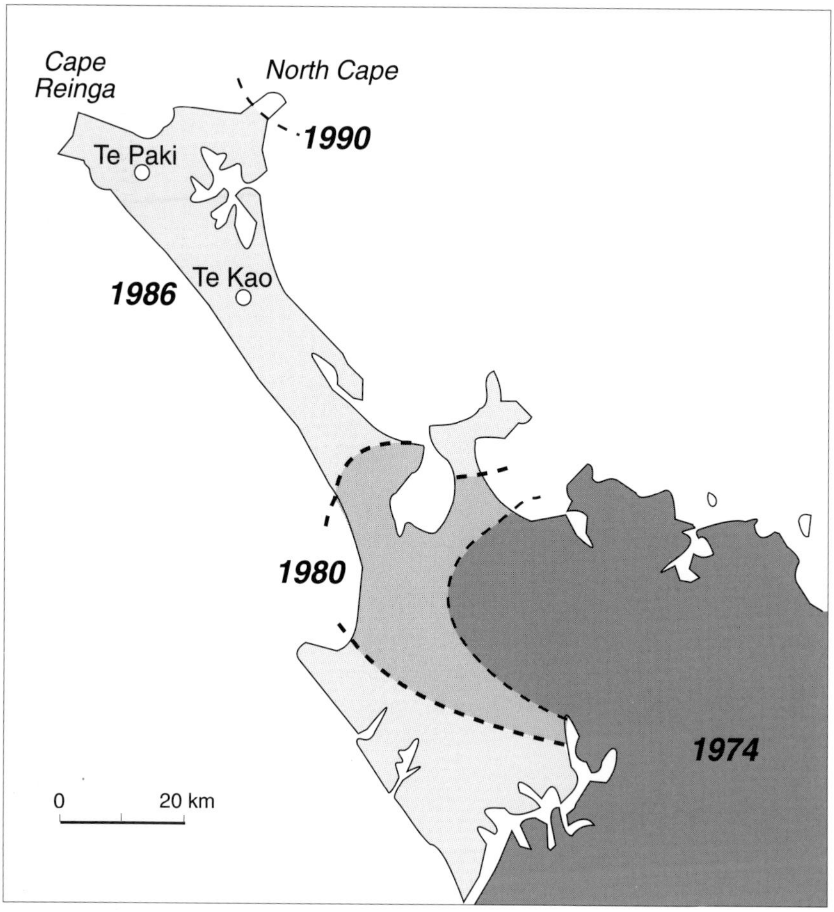

Fig. 1.2
Distribution of possums in Northland (after Cowan 1990a; D. M. McKenzie, DOC, Whangarei, pers. comm.).

reached its northern tip at Cape Colville by 1980 (Cowan 1990a,b). Here, as elsewhere, their spread was assisted over the years by a number of private liberations, the most northerly of which was at the base of Mt Moehau (R. Chappell, Department of Conservation, Coromandel, pers. comm.).

In Fiordland and South Westland it is possible that possums have been slowed in their spread both by the topography and by the wet and cold conditions. Cowan (1990a) noted that in the mid-1980s possums were still absent south of the Arawata River, apart from around Milford Sound. The range of possums in south-western New Zealand has expanded further since these previous reports were made (Fig. 1.3). Surveys carried out in southern Fiordland from 1995 to 1998 confirmed that a peninsula in Doubtful Sound and a large area around Long Sound remained possum free. Secretary Island is the largest offshore island in Fiordland known to be free of possums. The status of possums on Resolution Island is unconfirmed (P. Willemse, Department of Conservation, Invercargill, pers. comm.).

Possums have been recorded as present at some stage on at least 23 offshore or outlying islands. These include the 17 islands listed by Cowan (1990a), plus Motutapere Island, Uretara Island, Allports Island, and three small islands in Northland (Harakeke, Fortyseven, and Peach islands). Possums have also been recorded on Pig and Pigeon islands in Lake Wakatipu, and on Stevenson's Island in Lake Wanaka (Table 1.1). They have now been eradicated from most of these named islands, in addition to Rangitoto, Motutapu, Whanganui, Kapiti, Tarakaipa, and Codfish islands (Table 1.1).

The possum as an invasive species: could its success have been predicted?

The question of which factors determine whether or not a species will be a successful invader has been addressed by several authors, albeit without much success at any synthesis (Williamson 1996). Apart from the obvious characteristic of previous success as an invader, attributes which have commonly been proposed to distinguish successful invaders from others include high intrinsic rate of increase, extensive

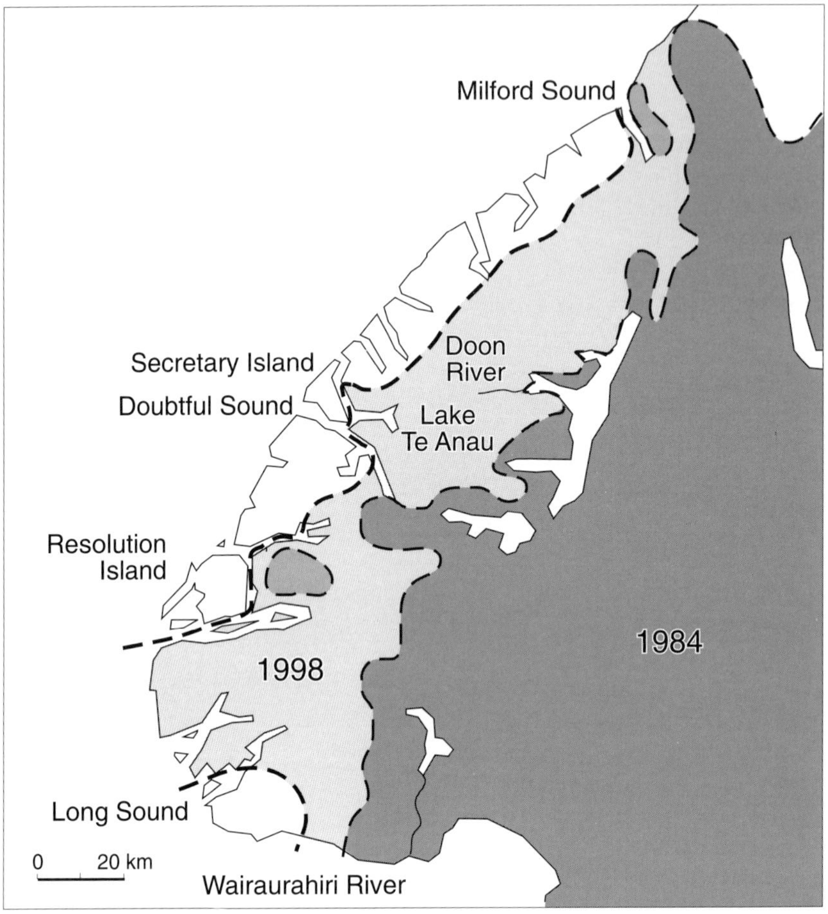

Fig. 1.3
Distribution of possums in Fiordland (after Cowan 1990a; P. Willemse, DOC, Invercargill, pers. comm.).

natural range, high abundance within this range, ecological distinctiveness in the invaded community, vacant niches in this community, and climatic matching with the natural range (Williamson 1996).

In the first respect the brushtail possum is an unlikely invasive species because it has an intrinsic rate of increase of only 0.2–0.3 per annum (Barlow 1991). Typical reproductive output is one young per female per annum, rising to an absolute maximum of two per annum under the most favourable conditions, such as in low-density populations during colonisation of new habitat or after control operations. There are, however, several other attributes which have contributed to the status of the brushtail possum as a highly successful invader in the New Zealand environment. Firstly, in its native Australia it is one of the most widely distributed native marsupials. It occurs naturally in a variety of habitat types in most parts of mainland Australia, although it is most abundant in dry woodlands of the south-east of the continent. It also occurs in Tasmania and on some offshore islands (Strahan 1983). This wide natural distribution suggests a broad environmental tolerance and was, in retrospect, an indication that the brushtail possum had the potential to become invasive in New Zealand. Brushtail possums are also more generalist and opportunistic in their feeding habits than most other arboreal marsupials, feeding not only on foliage, flowers, and fruits, but also insects, bird's eggs, and fungi (Strahan 1983; Cowan 1990a; Brown *et al.* 1993).

In addition to its broad environmental tolerance and generalist feeding habits, another factor that may have aided the success of brushtail possums as invasive species in New Zealand is the lack of competitors for food and shelter. In New Zealand it is ecologically distinctive and it has exploited a broad, essentially vacant niche for a large, nocturnal, arboreal herbivore. In Australia, brushtail possums have several competitors, including the marsupial gliders (*Petaurus, Petauraoides*) and other arboreal marsupials, about half of which use tree hollows similar to those required by brushtails and most of which have some dietary overlap with them in areas where they coexist (Menkhorst 1984; Smith & Hume 1984). The congeneric mountain brushtail possum (*Trichosurus caninus*) normally replaces *T. vulpecula* in wetter forests in Australia (Kerle 1984), but in New Zealand common brushtail possums inhabit a range of wet forest types right up to the tree line (Clout & Gaze 1984).

The relative lack of parasites and predators may have been yet another factor in the success of brushtail possums in New Zealand (see Chapter 7). In Australia, 31 species of endoparasites and 35 species of ectoparasites have been recorded from brushtail possums (Viggers & Spratt 1995), whereas in New Zealand only 8 endoparasites and 6 ectoparasites (two of these not recorded from Australia) have been found (Viggers & Spratt 1995; Clark *et al.* 1997). Predators of possums in Australia that are absent from New Zealand include large carnivores such as feral dogs and dingos (*Canis familiaris*), foxes (*Vulpes vulpes*), wedge-tailed eagles (*Aquila audax*), powerful owls (*Ninox strenua*), lace monitors (*Varanus varanus*), and carpet pythons (*Morelia spilota*) (Cowan 1990a). In New Zealand the largest feral predator (also present in Australia) is the domestic cat (*Felis catus*), which can prey on young possums (Fitzgerald & Karl 1979) and is also present in Australia. Introduced ferrets (*Mustela furo*) and stoats (*Mustela erminea*) also feed on possums in New Zealand, although it is uncertain how much of this involves scavenging on carcasses (King 1990). Domestic dogs kill some possums, but by far the most significant predators of possums in New Zealand are humans. Fur trappers continue to harvest some possums when skin prices are high (Parkes *et al.* 1996), but most killing (by shooting, trapping, and poisoning) is for pest control purposes, to prevent damage to crops, native forests, and wildlife, and to reduce risks of transmission of bovine tuberculosis to domestic livestock.

In addition to the lack of competitors and a reduced burden of parasites and predators, the New Zealand environment holds other advantages for the brushtail possum. New Zealand forest ecosystems are generally more productive than those in Australia, where aridity and the generally low nutrient status of soils and vegetation (Attiwill & Leeper 1987) are important factors limiting the abundance of arboreal herbivores (Pausus *et al.* 1995; Braithwaite 1996). Foliage in Australian eucalypt forests is notoriously sclerophyllous and low in nitrogen and phosphorus (Attiwill & Leeper 1987; Cork 1996). The vegetation of New Zealand is generally higher in nutrients than that in Australia and forests can support higher densities of browsing mammals such as possums. Furthermore, since New Zealand plants have evolved in the absence of any mammalian browsers, they tend to have fewer chemical defences than equivalent Australian species. *Eucalyptus* leaves form the bulk of the diet of brushtail possums and other arboreal herbivorous marsupials in Australia, but high levels of phenolics, terpenoids, and other chemical defences in eucalypt foliage limit the intake of any one species (Freeland & Winter 1975) or individual tree (Lawler *et al.* 1998). In New Zealand forests a higher proportion of the vegetation is palatable (Brockie 1992) and possum diets include a wide variety of foliage (Fitzgerald 1976; Allen *et al.* 1997) and fleshy fruits (Cowan 1990b). The New Zealand plants most favoured by possums tend to be those producing foliage or fruits high in carbohydrate (Brockie 1992). Some highly palatable and chemically "unprotected" plant species are so preferred by possums that their selective browsing can result in local plant extinctions (Campbell 1990).

The final factor that has been influential in the disastrous success of the brushtail possum as an invader of New Zealand ecosystems is the sheer number of introductions from Australia and Tasmania and the subsequent widespread releases around New Zealand of locally bred stock (Pracy 1974). The probability of establishment of populations of introduced vertebrates rises significantly with the size and number of liberations (Veltman *et al.* 1996). It is therefore entirely possible that without repeated introductions possums would never have established in New Zealand. The first recorded introduction was apparently unsuccessful, as were some other releases (Pracy 1974), especially in Fiordland and south Westland. The climate of these latter parts of New Zealand is wetter and cooler than the natural range of brushtail possums, which may have reduced the probability of them becoming established there. The multiple introductions of possums not only served to increase the demographic chances of successful establishment, they also broadened the genetic base of the New Zealand population through separate importations from Victoria, New South Wales, and Tasmania. Finally, without the later releases of New Zealand-bred stock, the spread of possums through the country would undoubtedly have been much slower. Continued human assistance, through repeated importations and liberations, has undoubtedly been a critical factor in the success of the brushtail possum as an invader of New Zealand ecosystems.

Conclusions

It is clear that brushtail possums in New Zealand have experienced an "ecological release" from some of their natural controls.The net result of this is that possums achieve and maintain much greater population densities in New Zealand ecosystems than in Australia. Densities of 10–12 brushtail possums per hectare are common in New Zealand podocarp/broadleaf forest (Cowan 1990a), which is an order of magnitude greater than the densities of <1/ha prevailing in most Australian forests (Dunnet 1964). The ecological consequences of these sorts of densities of arboreal mammals in forests that have not evolved to accommodate them are severe, as are the economic consequences of the damage these invaders cause as vectors of disease.

The lesson to be learned from the possum example is a very simple one: the breaching of biogeographic boundaries by introducing alien organisms is something that should never be undertaken lightly. New Zealand ecosystems and the New Zealand economy would certainly be much better off without the brushtail possum.

Summary

- Brushtail possums were introduced to New Zealand from Australia in the mid-nineteenth century, with the motive of establishing a fur trade.
- Multiple importations and liberations of possums were conducted by private individuals, acclimatisation societies, and government agencies, with a peak of introductions in the 1890s.
- After an initial period of legislative protection, growing realisation of the adverse effects of possums led (in 1947) to their official recognition as a pest.
- It is now recognised that possums not only damage crops and native forests, but also carry bovine tuberculosis and prey on nests of native birds.
- Possums have now spread throughout the main islands of New Zealand, but have recently been eradicated from several small islands.
- The success of the brushtail possum as an invasive species is partially attributable to its generalist habits; the lack of competitors, parasites or predators; and the abundance of palatable, nutritious vegetation in New Zealand. Human assistance has also been a major factor in its success.

Acknowledgements

We thank Jim Coleman, Tom Montague, John Parkes, Les Pracy, and an anonymous referee for their constructive comments on earlier versions of this manuscript. We also thank Chris Edkins for draughting the distribution maps.

References

Allen, R. B.; Fitzgerald, A. E.; Efford, M. G. 1997: Long-term changes and seasonal patterns in possum (*Trichosurus vulpecula*) leaf diet, Orongorongo Valley, Wellington, New Zealand. *New Zealand Journal of Ecology 21*: 181–186.

Attiwill, P. M.; Leeper G. W. 1987: Forest soils and nutrient cycles. Carlton, Victoria, Melbourne University Press. 202 p.

Barlow, N. D. 1991: A spatially aggregated disease/host model for bovine Tb in New Zealand possum populations. *Journal of Applied Ecology 28*: 777–793.

Braithwaite, L. W. 1996: Conservation of arboreal herbivores: The Australian scene. *Australian Journal of Ecology 21*: 21–30.

Brockie, R. 1992: A living New Zealand forest. Auckland, David Bateman. 172 p.

Brown, K.; Innes, J.; Shorten, R. 1993: Evidence that possums prey on and scavenge birds' eggs, birds and mammals. *Notornis 40*: 169–177.

Campbell, D. J. 1990: Changes in structure and composition of a New Zealand lowland forest inhabited by brushtail possums. *Pacific Science 44*: 277–296.

Clark, J. M.; Heath, D. D.; Stankiewicz, M. 1997: The ectoparasites of brushtail possum *Trichosurus vulpecula* in New Zealand. *New Zealand Journal of Zoology 24*: 199–204.

Clout, M. N.; Gaze, P. D. 1984: Brushtail possums (*Trichosurus vulpecula* Kerr) in a New Zealand beech (*Nothofagus*) forest. *New Zealand Journal of Ecology 7*: 147–155.

Cork, S. J. 1996: Optimal digestive strategies for arboreal herbivorous mammals in contrasting habitat types: Why koalas and colobines are different. *Australian Journal of Ecology 21*: 10–20.

Cowan, P. E. 1990a: Brushtail possum. *In:* King, C. M. *ed.* The handbook of New Zealand mammals. Auckland, Oxford University Press. Pp. 68–98.

Cowan, P. E. 1990b: Fruit, seeds, and flowers in the diet of brushtail possums, *Trichosurus vulpecula,* in lowland podocarp/mixed hardwood forest, Orongorongo Valley, New Zealand. *New Zealand Journal of Zoology 17*: 549–566.

Dunnet, G. M. 1964: A field study of local populations of the brush-tailed possum *Trichosurus vulpecula* in Eastern Australia. *Proceedings of the Zoological Society of London 142*: 665–695.

Ekdahl, M. O.; Smith, B. L.; Money, D. F.1970: Tuberculosis in some wild and feral animals in New Zealand. *New Zealand Veterinary Journal 18:* 44–45.

Fitzgerald, A. E. 1976: Diet of the opossum *Trichosurus vulpecula* (Kerr) in the Orongorongo Valley, Wellington, New Zealand, in relation to food-plant availability. *New Zealand Journal of Zoology* 3: 399–419.

Fitzgerald, B. M.; Karl, B. J. 1979: Foods of feral house cats (*Felis catus* L.) in forest of the Orongorongo Valley, Wellington. *New Zealand Journal of Zoology* 6: 107–126.

Freeland, W. J.; Winter, J. W. 1975: Evolutionary consequences of eating: *Trichosurus vulpecula* (Marsupialia) and the genus *Eucalyptus. Journal of Chemical Ecology 1:* 439–455.

Kean, R. I.; Pracy, L. 1953: Effects of the Australian opossum (*Trichosurus vulpecula* Kerr) on indigenous vegetation in New Zealand. *Proceedings of the Seventh Pacific Science Congress 4*: 696–705.

Kerle, J. A. 1984: Variation in the ecology of *Trichosurus*: its adaptive significance. *In:* Smith, A. P.; Hume, I. D. *ed.* Possums and gliders. Chipping Norton, NSW, Surrey Beatty in assoc. with the Australian Mammal Society. Pp. 115–128.

King, C. M. 1990: Stoat. *In:* King, C. M. *ed.* The handbook of New Zealand mammals. Auckland, Oxford University Press. Pp. 288–312.

Kirk, H. B. 1920: Opossums in New Zealand: Report on Australian opossums in New Zealand. *Appendix to the Journals of the House of Representatives of New Zealand H28:* 1–12.

Lawler, I. R.; Foley, W. J.; Eschler, B. M.; Pass, D. M.; Handasyde, K. 1998: Intraspecific variation in *Eucalyptus* secondary metabolites determines food intake by folivorous marsupials. *Oecologia 116:* 160–169.

Livingstone, P. G. 1993: Production pests: What values need protection? *New Zealand Journal of Zoology 20:* 273–277.

McDowall, R. 1994: Gamekeepers for the nation: the story of New Zealand's acclimatisation societies 1861–1990. Christchurch, Canterbury University Press. 508 p.

Menkhorst, P. W. 1984: The application of nest boxes in research and management of possums and gliders. *In:* Smith, A. P.; Hume, I. D. *ed.* Possums and gliders, Chipping Norton, NSW, Surrey Beatty in assoc. with the Australian Mammal Society. Pp. 517–525.

Ministry of Agriculture and Fisheries, Animal Health Division 1986: Possum research and cattle tuberculosis. *Surveillance 13(3)*: 18–37.

Parkes, J. P.; Nugent, G.; Warburton, B. 1996: Commercial exploitation as a pest control tool for introduced mammals in New Zealand. *Wildlife Biology 2(3)*: 171–177.

Parkes, J.; Baker, A. N.; Ericksen, K. 1997: Possum control by the Department of Conservation: Background, issues, and results from 1993 to 1995. Wellington, Department of Conservation. 40 p.

Pausas, J. G.; Braithwaite, L. W.; Austin, M. P. 1995: Modelling habitat quality for arboreal marsupials in the south coastal forests of New South Wales, Australia. *Forest Ecology and Management 78:* 39–49.

Pracy, L. T. 1974: Introduction and liberation of the opossum (*Trichosurus vulpecula*) into New Zealand. 2nd ed. *New Zealand Forest Service Information Series 45.* 28 p.

Smith A. P.; Hume I. D. *ed.* 1984: Possums and gliders. Chipping Norton, NSW, Surrey Beatty in assoc. with the Australian Mammal Society. 598 p.

Strahan, R. *ed.* 1983: The Australian Museum complete book of Australian mammals: the national photographic index of Australian wildlife. Sydney, Angus and Robertson. 530 p.

Veitch, C. R.; Bell, B. D. 1990: Eradication of introduced animals from the islands of New Zealand. *In:* Towns, D. R.; Daugherty, C. H.; Atkinson, I. A. E. *ed.* Ecological restoration of New Zealand islands. *Conservation Sciences Publication No. 2.* Wellington, Department of Conservation. Pp. 137–146.

Veltman, C. J.; Nee, S.; Crawley, M. J. 1996: Correlates of introduction success in exotic New Zealand birds. *American Naturalist 147:* 542–557.

Viggers, K. L.; Spratt, D. M. 1995: The parasites recorded from *Trichosurus* species (Marsupialia: Phalangeridae). *Wildlife Research 22*: 311–332.

Williamson, M. 1996: Biological invasions. London, Chapman and Hall. 244 p.

CHAPTER TWO

Possum Feeding Patterns: Dietary Tactics of a Reluctant Folivore

Graham Nugent, Peter Sweetapple, Jim Coleman and Phil Suisted

Traditionally possums were seen as conservation pests in New Zealand primarily because of their impacts as folivores; they were implicated in the widespread dieback of native forest because they were able to defoliate and kill some tree species (see Chapter 10). Increasingly, however, there is a realisation that, although the foliage of common canopy species usually forms most of their diet, they must obtain additional energy and/or nutrients from other sources to maintain the population densities seen in most New Zealand forests (Tyndale-Biscoe 1997). We postulate that this constraint explains, at least in part, the variable effect of possums in the defoliation of native forests, and that understanding the extent to which possums are constrained by the availability of non-foliar food sources may be crucial in any attempt to predict their abundance and impacts in New Zealand native ecosystems.

In this chapter, we outline the basic digestive capabilities and feeding strategies used by possums, before focusing more empirically on what is known of their diet and feeding preferences in New Zealand. We postulate a model of possum feeding behaviour specific to New Zealand native forest, and briefly explore its implications.

Digestive strategy

Possums are amongst the smaller folivorous mammals, and, as such, face severe problems in meeting their energy requirements from a leaf diet. Cork & Foley (1991) summarised the problems faced by animals that feed on woody plant foliage, a food source that is often low in protein, high in fibre, and contains an array of anti-herbivore chemicals such as phenols, terpenes, and tannins. Utilising fibrous plant tissue becomes less feasible as body size declines, as the two strategies for meeting energy requirements (maximising either food retention times or food intake) are ultimately limited by the capacity of the alimentary tract (which scales directly with body size whereas energy requirements per kilogram of body mass increase with declining body size). In theory, at least, "pure" folivores appear to need a body mass of tens of kilograms.

To overcome the constraints of small body size and fibrous diets, small mammals need to maintain a higher than average ratio of maximum food intake to energy requirement. Possums achieve this partly by having a low metabolic rate, and partly by modifications in their digestive and feeding strategies. Possums are hind-gut fermenters with a simple stomach but an expanded colon and caecum (Hume 1999). Most foods eaten are finely chewed, but, once ingested, there is little mixing of foods in the stomach, so much so that separate food items remain in discrete homogenous layers in the order they were eaten (Harvie 1973; Sweetapple & Nugent 1998). This is in stark contrast to the continual mixing (and regurgitation and remastication) of foods in the fore-stomachs of ruminants. For possums, mixing of foods and digestion occurs mainly in the hind-gut, particularly the expanded colon and caecum, an adaption that allows selective retention of smaller more digestible particles. This selective retention strategy is used even more effectively by small, strictly folivorous, hind-gut-fermenting herbivores such as the koala and ringtail possums (Hume 1999). Because it is not as highly developed in brushtail possums, however, it does not enable them to fully overcome the constraint of small gut capacity. As a consequence, possums require some high-energy/high-nutrient foods, and so have evolved as generalist and opportunistic feeders eating not only leaves, but fruit, flowers, insects, bird's eggs, and even meat as sources of protein and energy. This strategy in turn enables them to occupy a wide range of markedly different habitats (Cowan 1990a).

Intake rates

Possums have low metabolic rates and can maintain themselves on 30% less food than comparably sized eutherians such as rabbits. Captive adult possums weighing about 2.5 kg and housed indoors consumed 80 g digestible dry matter per day when fed either a diet of commercial pellets or one containing natural foliar foods, and digested approximately 55% (c. 850 Kj/day) of the available dietary energy; possums in the wild probably consume about twice as much (Fitzgerald *et al.* 1981). Reported daily nitrogen requirements vary from about 200 mg per kg $W^{0.75}$ for captive non-breeding adults fed highly digestible (89–94%) diets (Wellard & Hume 1981a) to 1000 mg per kg $W^{0.75}$ in the study referred to above (Fitzgerald *et al.* 1981). Mean retention times of digesta are 1.5–3.0 days for both synthetic and natural foliar diets (Wellard & Hume 1981b; Foley & Hume 1987).

Feeding behaviour

Possums in an Australian eucalypt woodland usually spent less than 25% of each night feeding (MacLennan 1984), typically in one to three bouts (depending on weather) with 2–3 hours between bouts. For a New Zealand forest, however, Ward (cited in MacLennan 1984) reported a bimodal pattern of feeding with a first peak in the second hour after dark followed by a lull of 2–3 hours, then a second peak in the sixth or seventh hour after dark. The low percentage of the night spent feeding suggests that the limiting factor in possum nutrition is likely to be their ability to extract sufficient energy and nutrients from the food eaten or to detoxify any anti-herbivore chemical ingested, rather than their ability to find sufficient food.

Foods eaten in New Zealand

Most of the studies in New Zealand that have produced quantitative estimates of possum diet are summarised in Appendix 2.1. Unless otherwise stated, all quantitative estimates of possum diet in this chapter are expressed as percent composition of stomach or faeces content, either by dry weight or some similar measure presumed to be equivalent. Inevitably for an animal with a wide range of feeding preferences, possum diet varies widely between studies, depending on what is locally available.

Foliage

Foliage comprises the bulk of the possum's diet, ranging from 50–95% of total diet in studies that have quantified diet by food type. In pasture-dominated sites, possums rely on introduced grasses and clover (*Trifolium* spp.; Gilmore 1967; Harvie 1973). However, such species are not always the main foods in these habitats. On farmland near Waverley, for example, these species contributed 32% of total diet (Harvie 1973). Possums living in native forest within foraging range of pasture generally obtain 20% or more of their food from pasture species (Coleman *et al.* 1985). In such situations, indigenous woody species usually provide the bulk of the non-pasture foliage diet, but introduced woody species such as willows (*Salix* spp.), macrocarpa (*Cupressus macrocarpa*), gums (*Eucalyptus* spp.), and broom (*Cytisus scoparius*), along with a wide range of introduced herbaceous species, are also eaten.

In exotic pine (*Pinus* spp.) forest, foliage of introduced broom, blackberry (*Rubus fruticosus*), grasses, and herbs are generally the main foods, although, when present, indigenous woody species such as tree fuchsia (*Fuchsia excorticata*), māhoe (*Melicytus ramiflorus*), pōhuehue (*Muehlenbeckia australis*), and wineberry (*Aristotelia serrata*) are also eaten. Pine needles are seldom eaten in any quantity, but some bark and large quantities of pine pollen, pollen cones, and (to a lesser extent) immature female cones are occasionally eaten (Harvie 1973; Clout 1977; Warburton 1978).

In native forests possums eat a wide range of foods, with 53 species eaten in the Orongorongo Valley (Mason 1958) and about 100 species in studies in both central Westland and central North Island (Coleman *et al.* 1985; Nugent *et al.* 1997). Typically, however, only 3–5 species make up the bulk of the diet. In mixed hardwood and podocarp–hardwood forests, the foliage of tree species usually dominates the possum's diet. Several of the most important diet components are dominant canopy species. Kāmahi (*Weinmannia racemosa*), or its northern congeneric tōwai (*W. silvicola*), is the most universally important food, ranking in the top five foods in 12 of the 20 studies of possum diet in native forest listed in Appendix 2.1. Northern and southern rātā and pōhutukawa (*Metrosideros robusta*, *M. umbellata*, and *M. excelsa*) are the other high-profile species commonly eaten by possums. In high-density post-peak possum populations they are often the most important foods (Appendix 2.1). Where locally abundant, tawa (*Beilschmiedia tawa*) and kohekohe (*Dysoxylum spectabile*) are also major components of possum diet (Appendix 2.1). Podocarps are generally not eaten by possums, with the key exceptions of Hall's tōtara (*Podocarpus hallii*) and pāhautea (*Libocedrus bidwillii*) (Nugent et al. 1997; Rogers 1997). Where commonly available, the leaves of Hall's tōtara can be the most important foliage food in summer and autumn.

Other important foliage foods are predominantly shrub hardwood species such as wineberry, tree fuchsia, māhoe, *Raukaua* spp., patē (*Schefflera digitata*), and several *Pseudopanax* spp., with the latter two foods often rapidly defoliated as possums frequently eat only the petiole of such leaves. Non-tree species taken in the canopy include epiphytes such as *Phymatosorus* spp., and parasitic mistletoes (e.g., *Peraxilla* spp., *Tupeia antarctica*), as well as vines such as pōhuehue, bush lawyer (*Rubus* spp.), and supplejack (*Ripogonum scandens*). "Ground foods" in native forest are sometimes important, and include herbs such as *Acaena, Hydrocotyle, Oxalis, Pratia,* and *Viola* spp., ferns, particularly shield fern (*Polystichum vestitum*), occasionally mosses, but seldom include native grasses.

Flowers, fruits, and fungi

Flowers and fruit are important but lesser components of possum diet. On farmland and in exotic forests, flowers of broom, gorse (*Ulex europaeus*), dandelions (flatweeds of the family Asteraceae), and clover, and the pollen or pollen cones of pines and macrocarpa are often eaten in large quantities, sometimes to the exclusion of other foods (Gilmore 1967; Harvie 1973; Clout 1977; Warburton 1978). Fruits of introduced species including walnuts (*Juglans* spp.), oaks (*Quercus* spp.), apples (*Malus* spp.), stone fruit (*Prunus* spp.), blackberry, boxthorn (*Lycium ferrocissimum*), and elderberry (*Sambucus nigra*) are also eaten (Gilmore 1967; Harvie 1973).

In native forest, flowers and flower buds are generally eaten when available (late spring or early summer), while fruits are eaten mostly in autumn but with a wider spread through the year, reflecting the more extended fruiting periods of most species (Cowan 1990b). The attractiveness of flowers as a food source is perhaps best demonstrated by their being the putative cause of many of the excursions outside their normal home range made by possums in a radio-tracking study (Ward 1978). In the Orongorongo Valley (Cowan 1990b), flowers of four native plants (hīnau (*Elaeocarpus dentatus*), rewarewa (*Knightia excelsa*), kawakawa (*Macropiper excelsum*), and nīkau (*Rhopalostylis sapida*)) were consistently eaten. Elsewhere in New Zealand, flowers of many other species are also eaten including kāmahi, tōwai, northern and southern rātā, five-finger (*Pseudopanax arborea*), tree fuchsia, kānuka (*Kunzea ericoides*), māhoe, *Astelia* spp., kiekie (*Freycinetia baueriana*), bush lawyer, wineberry, pōkākā (*Elaeocarpus hookerianus*), and tutu (*Coriaria arborea*), (Mason 1958; Gilmore 1967; Fitzgerald 1976; Sweetapple & Nugent 1998). At Waihaha in the central North Island, flower buds of bush lawyer comprised up to 87% of individual possum stomach contents (Nugent & Sweetapple unpubl. data).

In the Orongorongo Valley, possums eat the fleshy fruits of almost all the species available, with the large numbers of fruits consumed and the seasonally high proportion of possums eating them suggesting they are a highly sought after food (Cowan 1990b). By comparison, in Westland podocarp–hardwood forest, a wide range of fruits were eaten in small numbers, but only a few (pepper tree (*Pseudowintera colorata*), marbleleaf (*Carpodetus serratus*), and *Coprosma* spp.) were eaten in large quantities (Coleman *et al.* 1985). Possums also often eat large quantities of fruit of species whose foliage is rarely eaten. These species include some podocarps (rimu (*Dacrydium cupressinum*), kahikatea (*Dacrycarpus dacrydioides*), miro, (*Prumnopitys ferruginea*) and mataī (*P. taxifolia*)), and hardwood species such as hīnau, pepper tree, marbleleaf, pigeonwood (*Hedycarya arborea*), and black maire (*Nestegis cunninghamii*) (Coleman *et al.* 1985; Cowan 1990b; Nugent & Sweetapple unpubl. data). For large-seeded fruits, such as hīnau and miro, only the flesh is usually eaten (Nugent & Sweetapple unpubl. data).

Many of the earlier assessments of possum diet were based mainly on analyses of plant cuticle fragments in faeces or stomach contents, which made it difficult to assess accurately the relative contributions of fruit to the overall diet. Researchers either restricted their analyses to the leaf diet alone (e.g., Allen *et al.* 1997), or presented leaf and fruit data separately using different measures of abundance (e.g., Coleman *et al.* 1985). Nonetheless, these studies indicated large numbers of fruits were sometimes eaten. Fitzgerald & Wardle (1979) reported that 10% of diet of possums in Waiho (Westland) was "non-leaf" material compared with 35% in the Orongorongo Valley (Fitzgerald 1976). More recently, point sampling and layer separation techniques of stomach contents have been used to provide more comparable estimates of the proportions of fruit, flowers, and foliage eaten (Sweetapple & Nugent 1998). Studies using these techniques have shown that fruit formed 25% of the diet of possums in subalpine shrubland (Parkes & Thomson 1995), but in low-diversity beech forest, annual fruit use is typically low

(<5% of diet; Owen & Norton 1995; Sweetapple, Fraser, & Knightbridge, unpubl. data).

In podocarp–hardwood forest at Waihaha, fruit formed 4–42% of the annual diet of an uncontrolled possum population, but only about 4% of the total biomass of fruit, flowers, and leaves caught in litterfall traps (Nugent *et al.* 1997). Although the data are not strictly comparable (for a variety of reasons), they nonetheless suggest that fruits are generally much more preferred than foliage, presumably reflecting greater levels of available energy and/or nutrients in fruits (Williams 1982), and possibly lower levels of anti-herbivore chemicals. Consistent with that, use of fruit in the Waihaha forest increased from 9%, 2 years before possum control (Sweetapple & Nugent 1998) to 35% during the 2 years immediately after, and reached 86% of total diet in March 1996 (Nugent & Sweetapple unpubl. data).

Fungi are occasionally eaten by possums (Gilmore 1967; Harvie 1973; Warburton 1978; Nugent *et al.* 1997). They comprised 0.6% of possum diet at Waihaha before control (Nugent *et al.* 1997). Other items eaten occasionally or in minute quantities include bark (Fitzgerald & Wardle 1979; Clout 1977; Owen & Norton 1995), lichen, decomposing wood and litter, soil, and pumice (Nugent & Sweetapple unpubl. data; Sweetapple, Fraser, & Knightbridge unpubl. data).

Animal foods

Possum diets also routinely include insects and other animal material. In the most extensive study of possum consumption of invertebrates, Cowan & Moeed (1987) found 45% of monthly samples of possums faeces in the Orongorongo Valley contained invertebrate remains, with large, often slow-moving species such as stick insects, cicadas, wētā, and beetles dominating in summer and autumn, and fly larvae in winter. In that study, as in most previous studies, invertebrates formed a very small component of diet, such that some authors had presumed they had been eaten accidentally. A key exception was the presence of fly larvae in 30 of 31 possum stomachs taken in September in pine forest, representing 4.6% of stomach contents (Clout 1977). More recently, Rickard (1996) found large weevils and/or huhu beetles (*Prionoplus reticularis*) comprised over 10% of the stomach contents of each of three of 18 possums killed in rimu forest in January, with invertebrates comprising 2.4% of contents overall. Owen & Norton (1995) reported 28% of diet in August in beech forest consisted of insect larvae, though we suspect this is an overestimate caused by point-sampling bias. At Waihaha, insect larvae increased from 1% of the diet prior to control to 7% after the possum population was reduced by >90% (even though their diet after control contained a higher proportion of fruit; Nugent & Sweetapple unpubl. data). Intuitively, this increased use relative to other foods after a 10- to 12-fold increase (at least) in the per capita availability of all foods suggests that insects may be highly preferred items of food, and that the low occurrence usually reported in diet studies reflects either (a) the dietary assessment technique used, (b) low per capita availability, or (c) inefficient foraging ability by possums. Feeding trials have shown possums also eat native snails (K. Walker, cited in Innes 1995).

Possums sometimes eat eggs, nestlings, or adults of native bird species such as kōkako (*Callaeas cinerea*), kiwi (*Apteryx* spp.), kāhu (*Circus approximans*), fantail (*Rhipidura fuliginosa*), and kererū/kūkupa (*Hemiphaga novaeseelandiae*) (Brown *et al.* 1993; Innes 1995), and possums are increasingly being implicated as a major cause of decline of some of these species (e.g., James & Clout 1996). Although predation by possums is potentially crucial for any prey species in decline, vertebrates have seldom been reported as present in possum stomachs or faeces. This may partly reflect technical problems in identifying animal remains such as egg yolk, but even with "whole stomach" methods, their occurrence is low — usually just a few feathers indicate that a bird may have been eaten. Exceptions include one possum that ate a greenfinch (*Carduelis chloris*) (Parkes & Thomson 1995), and five of 23 possums that ate live or scavenged muttonbird chicks (*Puffinus griseus*) at Port Pegasus on Stewart Island; for one of these five, the chick eaten comprised about half the possum's stomach contents (Nugent unpubl. data).

Possums also readily eat raw and cooked meat. Possums in particularly poor condition have been observed feeding on deer (Thomas *et al.*, 1993), or possum carcasses (C. M. H. Clarke, retired ecologist, St. Arnaud, pers. comm.), suggesting that meat eating might be a response to starvation. However, they are routinely trapped in leg-hold traps baited with meat to catch ferrets, even when not apparently starving (Caley 1998).

Feeding preferences

Preference ranks for possum foods often differ from their ranking in the diet because the diet eaten in a particular area reflects not only what food is available, but how selectively possums feed. Preference rankings are calculated by comparing relative importance in the diet with some measure or index of the relative availability of potential foods. The technical and sampling difficulties in assessing forage availability for ephemeral foods such as insects, flowers, and fruit mean that there have been no quantitative studies of total forage availability. However, several studies have attempted to quantify the relative or total availability of foliage (Mason 1958; Fitzgerald 1976, 1978; Coleman *et al.* 1985; Owen & Norton 1995; Nugent *et al.* 1997). The foliage preference rankings derived are broadly similar, but are not fully consistent between areas and seasons. In all these studies the highest ranked species tend to be relatively uncommon. This suggests that possums make greater proportional use of some species when they are rare than when they are abundant. For example, kāmahi and toro (*Myrsine salicina*) rank ahead of māhoe in the Orongorongo Valley where the former two are rare but the latter is common (Fitzgerald 1976), but the rankings are reversed at Waihaha where relative abundance is reversed (Nugent *et al.* 1997).

Rarity aside, the general pattern is that possums most prefer the foliage of fast-growing seral and/or soft-leaved species: pōhuehue, tree fuchsia, wineberry, clover, kohekohe, and forest herbs. Some mistletoes may also belong to this group. This fits with the lack of any specialised adaption in possums to facilitate the clearance of indigestible bulk (fibre) from their hind-gut. Slightly less preferred are other seral species, small trees, and shrubs such as māhoe, five-finger, raukawa (*Raukaua edgerleyi*), haumakaroa (*R. simplex*), tītoki (*Alectryon excelsus*), tutu, and various large- and small-leaved *Coprosma* species. These intergrade in preference with slower-growing canopy and subcanopy species such as kāmahi, northern and southern rātā, tōtara, toro, and tawa that tend to form the bulk of the diet when locally abundant. The lianes (supplejack, scarlet rātā vine (*Metrosideros fulgens*), *Parsonsia* spp., and bush lawyer) probably fit within this group. The foliage of species that are seldom or never eaten by possums include many of the major forest canopy species: the podocarps other than Hall's tōtara and pāhautea, beech (*Nothofagus* spp.), hīnau, tōwheowheo (*Quintinia serrata*), rewarewa, and maire (*Nestegis* spp.), and shrubs and small trees such as broadleaf (*Griselinia littoralis*), marbleleaf, pigeonwood, and pepper tree. Ferns as a group are generally not preferred relative to woody species, but species such as *Blechnum penna-marina*, and the soft new growth of shield fern and mamaku tree fern (*Cyathea medullaris*) can sometimes form a substantial part of possums' diet.

Much has been made of the potential role of the herbivore-deterrent effects of plant secondary compounds in determining herbivore preferences (Hume 1999). Freeland & Winter (1975) suggest that possums in Australia are forced to be generalists because the high levels of secondary compounds in their main foods (*Eucalyptus* spp.) place limitations on the amount of such foliage they can consume. Consistent with this, Fitzgerald (1977, 1978) reported that possums fed single-species diets consumed lower than usual quantities of food, and that the main foliar foods of possums in the Orongorongo Valley (northern rātā, kāmahi) contained lower levels of secondary compounds than some less-preferred species. Likewise, Owen & Norton (1995) consider the high preference for seral species reflects a rapid growth – low defence strategy in these species that makes most of the energy and nutrients they contain readily available. However, possums in New Zealand do also regularly eat the foliage of some species known to contain toxic compounds such as saponins and alkaloids (e.g., māhoe, supplejack; Fitzgerald 1978), but typically only in conjunction with other foods. This, and the apparent tendency for possums to make greater proportional use of rare species, suggests deliberate diversification of diet.

Short-term diet changes – seasonality

The use of flowers, fruits, and invertebrates by possums appears to broadly track their availability. As already noted, flowers are mainly eaten in spring – early summer, and fruit in autumn. However, the ephemeral nature of flowers and fruits can result in rapid short-term diet changes. At Puketi in Northland, for example, the percentage of flowers and fruit in the autumn diet varied three and sevenfold respectively over six successive fortnightly periods (Table 2.1), demonstrating the dietary plasticity of possums.

Use of the foliage of deciduous or semi-deciduous species such as tree fuchsia, wineberry, and pōhuehue also appears to track their seasonal changes in availability. For example, in South Westland silver beech (*Nothofagus menziesii*) forest, the pattern is one of high use in spring and summer of soft-foliaged species such as pōhuehue and tree fuchsia, offset by

Table 2.1

Rapid changes in possum diet (percent dry weight of stomach contents) over 2.5 months in autumn 1993, Puketi, Northland. The predominant flowers eaten were nīkau and tōwai, the predominant fruit was hīnau, and the predominant leaves were kohekohe, northern rātā, and *Pinus radiata.*

	Date of collection and sample size					
	17 March (n = 41)	31 March (n = 31)	15 April (n = 45)	29 April (n = 42)	10 May (n = 23)	26 May (n = 32)
Flowers	32	16	23	10	9	13
Fruit	7	26	15	16	42	9
Leaf	58	51	57	68	32	78

(Source: P. Suisted, I. Payton, R. Pierce, Landcare Research, Lincoln, unpubl. data).

high late-autumn and winter use of the thicker-leaved species of intermediate or low preference (Owen & Norton 1995): regrouping of their data indicates that the latter species comprised just 2.5% of leaf diet in November, but 31% of diet in May and 21% in August. However, seasonal trends in use also exist for the foliage of evergreen species. For these, foliage is present in all seasons so seasonal variation in use presumably reflects seasonal changes in palatability, or at least in relative palatability. As an example, possum use of Hall's tōtara is consistently highest during late spring and summer when new foliage is available, and much lower in other seasons (Nugent *et al.* 1997; Rogers 1997). At Waihaha, the increased use of Hall's tōtara (the main food there) coincided with an equivalent fall in the use of kāmahi and toro foliage (the other two main foods), indicating that (a) there was a substantial seasonal change in the nutritive value of tōtara; (b) the change was greater than any similar variation for kāmahi and toro; and (c) that kāmahi and toro are less preferred foods than the summer foliage of tōtara. In Westland podocarp–hardwood forest, consumption of southern rātā and māhoe was usually lowest in autumn when kāmahi use was highest, at least for females (Coleman *et al.* 1985). In Orongorongo podocarp–hardwood forest, use of northern rātā was consistently highest in February, and much lower in June and September (Allen *et al.* 1997). In this instance the lower use of rātā in winter and spring was offset primarily by increased use of supplejack, māhoe, and rātā vine.

Long-term shifts in diet

When possums colonise an area, it is generally accepted that the population goes through some form of irruptive oscillation in density similar to that described for ungulates. Possum numbers increase to a peak, overshoot the carrying capacity of the area, and then decline to a more sustainable level (Thomas *et al.* 1993). Under this scenario, possums initially focus on the most highly preferred foods, but progressively switch to less preferred foods as the preferred foods are eaten out, as illustrated by the series of studies showing the disappearance or decline of highly preferred species from both forests and possum diet in the Orongorongo Valley between 1947 and 1976 (Mason 1958; Fitzgerald 1976; Campbell 1990; J. Alley, cited in Nugent 1995; Allen *et al.* 1997). This is also consistent with the differences in diet observed between colonising, near-peak, and post-peak populations in South Westland (Appendix 2.1).

Possum control effectively reverses these long-term changes in possum diet: as already noted, a > 90% reduction in possum density at Waihaha resulted in a marked decline in relative use of the major foliar foods (tōtara, kāmahi, and toro) from 47% to 17% of diet, while relative use of the foliage of highly preferred species (pōhuehue, wineberry, mistletoe, *Parsonsia* species, and forest herbs) rose from 9% to 19% of total diet (Nugent & Sweetapple unpubl. data).

Dietary model

Although foliage dominates the diet of possums in all habitats throughout New Zealand, the apparently heavy utilisation of flowers, fruits, and invertebrates whenever they are available suggest that foliage is the least preferred of these four food groups. The few studies of the chemical composition of possum foods in New Zealand suggest that the main foliar foods of post-peak possum populations in native forest are typically very low in protein and/or nitrogen

(Fitzgerald 1976; Williams 1982) and may be below the level required by possums to maintain a positive nitrogen balance (Tyndale-Biscoe 1997). Not surprisingly, Williams (1982) also found that the fruit from five species whose foliage is commonly eaten in the Orongorongo Valley were richer in available carbohydrates and lower in fibre compared with foliage. A logical inference from these studies is that the lower preference for foliage relative to the other three food groups reflects a generally lower nutritive value. There is some dietary evidence to support that. Kāmahi foliage, for example, is most commonly the main food of possums in native forest, but appears to be a "subsistence" food that is only eaten in large quantities when possum numbers are high and other more preferred foods are difficult to find; it increases in importance with increasing length of possum site occupation in South Westland (Appendix 2.1), and it drops quickly in importance when possum numbers are reduced (Nugent & Sweetapple unpubl. data). Clear evidence of the inadequacy of foliage of some commonly eaten native species as sole foods is provided by a series of feeding trials in which possums were offered leaves of either a single species (māhoe, five-finger, kāmahi, and northern rātā), a mixture of the first three, or commercial food (Fitzgerald 1977; Fitzgerald et al. 1981). In these trials, possums lost 6–18% of their body weight on leaf diets with dry-matter (DM) digestibilities of between 43% and 57%. On the commercial herbivore food (DM digestibility 57–60%), possums gained 2–4% of their body weight, while possums fed a 2:1 mixture of peas and bread (DM digestibility 91%) gained 28% of their body weight.

We postulate that the key to determining long-term local possum carrying capacities is not the availability of their main foliage foods, but rather the availability of energy-rich food sources such as fruit, or, as suggested by Tyndale-Biscoe (1997), the availability of nitrogen-rich species such as tree fuchsia, insects, bird's eggs or fungi. Supporting this argument, annual fluctuations in possum body weights and breeding success in the Orongorongo Valley are closely correlated to the size of the annual hīnau fruit crop, and few possum pouch young survive to independence in years when there are few hīnau fruit (Bell 1981). As a result the possum population fluctuates in size from year to year, but remains around a constant long-term mean. Paradoxically, this mean has remained unchanged since 1947 (Brockie 1992), despite the shift to less preferred, and presumably therefore less nutritious, foliage foods (Green 1984; Allen *et al.* 1997), which suggests the main foliage foods themselves cannot be the primary determinants of carrying capacity despite their overwhelming predominance in the diet. Paralleling that, possum-induced mortality of northern rātā in the Orongorongo Valley appears to be episodic, coinciding with periods of high possum numbers (Cowan *et al.* 1997), even though its importance in the diet appears to remain constant through time (Allen *et al.* 1997). This suggests that changes in possum numbers determine availability (survival) of northern rātā rather than the reverse.

In Australia, where possum diet is almost exclusively foliar (Freeland & Winter 1975; Fitzgerald 1984), possum densities are generally low compared with New Zealand forests (Fitzgerald 1984; Green 1984; Cowan 1990a). There, as in New Zealand, soft-leaved herbaceous foliage is an important addition to the predominant diet of leaves from woody species. We suggest that possum populations are able to attain the high densities seen in New Zealand native forests by topping up a subsistence diet of bulk foliar foods with high-energy and/or high-nutrient foods such as fruits, flowers, insects, eggs, or the soft highly digestible leaves of fast-growing seral or herbaceous species. Consistent with that, possum densities are generally low in low-diversity beech forests where there are few fleshy-fruit-producing species (Cowan 1990a) except where possums have only recently (within 20–30 years) invaded the area and highly preferred seral species (tree fuchsia, pōhuehue, wineberry, mistletoe, patē, five-finger) are still abundant.

On balance, the evidence suggests that possums are somewhat reluctant folivores that are nonetheless capable of subsisting on a pure leaf diet, but only if some component of it is highly digestible. Evidence also suggests that where high-energy/high-nutrient non-foliar foods are abundant, the site's possum-carrying capacity will be high. Further, impacts of possums at such sites will be more severe because the bulk of their diet will still consist of foliage, and result in the death of favoured species such as kāmahi and rātā (Cowan 1990a). Where highly nutritious foods are rare or absent, however, we would expect kāmahi and rātā to be largely undamaged. We argue that much of the temporal, regional, and local variation in possum impacts, for which a wide variety of explanations have been offered, may simply reflect variation in both the relative and absolute availability of high-quality foods.

One interesting consequence of this model is that where fruit-producing species such as hīnau are not themselves heavily affected by possums, they may be able to take advantage of the reduced competition for light and space that follows dieback of rātā and kāmahi (for example). Overall fruit production, and therefore possum carrying capacity, could conceivably increase through time rather than decrease. However, this scenario might not eventuate if increased possum use resulted in disproportionately increased suppression of fruit production. Cowan & Waddington (1990) document reduced fruiting in hīnau at high possum density.

Another implication is that reductions in possum density through possum control will instantly result in a greater per capita availability of highly preferred foods. The defoliation of species with intermediate or low preference rankings should therefore be particularly responsive to possum control efforts. At Waihaha, for example, the reduction in possum density was coupled with a 78% reduction in per capita consumption of kāmahi by the few remaining possums, magnifying the benefit of possum control for that species.

Summary

- Brushtail possums are generalist and opportunistic arboreal folivores that eat a wide range of foliar and non-foliar foods, both from trees and shrubs and on the ground. As a result they have successfully established in virtually every New Zealand terrestrial habitat, and their diets are as varied as the habitats they occupy.
- The digestive anatomy and physiology of possums is adapted to utilise the low-quality foliar foods that usually form the bulk of their diet, but they are less efficient at doing this than other more strictly folivorous species. They therefore tend to supplement their diet with high-energy/high-nutrient non-foliar foods whenever such foods are available.
- Where possums have access to pasture, pasture species (particularly clovers and some introduced grasses) are prominent diet components, but a wide range of exotic and native woody plants are also eaten. In exotic pine forests, broom, blackberry, grasses, and herbs are the main foods, with native woody species also eaten when available.
- In native forests, the plants most commonly eaten in large quantities are woody species including kāmahi, northern and southern rātā, and where present, kohekohe, pōhutukawa, tawa, and Hall's tōtara. Other species frequently eaten, but generally in smaller quantities, are tree fuchsia, wineberry, māhoe, patē, *Pseudopanax* and *Raukaua* spp., *Coprosma* spp., pōhuehue, bush lawyer, and forest herbs.
- The most highly preferred foods in native forest tend to be the less common and generally soft-leaved seral species such as tree fuchsia, wineberry, pōhuehue, and herbs, whereas the species that usually make up the bulk of the foliar diet are usually only intermediate in preference ranking.
- Flowers, fruit, and small quantities of invertebrates are eaten in all habitats whenever they are available. At times flowers or fruits may form the bulk of seasonal diets.
- Possum diet is often markedly seasonal. Seasonal shifts in diet reflect not only changing availability of non-foliar foods and foliage of preferred deciduous species, but also seasonal changes in the relative palatability of evergreen foliage. Long-term changes in possum diet can also occur where their browsing induces changes in vegetation composition.
- The apparently heavy utilisation of non-foliar foods whenever they are available suggests that possums prefer these to foliage. Limited evidence indicates that these non-foliar foods, particularly fruits, are nutritionally superior to foliage, and that possum reproductive success, and therefore local possum carrying capacity, is linked to their availability. The availability of these foods may therefore be the key to understanding variable possum densities and impacts in different habitats.

References

Allen, R. B.; Fitzgerald, A. E.; Efford, M. G. 1997: Long-term changes and seasonal patterns in possum (*Trichosurus vulpecula*) leaf diet, Orongorongo Valley, Wellington, New Zealand. *New Zealand Journal of Ecology 21*: 181–186.

Bell, B. D. 1981: Breeding and condition of possums *Trichosurus vulpecula* in the Orongorongo Valley , near Wellington, New Zealand, 1966–1975. *In:* Bell, B.D. *ed.* Proceedings of the first symposium on marsupials in New Zealand. *Zoology Publication from the Victoria University of Wellington No. 74*. Pp. 87–139.

Brockie, R. E. 1992: A living New Zealand forest. Auckland, David Bateman. 172 p.

Brown, K.; Innes, J.; Shorten, R. 1993: Evidence that possums prey on and scavenge birds' eggs, birds and mammals. *Notornis 40:* 169–177.

Caley, P. 1998: Broad-scale possum and ferret correlates of macroscopic *Mycobacterium bovis* infection in feral ferret populations. *New Zealand Veterinary Journal 46:* 157–162.

Campbell, D. J. 1990: Changes in structure and composition of a New Zealand lowland forest inhabited by brushtail possums. *Pacific Science 44:* 277–296.

Clout, M. N. 1977: The ecology of the possum (*Trichosurus vulpecula* Kerr) in *Pinus radiata* plantations. Unpublished PhD thesis, University of Auckland, Auckland, New Zealand.

Coleman, J. D.; Green, W. Q.; Polson, J. G. 1985: Diet of brushtail possums over a pasture-alpine gradient in Westland, New Zealand. *New Zealand Journal of Ecology 8:* 21–35.

Cork, S. J.; Foley, W. J. 1991: Digestive and metabolic strategies of arboreal mammalian folivores in relation to chemical defenses in temperate and tropical forests. *In:* Palo, R. T.; Robbins, C. T. *ed.* Plant defenses against mammalian herbivory. Boca Raton, Louisiana, USA. CRC Press. Pp. 133–166.

Cowan, P. E. 1990a: Brushtail possum. *In:* King, C. M. *ed.* The handbook of New Zealand mammals. Auckland, Oxford University Press. Pp. 68–98.

Cowan, P. E. 1990b: Fruits, seeds, and flowers in the diet of brushtail possums, *Trichosurus vulpecula,* in lowland podocarp/mixed hardwood forest, Orongorongo Valley, New Zealand. *New Zealand Journal of Zoology 17:* 549–566.

Cowan, P. E.; Moeed, A. 1987: Invertebrates in the diet of brushtail possums, *Trichosurus vulpecula,* in lowland podocarp/broadleaf forest, Orongorongo Valley, Wellington, New Zealand. *New Zealand Journal of Zoology 14:* 163–177.

Cowan, P. E.; Waddington, D. C. 1990: Suppression of fruit production of the endemic forest tree, *Elaeocarpus dentatus,* by introduced marsupial brushtail possums, *Trichosurus vulpecula. New Zealand Journal of Botany 28:* 217–224.

Cowan, P. E.; Chilvers, B. L.; Efford, M. G.; McElrea, G. J. 1997: Effects of possum browsing on northern rata, Orongorongo Valley, Wellington, New Zealand. *Journal of the Royal Society of New Zealand 27:* 173–179.

Fitzgerald, A. E. 1976: Diet of the opossum *Trichosurus vulpecula* (Kerr) in the Orongorongo Valley, Wellington, New Zealand, in relation to food-plant availability. *New Zealand Journal of Zoology 3:* 399–419.

Fitzgerald, A. E. 1977: Number and weight of faecal pellets produced by opossums. *Proceedings, New Zealand Ecological Society 24:* 76–78.

Fitzgerald, A. 1978: Aspects of the food and nutrition of the brush-tailed opossum, *Trichosurus vulpecula* (Kerr, 1792), Marsupialia: Phalangeridae, in New Zealand. *In:* Montgomery, G. G. *ed.* The ecology of arboreal folivores. *Symposia of the National Zoological Park.* Washington DC, Smithsonian Institution Press. Pp. 289–303.

Fitzgerald, A. E. 1984: Diet of the opossum *(Trichosurus vulpecula)* in three Tasmanian forest types and its relevance to the diet of possums in New Zealand forests. *In:* Smith, A. P.; Hume, I. D. *ed.* Possums and gliders. Chipping Norton, NSW, Surrey Beatty in assoc. with the Australian Mammal Society. Pp. 137–143.

Fitzgerald, A. E.; Wardle, P. 1979: Food of the opossum *Trichosurus vulpecula* (Kerr) in the Waiho Valley, South Westland. *New Zealand Journal of Zoology 6:* 339–345.

Fitzgerald, A. E.; Clarke, R. T. J.; Reid, C. S. W.; Charleston, W. A. G.; Tarttelin, M. F.; Wyburn, R. S. 1981: Physical and nutritional characteristics of the possum (*Trichosurus vulpecula*) in captivity. *New Zealand Journal of Zoology 8:* 551–562.

Foley, W. J.; Hume, I. D. 1987. Passage of digesta markers in two species of arboreal folivorous marsupials – the greater glider (*Petauroides volans*) and the brushtail possum (*Trichosurus vulpecula*). *Physiological Zoology 60:* 103–113.

Freeland, W. J; Winter, J. W. 1975: Evolutionary consequences of eating: *Trichosurus vulpecula* (Marsupialia) and the genus *Eucalyptus. Journal of Chemical Ecology 1:* 439–455.

Green, W. Q. 1984: A review of ecological studies relevant to management of the common brushtail possum. *In:* Smith, A. P.; Hume, I. D. *ed.* Possums and gliders. Chipping Norton, NSW, Surrey Beatty in assoc. with the Australian Mammal Society. Pp. 483–499.

Gilmore, D. P. 1967: Foods of the Australian opossum *(Trichosurus vulpecula* Kerr) on Banks Peninsula, Canterbury, and a comparison with other selected areas. *New Zealand Journal of Science 10:* 235–279.

Harvie, A. E. 1973: Diet of the opossum (*Trichosurus vulpecula* Kerr) on farmland northeast of Waverley, New Zealand. *Proceedings, New Zealand Ecological Society 20:* 48–52.

Hume, I. D. 1999: Marsupial nutrition. Cambridge, Cambridge University Press. 432 p.

Innes, J. 1995: The impacts of possums on native fauna. *In:* O'Donnell, C.F.J. *comp.* Possums as conservation pests. Proceedings of an NSSC Workshop . . . 29–30 November 1994. Wellington, Department of Conservation. Pp. 11–15.

James, R. E.; Clout, M. N. 1996: Nesting success of New Zealand pigeons (*Hemiphaga novaeseelandiae*) in response to a rat (*Rattus rattus*) poisoning programme at Wenderholm Regional Park. *New Zealand Journal of Ecology 20:* 45–51.

Leathwick, J. R.; Hay, J. R.; Fitzgerald, A. E. 1983: The influence of browsing by introduced mammals on the decline of North Island kokako. *New Zealand Journal of Ecology 6:* 55–70.

MacLennan, D. G. 1984: The feeding behaviour and activity patterns of the brushtail possum, *Trichosurus vulpecula,* in an open eucalypt woodland in Southeast Queensland. *In:* Smith, A. P.; Hume, I. D. *ed.* Possums and gliders. Chipping Norton, NSW, Surrey Beatty in assoc. with the Australian Mammal Society. Pp. 155–161.

Mason, R. 1958: Foods of the Australian opossum (*Trichosurus vulpecula,* Kerr) in New Zealand indigenous forest in the Orongorongo Valley, Wellington. *New Zealand Journal of Science 1:* 590–613.

Nugent, G. 1995: Effects of possums on the native flora. *In:* O'Donnell, C. F. J. *comp.* Possums as conservation pests. Proceedings of an NSSC workshop . . . 29–30 November 1994. Wellington, Department of Conservation. Pp. 5–10.

Nugent, G.; Fraser, K. W.; Sweetapple, P. J. 1997: Comparison of red deer and possum diets and impacts in podocarp-hardwood forest, Waihaha Catchment, Pureora Conservation Park. *Science for Conservation 50.* Wellington, Department of Conservation. 61 p.

Owen, H. J.; Norton, D. A. 1995: The diet of introduced brushtail possums *Trichosurus vulpecula* in a low-diversity New Zealand *Nothofagus* forest and possible implications for conservation management. *Biological Conservation 71:* 339–345.

O'Cain, M. J. 1997: The role of possums in forest regeneration, Hoon Hay Valley, Port Hills, Canterbury, New Zealand. Unpublished MForSc thesis, University of Canterbury, Christchurch, New Zealand.

Olds, C. B.1987: Aspects of the ecology of the brush-tail possum (*Trichosurus vulpecula* Kerr, 1792) on Rangitoto Island. Unpublished MSc thesis, University of Auckland, Auckland, New Zealand.

Parkes, J. P.; Thomson, C. 1995: Management of thar. Part II: Diet of thar, chamois, and possums. *Science for Conservation 7.* Wellington, Department of Conservation. Pp. 22–42.

Rickard, C. G. 1996: Introduced small mammals and invertebrate conservation in a lowland podocarp forest, South Westland, New Zealand. Unpublished MForSc thesis, University of Canterbury, Christchurch, New Zealand.

Rogers, G. 1997: Trends in health of pahautea and Hall's totara in relation to possum control in central North Island. *Science for Conservation 52.* Wellington, Department of Conservation. 49 p.

Sweetapple, P. J.; Nugent, G. 1998: Comparison of two techniques for assessing possum (*Trichosurus vulpecula*) diet from stomach contents. *New Zealand Journal of Ecology 22:* 181–188.

Thomas, M. D.; Hickling, G. J.; Coleman, J. D.; Pracy. L. T. 1993: Long-term trends in possum numbers at Pararaki: Evidence of an irruptive fluctuation. *New Zealand Journal of Ecology 17:* 29–34.

Tyndale-Biscoe, C. H. 1997: Possums and Westland forests – a 45 year perspective. Manaaki Whenua – Landcare Research 1996 Hayward Fellowship final report (unpublished). 46 p.

Warburton, B. 1978: Foods of the Australian brush-tailed opossum (*Trichosurus vulpecula*) in an exotic forest. *New Zealand Journal of Ecology 1:* 126–131.

Ward, D. G. 1978: Habitat use and home range of radio-tagged opossums *Trichosurus vulpecula* (Kerr) in New Zealand lowland forest. *In:* Montgomery, G. G. *ed.* The ecology of arboreal folivores. *Symposia of the National Zoological Park.* Washington DC, Smithsonian Institution Press. Pp. 267–287.

Wellard, G. A.; Hume, I. D. 1981a: Nitrogen metabolism and nitrogen requirements of the brushtail possum, *Trichosurus vulpecula* (Kerr). *Australian Journal of Zoology 29:* 147–156.

Wellard, G. A.; Hume, I. D. 1981b: Digestion and digesta passage in the brushtail possum, *Trichosurus vulpecula* (Kerr). *Australian Journal of Zoology 29:* 157–166.

Williams, C. K. 1982: Nutritional properties of some fruits eaten by the possum *Trichosurus vulpecula* in a New Zealand broadleaf-podocarp forest. *New Zealand Journal of Ecology 5:* 16–20.

Appendix 2.1

Summary of major possum diet studies in New Zealand including unpublished studies known to the authors. The eight most dominant foods reported in each study in terms of percentages of total dietary composition are listed. For those studies marked with asterisks the percentages are for the leaf component of diet only, whereas the others include fruits as well. Data are directly measured percent dry weights or calculated equivalents, except for the two earliest Orongorongo studies where the data are percentages of the total number (across all species).

	1	2	3	4	5	6	7	8
Silver beech forest, S. Westland. Owen & Norton 1995	*Aristotelia serrata* 31%	*Muehlenbeckia australis* 24%	Insect larvae 8%	*Rubus* spp. 5%	*Fuchsia excorticata* 4%	*Pseudopanax* spp. 4%	*Neomyrtus pedunculata* 3%	Fruit 2%
Silver beech forest , S. Westland. (Invading). Sweetapple unpubl.[1]	*Muehlenbeckia australis* 21%	*Aristotelia serrata* 18%	*Weinmannia racemosa* 13%	*Fuchsia excorticata* 8%	Ferns 8%	*Peraxilla colensoi* 4%	*Rubus* spp. 3%	*Pennantia corymbosa* 1%
Silver beech forest S. Westland. (Peak). Sweetapple unpubl.[1]	*Fuchsia excorticata* 40%	*Aristotelia serrata* 13%	*Muehlenbeckia australis* 12%	Ferns 9%	*Weinmannia racemosa* 7%	*Raukaua simplex* 4%	Herbs 3%	*Melicytus ramiflorus* 2%
Silver beech forest S. Westland. (Post-peak). Sweetapple unpubl.[1]	*Weinmannia racemosa* 21%	Ferns 10%	*Fuchsia excorticata* 9%	*Aristotelia serrata* 9%	*Coprosma foetidissma* 7%	*Muehlenbeckia australis* 5%	*Melicytus ramiflorus* 5%	*Hoheria glabrata* 3%
Podocarp–hardwood forest, Stewart Island. March 1993. Nugent unpubl.[3]	*Podocarpus hallii* 24%	*Metrosideros umbellata* 20%	*Pseudopanax simplex* 18%	*Senecio reinoldii* 8%	Ferns 7%	*Prumnopitys ferruginea* 4%	Bird 3%	Fungi 1%
Mixed hardwood–pasture Mt Bryan O'Lynn, Westland. Coleman *et al.* 1985*	*Weinmannia racemosa* 33%	*Metrosideros umbellata* 24%	*Melicytus ramiflorus* 12%	*Pseudopanax* spp. 4%	*Coprosma foetidissima* 3%	Ferns 2%	Herbs and grasses 2%	*Raukaua simplex* 2%
Hardwood forest Alex Knob, Westland. Fitzgerald & Wardle 1979*	*Weinmannia racemosa* 27%	Ferns 16%	*Melicytus ramiflorus* 12%	*Metrosideros umbellata* 12%	*Fuchsia excorticata* 8%	*Schefflera digitata* 6%	*Ripogonum scandens* 4%	Unidentified sp. 3%

Appendix 2.1 continued

	1	2	3	4	5	6	7	8
Hardwood forest Douglas Track, Westland Fitzgerald & Wardle 1979*	*Metrosideros umbellata* 20%	*Weinmannia racemosa* 13%	*Schefflera digitata* 13%	*Fuchsia excorticata* 13%	Ferns 10%	*Melicytus ramiflorus* 10%	*Hoheria glabrata* 6%	Unidentified sp. 4%
Podocarp–hardwood forest remnant Banks Peninsula, 1993. Suisted unpubl. [2]	*Podocarpus hallii* 29%	*Fuchsia excorticata* 27%	Herbs 12%	*Rubus* spp. 8%	*Griselinia littoralis* 5%	*Muehlenbeckia australis* 4%	*Pittosporum eugenioides* 4%	*Pennantia corymbosa* 3%
Māhoe older forest remnant Banks Peninsula. O'Cain 1997	*Myoporum laetum* 24%	*Muehlenbeckia australis* 20%	*Sambucus nigra* 14%	*Melicytus ramiflorus* 11%	*Solanum aviculare* 9%	*Cytisus scoparius* 7%	*Rubus fruticosus* 4%	Herbs 3%
Mixed beech podocarp Orongorongo Wellington 1946–47. Mason 1958.*	*Fuchsia excorticata* 14%	*Metrosideros robusta* 13%	*Alectryon excelsus* 12%	*Weinmannia racemosa* 9%	*Pseudopanax arboreus* 9%	*Brachyglottis repanda* 4%	*Leptospermum* spp. 4%	*Metrosideros fulgens* 2%
Mixed beech podocarp Orongorongo, Wellington 1969–73. Fitzgerald 1976*	*Weinmannia racemosa* 33%	*Metrosideros robusta* 29%	*Metrosideros fulgens* 9%	*Melicytus ramiflorus* 8%	*Ripogonum scandens* 7%	*Geniostoma ligustrifolium* 5%	*Pseudopanax arboreus* 5%	*Elaeocarpus dentatus* 2%
Mixed beech podocarp Orongorongo, Wellington 1976–89. Allen *et al.* 1997*	*Metrosideros robusta* 36%	*Ripogonum scandens* 20%	*Weinmannia racemosa* 16%	*Melicytus ramiflorus* 16%	*Metrosideros fulgens* 6%	*Geniostoma ligustrifolium* 3%	*Laurelia novae-zelandiae* 2%	*Elaeocarpus dentatus* 1%
Podocarp–hardwood Hihitahi, Central N.I. Rogers 1997	*Aristotelia serrata* 21%	Ferns 20%	*Podocarpus hallii* 17%	Herbs 9%	*Libocedrus bidwillii* 8%	*Carpodetus serratus* 6%	*Rubus* spp. 6%	Insects 2%
Podocarp–hardwood Waihaha, W. Taupo (pre-control) Nugent *et al.* 1997.	*Podocarpus halli* 19%	*Weinmannia racemosa* 18%	*Myrsine salicina* 10%	*Prumnopitys ferruginea* 8%	*Rubus* spp. 4%	*Elaeocarpus hookerianus* 4%	*Carpodetus serratus* 4%	*Aristotelia serrata* 3%

Appendix 2.1 continued

	1	2	3	4	5	6	7	8
Podocarp–hardwood Waihaha, W. Taupo (post-control). Nugent unpubl.[3]	*Rubus* spp. 17%	*Elaeocarpus hookerianus* 14%	*Podocarpus hallii* 9%	Fly larvae 7%	*Aristotelia serrata* 7%	*Muehlenbeckia australis* 7%	*Weinmannia racemosa* 4%	*Myrsine salicina* 4%
Podocarp–hardwood Pureora, Central N.I. Leathwick *et al.* 1983*	*Weinmannia racemosa* 20%	*Beilschmiedia tawa* 10%	*Melicytus ramiflorus* 9%	*Pseudopanax arboreus* 8%	*Pseudopanax edgerleyi* 7%	Ferns 6%	*Ripogonum scandens* 6%	*Rubus* spp. 6%
Podocarp–hardwood Mokau, Northland. Suisted unpubl.[4]	*Dysoxylum spectabile* 21%	*Elaeocarpus dentatus* 11%	*Rophalostylis sapida* 9%	*Weinmannia silvicola* 8%	*Metrosideros robusta* 7%	*Pinus radiata* 5%	Ferns 8%	Herbs 3%
Tawa–hardwood Mapara, Central N.I. Leathwick *et al.* 1983*	*Beilschmiedia tawa* 24%	*Melicytus ramiflorus* 18%	*Alectryon excelsus* 14%	Ferns 13%	*Fuchsia excorticata* 7%	*Aristotelia serrata* 6%	Grasses 5%	*Geniostoma ligustrifolium* 3%
Kāmahi–tawa forest, Mamaku Plateau, Central. N.I. Clout 1977	Dicotyledons 68%	Fruit 10%	Ferns 4%	Flowers 3%	Monocoty-ledons 3%	Invertebrates 2%	Bark <1%	*Pinus radiata* <1%
Pōhutukawa Rangitoto Island, Auckland. Olds 1987*	*Astelia banksii* 23%	*Myrsine australis* 21%	*Metrosideros* spp. 19%	*Cyathodes juniperina* 13%	*Griselinia lucida* 7%	*Coprosma* spp. 6%	*Coriaria arborea* 2%	*Trichomanes reniformes* 2%
Alpine shrubland Rangitata, Central S.I. Parkes & Thomson 1995	*Podocarpus nivalis* 22%	*Muehlenbeckia axillaris* 11%	*Coriaria angustissima* 10%	*Hieracium* spp. 8%	*Hoheria lyallii* 6%	*Trifolium repens* 5%	*Raoulia tenuicaulis* 4%	*Aristotelia fruticosa* 3%
Pine plantation – 14-yr-old stand Tokoroa, Central. N.I. Clout 1977	Dicotyledons 45%	*Pinus radiata* 31%	Ferns 9%	Invertebrates 2%	Monocoty-ledons 1%	Flowers 1%	Fruit 1%	Bark <1%

Appendix 2.1 continued

	1	2	3	4	5	6	7	8
Pine plantation – 3-yr-old stand Tokoroa, Central. N.I. Clout 1977.	Dicotyledons 40%	Monocotyledons 30%	Ferns 14%	Fruit 2%	*Pinus radiata* 1%	Bark <1%	Flowers <1%	Invertebrates <1%
Pine plantation Ashley Forest, Canterbury Warburton 1978	*Cystisus scoparius* 27%	*Pinus radiata* 22%	Grasses 12%	*Rubus fruticosa* 11%	*Ulex europaeus* 4%	*Muehlenbeckia australis* 3%	*Rumex acetosa* 3%	*Pseudotsuga menziesii* 2%
Pastureland Banks Peninsula, Canterbury Gilmore 1967	Clovers c. 37%	Grasses c. 26%	NA	NA	NA	NA	NA	NA

Authors of unpublished Landcare Research, Lincoln, studies
[1] P. J. Sweetapple, K. W. Fraser & P. Knightbridge
[2] P. Suisted & I. Payton
[3] G. Nugent & P. Sweetapple
[4] P. Suisted, I. Payton & R. Pierce

CHAPTER THREE

Possums on the Move: Activity Patterns, Home Ranges, and Dispersal

Phil Cowan and Mick Clout

Brushtail possums, *Trichosurus vulpecula*, despite their relatively low reproductive rate, have proven to be highly successful colonists of New Zealand and formidable pests (Cowan 1990). They transmit livestock diseases, damage crops, and threaten native forests and wildlife, despite persistent efforts to control them (Green 1984; Cowan 1990). To control them efficiently it is necessary to understand not only their population dynamics, but also their behaviour. The daily and seasonal movements of individuals influence their susceptibility to control, and dispersal movements affect the rate of recovery after populations have been reduced.

Knowledge of possum movement patterns has three important management applications. First, the most effective spacing of traps, bait stations, and poison lines for possum control can be determined from movement information. Second, the rate and scale of movements by possums recolonising controlled areas (along with reproductive and survival rates) dictate the frequency with which control must be carried out to keep possum damage or disease transmission risks at acceptable levels. Third, long-distance dispersal by possums may be responsible for the continuing spread of bovine Tb or reinfection of areas under control. In this chapter we review current information about home range, activity patterns, den use, and movements of brushtail possums, placing these findings in the context of control efforts.

Activity patterns

Possums are strictly nocturnal in their activity, spending the daytime in dens. Dens are usually located in sheltered sites, typically above ground in tree hollows, other cavities, or perching epiphytes (Cowan 1989). Otherwise they are under logs, in dense clumps of ground vegetation, underground in the burrows of other animals, or under tree roots. Males and females use similar dens, but males use more dens than females (Green & Coleman 1987; Cowan 1989). Individual possums use 11–15 different dens per year (most of them only occasionally) and change dens frequently, on average two nights in three (Cowan 1989). Den sharing is uncommon, but dens are often used sequentially by several different possums. Exceptions to this general lack of den sharing are co-occupation by females and their young, and unusual situations where dens are few and possum density is high (Fairweather *et al.* 1987; Cowan 1990). Where possum numbers have been reduced by control, den sharing is greatly reduced (Caley *et al.* 1998).

Possums awaken and start to become active in their dens about 1–2 hours before sunset, but do not usually emerge until about 30 minutes after sunset (Winter 1976; Ward 1978; MacLennan 1984). The first 1–2 hours after emergence are usually spent in the den tree, if present, grooming, sitting, or moving about (Ward 1978; MacLennan 1984) (Fig. 3.1). Heavy rain may delay emergence for up to 5 hours (Ward 1978). Exceptions to this pattern are starving or sick animals, which may emerge to feed in the afternoon, especially during winter.

Once out of the den, feeding occupies only about 1–2 hours (10–15%) of the nocturnal activity period, starting from about 2 hours after sunset (Fig. 3.1). Individuals feed mostly at two to four different sites during the night, in two to three sessions, separated by episodes of 2–3 hours of relative inactivity, and together these occupy about 45% of the time spent out of the den (Winter 1976; Ward 1978; MacLennan 1984). About 10–20% of the time is spent grooming and 20–30% on other activities, such as travelling between feeding sites. Little time is spent on active social interaction, except during the breeding season (Winter 1976; Ward 1978; MacLennan 1984; Paterson *et al.* 1995). Paterson *et al.* (1995) found that peak numbers of possums were observed between 2300 and 0230 hours. Unless they are driven to shelter early by particularly bad weather conditions, possums return to their dens just before dawn in summer, but often several hours earlier in

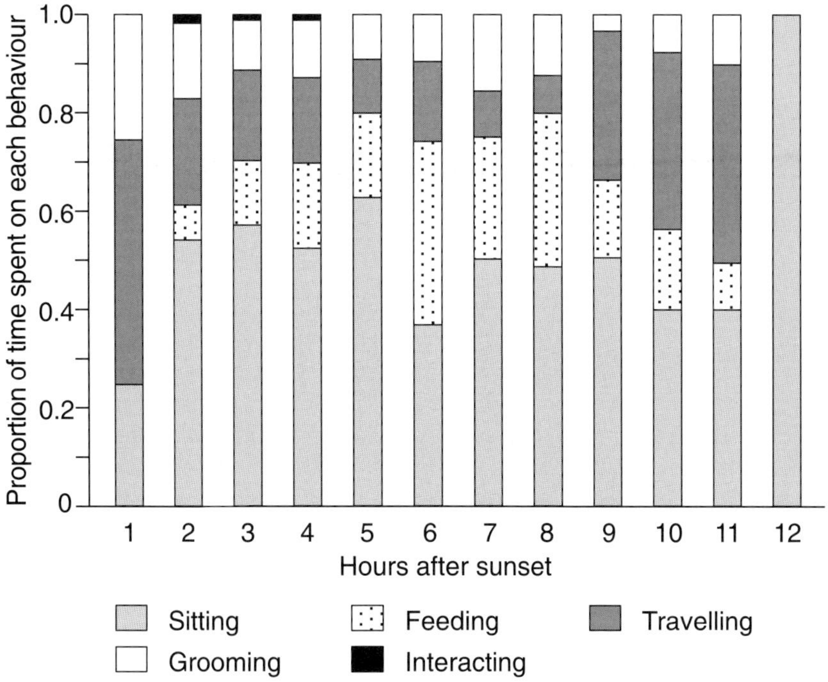

Fig. 3.1
Proportions of each hour from sunset to denning spent by *Trichosurus vulpecula* sitting, feeding, travelling, grooming, and interacting. (Derived from consolidated data from observations of 50 brushtail possums) (from MacLennan 1984).

winter (Ward 1978). Commonly, males return later than females (Winter 1976).

Although possums are largely arboreal, they spend about 10–15% of their time on the ground in forest habitats, feeding and moving about. They are more active on the ground on moonlit nights and less so in heavy or persistent rain (Winter 1976; Ward 1978; MacLennan 1984). Strong winds may also depress activity (Paterson *et al.* 1995). In Australia, predators such as foxes influence possum activity, such that when fox numbers are reduced, possums significantly increase the time they spend feeding and moving on the ground (Pickett 1997). Such an effect is probably unlikely in New Zealand because of the lack of predators.

Home range and movements

By *home range* we mean the area within which an animal lives, feeds, and breeds (Burt 1940). Although home ranges tend to be stable over time, there may be seasonal differences in the way in which they are used. *Home-range length* is defined as the maximum distance across the range, and gives a useful estimate of the maximum distance a possum might move to traps, bait stations, or poison lines. *Home-range area* is a more difficult parameter to define, because its size depends on how animals are located (e.g., trapping, spotlighting, radio-tracking), on the method of calculation from the points at which the animal has been located, and the time over which movements are considered (Ward 1984, 1978; Jolly 1976). A conservative method that has been commonly applied in estimation of possum home ranges is the minimum convex polygon, which encompasses the known locations of an individual.

The size of brushtail possums' home ranges in New Zealand and Australia has been studied in a variety of habitats (native forests, scrublands, plantations, swamps, farmland) using both live trapping and radiotelemetry (Table 3.1). Range areas revealed by radiotelemetry are typically much larger than those revealed by live trapping (Ward 1984). Most published estimates of possum home ranges (Table 3.1), which are largely based on live trapping alone and often over periods of 1 year or less, are therefore likely to be underestimates. Nevertheless, some consistent patterns emerge. There is a clear tendency for male possums to have larger home ranges (mean 1.9 ha, range 0.7–3.4 ha, length 295 m) than females (mean 1.3 ha, range 0.6–2.7 ha, length 245 m) (Green 1984), and to move about most extensively in summer and autumn, possibly in search of females during the breeding season.

Home-range sizes tend to be larger in habitats where possums are at low density, or where they have access to farmland. Within high-altitude beech (*Nothofagus*) forest, where possum densities were less than 0.5/ha (Clout & Gaze 1984), trap-revealed adult range lengths were significantly larger than

Table 3.1
Summary of Australian and New Zealand information on the mean size and length of home ranges of brushtail possums in various habitats, updated from Green (1984) and Kerle (1984). Figures in parentheses indicate range of values. Because studies differed both in practical and statistical methods of home-range estimation, comparisons between studies must be treated with caution. (T = trapping; R = radio-tracking; S = spotlight).

Vegetation type	Home range area (ha)		Home range length(m)		Reference	Methods
	male	female	male	female		
Australia						
Urban, wooded grassland	3.0 (2.0–4.2)	1.1 (0.3–2.4)	320	200	Dunnet 1956	T
Open *Eucalyptus* woodland	5.0 (0.3–20.6)	1.1 (0.1–3.9)	340	180	Dunnet 1964	T
Eucalyptus/*Casuarina* woodland	2.6	4.6			Sampson 1971	T
Open *Eucalyptus* forest	3.7 (2.7–5.3)	1.7 (1.1–2.6)	340	210	Winter 1976	T
Eucalyptus forest	7.4 (6.0–8.3)	4.7 (3.3–6.5)	460	350	How 1981	T
Eucalyptus rain forest	7.0	4.2	300	230	Hocking 1981	S
Eucalyptus forest	6.3 (3.8–9.4)	1.8	500	320	Allen 1982	R
Eucalyptus woodland (tropical)	1.1	0.9	165	155	Kerle 1984	T
Urban	10.9 (0.3–42.1)	1.9 (0.4–4.6)			Statham & Statham 1997	T, R
New Zealand						
Modified forest, urban	1.5 (0.5–3.6)	2.7 (1.7–4.5)	260	210	Winter 1963	T
Podocarp–mixed broadleaf forest	0.8 (0.1–3.0)	0.5 (0.03–3.8)			Crawley 1973	T+S+R
Pasture/scrub/remnant forest	3.1 (2.4–3.6)	0.9 (0.3–1.2)	435 (407–550)	295 (110–516)	Jolly 1976	T
Pasture/scrub/willows	29.9 (10.4–61.3)	31.0 (2.2–105.2)	883 (480–1370)	784 (250–1650)	Brockie *et al.* 1987	R
Pasture/scrub/remnant forest	1.4 (0.1–8.8)	0.9 (0.05–2.32)	359 (100–652)	291 (66–616)	Paterson *et al.* 1995	R
Pasture/remnant forest	5.6	1.7	544	292	D. Ramsey unpubl.	R
Pine plantation	0.7 (0.4–0.8)	0.7 (0.2–1.4)	310	230	Warburton 1977	T
Pine plantation	1.4	1.0	280	220	Clout 1977	T
Pine plantation			296–317	163–272	Keber 1988	T
Beech forest			529	390	Clout & Gaze 1984	T
Pine, scrub, forest areas	1.9	1.3	210	190	Triggs 1982	T
Podocarp–mixed broadleaf forest	3.9 (3.1–4.8)	2.6 (2.2–3.0)	319 (278–360)	262 260–263)	Ward 1978	R
Podocarp–mixed broadleaf forest	24.6 (2.5–65.0)	18.3 (4.2–45.8)	880	820	Green 1984 Green & Coleman 1986	R

those in a lowland podocarp–hardwood forest where possum densities exceeded 7/ha (Crawley 1973; Ward 1978).

Possums living wholly within lowland forests tend to have small home ranges of about 1–4 ha, and they range over distances of only 200–500 m. In modified lowland habitats and along native forest–pasture margins possums may range over much larger distances. Some possums feeding out on pasture in Westland came from dens up to 1200 m into the adjacent forest (Green & Coleman 1986). The nearer to the forest–pasture margin that the possums denned, the more likely they were to move out onto the pasture to feed; about 95% of possums denning within 200 m of the forest–pasture margin fed out on pasture compared with fewer than 5% denning 1000 m away (Green & Coleman 1986). Some of these possums fed up to 1000 m out onto farmland beyond the forest margin.

Where possums live on farmland with scattered, remnant patches of forest or scrub, they show two types of ranging behaviour. Some have small ranges centred on preferred habitats, such as stream-side willows or swamps, and never venture far out onto farmland, while others range up to 1600 m over open pasture (Jolly 1976; Brockie *et al.* 1997) (Fig. 3.2), and have annual home ranges of up to 60 ha in area. This variation in ranging behaviour suggests that not all possums are likely to encounter a trap or bait station if control is restricted to the margins of forest patches greater than 500 m wide.

Bimodal or "dumbell-shaped" home ranges, where possums den in one area and move significant distances to another area to forage, have not only been observed where possums move out to feed on pasture, but are also displayed by some possums within more homogeneous habitats. Efford *et al.* (1994) found distinct bimodal ranges for some of the possums they radio-tracked within a 13-ha forest remnant. Similar behaviour is shown in the trap-revealed ranges of some possums in long-term mark-recapture studies in lowland podocarp–hardwood forest (Cowan, Efford & Ramsey unpubl. data). Such behaviour may allow possums to take advantage of patches of higher quality or seasonally available foods.

These observations on possum home ranges suggest that both size and shape are influenced by

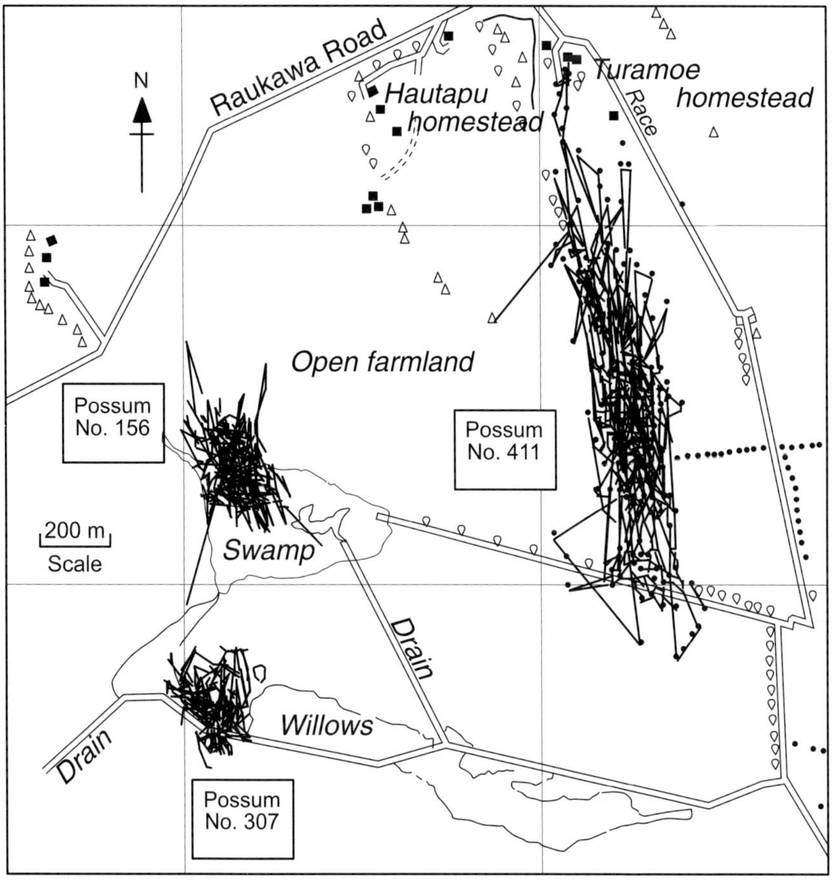

Fig. 3.2
Radio-tracking-revealed annual home ranges of three female possums on Hawke's Bay farmland. Home-range areas were 23 ha (No. 156); 6 ha (No. 307); 105 ha (No. 411). (from Brockie *et al.* 1987).

habitat, density relative to carrying capacity, and absolute density, but the relative importance of these influences is unknown.

Possum home ranges typically overlap extensively, both within and between the sexes (Crawley 1973; Green 1984; Paterson *et al.* 1995). Dunnet (1964 p. 665) concluded that (in Australia) "Adult resident males are clearly territorial while females occupy smaller individual ranges which overlap extensively where they are dense." Green (1984) reanalysed Dunnet's data and commented that both sexes appeared equally likely to hold non-overlapping home ranges. Based on his observations of social interactions of possums in open *Eucalyptus* woodland, Winter (1976) proposed a system of home ranges based on mutual avoidance between co-dominant possums of each sex.

In New Zealand, no evidence of territorial behaviour has been observed in podocarp–broadleaf forest (Crawley 1973; Ward 1978; Green & Coleman 1986), although Clout (1977) found evidence of lower than expected amounts of overlap between male possum home ranges in pine forest. Rather, the exclusive areas of high-ranking males and females extensively overlap the ranges of lower ranking individuals (Winter 1976; Clout 1977; Triggs 1982). Female young, in particular, tend to establish home ranges close to, or overlapping with, those of their mother (Clout & Efford 1984). Because of this behaviour, home ranges of females are effectively inherited (Efford 1991b and unpubl. data). In undisturbed populations this gives rise to spatial groupings of related females, which is reflected in the higher level of inbreeding found in female possums compared to males in genetic studies (Triggs 1987). The few juvenile males that stay where they were born do not, however, show this behaviour, but over the first 2–3 years after independence their home ranges gradually shift away from those of their mothers to lie, on average, several range lengths away (Efford 1991b).

Long-term trapping records from sites such as the Orongorongo Valley (Crawley 1973; M.G. Efford, Landcare Research, Dunedin, pers. comm.) show that once an adult possum has established a home range, it is likely to occupy it for life. A shift of more than a few hundred metres is highly unusual, even if immediate neighbours die or are killed in a control operation (Cowan 1993; Efford *et al.* 2000). However, individual possums do not use all of their home range every night. On any one night only part of it will be used, and it may take 3–5 nights for a possum to completely cover its current range (Ward 1978; Brockie *et al.* 1987; Paterson *et al.* 1995). Although home range size tends to be relatively stable from year to year, it may fluctuate seasonally by 10–20% (Ward 1978) as possums move more extensively to take advantage of seasonally available foods.

Movements to seasonal food sources

Possums forage extensively on seasonally available foods, and have been recorded moving 200–1600 m from their den sites to feed on such things as apples, walnuts, flowers of native trees, riverbed vegetation, pine pollen cones, and the flush of new leaves on poplars and willows (Jolly 1976; Ward 1978; Thomas *et al.* 1984). Seasonal food sources probably concentrate animals whose home ranges normally include those sites, rather than attracting animals from long distances. In one study, most possums attracted to an experimental planting of palatable poplars had travelled less than 500 m from the sites where they were originally trapped (Thomas *et al.* 1984), and on a Hawke's Bay farm with brassica crops, lucerne, pines with pollen cones, willows, and flowers and vegetables available at different times, none of the possums radio-tracked over the course of a whole year moved outside their normal home ranges to forage on these seasonal foods (Brockie *et al.* 1987).

Movements in response to control measures

A number of studies have looked at the distances possums will move to bait stations or to poison lines. A single permanent bait station was established and supplied with food for 14 weeks in an area of pine forest where most of the possums had been individually marked; poison and traps were then set around the area for 13 nights. None of the marked possums killed at the bait station had moved further than 300 m to feed there (Keber 1987), and only 10% of possums killed on cyanide lines laid through an area of swamp and willows on Hawke's Bay farmland had moved more than 300 m to find a poison bait (Brockie *et al.* 1989, 1997). Bait stations established in forest, scrub, swamp, and gully habitats were pre-fed with rhodamine-dyed bait for 2 weeks, and possums were then trapped at various distances from the bait stations (Hickling *et al.* 1990). The proportion of possums trapped that had fed at the bait stations decreased gradually with distance, with significantly fewer possums with internal or external evidence of dye caught more than 400 m from the bait stations (Fig. 3.3). When bait stations were placed 200 m apart on a grid, only about half as many possums were marked as when the bait stations were only 100 m apart (Thomas & Fitzgerald 1994).

The distances possums are likely to move to bait stations and poison lines within a 1- to 2-week period thus appears to be limited to 200–400 m in these habitats. This roughly matches estimates of home-range lengths in similar habitats (Table 3.1). The operating distance for bait stations can be expected to increase, however, if they are left deployed for prolonged periods. This is partly because possums move over different parts of their range with time (Ward 1978; Paterson *et al.* 1995) and partly because there is likely to be some movement into the area by adjacent possums (Efford *et al.* 2000).

Recolonisation after control (the "vacuum" effect)

After any control operation, the possum population eventually recovers through a mixture of breeding by survivors and recruitment of their offspring, and immigration from surrounding areas. In the 2–3 years after an initial control operation, or where maintenance control is carried out on a regular basis, immigration is likely to be more important than breeding. Control areas have often been likened to "vacuums", supposedly attracting possums in from immediately surrounding areas to take advantage of the excess of food and den sites available in the controlled area. This "vacuum effect" has frequently

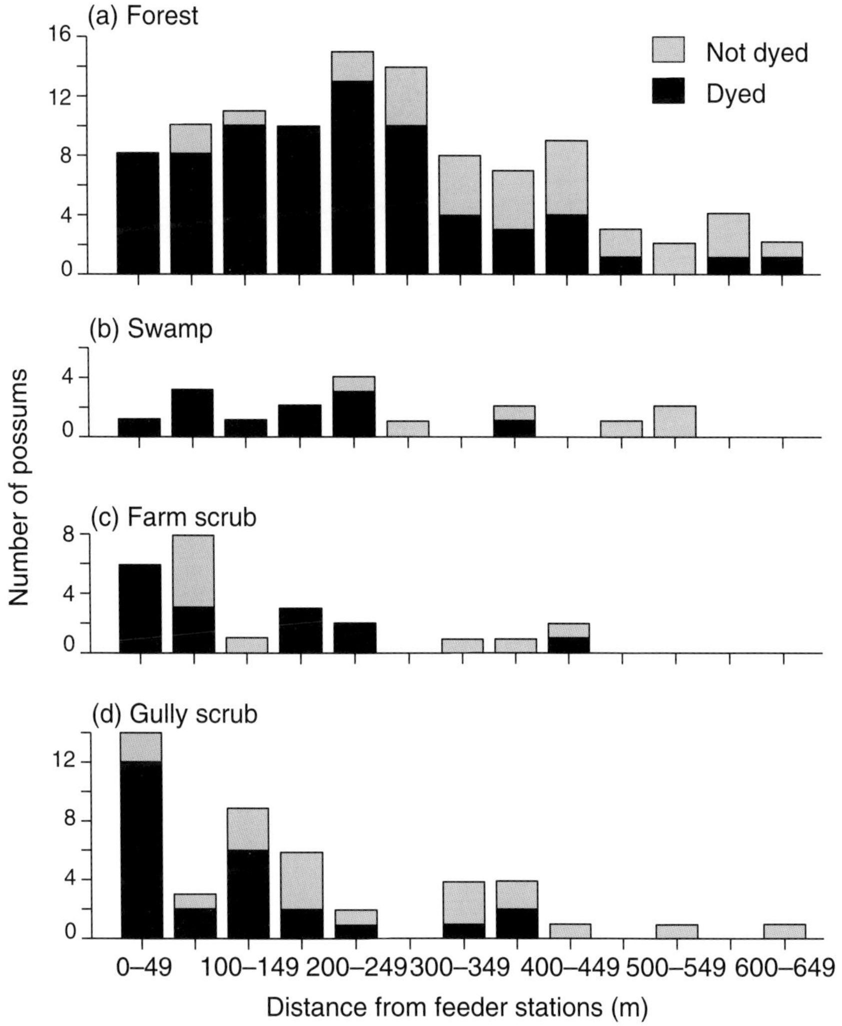

Fig. 3.3
Numbers of possums caught that were dyed or not dyed with Rhodamine B dye from baits on the four study blocks (from Hickling *et al.* 1990).

been suggested as a flaw in the concept of buffer zones to control the spread of bovine Tb, and as a reason for the apparent rapid recolonisation of areas controlled for conservation.

Several studies have tested this hypothesis. In Westland, few possums moved into a 100-ha area of forest in the 3 years after possums had been all but eradicated there (Green & Coleman 1984). On Kapiti Island, adult possums whose home ranges were known from repeated live-trapping and releasing, did not move their ranges even after the average density of possums on the island had been reduced by 80% (Cowan 1993). In a central King Country valley catchment, none of 65 adult possums fitted with radio transmitters shifted their ranges in the 4–6 months after a poison buffer zone was established on immediately adjacent land (Cowan & Rhodes 1992).

Brockie *et al.* (1991) tagged 141 possums at bush patches within 3 km of two adjacent farms where about 90% of possums had been killed. Only one of these, a juvenile male on first capture, was trapped on the controlled farms over the subsequent 2 years. In the same study, only one out of 18 survivors of the control operation shifted its home range in the following 12 months (Brockie *et al.* 1997). Efford *et al.* (2000) found very few new possums arriving in a 13-ha bush remnant on farmland in the 12 months after possum numbers in part of the patch had been controlled. There was, however, some local shifting of home ranges by survivors adjacent to the controlled areas during that time, but the effect was only detectable over a distance of 200–300 m from the edge of the control zone and seemed to involve animals whose ranges overlapped the control zone (Efford *et al.* 2000).

Thus there appears to be little movement of adult possums, even among the survivors of control operations, or animals adjacent to control zones. This conclusion is supported by studies of the structure of possum populations recolonising controlled areas. Most of the immigrants are young animals, often recently matured juveniles (Clout 1977; Clout & Efford 1984). Only a few older animals are also found to have immigrated (Clout 1977).

Long-distance dispersal

By *dispersal* we mean the movement of an individual animal from its place of birth to the place where it reproduces (Gaines & McGlenaghan 1980). It is not to be confused with local movement within a home range, or migration between seasonal ranges. By definition, dispersal movements are made by young animals, although adult possums occasionally shift their home ranges permanently (Efford *et al.* 2000). For possums, dispersal may be over either short or long distances, but interest focuses particularly on long-distance dispersal because of its greater significance in recolonisation and spread of possum-borne diseases.

Long-distance dispersal is the key factor influencing initial rates of recovery after control operations, and thus dictating the frequency of maintenance control, and the major mechanism by which possums spread bovine Tb into new areas. Attempts are usually made in control operations to limit immigration by use of natural boundaries, such as rivers and altitudinal barriers, although how well that approach works has not been formally evaluated.

Possums that disperse long distances are almost always animals that are undergoing, or have recently undergone, sexual maturation (often referred to as sub-adults), and include about four times as many males as females (Ward 1985; Efford 1991a; Cowan *et al.* 1996). About 20–30% of juvenile possums disperse, with most dispersals occurring in the 4 months before an animal's first birthday. In forest populations where maturation is delayed, some possums may disperse later, between the ages of 18 and 24 months (Efford 1998). Dispersal occurs most often about the time of peak breeding in late summer and early autumn, but will also occur in the spring if the population breeds then.

The average distance dispersed is about 5 km, although one possum was recorded moving 41 km. Dispersing possums can cover long distances in a short time, up to 3 km in a night, and 10 km in a week (Cowan & Rhodes 1992, 1993; Cowan *et al.* 1996, 1997) (Fig. 3.4). They also may make several consecutive moves before finally settling in a new area; for example, a juvenile female in the central King Country moving five times in 72 days (Cowan & Rhodes 1993). Dispersing female possums behave differently from males. They tend to move further (the two longest recorded moves, 32 and 41 km, are both by females) and make more moves before settling (Efford 1991a). So, although females disperse much less frequently, they may be disproportionately important in spreading Tb.

Dispersal of possums from their natal areas appears to be independent of density. After a control operation that killed more than 90% of possums, the absolute number of juvenile possums that dispersed decreased but the proportion that dispersed remained unchanged (Cowan *et al.* 1997). Control thus reduces but does not eliminate juvenile dispersal. The distances

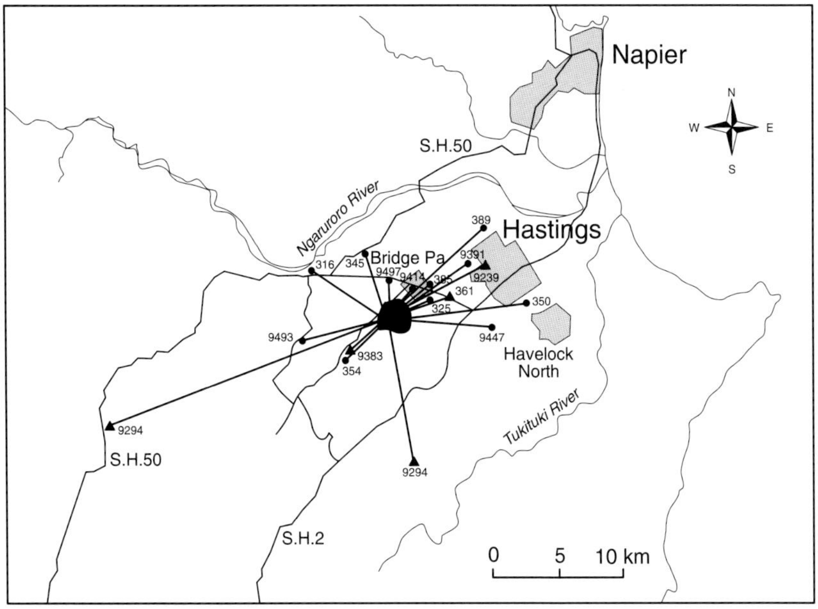

Fig. 3.4
Movements of radio-tagged (●) and ear-tagged (▲) possums that dispersed more than 2 km. Numbers are tag numbers. Lines join origin and final settlement location or last known location. Urban areas indicated by light shading. (from Cowan *et al.* 1996).

moved by juveniles dispersing from the controlled population also did not change significantly, so that, even after control, some juvenile possums are still likely to disperse through areas of reduced possum density (buffer zones) established for Tb management.

Areas of farmland and forest where Tb is endemic in possums and other wildlife are surrounded by buffer zones to contain the disease. It was thought originally that possums would disperse into the buffer zones from the endemic areas and settle there preferentially, because of the lower density and excess of food and nest sites. This hypothesis was not confirmed by a study adjacent to a central King Country buffer zone. There, the distances dispersed by possums were not reduced on average and some possums dispersed right across the buffer zone (Cowan & Rhodes 1992, 1993).

Homing behaviour

Possums are also capable of homing over distances significantly greater than their normal nightly movements would suggest. How (1972) translocated both brushtail possums and mountain brushtail possums (*T. caninus*) various distances to the north or south of their normal home ranges in a pine plantation into surrounding sclerophyll forest. Ten out of 15 (67%) returned from 2.3 km away, 10 out of 20 (50%) from 3.8 km away, 4 out of 10 (40%) returned after being moved 5.5 km south, and 1 out of 2 after being moved 7 km south. Some return movements were quite rapid. One individual returned from 3.8 km away in 2 days, and another from 7 km away in 19 days. Cowan (2000) observed similar behaviour in possums translocated on Manawatu farmland. Four possums, all from the same location, were moved about 4 km and returned successfully to their original home areas. The translocations were repeated for two of them and they again homed to their original capture points. The exact times taken to return were not known, but the maximum times were between 3 and 19 days. The mechanism by which possums are able to achieve this is unknown.

Conclusions

Activity patterns, home ranges, and dispersal of possums clearly demonstrate the flexible behaviour that has been one of the key factors in their success as colonists of New Zealand (Green 1984; Kerle 1984). Activity patterns vary in response to weather; denning varies in response to density; home range movements vary in response to season, habitat, local density, and available resources; and the sexes vary both in their tendency to disperse and in settlement behaviour if recruited into their natal population. Once an adult possum has established a home range, that range is likely to remain relatively stable for its lifetime. Even the removal of most of their immediate neighbours does not appear to evoke major range reorganisations among adjacent possums, affecting only those possums within about one to two home-range lengths of the removal.

Not all of a possum's home range is used at any one time, and movements to seasonal food sources probably mostly represent use of specific parts of that range at particular times, rather than "unusual" movements to particularly attractive resources. This suggestion is supported by the observation that most possums attracted to seasonal food sources or bait stations seem to have moved distances to them roughly equivalent to their home-range lengths. But the concentration of possums at seasonal food sources and bait stations makes their use as foci for control efforts an effective management strategy.

Managing possum dispersal is perhaps the most important requirement in limiting the spread of bovine Tb (see Chapter 8) and the recovery of possum numbers after control (see Chapter 13). The large distances moved by juvenile possums and the tendency of possums to disperse even at very low densities mean that buffer zones around control areas are always likely to be "leaky". Improving our ability to manage these problems will require a better understanding of the factors driving dispersal at source and ultimate settlement in a new area, such as hormonal effects on behaviour, density, sex ratio, and competition for resources.

Summary

- Possums are nocturnal. Daytime dens are usually under cover, either above or below ground. Many dens are used, but only a few are used repeatedly, and dens may occasionally be shared by up to five possums. Den sites may be a limiting resource in some habitats.
- Possums are usually active from about 30 minutes after sunset, returning to their dens just before dawn in summer but much earlier in winter. Rain at dusk may delay emergence and reduce time spent out of the den during the night. Possums spend about 1–2 hours per night feeding in two or three separate sessions, both on the ground and in trees. The rest of the time is spent grooming, sitting, or moving about.
- Home ranges are typically about 1–4 ha in area in forested areas in New Zealand, overlap each other extensively, with males having larger ranges than females, and ranges being larger in late summer/early autumn than in other seasons. Possums living in pastoral grazing areas or emerging from forest to feed on pasture may have much larger home ranges (up to 60 ha) or "dumbell-shaped" ranges. Once established, possum home ranges are largely stable in both space and time.
- Females born in an area mostly take up home ranges that largely overlap those of their mothers, thus effectively "inheriting" their mothers' home ranges. Male possums gradually shift their ranges away from those of their mothers.
- Removal of possums by control results in some rearrangement of the home ranges of immediately adjacent possums, but mostly only those within one to two range lengths of the controlled area.
- Use by possums of seasonal food sources and bait stations is largely restricted to animals from within a radius of 300–500 m around the food source or bait station.
- Juvenile possums, particularly males, disperse about the time of sexual maturity. The average dispersal distance is about 5 km, with the longest record being 41 km. Dispersing possums may move 3 km or more in a night. Juvenile possums disperse from their natal areas even when densities have been reduced by more than 90%.
- Possums are capable of homing over distances of three or more kilometres. How they do that is unknown.
- Improved management of possum dispersal is an urgent need because of the contribution of dispersal to the spread of bovine Tb and to the recovery of possum numbers after control.

References

Allen, N. T. 1982: A study of the hormonal control, chemical constituents, and functional significance of the paracloacal glands in *Trichosurus vulpecula* (including comparisons with other marsupials). Unpublished MSc thesis, University of Western Australia, Perth, Australia.

Brockie, R. E.; Fairweather, A. A. C.; Ward, G. D.; Porter, R. E. R. 1987: Field biology of Hawke's Bay farmland possums, *Trichosurus vulpecula. Ecology Division Report 10.* Lower Hutt, Department of Scientific and Industrial Research (unpublished). 40 p.

Brockie, R. E.; Herritty, P. J.; Ward, G. D.; Fairweather, A. A. C. 1989: A population study of Hawke's Bay farmland possums, *Trichosurus vulpecula. Ecology Division Report 26.* Lower Hutt, Department of Scientific and Industrial Research (unpublished). 43 p.

Brockie, R. E.; Ward, G. D.; Herritty, P. J.; Smith, R. N. 1991: Recolonisation and dispersal of possums on Hawke's Bay farmland 1988-1991: a progress report. *DSIR Land Resources Technical Record 51.*

Brockie, R. E.; Ward, G. D.; Cowan, P. E. 1997: Possums (*Trichosurus vulpecula*) on Hawke's Bay farmland: Spatial distribution and population structure before and after a control operation. *Journal of the Royal Society of New Zealand 27*: 181–191.

Burt, W. H. 1940: Territorial behavior and populations of some small mammals in southern Michigan. *University of Michigan Museum of Zoology Miscellaneous Publications 45.* 58 p.

Caley, P.; Spencer, N. J.; Cole, R. A.; Efford, M. G. 1998: The effect of manipulating population density on the probability of den-sharing among common brushtail possums, and the implications for transmission of bovine tuberculosis. *Wildlife Research 25*: 383-392.

Clout, M. N. 1977: The ecology of the possum (*Trichosurus vulpecula* Kerr) in *Pinus radiata* plantations. Unpublished Ph.D. thesis, University of Auckland, Auckland, New Zealand.

Clout, M. N.; Efford, M. G. 1984: Sex differences in the dispersal and settlement of brushtail possums (*Trichosurus vulpecula*). *Journal of Animal Ecology 53*: 737–749.

Clout, M. N.; Gaze. P. D. 1984: Brushtail possums (*Trichosurus vulpecula* Kerr) in a New Zealand beech (*Nothofagus*) forest. *New Zealand Journal of Ecology 7*: 147–155.

Cowan, P. E. 1989: Denning habits of common brushtail possums, *Trichosurus vulpecula,* in New Zealand lowland forest. *Australian Wildlife Research 16*: 63–78.

Cowan, P. E. 1990: Brushtail possum. *In:* King, C. M. *ed.* The handbook of New Zealand mammals. Auckland, Oxford University Press. Pp. 68–98.

Cowan, P. E. 1993: Effects of intensive trapping on breeding and age structure of brushtail possums, *Trichosurus vulpecula,* on Kapiti Island , New Zealand. *New Zealand Journal of Zoology 20*: 1–11.

Cowan, P. E. 2000: Responses of common brushtail possums (*Trichosurus vulpecula*) to translocation on farmland, southern North Island, New Zealand. *Wildlife Research* (submitted).

Cowan, P. E.; Rhodes, D. S. 1992: Restricting the movements of brushtail possums (*Trichosurus vulpecula*) on farmland with electric fencing. *Wildlife Research 19*: 47–58.

Cowan, P. E.; Rhodes, D. S. 1993: Electric fences and poison buffers as barriers to movements and dispersal of brushtail possums (*Trichosurus vulpecula*) on farmland. *Wildlife Research 20*: 671–686.

Cowan, P. E.; Brockie, R. E.; Ward, G. D.; Efford, M. G. 1996: Long-distance movements of juvenile brushtail possums (*Trichosurus vulpecula*) on farmland, Hawke's Bay, New Zealand. *Wildlife Research 23*: 237–244.

Cowan, P. E.; Brockie, R. E.; Smith, R. N.; Hearfield, M. E. 1997: Dispersal of juvenile brushtail possums, *Trichosurus vulpecula,* after a control operation. *Wildlife Research 24*: 279–288.

Crawley, M. C. 1973: A live-trapping study of Australian brush-tailed possums, *Trichosurus vulpecula* (Kerr), in the Orongorongo Valley, Wellington, New Zealand. *Australian Journal of Zoology 21*: 75–90.

Dunnet, G. M. 1956: A live-trapping study of the brush-tailed possum *Trichosurus vulpecula* Kerr (Marsupialia). *CSIRO Wildlife Research 1*: 1–18.

Dunnet, G. M. 1964: A field study of local populations of the brush-tailed possum *Trichosurus vulpecula* in Eastern Australia. *Proceedings of the Zoological Society of London 142*: 665–695.

Efford, M. G. 1991a: A review of possum dispersal. *DSIR Land Resources Scientific Report 23.* 70 p.

Efford, M. G. 1991b: The ecology of an uninfected forest possum population. *In:* Jackson, R. *convenor.* Symposium on tuberculosis. *Publication No. 132, Veterinary Continuing Education.* Palmerston North, New Zealand, Massey University. Pp. 41–51.

Efford, M. 1998: Demographic consequences of sex-biased dispersal in a population of brushtail possums. *Journal of Animal Ecology 67*: 503–517.

Efford, M.; Spencer, N.; Warburton, B.; Drew, K. 1994: Possum movements after control – assessment of the "vacuum" effect. Landcare Research Contract Report LC9495/57. (unpublished) 21 p.

Efford, M. G.; Warburton, B. W.; Spencer, N. J. 2000: Home-range changes by brushtail possums in response to control. *Wildlife Research* (in press).

Fairweather, A. A. C.; Brockie, R. E.; Ward, G. D. 1987: Possums (*Trichosurus vulpecula*) sharing dens: a potential infection route for bovine tuberculosis. *New Zealand Veterinary Journal 35*: 15–16.

Gaines, M. G.; McGlenaghan, L. R., Jr. 1980: Dispersal in small mammals. *Annual Review of Ecology and Systematics 11*: 163–196.

Green, W. Q, 1984: A review of ecological studies relevant to management of the common brushtail possum. *In:* Smith, A. P.; Hume, I. D. *ed.* Possums and gliders. Chipping Norton, NSW, Surrey Beatty in assoc. with the Australian Mammal Society. Pp. 483–499.

Green, W. Q.; Coleman, J. D. 1984: Response of a brush-tailed possum population to intensive trapping. *New Zealand Journal of Zoology 11*: 319–328.

Green, W. Q.; Coleman, J. D. 1986: Movement of possums (*Trichosurus vulpecula*) between forest and pasture in Westland, New Zealand: Implications for bovine tuberculosis transmission. *New Zealand Journal of Ecology 9*: 57–69.

Green, W. Q.; Coleman, J. D. 1987: Den sites of possums, *Trichosurus vulpecula*, and frequency of use in mixed hardwood forest in Westland, New Zealand. *Australian Wildlife Research 14*: 285–292.

Hickling, G. J.; Thomas, M. D.; Grueber, L.S.; Walker. R. 1990: Possum movements and behaviour in response to self-feeding bait stations . Forest Research Institute Contract Report FWE 90/9 (unpublished). 17 p.

Hocking, G. J. 1981: The population ecology of the brush-tailed possum, *Trichosurus vulpecula* (Kerr), in Tasmania. Unpublished MSc thesis, University of Tasmania, Hobart, Australia.

How, R. A. 1972: The ecology and management of *Trichosurus* species (Marsupialia) in N.S.W. Unpublished PhD thesis, University of New England, Armidale, NSW, Australia.

How, R. A. 1981: Population parameters of two congeneric possums, *Trichosurus* spp., in north-eastern New South Wales. *Australian Journal of Zoology 29*: 205–215.

Jolly, J. N. 1976: Habitat use and movements of the opossum (*Trichosurus vulpecula*) in a pastoral habitat on Banks Peninsula. *Proceedings, New Zealand Ecological Society 23*: 70–78.

Keber, A. 1987: The efficacy of permanent bait stations. *Fur Facts 8(29)*: 34–37.

Keber, A. W. 1988: An enquiry into the economic significance of possum damage in an exotic forest near Taupo. Unpublished PhD thesis, 2 vols. University of Auckland, Auckland, New Zealand.

Kerle, J. A. 1994: Variation in the ecology of *Trichosurus*: its adaptive significance. *In:* Smith, A. P.; Hume, I. D. *ed.* Possums and gliders.

Chipping Norton, NSW, Surrey Beatty in assoc. with the Australian Mammal Society. Pp. 115–128.

MacLennan, D. G. 1984: The feeding behaviour and activity patterns of the brushtail possum, *Trichosurus vulpecula*, in an open eucalypt woodland in southeast Queensland. *In:* Smith, A. P.; Hume, I. D. *ed.* Possums and gliders. Chipping Norton, NSW, Surrey Beatty in assoc. with the Australian Mammal Society. Pp. 155–161.

Paterson, B. M.; Morris, R. S.; Weston, J.; Cowan, P. E. 1995: Foraging and denning patterns of brushtail possums, and their possible relationship to contact with cattle and the transmission of bovine tuberculosis. *New Zealand Veterinary Journal 43*: 281–288.

Pickett, K. 1997: The sublethal effects of predation on the common brushtail possum (*Trichosurus vulpecula*) at Lake Burrendong, New South Wales. Poster presentation, 10th Annual Conference, Australasian Wildlife Management Society, University of New England, 25–27 November 1997.

Sampson, J. C. 1971: The biology of *Bettongia penicillata* Gray, 1837. Unpublished PhD thesis, University of Western Australia, Perth, Australia.

Statham, M.; Statham, H. L. 1997: Movements and habits of brushtail possums (*Trichosurus vulpecula* Kerr) in an urban area. *Wildlife Research 24*: 715–726.

Thomas, M. D.; Fitzgerald, H. 1994: Bait-station spacing for possum control in forest. Landcare Research Contract Report LC9394/118 (unpublished) 9 p.

Thomas, M. D.; Warburton, B.; Coleman, J. D. 1984: Brush-tailed possum (*Trichosurus vulpecula*) movements about an erosion-control planting of poplars. *New Zealand Journal of Zoology 11*: 429–436.

Triggs, S. J. 1982: Comparative ecology of the possum, *Trichosurus vulpecula* in three pastoral habitats. Unpublished MSc thesis, University of Auckland, Auckland, New Zealand.

Triggs, S. J. 1987: Population and ecological genetics of the brush-tailed possum (*Trichosurus vulpecula*) in New Zealand. Unpublished PhD thesis, Victoria University of Wellington, Wellington, New Zealand.

Warburton, B. 1977: Ecology of the Australian brush-tailed possum (*Trichosurus vulpecula* Kerr) in an exotic forest. Unpublished MSc thesis, University of Canterbury, Christchurch, New Zealand.

Ward, G. D. 1978: Habitat use and home range of radio-tagged opossums, *Trichosurus vulpecula* (Kerr) in New Zealand lowland forest. *In:* Montgomery, G.G. *ed.* The ecology of arboreal folivores. *Symposia of the National Zoological Park.* Washington DC, Smithsonian Institution Press. Pp. 267–287.

Ward, G. D. 1984: Comparison of trap- and radio-revealed home ranges of the brush-tailed possum (*Trichosurus vulpecula* Kerr) in New Zealand lowland forest. *New Zealand Journal of Zoology 11*: 85–92.

Ward, G. D. 1985: The fate of young radiotagged common brushtail possums, *Trichosurus vulpecula*, in New Zealand lowland forest. *Australian Wildlife Research 12*: 145–150.

Winter, J. W. 1963: Observations on a population of the brush-tailed opossum (*Trichosurus vulpecula* Kerr). Unpublished MSc thesis, University of Otago, Dunedin, New Zealand.

Winter, J. W. 1976: The behaviour and social organisation of the brush-tail possum *(Trichosurus vulpecula* Kerr). Unpublished PhD thesis, University of Queensland, Brisbane, Australia.

CHAPTER FOUR

Possum Social Behaviour

Tim Day, Cheryl O'Connor and Lindsay Matthews

Social behaviour is the behaviour displayed by animals when they interact with other animals of the same species. Repeated social interactions lead to social relationships, and the outcome of a consistent set of these relationships is social organisation (Hinde 1976). Within a species, social organisation is not necessarily a fixed attribute and may vary in different circumstances or populations (Lott 1991). The ability of species to vary their social structure to cope with their current situation is thought to be an adaptive strategy that allows individuals to maximise the benefits they can gain from a given situation at a minimal cost (Eisenberg 1966). Species that can readily adapt may have an advantage when faced with new or modified conditions.

Brushtail possums (*Trichosurus vulpecula*) are recognised as being the most adaptable and most widely distributed of the Australian marsupials (Kerle 1984). They have been able to colonise many habitats and, since being introduced to New Zealand, have thrived. With few limitations in New Zealand environments, possums have established so successfully that they have become one of the country's most serious vertebrate pests (Cowan & Tyndale-Biscoe 1997). One of the factors that has probably contributed to this success has been the flexibility of the possum's behaviour and social organisation in different ecological circumstances.

Ecological conditions in New Zealand differ markedly to those found in Australia. Factors such as food supply (species, distribution, stability, abundance), habitat type (shelter, alternative resources, etc.), and climate may make significant contributions to variations observed in the social organisation of possums in both countries. The factors that determine variation in social structure have rarely been systematically examined, but can be generalised into four main types: ecological circumstances, demographics, kinship, and the current pattern of social behaviour (Lott 1991). One or more of these factors may combine to influence the nature of social organisation present in a given situation. Demographics, kinship, and the nature of social structure among possums in Australia will also underpin the social structure of possums living in New Zealand.

A range of social structures have been described within the different species of Australian possums and gliders (Smith & Lee 1984), and these structures vary for different activities and in response to habitat type and predation risk. The majority of these species, including brushtail possums, tend to be solitary when feeding, unless localised abundant food sources are available (e.g., *Petaurus* spp. at sap-feeding sites). About half of the possum and glider species are also solitary when moving at night or when sleeping in dens. Interestingly, the larger species (including possums) tend to be solitary for sleeping and the smaller species tend to form mixed, matriarchal or maternal groups for this activity (Smith & Lee 1984). Group living reduces predation risk for some animals, and this risk may be greater for the smaller Australian marsupials than it is for the larger species. Dominance relationships are important for maintaining individual space in many of the possums and gliders. Most species appear to have a polygynous mating system, although the mountain brushtail possum (*Trichosurus caninus*), which is closely related to *T. vulpecula*, is thought to be monogamous (How 1978; Kerle 1984).

In New Zealand, the social organisation of brushtail possums appears to vary from place to place, and may differ from the social structures typical for possums in Australia (Green 1984). This chapter outlines our understanding of the social behaviour and organisation of brushtail possums in New Zealand, drawing on research collected from both wild and captive possums in New Zealand and Australia. It also serves to highlight areas of possum social behaviour that are poorly understood. The relevance of social behaviour to the development of sustainable possum control techniques is discussed.

Measuring social behaviour

The nocturnal, arboreal, and solitary habits of possums make the study of their social behaviour difficult. Very few quantitative data are available to describe the influence of factors such as habitat type, population density, or demographics on social organisation. The effects of current control strategies on possum social behaviour, and the effects of different social structures on the efficacy of possum control, are also poorly understood. The most detailed studies of social behaviour have been conducted using captive groups of possums, as the animals are easier to observe in a restricted area. Research has typically focused on dominance relationships and mating systems (e.g., Biggins & Overstreet 1978; Oldham 1986; McAllum 1996; Jolly *et al.* 1995) or specific animal-to-animal interaction patterns (e.g., Day *et al.* 1998b). Recently, some studies have investigated the effects of sterilisation on social behaviour (e.g., Jolly *et al.* 1996; Jolly & Spurr 1996) or the effect of social behaviour on the spread of infectious organisms (e.g., Day *et al.* 1998a). One of the concerns with studying social behaviour in captivity is that the social structure observed in captive animals may not be representative of that found in wild possums. Captive groups of possums are usually kept at densities many times greater than that in the wild (e.g., equivalent to 200 possums/ha; Day *et al.* 2000), in much less complex environments than their natural habitat. In captivity, subordinate possums usually have no use of an exclusive area or place to escape from more dominant animals, and competition for resources such as food and dens is high. In many other species, social organisation varies with population density. For example, the house mouse (*Mus musculus*) uses a system of territoriality to defend resources when density is low, and a dominance hierarchy system in high-density groups where individuals are unable to defend exclusive areas (Alleva *et al.* 1995).

Only a few studies have investigated the social behaviour or organisation of possums in the wild. These studies have also tended to be conducted in areas of relatively high possum density (e.g., scrubby farmland, which has densities above 1 possum/ha; Jolly 1976), which may influence the type of social structure that develops. Evidence from Australia suggests that possums living in simple forest habitats defend exclusive areas in low-density populations (Dunnet 1956; Winter 1976), but that defended areas are rare in high-density populations in more complex forest types (Green 1984). This suggests that, to further our understanding of social organisation in possums, we will need to investigate social behaviour in a range of habitats at different possum densities.

Social behaviour studies with wild possum populations have often been limited by the methods available to collect data. Traditionally, most studies of wild possum behaviour have relied on observers following individual animals throughout the night with binoculars and dimmed spotlights, which may sometimes disturb the possums from their normal activities (Winter 1976; Jolly 1976; MacLennan 1984). More recently, aspects of possum behaviour have been examined using remote infrared or low-light cameras (Day 1996) or light-intensifying telescopes (Paterson *et al.* 1995). Using operant conditioning, it has been shown that possums do not see infrared light (J. Chandler, University of Waikato, Hamilton, pers. comm.), so conducting observations of possum behaviour using this lighting regime and infrared-sensitive equipment may be less intrusive to possums than techniques used in the past. New technologies, such as proximity-measuring sensors, may become useful for advancing our understanding of social structure in possums in the future.

Social structure

Wild possums in New Zealand are essentially solitary animals, and when active at night, spend the majority of their time alone, either sitting, feeding, grooming, or travelling between trees or ground-feeding sites (MacLennan 1984). Possums often have widely overlapping home ranges, both between and within sexes (Crawley 1973; Green 1984). Males usually have larger home ranges than females (Winter 1976; Jolly 1976; Ward 1978) and travel more extensively each night (Jolly 1976; Green 1984). The home ranges of possums may shift spatially, both seasonally (Jolly 1976; Ward 1978) and during their lifetime (Green 1984). All social interactions occur in overlapping areas, with priority for resources primarily being based on dominance when possums meet (Winter 1976; Green 1984). The home ranges of some subordinate individuals may be entirely overlapped by more dominant animals (Clout 1977; Triggs 1982).

Possums spend less than 1% of their time directly interacting, except during the breeding season (Winter 1976; Jolly 1976; MacLennan 1984; Herbert 1996). Typically, when two possums interact, a dominance relationship is evident. One of the animals "wins" the interaction, either by maintaining its position when approached, or if it is the approacher, by displacing the other (Jolly & Spurr 1996; Henderson & Hickling 1997). Older and heavier animals are usually dominant to smaller, younger possums (Jolly 1976; Biggins & Overstreet 1978). Females are generally dominant over males (Winter 1976; Jolly & Spurr 1996), although some large old males are dominant over smaller young females (Winter 1976; Henderson & Hickling 1997). Triggs (1982) found that possums that are more than 100 g heavier than their standard weight for length are usually dominant. Dominance status between two possums appears to be maintained after initial encounters without further interactions, by individual recognition and avoidance (Winter 1976; Biggins & Overstreet 1978).

When two possums are of a similar size, age, and sex, there may not be a clearly dominant individual. Winter (1976) described mutual avoidance between co-dominants, where two possums of similar social status avoided using the same area at the same time. In both wild and captive possums, animals that are subordinate to others will also temporally or spatially separate (Jolly 1973; Winter 1976). Thus, the movements of subordinate animals may be constrained by the presence of dominant possums (Henderson & Hickling 1997). Scent marking (described in the communication section) may play an important role in regulating temporal or spatial avoidance.

In Australia, dominance hierarchies in overlapping areas, coupled with defence of core areas of the home range, have been suggested as the primary social structure among wild possums (Green 1984). Territorial defence has previously been described for possums in Australia by Dunnet (1964). However, reanalysis of Dunnet's data, and more recent research (How 1981), has found that, in accordance with Green (1984), only limited parts of possums' home ranges are defended. Exclusively defended areas are most evident in low-density populations occupying simple forest habitats (Dunnet 1956; Winter 1976). In a recent study of the northern brushtail possum (*T. vulpecula arnhemensis*) in Australia, Kerle (1998) found that possums in a high-density population were not territorial and allowed other possums to live in close proximity. It is unknown whether these possums exhibited dominance relationships. Social dominance has been suggested to be one of the major factors controlling possum density in Australia, as dominant animals are able to exclude subordinates from access to limiting resources, such as food or dens (Green 1984).

In New Zealand, wild possums of similar dominance status and the same sex tend to avoid each other (Clout 1977). If challenged, both male and female possums will actively defend their den or denning tree. While feeding, possums may defend a high-quality food source (e.g., a bait station), but will not continue to defend it after they cease feeding (Hickling & Sun unpubl. data). Social structure in New Zealand appears to vary with habitat type, although there has been no systematic study conducted to determine this relationship. In less complex habitats, dominant animals may have small exclusive areas in their home range (Clout 1977; Triggs 1982), but there is no conclusive evidence that they defend these areas. In complex broadleaf–podocarp forest no exclusive areas have been detected (Ward 1978). Green (1984) hypothesised that territorial defence in possums evolved in the open *Eucalyptus* woodlands of Australia, is primarily adaptive in a two-dimensional space, and is ineffective in complex New Zealand forests. He concluded that social behaviour is unlikely to be a factor regulating possum density in New Zealand, because individual possums would be unlikely to be able to defend limited resources (e.g., food and dens) in New Zealand forests. Dominance structures are commonly found in animal populations where resources, such as food and shelter, can not readily be defended through territoriality (Brown 1964). The social organisation of possums in New Zealand, as in Australia, appears to be based primarily on dominance relationships between possums in areas of overlapping habitat use or home ranges (Triggs 1982; Ward 1978). There is no strong evidence of territorial defence. However, the effects of possum density, habitat type, demographic factors, and climate on social organisation need to be critically examined before we will fully understand the variability of social organisation and controlling factors among wild possums in New Zealand.

Interaction patterns

Other than in the maintenance of the mother-juvenile bond, a possum's social interactions are oriented mostly towards forming reproductive relationships or maintaining dominance relationships (Green 1984). *Interactions* can be defined as any behaviour performed by one animal that elicits a behavioural response from another animal, and may be indirect or direct. *Indirect interactions* occur when an animal encounters evidence of another animal (e.g., scent marking) and responds in the absence of the other animal. *Direct interactions* begin when one animal approaches another (either deliberately or not) and end when one animal moves away from the other. Generally, a possum responds to another when it comes within a few metres of the other animal, and (in captivity) almost always responds when within 1 m of another (Day 1996). Direct interactions usually occur between two possums only, at a time. There are few quantitative data available on the frequency with which possums directly interact, although MacLennan (1984) suggested that in his Australian study site most possums interact only once or twice per night. Possums display a variety of social behaviours when interacting (Table 4.1), including agonistic, affiliative, or sexual behaviour (Winter 1976; Jolly 1976; Biggins & Overstreet 1978).

Agonistic interactions range in form and intensity from submissive "give-ways" or low-intensity "threats", involving no physical contact, to "fights", which always involve bodily contact (Winter 1976; Day *et al.* 1998b). Threats are the most frequently observed type of interaction between possums (Day *et al.* 1998b), and include: a range of vocalisations, the "glare", the "raised-paw threat", and the "bipedal threat" (Winter 1976; Jolly 1976; Biggins & Overstreet 1978). While performing any of the threat postures, the possum may "swipe" or "lunge" at the other possum without making contact (Day 1996). In response to these threats one possum may submit to the other by "giving-way" (Winter 1976) or "rolling" onto its back, or it may retaliate with another threat (e.g., raised-paw threat), a "chase", a "pounce", a "box", or a "fight". Some actions, such as the "bipedal threat", may also function as a defensive response by a possum (Spurr & Jolly unpubl. data). Fights between possums are less common than threats, and usually result from a series of increasingly intense threat displays (Winter 1976; Biggins & Overstreet 1978; Day 1996). During intense fights, the possums often "back-roll" on the ground while "swiping" at each other with their paws and biting. Often much fur is lost and blood is drawn from scratch and bite wounds (Day *et al.* 1998b). Fights end only when one possum submits (Winter 1976; Day 1996). The highest-intensity fights tend to occur between possums of similar dominance status (Winter 1976), as neither possum readily submits to the other.

Affiliative interactions or "associations" are non-aggressive and are rare outside the breeding season (Day 1996). They occur between den-sharing possums (Day *et al.* 2000) and include activities such as "nose-to-nose sniffing", "touching", "allogrooming" and "food sharing" (Table 4.1). Winter (1976) suggested that "nose-to-nose sniffing" is used for individual recognition, but there has been no systematic study of the function of any of the affiliative behaviours. "Allogrooming" between adult possums is directed towards the dorsal part of the possum, either on the head, neck, back, tail, or ears, but not towards the underbelly, mouth, or anal area. It is not clear whether one possum grooming another removes or consumes anything from the fur or skin of the other animal (Day 1996). "Food sharing" has only been described in captive animals (Day 1996), and occurs when one animal joins a feeding possum and begins eating the same item. While the specific functions of affiliative behaviours are not obvious, they may serve to reinforce bonds between possums, as they occur most frequently between mothers and their offspring, male–female consort pairs during the breeding season, and pairs of mature females (Winter 1976; Day 1996). Affiliative behaviours between males (captive or wild) have rarely been reported.

Sexual interactions occur exclusively between males and females, with no evidence of homosexual behaviour. These interactions involve affiliative, agonistic, and sexual behaviours and are discussed further in the section on reproductive behaviour.

Common interactions

Mother–juvenile interactions

Juvenile possums accompany their mother continuously for the first 8 months of life (Winter 1976). Affiliative behaviour is common during the first 160 days, including "touching", "allogrooming", "food sharing", "nose-to-nose sniffing" and "play" behaviour. Play behaviour by the juvenile involves "patting" the mother with the paws and "play biting" (Table 4.1; Winter 1976). Between days 160 and 200, the juvenile gradually becomes more independent and will climb off its mother's back, but continue to follow her closely. At this time the mother begins to display agonistic behaviour towards the juvenile, occasionally preventing it from climbing onto her back by threatening, striking, or biting (Winter 1976). The juvenile does not show any submissive behaviour towards its mother until about 230 days after birth, when the maternal behaviour becomes more aggressive in nature, resulting in the end of the back-riding phase and weaning. Once the juvenile has ceased back-riding, a distance of more than a metre is maintained between mother and juvenile, except when sharing dens (Winter 1976).

Table 4.1
Definitions of social behaviours observed between interacting possums. Behaviours are grouped by type and listed in increasing intensity.

Behaviour type	Behaviour	Definition
	Approach[1]	Possum moves toward another possum on all four limbs, either deliberately or by chance. All social interactions began with an approach by one possum.
	Leave[4]	Possum moves away (using any gait) from another possum in response to another possum's behaviour. All social interactions end with one possum "leaving" the other.
Agonistic behaviours	Glare[2,4]	Stationary possum sits on all four limbs and stares fixedly at another possum with body held hunched and rigid.
	Raised paw threat[1]	Possum raises one front paw off the ground towards another possum from a quadrupedal stance, with its body lowered and head extended forward.
	Bipedal threat[3,4]	Possum raises its body vertically into a bipedal posture, with front paws outstretched, ears flattened, and mouth slightly open.
	Swipe[4]	Possum rapidly swings one front paw toward another possum, without making contact with any part of the other possum.
	Lunge[4]	Possum moves its body (> 0.5 body length) rapidly towards another possum, without contacting it, from a quadrupedal or bipedal posture.
	Chase[1,4]	Possum follows rapidly behind a "leaving" possum (runs), with the distance between possums not exceeding 3 metres.
	Pounce[4]	Possum lunges or jumps rapidly towards another possum, contacting it with its paws or body.
	Box[3]	Possum rapidly swings its front paws toward another possum, making contact with other possum's body or head.
	Fight[1,3,4]	Possum maintains physical contact with another possum, usually after a "pounce", and may "box", "kick", or "bite" the other animal.
	Give-way[1,2]	Possum turns or moves part or all of its body away from another possum, without there being a "chase", "box", or "fight".
	Back roll[1]	Possum lies on its back with its front paws extended towards another possum and head raised above the ground.
Affiliative behaviours	Nose-to-nose sniff[1,3]	Both possums, while in quadrupedal postures, stretch their noses forward to within 30 cm of each other, possibly sniffing.
	Touch[1,4]	Possum contacts some part of another possum (for >2 seconds) with part of its body, face or paws, without displaying "box", "pounce", or "fight" behaviours. Includes sleeping, sitting, and lying in contact.
	Pat[1]	Mother or juvenile gently touches the others body or head several times with its front paws.
	Play-bite[1]	Mother or juvenile gently bites at the head or body of the other.
	Food share[4]	Both possums in a pair consume the same food item at the same time. Possums' mouths are < 10 cm apart.
	Allogroom[1,4]	One possum uses its front paws and teeth to manipulate the fur or skin of another possum, without performing any "box" or "fight" behaviour.
Sexual behaviours	Mate[1]	Male possum approaches a female and "mounts" her (climbs onto her back), moves his pelvis forward for intromission and then "pelvic thrusts" in bouts of 10–12 thrusts. Female may threaten male or move under him. After ejaculation the male dismounts and "leaves".

Definitions from: [1]Winter 1976; [2]Hickling & Sun unpubl. data; [3]Biggins & Overstreet 1978; [4]Day 1996.

All juveniles become independent and establish their own home ranges (which may overlap their mother's range) by 12–17 months of age.

Female–female interactions

Interactions between wild female possums are rare (Winter 1976; Oldham 1986). Female home ranges are generally close to their mother's range (Efford 1991) and most females would not encounter many other females during their nightly movements. Although rare, agonistic interactions between wild females are more aggressive than between mixed-sex or male–male pairs of possums (Biggins & Overstreet 1978). In both captivity and the wild, however, pairs of female possums will readily share dens and perform affiliative as well as agonistic behaviours (Day 1996 unpubl. data). Female-only groups of captive possums interact more frequently than either mixed-sex or male-only captive groups, with little evidence of temporal separation of individuals (Day *et al.* 1998b).

Male–male interactions

The majority of interactions between males in the wild occur in the vicinity of oestrous females (Winter 1976) and are predominately agonistic in nature (Biggins & Overstreet 1978; Day *et al.* 1998b). Winter (1976) found that, in the absence of females, threat interactions resulting in one possum "giving-way" are typically between males. When females were present, interactions between males were more aggressive (Winter 1976). While male possums travel further than females each night (Crawley 1973; Jolly 1976; Green 1984), they tend to avoid each other temporally (Jolly 1973; Day et al. 1998b), reducing the number of contacts between them.

Male–female interactions

Most interactions between male and female possums in the wild occur during the breeding season (Winter 1976). Threats are the most commonly observed behaviours in wild and captive animals (Winter 1976; Day *et al.* 1998b), although most interactions are associated with the formation of reproductive relationships. Fights and affiliative behaviours are also frequently observed between males and females in the breeding season (Winter 1976; Day 1996). Interactions between males and females occur less frequently in the non-breeding season (Winter 1976; Day *et al.* 1998b) and involve mainly low intensity threats, such as "give-ways" and "glares".

Reproductive behaviour

During the breeding season, in both the wild and captivity, some female possums are accompanied by one or more males, who "consort" with her for 30 to 40 days before mating (Jolly 1976; Winter 1976). Consort males are usually males whose home ranges overlap with the female (Smith & Lee 1984). A male will approach or follow the female repeatedly during this "consort" period (usually staying within 3 m), until she becomes more tolerant of him (Jolly 1981). Some "consort" males share the den of the female during the daytime (Winter 1976) and both affiliative and agonistic behaviours occur. Consort males are usually dominant animals, and one or more subordinate males may also follow the female (Winter 1976). After the consort period, mating usually occurs between the most dominant male and the female. Based on visual observation, mating always appears to be initiated by the male, when he approaches the female, either from in front or behind, and attempts to mount her. It is unknown whether the female signals her readiness to mate with the male via other means (e.g., scent or subtle postural changes). The female is usually not physically aggressive during mounting, but may threaten the male with vocalisations before and during the mounting, or may move from under him (Winter 1976). Mating is typically of 2–4 minutes duration (Winter 1976). The male may mount the female several times, but after successful mating there is no further "consort" behaviour. Mating without a consort period (described by Cowan (1990) as "casual" mating) is also observed. It is usually aggressive, with the male repeatedly trying to overpower the female and mount her. Such casual matings (or "rape") often attract other males to the area, who may also attempt to mate with the female (Ward 1978; Winter 1976). Day *et al.* (1998b) found that, in captivity, offspring were produced after either consort or casual matings.

Oldham (1986) suggested that breeding success in captive female possums is not related to the female's social dominance status. However, Jolly *et al.* (1995) found that heavier females (> 3 kg), which are usually dominant over lighter females, produce more offspring than light possums. For wild and captive male possums, the highest-ranking animals have the greatest reproductive opportunity, as they are able to exclude subordinate males from access to females (Winter 1976; Oldham 1986; Jolly *et al.* 1998). McAllum (1996) proposed that female possums may also "choose" dominant sires. It is unknown whether the increased mating opportunity of dominant males results in them having greater overall reproductive success than subordinate males. Dominant males will mate with females of any social rank, and may mate with different females each season (Jolly 1981).

The mating systems of wild possums are not well understood, although males will mate with several females per season in captive colonies (Oldham 1986; Jolly *et al.* 1995). Many of the mammalian mating systems that have been examined are described as polygynous, with a relatively small proportion of males monopolising breeding with the available females (Greenwood 1980; Clutton-Brock 1989). Possums in Australia have previously been suggested to be polygynous by Smith & Lee (1984). Current research is attempting to determine the nature of mating systems in different possum populations in New Zealand. To date, it appears that the populations examined are not monogamous (Sarre & Clout 1998), and may be polygynous. However, mating systems in other species vary with habitat type (Clutton-Brock 1989) and the mating systems of possums in New Zealand may also vary in a similar way. Taylor *et al.* (1998) suggested that the mating system exhibited by possums may, to a certain extent, determine the degree and nature of contact between individuals in different populations. Therefore, understanding the variation in mating systems in New Zealand possum populations may also help us to understand social organisation.

Denning behaviour

During the daytime possums sleep in dens. These may be in hollow trees, clumps of epiphytes, vines or flax, under logs, or underground in burrows of other animals such as kiwi (Cowan 1990). Possums will also den in the ceiling or wall cavities of buildings or any other place that affords cover from the sun and rain. When suitable dens are available (e.g., in large epiphytes), possums tend to den above ground in preference to ground level or below-ground sites (Green & Coleman 1987; Cowan 1989). Possums regularly use between five and ten dens within their home range, although dens are not exclusive to one possum and some dens are used regularly by several individuals (Cowan 1990). In high-density populations or areas with few potential dens, den sharing (two or more possums using the same den at the same time) may occur (Cowan 1990). Sharing occurs most commonly between mothers and their young or between consort pairs of possums (Winter 1976), but up to five adult possums have been observed in one den at the same time (Fairweather *et al.* 1987). Sharing of dens is most likely to occur in deep sheltered dens (Caley 1996) and has a higher prevalence on farmland (Fairweather *et al.* 1987). Several estimates have been made of how often possums share dens in specific situations, ranging from very rarely (Paterson *et al.* 1995) to a daily probability of sharing dens of 0.07 (Caley *et al.* 1998). However, estimates of the incidence and frequency of den sharing are difficult to make, as dens are not usually open to inspection (Fairweather *et al.* 1987). In captivity, den sharing can occur up to 90% of the time between some animals and is most common between pairs of females or male and female "consort" pairs during the breeding season (Day *et al.* 2000; Kerle 1998). When captive possums share dens, affiliative interactions occur throughout the period of sharing. Long periods of "touching" are observed between sharing possums, and when active in the den, "food sharing" and "allogrooming" are also observed (Day *et al.* 2000).

Communication

Possums have a well-developed system for conveying information to other possums, either during direct interactions or indirectly (in the absence of another animal). This communication system is based mainly on smell, although visual and auditory stimuli are also used (Biggins 1984). Possums have several scent glands, including the labial and chin glands, sternal gland, and paracloacal glands (Biggins 1979; Russell 1984). Scent-marking behaviour is performed regularly by most possums, with stereotyped patterns of sternal rubbing, labial rubbing, and cloacal dragging evident (Biggins 1984). Sternal and chin gland rubbing on branches and tree trunks are the most commonly observed marking behaviours (Winter 1976), with most animals marking parts of their home range every night. Biggins (1984) suggested that scent marks may provide information to other possums about the marker's individual identity, social status, location, sex, age, and time of marking. Dominant possums appear to scent mark more often than subordinates (Biggins & Overstreet 1978), and these markings may play a role in reducing the number of contacts between possums in overlapping home ranges. Winter (1976) observed that subordinate possums avoided areas that had been recently marked by animals they recognised to be dominant.

Possums also use visual cues to communicate, mainly during agonistic interactions (Biggins 1984).

They display threatening and submissive postures that convey information to the approaching possum regarding the likelihood of the signaller attacking (Biggins 1984). The "bipedal threat" posture exposes the light colouring of the paws and ventral surface, which is thought to provide an important visual signal for nocturnal animals (Eisenberg & Golani 1977). This posture also exposes the stained sternal-gland patch. Biggins (1984) suggested that the size, secretory activity, and colouration of the sternal gland region reflects the possum's dominance status. A submissive display typically involves the possum rolling onto its back with its paws extended. As this posture is not observed in any other situation, it may be easily identified by other possums. Several other behaviours of possums are also thought to coincidentally provide important visual signals that demonstrate identity or status to others, including stereotyped scent-marking behaviour and the "upright alert" posture (Biggins 1984). Possums also leave visual marks around their home range, by chewing and gnawing repeatedly on trees and scratching the ground. Kean (1967) suggested that these visual signals indicate to other possums the extent to which that area is being used by the signalling animal.

Vocal communication between possums mainly occurs during close interactions, with little evidence of deliberate long-range signalling (Winter 1976). Winter (1976) proposed that the vocal repertoire of possums contains at least 22 sounds of two broad types: threats and appeasement calls. Threat vocalisations include screeches, grunts, growls, hisses, and chatters, all of which serve to warn other possums during aggressive interactions. The different types of threat vocalisations represent a graded series of calls that signal the likelihood the caller will initiate an agonistic attack on another possum (Winter 1976). The possum has a dilated thyroid cartilage (Winter 1976) that enables it to markedly increase the loudness and presumably the degree of threat, or seriousness, of some vocalisations. The appeasement calls include sounds like "zook-zooks" and "squeaks" (made by dependent juveniles) and "shook-shooks" and "clicks" made by males during courtship. Females also use "clicking" vocalisations to communicate with their offspring (Winter 1976; Biggins 1984). The "shook-shook" call of the male was suggested by Wemmer & Collins (1978) to imitate juvenile distress or appeasement calls, and may serve to reduce female aggression. Further research is required to fully determine the social-signalling function of each of the sounds made by possums.

Implications of social behaviour for conventional possum control

Possum control operations using poison baits rely on possums being able to locate the baits and then consume a lethal dose. Bait presentation method and possum social behaviour will combine to influence a possum's ability to both locate and consume a lethal dose of poison. Male possums have larger home ranges and move greater distances than females each night, increasing their probability of encountering baits (Hickling & Sun unpubl. data). Some females have very restricted home ranges (less than 0.5 ha), so will not encounter baits in bait stations unless the bait stations are at high density (Henderson & Hickling 1997). Movement patterns will also affect possums' ability to find traps. Possums generally climb preferred, and often heavily marked, trees (Ward 1978; MacLennan 1984), so above-ground bait stations may be most effective if placed in these "focal" sites. Possums consume more bait from bait stations placed at ground level than from stations placed in trees (Henderson & Frampton 1999). The reasons for this are not clear, although it is likely that more possums encounter bait stations at ground-level than in trees. Dominant female possums appear to restrict some subordinate animals from using bait stations, either through direct interactions or by regularly scent marking the bait station (Hickling & Sun unpubl. data). Some subordinate animals may also be displaced from bait stations while feeding (Hickling & Sun unpubl. data), which may lead to sub-lethal poisoning and thereby set up ideal conditions for the development of bait aversions (Clapperton *et al.* 1996; O'Connor & Matthews 1996).

Implications of social behaviour for biological possum control

It has been suggested that the "ideal" biological control system for possums would be based around an infectious agent that induces permanent sterility without suppressing reproductive cycles (Cowan 1996). This is because the possum contact rate should be maintained as possum density declines (possums actively seek each other for mating) and multiple matings following sterilisation could actually increase contact rates, thereby spreading the agent more quickly (Barlow 1997). Recent research has demonstrated that disruption of sex steroid levels by ovariectomy (Jolly & Spurr 1996), vaccination against gonadotrophin-releasing hormone (Jolly *et al.* 1996), or castration (McAllum 1996) do not alter established dominance status in groups of captive possums. Therefore, biological control methods that disrupt sex steroid production may not be inhibited by changes to dominance relationships and may be suitable for use in biological control.

A biological control system will require an effective vector for transmitting the control agent between possums (Cowan 1996). A vector may be either non-disseminating (e.g., a bait that each individual must eat) or self-disseminating (e.g., an infectious organism that spreads between possums). If the vector is in bait form, possum behaviour will affect bait consumption, and therefore control efficacy, in a similar way to that described for toxic baits. However, if dominant possums excluded subordinates from eating the non-toxic bait used to deliver the sterilising agent, then the subordinate would remain fertile and the effectiveness of control would be reduced. In a non-disseminating system (such as bait delivery), there would need to be procedures put in place (e.g., high-density bait stations) to ensure that all breeding-age animals were exposed to the control agent.

The spread of a self-disseminating organism will also be affected by social behaviour and the routes by which the agent is transmitted (e.g., *Leptospira interrogans* serovar *balcanica*; Day *et al.* 1998a,b). A self-disseminating organism may be transmitted either indirectly (requiring no contact between individuals for transmission) or directly, relying on social contact for transmission. Indirect transmission will occur in overlapping home ranges and will be reliant on suitable environmental conditions for persistence of the vector organism. Dominance relationships and possum density will help determine the degree of home range overlap and shared habitat use in a population, both of which may vary seasonally. The greater the sharing of habitat, the greater the chance of uninfected animals contacting infectious material. Scent marks left by infected possums may provide a source of infectious material. Agents spread by indirect transmission would not require social contact to disseminate, so may spread more rapidly than directly disseminated agents.

Directly disseminating vectors will be most strongly affected by patterns of possum social behaviour. The rate of social contact between possums will affect the speed with which the organism is transmitted. Little is known about the effects of possum density or dominance status on social contact rate, although in captivity dominant animals interact most frequently (Biggins & Overstreet 1978). Most interactions between wild possums occur in the breeding season (Winter 1976), suggesting that transmission may occur most quickly during this period.

The type of social interaction between possums may determine the likelihood of a control agent disseminating. For example, *Leptospira interrogans* serovar *balcanica* is transmitted between possums only during affiliative or sexual interactions, but not during agonistic interactions that involve equally as close contact (Day *et al.* 1998b). Sexual transmission has been suggested as the ideal method for disseminating a biological control agent through possum populations (Cowan 1996). Possums actively seek mates in the breeding season, so as possum density reduces, sexual contact would be maintained (Barlow 1997). Sexual transmission is also likely to be highly species specific. The disadvantage of sexual transmission is that juvenile possums would not be infected. As juveniles are the primary dispersers from possum populations (Clout & Efford 1984), a sexually transmitted agent may not be readily disseminated over a wide area. Some subordinate animals may also be restricted from mating opportunities by more dominant possums, so may not easily be infected with a sexually transmitted agent. Jolly *et al.* (1996, 1999) have shown that, in captivity, sterilised dominant possums are able to restrict subordinate animals from mating. From a behavioural perspective, dissemination during non-sexual behaviours that occur both at mating and also between adult and juvenile possums at other times (e.g., agonistic or affiliative behaviours) may be most appropriate for widely disseminating a control agent.

Social factors will influence the best time for the release of an agent and the ideal animals to disseminate the infection. If dominant animals only were infected, some subordinate animals may never interact with them and continue to reproduce with other subordinates, thereby avoiding the control measure. The nature and variation of the mating system may have a considerable impact on the success of sterilisation, with polygamous populations (as is likely for possum populations) being least affected by fertility control (Barlow *et al.* 1997). Mathematical modelling has suggested that up to 80% of infected possums would need to be sterile at mating before biological control would become effective (Barlow 1997). Current research in experimentally sterilised populations is being conducted to determine the influence of potential biological control strategies on possum populations. Further research will be required to examine the effects of social behaviour on the spread of self-disseminating agents.

Conclusions

The ability of possums to adapt to a wide range of conditions has been the primary reason for their phenomenal success, and subsequent pest status, in New Zealand. Their nocturnal, arboreal, and predominantly solitary existence has made the study of their behaviour in the wild difficult, often leaving pest managers and researchers guessing as to the best methods for controlling them. Currently, very little is known about the effects of control operations on the social behaviour of possums and the implications of any changes in behaviour that control operations may have for future control strategies. There is ample evidence to show that, if left untouched, possum populations will readily recover after highly successful conventional poisoning operations. This suggests that possums' feeding, reproductive, and social strategies are sufficiently flexible to cope with rapidly altered and very low density situations. It is possible that possums may adjust their social structure and mating patterns to compensate for changes in their populations caused by fertility-based control techniques. These possible changes need to be considered in the development of biological control. Only through a better understanding of the factors controlling the variability and nature of social organisation among possums in New Zealand will we be able to develop effective methods for controlling them in the long term. Until these social factors are identified and manipulated, possums will remain one of the most successful and devastating vertebrate pests in New Zealand.

Summary

- Possums in New Zealand are generally solitary, but they have extensively overlapping home ranges. Possums do not appear to defend a territory, but at low density may have exclusive areas.
- The social organisation of wild possums in New Zealand is not well described, but appears to be based on a system of dominance and mutual avoidance. The influences of possum density, habitat type, and demographic factors on social organisation are poorly understood.
- Possums spend little time in direct social interactions, except during the breeding season. Interactions are of three main types: agonistic, affiliative, or sexual.
- Male possums will consort with a female possum for 30 to 40 days before mating, although mating without a consort period also occurs. Dominant males have greater reproductive opportunity than subordinate males.
- Den sharing is rare between possums, although few dens are exclusive to one possum. Within dens, affiliative interactions such as touching and allogrooming are common.
- Possums communicate in the absence of other animals by scent marking. Visual and vocal communication are used predominately when possums are interacting directly.
- The social organisation of possums and its variation will influence the success of control operations, and should be considered in the future development of conventional and biological control strategies.

References

Alleva, E.; Petruzzi, S.; Ricceri, L. 1995: Evaluating the social behaviour of rodents: Laboratory, semi-naturalistic and naturalistic approaches. *In:* Alleva, E.; Fasolo. A.; Lipp, H.; Nadel, L.; Ricceri, L. *ed.* Behavioural brain research in naturalistic and semi-naturalistic settings. Dordrecht, Kluwer Academic Publishers. Pp. 375–393.

Barlow, N. D. 1997: Modelling immunocontraception in disseminating systems. *Reproduction, Fertility and Development 9*: 51–60.

Barlow, N. D.; Kean, J. M.; Briggs, C. J. 1997: Modelling the relative efficacy of culling and sterilisation for controlling populations. *Wildlife Research 24*: 129–141.

Biggins, J. G. 1979: Olfactory communication in the brush-tailed possum, *Trichosurus vulpecula* Kerr, 1972, (Marsupialia: Phalangeridae). Unpublished PhD thesis, Monash University, Melbourne, Australia.

Biggins, J. G. 1984: Communications in possums: a review. *In:* Smith, A. P.; Hume, I. D. *ed.* Possums and gliders. Chipping Norton, NSW, Surrey Beatty in assoc. with the Australian Mammal Society. Pp. 35–57.

Biggins, J. G.; Overstreet, D. H. 1978: Aggressive and nonaggressive interactions among captive populations of the brush-tail possum, *Trichosurus vulpecula* (Marsupialia: Phalangeridae). *Journal of Mammalogy 59*: 149–159.

Brown, J. L. 1964: The evolution of diversity in avian territorial systems. *Wilson Bulletin 76*: 160–169.

Caley, P. 1996: Is the spatial distribution of tuberculous possums influenced by den "quality"? *New Zealand Veterinary Journal 44*: 175–178.

Caley, P.; Spencer, N. J.; Cole, R. A.; Efford, M. G. 1998: The effect of manipulating population density on the probability of den-sharing among common brushtail possums and the implications for transmission of bovine tuberculosis. *Wildlife Research 25*: 383–392.

Clapperton, B. K.; Matthews, L. R.; Fawkes, M. S.; Pearson. A. J. 1996: Lithium and cyanide-induced conditioned food aversions in brushtail possums. *Journal of Wildlife Management 60*: 195–201.

Clout, M. N. 1977: The ecology of the possum (*Trichosurus vulpecula* Kerr) in *Pinus radiata* plantations. Unpublished PhD thesis, University of Auckland, Auckland, New Zealand.

Clout, M. N.; Efford, M.G. 1984: Sex differences in the dispersal and settlement of brushtail possums (*Trichosurus vulpecula*). *Journal of Animal Ecology 53*: 737–749.

Clutton-Brock, T. H. 1989: Mammalian mating systems. *Proceedings of the Royal Society of London B 236*: 339–372.

Cowan, P. E. 1989: Denning habits of common brushtail possums, *Trichosurus vulpecula*, in New Zealand lowland forest. *Australian Wildlife Research 16*: 63–78.

Cowan, P. E. 1990: Brushtail possum. *In:* King, C. M. *ed.* The handbook of New Zealand mammals. Auckland, Oxford University Press. Pp. 68–98.

Cowan, P. E. 1996: Possum biocontrol: Prospects for fertility regulation. *Reproduction, Fertility and Development 8*: 655–660.

Cowan, P. E.; Tyndale-Biscoe, C. H. 1997: Australian and New Zealand mammal species considered to be pests or problems. *Reproduction, Fertility and Development 9*: 27–36.

Crawley, M. C. 1973: A live-trapping study of Australian brush-tailed possums, *Trichosurus vulpecula* (Kerr), in the Orongorongo Valley, Wellington, New Zealand. *Australian Journal of Zoology 21*: 75–90.

Day, T. D. 1996: *Leptospira interrogans* serovar *balcanica* transmission in the brushtail possum (*Trichosurus vulpecula*). Unpublished MSc thesis, University of Waikato, Hamilton, New Zealand.

Day, T. D.; O'Connor, C. E.; Matthews, L. R. 1998a: Effects of possum social behaviour on potential biological control strategies. Proceedings: 11th Australian Vertebrate Pest Conference, Bunbury, Western Australia. Pp. 221–225.

Day, T. D.; O'Connor, C. E.; Waas, J. R.; Pearson, A. J.; Matthews, L. R. 1998b: Social behaviour and leptospirosis transmission in possums. *In:* Biological control of possums. *The Royal Society of New Zealand Miscellaneous Series 45*: 115–121.

Day, T. D.; O'Connor, C. E.; Waas, J. R. 2000: Den sharing behaviour in captive brushtail possums (*Trichosurus vulpecula*). *New Zealand Journal of Zoology*. (in press).

Dunnet, G. M. 1956: A live-trapping study of the brush-tailed possum *Trichosurus vulpecula* Kerr (Marsupialia). *CSIRO Wildlife Research 1*: 1–18.

Dunnet, G. M. 1964: A field study of local populations of the brush-tailed possum *Trichosurus vulpecula* in Eastern Australia. *Proceedings of the Zoological Society of London 142*: 665–695.

Efford, M. G. 1991: The ecology of an uninfected forest population *In:* Jackson, R. *ed.* Symposium on tuberculosis. *Publication No. 132, Veterinary Continuing Education*. Palmerston North, New Zealand, Massey University. Pp. 41–51.

Eisenberg, J. F. 1966: The social organisation of mammals. *In:* Eisenberg, J. F. *ed.* Handbuck der Zoologie. Berlin, Walter de Gruyter.

Eisenberg, J. R.; Golani, I. 1977: Communication in Metatheria. *In:* Sebeok, T. A. *ed.* How animals communicate. Bloomington, Indiana University Press. Pp. 575–599.

Fairweather, A. A. C.; Brockie, R. E.; Ward, G. D. 1987: Possums (*Trichosurus vulpecula*) sharing dens: a potential infection route for bovine tuberculosis. *New Zealand Veterinary Journal 35*: 15–16.

Green, W. Q. 1984: A review of ecological studies relevant to management of the common brushtail possum. *In:* Smith, A. P.; Hume, I. D. *ed.* Possums and gliders. Chipping Norton, NSW, Surrey Beatty in assoc. with the Australian Mammal Society. Pp. 483–499.

Green, W. Q.; Coleman, J. D. 1987: Den sites of possums, *Trichosurus vulpecula*, and frequency of use in mixed hardwood forest in Westland, New Zealand. *Australian Wildlife Research 14*: 285–292.

Greenwood, P. J. 1980: Mating systems, philopatry and dispersal in birds and mammals. *Animal Behaviour 28*: 1140–1162.

Henderson, R.J.; Frampton, C.M. 1999: Avoiding bait shyness in possums by improved bait standards. Landcare Research Contact Report LC9899/60 (unpublished) 54 p.

Henderson, R. J.; Hickling, G. J. 1997: Possum behaviour as a factor in sublethal poisoning during control operations using cereal baits. Landcare Research Contract Report LC9798/03 (unpublished) 26 p.

Herbert, P. A. 1996: The chronobiology of the common brushtail possum *Trichosurus vulpecula* (Kerr 1972). Unpublished MSc thesis, University of Auckland, Auckland, New Zealand.

Hinde, R. A. 1976: Interactions, relationships and social structure. *Man 11*: 1–17.

How, R. A. 1978: Population strategies of four species of Australian possums. *In:* Montgomery, G. G. *ed.* The ecology of arboreal folivores. *Symposia of the National Zoological Park*. Washington DC, Smithsonian Institution Press. Pp. 305–313.

How, R. A. 1981: Population parameters of two congeneric possums, *Trichosurus* spp., in north-eastern New South Wales. *Australian Journal of Zoology 29*: 205–215.

Jolly, J. N. 1973: Movements and social behaviour of the opossum, *Trichosurus vulpecula* Kerr, in a mixed scrub and pasture habitat. *Mauri ora 1*: 65–71.

Jolly, J. N. 1976: Habitat use and movements of the opossum (*Trichosurus vulpecula*) in a pastoral habitat on Banks Peninsula. *Proceedings, New Zealand Ecological Society 23*: 70–78.

Jolly, J. N. 1981: Aspects of the social behaviour of the possum *Trichosurus vulpecula*. *In:* Bell, B. D. *ed.* Proceedings of the first symposium on marsupials in New Zealand. *Zoology Publications from the Victoria University of Wellington 74*. Pp. 141–142.

Jolly, S. E.; Spurr, E. B. 1996: Effect of ovariectomy on the social status of brushtail possums *Trichosurus vulpecula* in captivity. *New Zealand Journal of Zoology 23*: 27–32.

Jolly, S. E.; Scobie, S.; Coleman, M. C. 1995: Breeding capacity of female brushtail possums *Trichosurus vulpecula* in captivity. *New Zealand Journal of Zoology 22*: 325–330.

Jolly, S. E.; Scobie, S.; Cowan, P. E. 1996: Effects of vaccination against gonadotrophin-releasing hormone (GnRH) on the social status of brushtail possums in captivity. *New Zealand Journal of Zoology 23*: 325–330.

Jolly, S. E.; Scobie, S.; Spurr, E. B.; McAllum, C.; Cowan, P. E. 1998: Behavioural effects of reproductive inhibition in brushtail possums. *In:* Biological control of possums. *The Royal Society of New Zealand Miscellaneous Series 45*: 125–127.

Jolly, S.E.; Spurr, E.B.; Cowan, P.E. 1999: Social dominance and breeding success in captive brushtail possums, *Trichosurus vulpecula*. *New Zealand Journal of Zoology 26*: 21–25.

Kean, R. I. 1967: Behaviour and territorialism in *Trichosurus vulpecula* (Marsupialia). *Proceedings, New Zealand Ecological Society 14*: 71–78.

Kerle, J. A. 1984: Variation in the ecology of *Trichosurus*: its adaptive significance. *In:* Smith, A. P.; Hume, I. D. *ed.* Possums and gliders. Chipping Norton, NSW, Surrey Beatty in assoc. with the Australian Mammal Society. Pp. 115–128.

Kerle, J. A. 1998: The population dynamics of a tropical possum, *Trichosurus vulpecula arnhemensis* Collett. *Wildlife Research 25*: 171–181.

Lott, D. F. 1991: Intraspecific variation in the social systems of wild vertebrates. Cambridge, Cambridge University Press. 238 p.

MacLennan, D. G. 1984: The feeding behaviour and activity patterns of the brushtail possum, *Trichosurus vulpecula*, in an open eucalypt woodland in southeast Queensland. *In:* Smith, A. P.; Hume, I. D. *ed.* Possums and gliders. Chipping Norton, NSW, Surrey Beatty in assoc. with the Australian Mammal Society. Pp. 155–161.

McAllum, P. 1996: Social rank, hormones and reproductive behaviour of male brushtail possums (*Trichosurus vulpecula*): Implications for biocontrol. Unpublished MSc thesis, Lincoln University, Lincoln, New Zealand.

O'Connor, C.E.; Matthews, L.R. 1996: Behavioural mechanisms of bait and poison avoidance. *In:* Improving conventional control of possums. *The Royal Society of New Zealand Miscellaneous Series 35*: 51–53.

Oldham, J.M., 1986. Aspects of reproductive biology in the male brush-tailed possum, *Trichosurus vulpecula*. Unpublished MSc thesis, University of Waikato, Hamilton, New Zealand.

Paterson, B. M.; Morris, R. S.; Weston, J.; Cowan, P. E. 1995: Foraging and denning patterns of brushtail possums, and their possible relationship to contact with cattle and the transmission of bovine tuberculosis. *New Zealand Veterinary Journal 43*: 281–288.

Russell, E. M. 1984: Social behaviour and social organization of marsupials. *Mammal Review 14*: 101–154.

Sarre, S.; Clout, M. 1998: Mating patterns of possums. *In:* Biological control of possums. *The Royal Society of New Zealand Miscellaneous Series 45*: 128–129.

Smith, A.; Lee, A. 1984: The evolution of strategies for survival and reproduction in possums and gliders. *In:* Smith, A. P.; Hume, I. D. *ed.* Possums and gliders. Chipping Norton, NSW, Surrey Beatty in assoc. with the Australian Mammal Society. Pp. 17–33.

Taylor, A.; Cooper, D.; Fricke, B.; Cowan, P. 1998: Population genetic structure and mating system of brushtail possums in New Zealand — implications for biocontrol. *In:* Biological control of possums. *The Royal Society of New Zealand Miscellaneous Series 45*: 122–124.

Triggs, S. J. 1982: Comparative ecology of the possum, *Trichosurus vulpecula*, in three pastoral habitats. Unpublished MSc thesis, University of Auckland, Auckland, New Zealand.

Ward, G. D. 1978: Habitat use and home range of radio-tagged opossums *Trichosurus vulpecula* (Kerr) in New Zealand lowland forest. *In:* Montgomery, G. G. *ed.* The ecology of arboreal folivores. *Symposia of the National Zoological Park.* Washington DC, Smithsonian Institution Press. Pp. 267–287.

Wemmer, C.; Collins, L. 1978: Communication patterns in two phalangerid marsupials, the grey cuscus (*Phalanger gymnotis*) and the brush possum (*Trichosurus vulpecula*). *Säugetierkundliche Mitteilungen 3*: 161–172.

Winter, J. W. 1976: The behaviour and social organisation of the brush-tail possum (*Trichosurus vulpecula* Kerr). Unpublished PhD thesis, University of Queensland, Brisbane, Australia.

CHAPTER FIVE

Possum Density, Population Structure, and Dynamics

Murray Efford

Population density is probably the single most important predictor of possum impacts on other biota and on the values that people hold dear, from Tb-free livestock to birds in our forests. Foliage consumption, nest predation, and disease transmission will all be greater in absolute terms at greater possum densities. The understanding of broad patterns of possum density and the processes controlling these is therefore fundamental to determining and managing the pest status of possums. This chapter outlines what is known about possum density and population dynamics in New Zealand. It draws particularly on the intensive and long-term Orongorongo Valley study that was started in 1966 by Crawley (1973) and continued by a succession of workers to the present.

How to count?

Simple as it is in concept, the density of a possum population remains a difficult property to measure with much precision. Index methods for monitoring *relative density* are detailed in Chapter 12. They do not allow density to be compared between studies without assumptions (e.g., equal capture probability and equal home-range size). *Absolute density* refers to the number of animals per hectare. Early workers were optimistic that absolute density might be inferred from trap catch (Batcheler *et al.* 1967). However, reliable calibration of trap catch is elusive as we have little reason to believe that home-range size and capture probability are constant between habitats and over time. This chapter focuses on absolute estimates of population density because their calculation includes explicit adjustments for capture probability and home-range size, and they are therefore more likely to be comparable between studies.

Absolute density has usually been estimated by surveying a defined local population (e.g., the animals overlapping a trapping grid) and dividing by an estimate of its extent – usually called the effective trapping area. Census by mark-recapture or removal methods (Otis *et al.* 1978; Pollock *et al.* 1990) can be problematic when animals vary in their capture probabilities either individually or over time. Both these effects apply to possums in New Zealand. For example, individual nightly trappability had a coefficient of variation of 75% between Orongorongo Valley possums (Pledger & Efford 1998). Nevertheless, new techniques and software make the goal of reliable estimation more achievable (S. Pledger, Victoria University of Wellington, pers. comm.; White 1998). Simple pooling of data across nights may be enough to remove most of the bias due to individual and temporal variation (e.g., Efford 1998).

To estimate effective trapping area we must know (or make assumptions) about the movements of members of the target population. This frequently requires expensive radio-tracking of many animals. A common compromise is to treat the range of each animal as a uniform disc with a fixed radius, and to estimate the effective trapping area as the area of the trapping grid plus a boundary strip of width equal to the range radius. Looking beyond traditional approaches to density estimation, trapping webs (Link & Barker 1994) potentially avoid the need for separate census and area estimates, but they have yet to be tested on possums.

How many possums?

Kean & Pracy (1953 p. 702) reported that "... in badly infested country in Poverty Bay, concentrations in isolated patches of forest or scrub [of] opossums have been taken ... at the rate of 20 or more per acre." Such extreme densities (>45.0/ha) have not been verified by formal studies. Population density has been measured by live trapping or intensive removal in several forest and farmland habitats (Fig. 5.1. See Appendix 5.1 for raw data). These studies sampled the long-term average density of possums in the selected habitats, which may be interpreted as an equilibrium density or "carrying capacity" because the studies were in areas

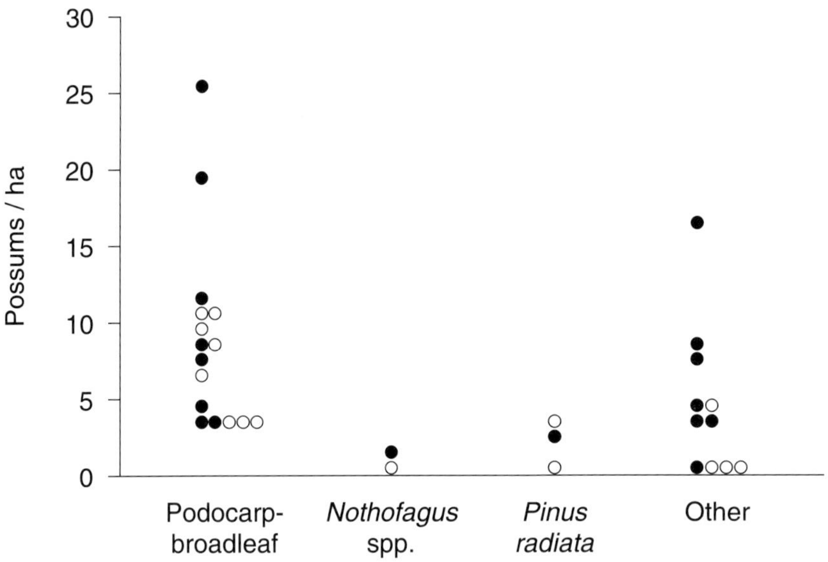

Fig. 5.1
Population densities of brushtail possums in different New Zealand habitats. Each symbol represents the average density from an intensive study of an undisturbed population. Mark-recapture or removal estimates as reported by the original authors. Sites adjoining pasture are shown as filled circles; no boundary strip was added to the effective trapping area on pasture margins.

"Podocarp-broadleaf" refers to mixed, mostly lowland, forest with dominant *Metrosideros robusta, M. umbellata, Beilschmiedia tawa, Weinmannia racemosa,* or *Fuchsia excorticata* in combination with one or more podocarp species. (Triggs 1982; Ji Weihong & M. Clout, University of Auckland, pers. comm. [2 points]; Ramsey *et al.* 1997 [7 points]; Efford 1998 and unpubl. results; Coleman *et al.* 1980 [3 points]; Efford *et al.* 2000; J. Coleman, Landcare Research, Lincoln, pers. comm.).

"*Nothofagus*" refers to forest composed principally of one or more *Nothofagus* species (*N. fusca, N. solandri* var. *cliffortioides*) at Mt Misery, Nelson Lakes (Clout & Gaze 1984) and Burwood Bush, Te Anau (R. van Mierlo, Department of Conservation, Hokitika, pers. comm.).

"*Pinus radiata*" refers to maturing (>10 year old) pine plantations at Silverdale (Triggs 1982), Tokoroa (Clout 1977), and Ashley Forest (Warburton 1977).

"Other" includes willows (*Salix* spp.) and a variety of grassland, scrub and scrub-pasture habitats. (Jolly 1976; Triggs 1982; Olds 1987; Pfeiffer 1994; Patterson *et al.* 1995; Brockie *et al.* 1997 [3 points]; G. Hickling, Lincoln University, Lincoln, pers. comm [2 points]; Efford unpubl. data).

---

not directly affected by possum control operations (see below). Most of the populations monitored (16/31 = 52%) have been in lowland podocarp-broadleaved forest adjacent to pasture, where an important research objective has been to understand the role of possums in transmitting Tb to stock.

The tree species most at risk to possum browsing commonly occur as components of podocarp-broadleaved forests, and the highest possum densities have been recorded in them (Fig. 5.1). The maximum density measured so far, 25.4/ha, occurred in a previously logged forest of kāmahi (*Weinmannia racemosa*) and tree fern (*Cyathea smithii*), adjoining pasture near the Haupiri River, Westland (Coleman *et al.* 1980). The second-ranked at 19.8/ha was also in modified forest comprising regenerating tree fuchsia (*Fuchsia excorticata*), māhoe (*Melicytus ramiflorus*), and broadleaf (*Griselinia littoralis*), with occasional emergent podocarps, at Pigeon Flat near Dunedin (Efford *et al.* 2000). The remaining estimates for mixed podocarp-broadleaved forest, including all those not adjoining pasture, lie between 3.0 and 13.0 per hectare (Fig. 5.1).

Possums have been surveyed in beech (*Nothofagus* spp.) forests by cyanide removal on a grid at Mt Misery, Nelson Lakes (0.5/ha; Clout & Gaze 1984) and by exhaustive trapping of a 24-ha fenced enclosure at Burwood near Te Anau (1.7/ha; R. van Mierlo, Department of Conservation, Hokitika, pers. comm.). The population density in South Westland beech forest was similar to that at Mt Misery, judging from the very low kill rates on cyanide baits (Owen & Norton 1995). Measured densities in radiata pine (*Pinus radiata*) plantations were also low, ranging from 0.9/ha to 3.0/ha (Fig. 5.1). Low densities in pine and beech forests seem to reflect the unpalatability of their foliage to possums (although flowers, buds, and pollen cones are eaten when available), and the low diversity of alternative food species. The possum populations of plantation forests initially increase with time since planting (Keber 1988), reflecting an initial increase in the diversity of shrub-layer vegetation; numbers later decline as the canopy shades out the shrub layer.

Seral woody vegetation (scrub) and plantings for shelter or erosion control provide suitable habitat for possums in many modified New Zealand landscapes. Possum densities in these habitats can be high (>8.0/ha), as found by Brockie *et al.* (1997) in the parts of their Hawke's Bay study area dominated by willow (*Salix* spp.) and blackberry (*Rubus fruticosus*).

Possums occur in extensive tracts of tussock grassland and other open country, but there are few measurements of their abundance. No possums were caught in a 3-day survey by cage trapping in 11 ha of red tussock (*Chionochloa rubra*) at Burwood near Te Anau in 1991, although they were present in nearby radiata pine shelter belts (Efford unpubl. data). Possums may occur more commonly in native grassland where interspersed shrubs and rocks provide daytime refuges (G. J. Hickling, Lincoln University, Lincoln, pers. comm.).

From the ad hoc and uneven sampling of habitats represented in Figure 5.1, it is difficult to determine any overall patterns of variation in density with, for example, latitude or altitude. Among sites in mixed podocarp-broadleaved forest, there was no statistically significant relationship between density and latitude ($r = 0.35$, df=14, P=0.18).

Possum densities have been studied on altitudinal transects at Mt Misery, Nelson Lakes (Clout & Gaze 1984), and at Mt Bryan O'Lynn, Westland (Coleman *et al.* 1980). The capture rate in live traps in beech forest at Mt Misery declined from 0.114 captures/trap night at 460–650 m a.s.l. to 0.065 captures/trap night at 1100–1400 m; a higher capture rate (0.214 captures/trap night) was recorded in beech-podocarp forest on an alluvial fan at low altitude (455–460 m), while there were no captures in traps set above the bushline (>1400 m). Density also declined with altitude at Mt Bryan O'Lynn, except that density was high in a narrow zone at the high-altitude-forest scrub interface (900–1000 m). From these intensive studies it seems plausible that possum densities generally decline with altitude. However, the pattern is more likely driven by altitudinal zonation of palatable vegetation than by physical variables such as temperature and rainfall. Strong, direct climatic effects on density are contradicted by the occurrence of high densities at both high and low latitudes, and in both the wet west and the drier east.

Possums reach population densities in New Zealand (Fig. 5.1) that are much higher than any recorded in Australia. The most dense Australian population tabulated by Kerle (1984) was 4.0/ha from Tasmanian dry sclerophyll forest adjacent to pasture. Although several populations studied in eucalypt woodland have been at densities of 1.0–3.0/ha, Australian populations are typically sparse (<1/ha) (Kerle 1984). Comparison is difficult because few New Zealand habitats have direct equivalents in Australia and vice versa. A possible exception is beech forest, which occurs extensively both in New Zealand and Tasmania, but in neither place does it support high densities of possums (Fig. 5.1; Green 1973, cited by Kerle 1984). Lower densities in Australia may be attributed to the combined effects of numerous factors: the presence of large predators (e.g., dingos (*Canis familiaris*) and powerful owls (*Ninox strenua*)), a wider range of parasites (Viggers & Spratt 1995), lower digestibility or nutrient content of foliage and, locally, competition with other folivorous marsupials.

In summary, possum density varies greatly between major habitat types in New Zealand, but within each broad vegetation type there is much unexplained variation. Future research should focus on understanding in more detail how possum densities are influenced by the local vegetation (e.g., species diversity, community structure, tree cavities) and site attributes (e.g., soil fertility, rainfall). Without this perspective, we cannot extrapolate reliably from intensive studies to the regional and national scales necessary for effective control planning.

Spatial distribution within habitat patches

Possums may not be uniformly distributed within habitats for a variety of reasons, including fine-scale heterogeneity of the same vegetation and site factors that cause density to vary between habitats. Two further types of unevenness are habitat ecotones or edge effects, and social clustering.

Edge effects

The highest accurately measured possum densities (>15.0/ha) have all been within 250 m of pasture – at Mt Bryan O'Lynn (Coleman *et al.* 1980), Bridge Pa (Brockie *et al.* 1997), and Pigeon Flat (Efford *et al.* 2000). In each situation, at least some possums within this zone were known to feed on pasture plants, particularly clovers and grasses. Nutritional benefits from these foods may lead to higher carrying capacity, but the role of pasture in supporting elevated possum densities in adjoining forest is not yet clear.

Possum feeding on pasture across the forest/pasture ecotone was best documented at Mt Bryan O'Lynn. There, females living within about 200 m of the forest edge found the great majority of their food (about 93%) within the forest (Coleman *et al.* 1985 table 4). Pasture species were more than twice as important in the diet of males compared to females, but remained a small fraction of the total diet. Even allowing for the greater nutritional value of pasture, it is doubtful whether the direct nutritional gain from pasture is sufficient to drive the greater than twofold increase in population observed in this edge zone. If proximity to pasture is inadequate to explain the observed high densities, how do they arise? At Mt Bryan O'Lynn the milled forest close to pasture differed greatly in species composition from the unmilled forest on sites at higher altitude more distant from pasture (Coleman *et al.* 1980), and this may be sufficient to explain the high edge densities.

A general tendency towards higher possum densities on edges may be caused by greater diversity and biomass of understorey species along forest margins, particularly where livestock are excluded. The location of forest-pasture edges is likely to correlate with discontinuities in soil fertility and forest composition, for historical and economic reasons. Edge forests are also more likely to have been disturbed and to include seral species thought to be highly nutritious to and favoured by possums (e.g., Owen & Norton 1995). The empirical evidence for high possum abundance along forest edges needs further consideration.

Clustering

The home ranges of both males and females overlap extensively (Chapter 3), and it is therefore clear that possums in New Zealand are not rigidly territorial. However, little is known of the precise spatial structure of local populations, such as whether they show some weaker form of territoriality (Green 1984; Efford *et al.* 2000) or even social aggregation. Hickling (1995) reported clustering of captures along possum trap lines in kill-sampling of several possum populations. Intensive grid-based live trapping data from the Orongorongo Valley show that the distribution of individuals is spatially random or dispersed rather than aggregated in this moderately high density population (Fig. 5.2). The number of captures per live trap also showed no autocorrelation at any spatial lag (data from 1997–1998; Efford, unpubl. results). Removal of neighbours can induce some individual female possums to immediately shift their home range towards the vacant space (Efford *et al.* 2000), suggesting that they actively space out from each other. Where it occurs, clustering would seem to reflect unevenness in the habitat or the transient effects of control (Hickling 1995) rather than any tendency of possums to form social aggregations. Highly attractive and seasonally abundant food sources such as flowering or fruiting trees may cause aggregations of foraging possums outside their usual home ranges (e.g., Jolly 1976; Ward 1978; Chapter 3). The issue of clustering requires more investigation, particularly in view of its importance for disease transmission (e.g., Barlow 1991; Hickling 1995).

Population turnover: the predictable effects of age, sex, and season

Possum populations are subject to continuous recruitment and loss, resulting from both *in situ* births and deaths, and movement. Adult possums are typically sedentary, i.e., they remain within a stationary home range. However, even in the absence of control or other disturbance, movement of young animals contributes strongly to the turnover of local populations. In the small (22.2 ha) Orongorongo Valley study area, for example, about a quarter of the male population had immigrated in the last 12 months (Fig. 5.3), and the majority are believed to be young possums.

The contributions of different population processes to the overall flux are strongly seasonal. This is driven in part by the seasonality of breeding and subsequent recruitment (Chapter 6). Mortality of adult males is particularly high in winter (June–September), following their loss of fat reserves during the autumn breeding season (Efford 1998). Conversely, immigration and, by inference, emigration are curtailed in winter.

Recruitment almost exactly balanced losses over 15 years in the Orongorongo Valley study population

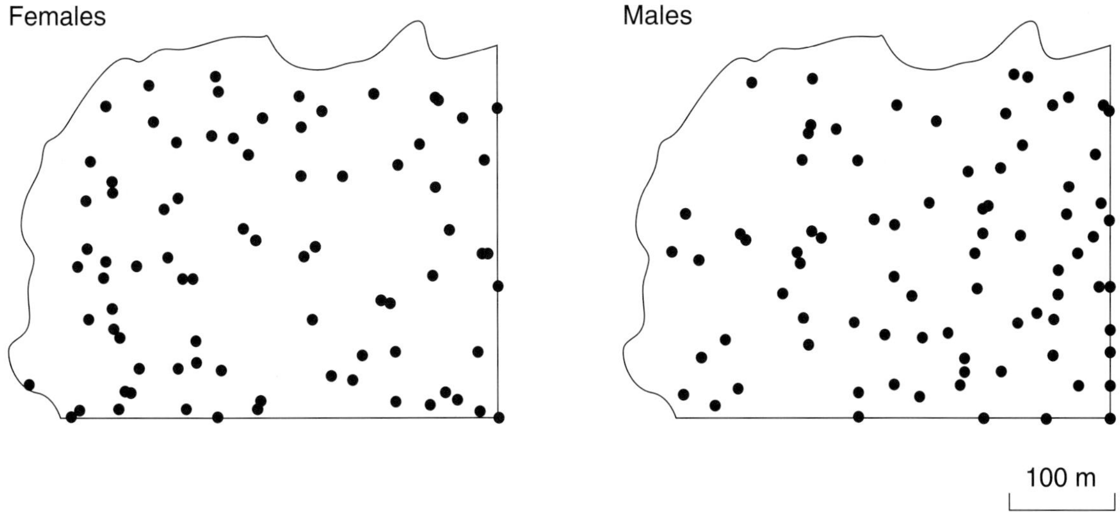

Fig. 5.2
Spatial distribution of possums in an intensively live-trapped local population. Each point represents the averaged lifetime adult trap locations of a possum present on the main Orongorongo Valley study grid in 1997; females ($n = 80$) and males ($n = 81$) are plotted separately. The median number of captures per individual was seven for both sexes. Traps were set on a 30-m grid between the river bank and the boundary line shown. Analysis of the spatial point patterns in a rectangular subarea with Ripley's (1976) "k" functions shows that females tend to be more evenly spaced than expected at scales of 100–250 m; no spacing of females can be distinguished at scales <100 m; and males are more or less randomly located at all these scales (Efford unpubl. data).

(Fig. 5.3). In other areas, females may mature earlier (often in their first year of independence) and some females bear a second young in spring (Chapter 6). Taken together, these additional breeding inputs might be expected to generate a considerable surplus of recruitment and steadily increasing populations. This does not appear to be the case, so higher breeding productivity must be balanced by greater *in situ* mortality, by higher rates of emigration, or by lower rates of immigration. On our present understanding, increased mortality of adults in high-productivity populations seems unlikely. The nutritional conditions that favour breeding by one-year-olds in spring tend to enhance rather than diminish survival prospects, especially as winter is the critical season for both breeding and survival. On land that is farmed intensively, possums face additional risks from vehicles, dogs, and shooting, but these causes of mortality are usually quite localised. It is possible that young born to one-year-old mothers, or in spring, suffer exceptionally high juvenile mortality, but this has yet to be shown. Nor is there yet any evidence for greater per capita rates of emigration from fast-breeding populations: natal dispersal seems to occur at a more or less constant rate (Cowan & Rhodes 1993; Chapter 3).

This leads to the hypothesis that a low rate of immigration relative to emigration is the critical variable compensating for high productivity of young. Why the asymmetry? In a patchy habitat, source areas may receive fewer dispersers than they export simply because there are few source areas nearby, or because more dispersers die in transit in these habitats.

The prospects of a possum surviving through the next year depend strongly on its age (Fig. 5.4). Survival of each sex is greatest soon after they reach maturity, usually at 1–2 years old (Chapter 6). Among females, survival remains high for a further 6–7 years. The estimated annual survival rate of Orongorongo females averaged 89.1% in the prime years 2–5 and, making allowance for probable negative bias in the estimates (Carothers 1979) and occasional emigration, actual survival most likely exceeded 90%. This is a moderately dense population, which has eaten out some preferred food species; in more favourable conditions, the survival of young adult possums may approach 100%.

Beyond 7 years of age there is an increasing decline in the annual survival rate of Orongorongo females. The oldest recorded female reached 14 years. Survival of males was generally lower than that of females, although at its peak (3–4 years old) the difference was small; thereafter, male survival declined steadily with age. Few comparable mark-recapture data exist for other areas, but the age composition of many populations has been sampled by kill trapping and the counting of annual growth lines in tooth cementum. These data also indicate a maximum life span of around 14 years, and an increasing dominance of females in older age classes that can be attributed to male-biased mortality (e.g., Brockie *et al.* 1981; Coleman & Green 1984).

More male possums are born than females: in the Orongorongo Valley, for example, 56% of pouch young born are males (Efford 1998). Mortality in the pouch and post-weaning stages is probably slightly greater among males than females, bringing the sex ratio to parity during the first year of independence (Hope 1972; Crawley 1973; Efford 1998). The sex ratio of adult possum populations varies between moderately male-biased and moderately female-biased. Precision here is difficult because the greater susceptibility of males to trapping or poisoning leads to sampling biases (Coleman & Green 1984). The faster turnover of the male fraction of the population (Fig. 5.3) also leads to a male bias in any sample accumulated over time.

Sex ratio need not be constant through time even within one population. For example, the ratio in the long-established population in the Orongorongo Valley shifted from near parity prior to 1978 (e.g., Crawley 1973) to a distinct male bias (57% males) in 1980–1994 (Efford 1998). By comparison, the colonising possums of depopulated areas usually include an excess of males, which declines with time (Clout & Efford 1984), although Little & Cowan (1992) reported an almost even sex ratio of colonists to a small island connected to the mainland by a 200 m causeway.

How do possum numbers change over time?

Local possum populations in New Zealand were founded either by a small number of liberated animals or by small numbers of animals that dispersed from liberation sites (Pracy 1974). Several decades of steady population growth must have been required to reach present densities. For example, accepting the suggested upper limit to the number of introductions from Australia (300; Pracy 1974), and a generous estimate of the intrinsic growth rate (r_m = 0.3; Clout & Barlow 1982; Barlow 1991), four decades of unrestrained population growth are required to reach the conservative estimate of the national possum population in the 1980s (69 million; Brockie 1992). The phase of initial population

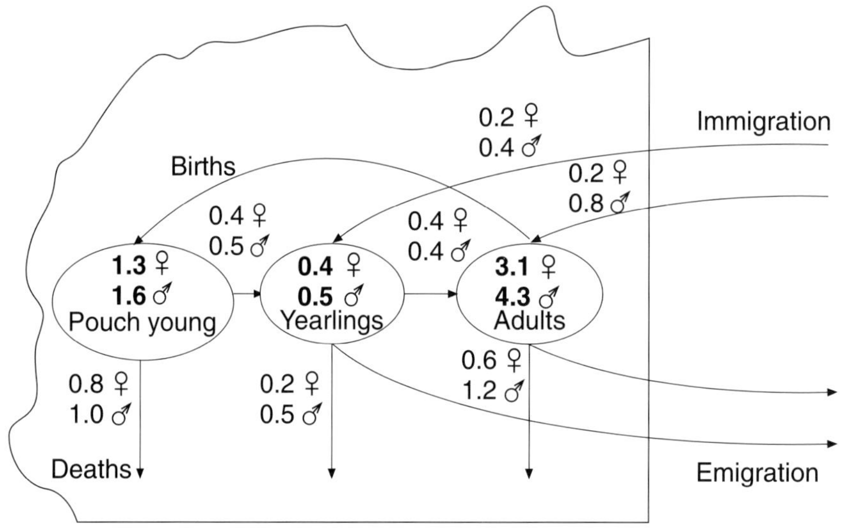

Fig. 5.3
Annual turnover in a possum population. Average of mark-recapture estimates (per hectare) from the Orongorongo Valley, 1980–1994 (Efford 1998). Bold values in circles indicate the population density of newborn young in June, and of independent yearlings and older animals in February. Values on arrows indicate the absolute population flux due to immigration, and to mortality and emigration combined. An effective trapping area of 22.2 ha is assumed.

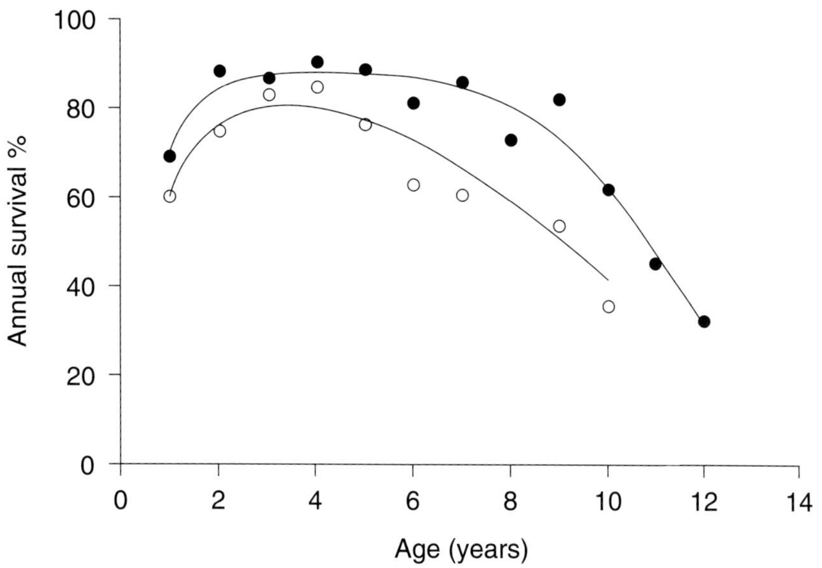

Fig. 5.4
Age-dependent survival of male (o) and female (•) possums. Mark-recapture estimates from the Orongorongo Valley, 1967–1997, and double Weibull curves fitted by non-linear least squares. Disappearance of 1- and 2-year-olds includes natal dispersal, so the plotted points for these ages are underestimates of the true survival rate. Sample restricted to animals that had been first caught as pouch young or one-year-olds, and hence were of known age.

growth is now considered to have passed for almost all New Zealand populations, and in most areas was probably completed several decades ago. The long-term trajectory of these "established" populations is less certain, but it is of critical concern to population managers. Long-term ecological studies allow us to address several questions. Firstly, are established populations regulated about some short-term equilibrium density? Secondly, is any equilibrium density itself stable and if so over what time span? Thirdly, what is the nature of the negative feedback mechanisms that underlie any equilibrium and how should they be modelled? By analysing the results of long-term population monitoring we can currently provide at least tentative answers to these questions.

Long-term studies

Possum populations have been monitored annually over a number of generations in two areas: the Pararaki catchment in the southern Wairarapa and the Orongorongo Valley near Wellington. The Pararaki study started in 1946 as a commercial trapping operation (Batcheler *et al.* 1967), but the same area has been monitored annually by kill trapping on standard trap lines since 1965. The kill rate on the initial trap lines at Pararaki fell by 80% between 1946 and 1965 (= 8% p.a.) (Thomas *et al.* 1993). Subsequent trapping from1965 to the present suggests a stepped decline in density at an overall rate of 2% per annum, with periods of stasis interrupted by episodes of unusually heavy mortality, as in 1977 (Thomas *et al.* 1993; Coleman *et al.* 1999). The Pararaki data lend support to an "irruptive fluctuation" model for the coupled dynamics of possums and their food supply (Thomas *et al.* 1993). Under this interpretation, possums initially increased in numbers well beyond a level that could be sustained by the vegetation.

The Orongorongo pattern is rather different. Intensive kill trapping in 1946–47 gave density estimates of 11.1/ha (Mason 1958) and 6.5/ha (Batcheler *et al.* 1967) and these lie comfortably within the range of densities later monitored by live trapping over 1967–1998 (Fig. 5.5). Evidence for initial population irruption and subsequent decline in the Orongorongo Valley is weak, although Mason (1958 p. 593) noted that "from information gleaned from trappers density had been heavier" before 1946–47. The overall trend in density over the last 30 years is slight, and not significantly different from zero (slope ± 95% confidence interval: + 0.04 ± 0.06/ha/yr). However, densities were relatively low in the middle years of the study, leading to a significant curvilinear trend ($P < 0.05$).

Population stability

The occurrence of characteristic possum densities in different habitat types (Fig. 5.1) in itself suggests that possum populations are regulated around a habitat-determined equilibrium density. Segments of the post-1965 Pararaki data can be interpreted as varying around an equilibrium (Thomas *et al.* 1993) although this level may decline over time (Coleman *et al.* 1998). The Orongorongo time series is even more convincing on this, primarily on account of the high precision of the density estimates (Fig. 5.5). Between 1967 and 1998 this population fluctuated between 5.5 and 13.3 per hectare (9.2 ± 1.47, mean ± SD). Major departures from the mean in 1971–72, 1977, and 1995–96, apparently caused by extreme weather patterns, were in each case followed by an abrupt correction. The natural behaviour of possum populations in areas that have had possums for decades is therefore probably a modest annual fluctuation around the mean.

In statistical terms, a population is "regulated" if its rate of change varies inversely with population density. This is clearly the case in the long-term Orongorongo Valley study (Fig. 5.6). Mechanisms of regulation are discussed below.

Comparisons between populations can be undertaken using *s*, a standard index of population variance.[1] The Orongorongo possum population ($s = 0.07$) was substantially more stable than populations of most other mammals that have been studied, particularly northern hemisphere rodents (Ostfeld 1988). The effect persists even when allowance is made for the effect of body size (Sinclair 1997). Nor is there any indication of regular cycles in possum population density such as are found in many northern hemisphere small mammals at high latitudes (Finerty 1980): population density showed no significant positive serial correlation at lags greater than 1 year (Efford, unpubl. results).

High population stability is an unusual attribute for an introduced pest. For example, Sinclair (1997) suggested that populations of introduced mammals are typically more variable than those of the same or similar species in their country of origin. Possums in New Zealand appear to be an exception to this rule – Orongorongo Valley possums appear as the least variable marsupial or eutherian population on

1 $s = SD(\log 10(N))$ where N is the annual density (Ostfeld 1988)

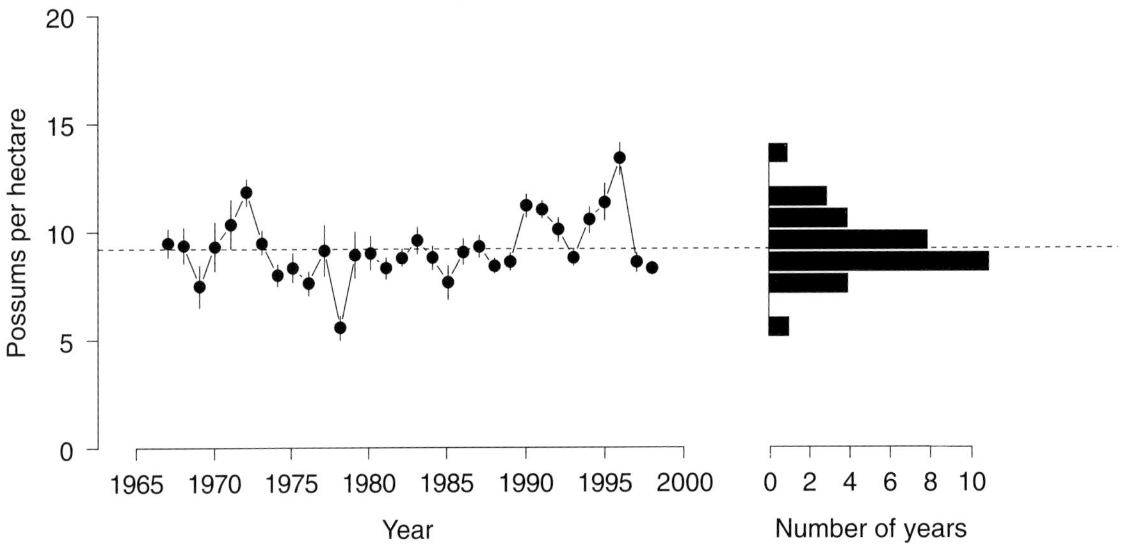

Fig. 5.5
Variation in population density over 1967–1998 in one intensively studied possum population, Orongorongo Valley, Wellington. The population on a 15-ha grid in mixed lowland podocarp-broadleaved forest was monitored by cage trapping for 5 nights at least three times each year (Efford 1998). Vertical bars show 95% confidence intervals of age-structured Jolly-Seber estimates for the population density in late summer (summed yearlings and adults; 83-m boundary strip). Horizontal line indicates long-term mean of possums/ha (9.2 ± 1.47 (SD)). Histogram shows frequency distribution of annual density (class width 1 possum per hectare).

Sinclair's graph (1997: fig. 7), although the only comparative long-term possum data from Australia are relative indices that may have large sampling errors. Sinclair concluded that fertility control may be an effective way of preventing outbreaks of variable exotic pests. It is unclear whether the converse argument holds (i.e., fertility control will be ineffective for possums).

Long-term stability of equilibrium

Possum populations appear stable when studied on time scales of 10–30 years, but it is an open question whether relatively constant possum densities will be maintained in the face of longer-term vegetation change. Many New Zealand forests continue to suffer incremental loss of long-lived possum-palatable tree species decades after their initial colonisation by possums (Campbell 1990). An example is the episodic mortality of emergent northern rātā (*Metrosideros robusta*) in the Orongorongo Valley (Meads 1976; Cowan *et al.* 1997). As leaves of these "preferred" trees seasonally contribute a large part of the possum's diet, we might expect equilibrium possum density in the altered forest to trend downwards over time (Bell 1981). Alternatively, possums may be able to meet their nutritional needs from replacement opportunist or subcanopy species and thereby maintain more or less constant density in the long term. Total possum density did not trend upwards or downwards in the Orongorongo Valley over 1967–1998, despite changes in forest composition (Campbell 1990). For this site the second interpretation is therefore more plausible, and is supported by faecal analysis of food habits. Between 1969 and 1989 the diet shifted subtly away from northern rātā and kāmahi and towards lianes (*Ripogonum scandens* and *Metrosideros fulgens*) and māhoe (Allen *et al.* 1997).

Can we predict the future trajectory of these lowland-mixed-forest possum ecosystems? Incremental loss of vulnerable, long-lived forest trees seems likely to continue, and these species must diminish in importance in possum diet. However, the species that replace them are often both palatable and resilient to browsing, and we should therefore not expect a natural decline in the number of possums. We already see instances in which low forest supports high possum densities on a diet that is largely composed of rapidly regenerating opportunist species (e.g., Pigeon Flat, Dunedin. Efford *et al.* 2000; R. B. Allen, Landcare Research, Dunedin, pers. comm.).

Mechanisms of regulation

Regulation of mammalian populations is the subject of a large and generally inconclusive literature. Census data may indicate that regulation occurs (e.g., Fig. 5.6), although even here statistical testing remains controversial (Shenk *et al.* 1998). The search for underlying mechanisms of regulation often reflects theoretical prejudices (Krebs 1995) and we should proceed with caution.

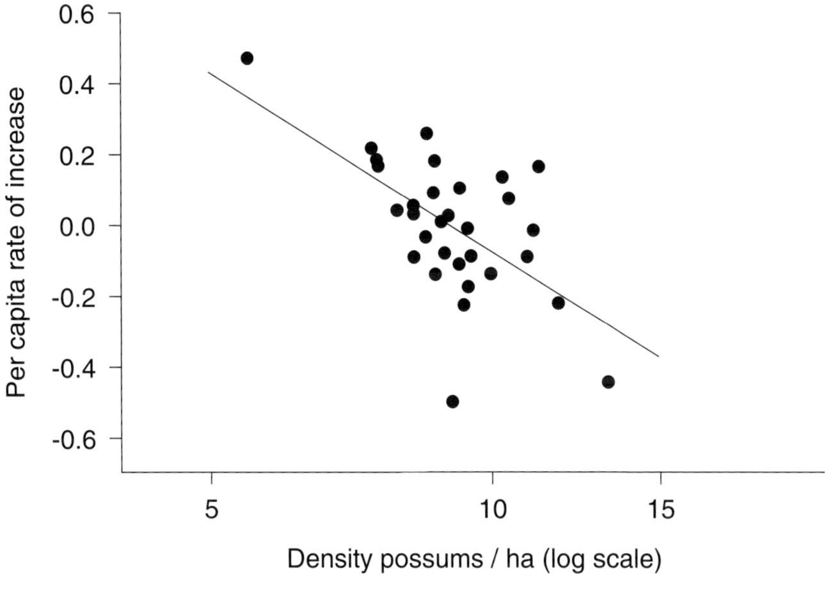

Fig. 5.6
Regulation of a possum population. Rate of increase ($\log(N_{t+1}/N_t)$) as a function of population size ($\log(N_t)$). Annual (February) mark-recapture estimates Orongorongo Valley 1967–1998. Least squares regression fit $r = 1.60 - 0.728 \log(N_t)$; the randomisation test of Pollard *et al.* (1987) shows the relationship is unlikely to have arisen by chance ($P = 0.009$; 10 000 randomisations).

Possums in New Zealand lack significant non-human predators, and it is natural therefore to think of them reaching an equilibrium with their food supply as documented for ungulate-grassland systems (Riney 1964; Caughley 1970). The ungulate model is appealing because it potentially accounts for both the irruptive dynamics of the colonisation phase (steady increase in conditions of abundant food followed by an abrupt impact on vegetation when density first overshoots carrying capacity), and the later equilibrium. Simple plant–herbivore interactive models describe these dynamics well, and in these models both colonising and established populations follow the same rule: a time-lagged feedback through the biomass of edible vegetation (Caughley 1970, 1976). In principle, a new irruption may be triggered by drastically reducing population density.

How closely do established possum populations conform to a simple plant–herbivore interaction? The diversity of foods used by possums, and their frequent inaccessibility, has inhibited any attempt to measure total food availability. However, it has been argued (e.g., Bell 1981; Williams 1982; Nugent *et al.* in Chapter 2) that seasonal fruits are an important high-energy food resource for possums, critical for breeding and overwinter survival. Annual production of one such fruit (hīnau *Elaeocarpus dentatus*) has been measured on the Orongorongo Valley study area (P. E. Cowan, Landcare Research, Palmerston North, pers. comm.). The rate of increase of the Orongorongo population was only weakly related to the hīnau crop ($r^2 = 0.12$, $F_{1,28} = 3.76$, $P = 0.06$; data 1968–1997). Although feeding by possums tends to suppress the production of hīnau fruit (Cowan & Waddington 1990), there is little evidence that depression of the annual fruit crop is a critical feedback regulating year-to-year variation in possum populations.

Although the original Pararaki "irruptive fluctuation" entailed an initial overshoot of the later carrying capacity, severe (50%) natural reductions in population density in both the Pararaki and Orongorongo populations in 1977 were followed by a direct rather than oscillatory return to equilibrium. This suggests that if a plant-herbivore dynamic is the mechanism of the ultimate equilibrium, it is substantially different to the dynamic in the colonising phase. Such a change is not surprising given the effectively permanent loss of some long-lived tree species.

Plant–herbivore dynamics belong to the class of regulatory mechanisms described as "extrinsic" (Caughley & Krebs 1983). Except for diseases and pathogens, about which very little is known, alternative mechanisms of regulation are mostly "intrinsic", i.e., acting by direct competition between individuals (Caughley & Krebs 1983). Direct competition at high density may act through any of several mechanisms, for example, lowered feeding efficiency in the presence of conspecifics, territorial behaviour of residents causing dispersers to settle in less favourable habitats, or increased use of substandard refuges once the best are occupied. Experimental work is needed to test these hypothesised mechanisms.

Summary

- The equilibrium density of an undisturbed possum population depends to a large extent upon the habitat. Podocarp-broadleaved forests carry higher densities (3.0–25.0/ha) than beech forests (<0.5–1.7/ha) or pine plantations (0.7–3.0/ha).
- Very high possum density populations (>15.0/ha) have been recorded only in forest near pasture edges, but it is unclear whether this is due to the nutritional benefits of feeding on pasture or the particular suitability of these forests per se.
- Sampling with removal trap lines appears to show spatial clustering of possums on scales of about 100 m, but live-trapping studies tend to show the reverse.
- The annual rate of survival varies strongly with age. Females live longer than males. Annual survival peaks at about 90% for females aged 2–5 years old, compared with 80% for similar-aged males.
- Left to themselves, possum populations fluctuate within rather narrow bounds that are roughly equal to ± 50% of the long-term average density.
- The exact mechanisms by which established possum populations are regulated remain obscure. Simple plant–herbivore dynamics do not seem to apply, but nor is there yet clear evidence for direct competition.

Acknowledgements

I am grateful to Mick Clout, Jim Coleman, Phil Cowan, Graham Hickling, Dave Ramsey, Ron van Mierlo, and Ji Weihong for unpublished results; and to Jim Coleman, Phil Cowan, David Lindenmayer, Tony Pople and Dave Ramsey for their helpful comments on drafts. Discussions with Ralph Allen and Bruce Warburton were also helpful.

References

Allen, R. B.; Fitzgerald, A. E.; Efford, M. G. 1997: Long-term changes and seasonal patterns in possum (*Trichosurus vulpecula*) leaf diet, Orongorongo Valley, Wellington, New Zealand. *New Zealand Journal of Ecology 21*: 181–186.

Barlow, N. D. 1991: A spatially aggregated disease/host model for bovine Tb in New Zealand possum populations. *Journal of Applied Ecology 28*: 777–793.

Batcheler, C. L.; Darwin, J. H.; Pracy, L. T. 1967: Estimation of opossum (*Trichosurus vulpecula*) populations and results of poison trials from trapping data. *New Zealand Journal of Science 10*: 97–114.

Bell, B. D. 1981: Breeding and condition of possums *Trichosurus vulpecula* in the Orongorongo Valley, near Wellington, New Zealand, 1966-1975. *In:* Bell, B. D. *ed.* Proceedings of the first symposium on marsupials in New Zealand. *Zoological Publications from Victoria University of Wellington No. 74.* Pp. 87–139.

Brockie, R. E. 1992: A living New Zealand forest. Auckland, Bateman. P. 147.

Brockie, R. E.; Bell, B. D.; White, A. J. 1981. Age structure and mortality of possum *Trichosurus vulpecula* populations from New Zealand. *In*: Bell, B. D. *ed.* Proceedings of the first symposium on marsupials in New Zealand. *Zoological Publications from Victoria University of Wellington No. 74.* Pp. 63–83.

Brockie, R. E.; Ward, G. D.; Cowan, P. E. 1997: Possums (*Trichosurus vulpecula*) on Hawke's Bay farmland: spatial distribution and population structure before and after a control operation. *Journal of the Royal Society of New Zealand 27*: 181–191

Campbell, D. J. 1990: Changes in structure and composition of a New Zealand lowland forest inhabited by brushtail possums. *Pacific Science 44*: 277–296.

Carothers, A. D. 1979: Quantifying unequal catchability and its effect on survival estimates in an actual population. *Journal of Animal Ecology 48*: 863–869.

Caughley, G. 1970: Eruption of ungulate populations, with emphasis on Himalayan thar in New Zealand. *Ecology 51*: 53–72.

Caughley, G. 1976: Wildlife management and the dynamics of ungulate populations. *Applied Biology 1*: 183–246.

Caughley, G.; Krebs, C. J. 1983: Are big mammals simply little mammals writ large? *Oecologia 59*: 7–17.

Clout, M. N. 1977: The ecology of the possum (*Trichosurus vulpecula* Kerr) in *Pinus radiata* plantations. Unpublished PhD thesis, University of Auckland, Auckland, New Zealand.

Clout, M. N.; Barlow, N. D. 1982. Exploitation of brushtail possum populations in theory and practice. *New Zealand Journal of Ecology 5*: 29–35.

Clout, M. N.; Efford, M. G. 1984: Sex differences in the dispersal and settlement of brushtail possums (*Trichosurus vulpecula*). *Journal of Animal Ecology 53*: 737–749.

Clout, M. N.; Gaze, P. D. 1984: Brushtail possums (*Trichosurus vulpecula* Kerr) in a New Zealand beech (*Nothofagus*) forest. *New Zealand Journal of Ecology 7*: 147–155.

Coleman, J. D.; Green, W. Q. 1984: Variations in the age and sex distributions of brush-tailed possum populations. *New Zealand Journal of Zoology 11*: 313–318.

Coleman, J. D.; Gillman, A.; Green, W. Q. 1980: Forest patterns and possum densities within podocarp/mixed hardwood forests on Mt Bryan O'Lynn, Westland. *New Zealand Journal of Ecology 3*: 69–84.

Coleman, J. D.; Green, W. Q.; Polson, J. G. 1985: Diet of brushtail possums over a pasture-alpine gradient in Westland, New Zealand. *New Zealand Journal of Ecology 8*: 21–35.

Coleman, J. D.; Thomas, M. D.; Pracy, L. T.; Hansen, Q. 1999: Fluctuations in possum numbers in the Pararaki Valley, Haurangi State Forest Park. *Science for Conservation 128*. Wellington, Department of Conservation. 16 p.

Cowan, P. E.; Chilvers, B. L.; Efford, M. G.; McElrea, G. J. 1997: Effects of possum browsing on northern rata, Orongorongo Valley, Wellington, New Zealand. *Journal of the Royal Society of New Zealand 27*: 173–179.

Cowan, P. E.; Rhodes, D. S. 1993: Electric fences and poison buffers as barriers to movements and dispersal of brushtail possums (*Trichosurus vulpecula*) on farmland. *Wildlife Research 20*: 671–686.

Cowan, P. E.; Waddington, D. C. 1990: Suppression of fruit production of the endemic forest tree, *Elaeocarpus dentatus*, by introduced marsupial brushtail possums, *Trichosurus vulpecula*. *New Zealand Journal of Botany 28*: 217–224.

Crawley, M. C. 1973: A live-trapping study of Australian brush-tailed possums, *Trichosurus vulpecula* (Kerr), in the Orongorongo Valley, Wellington, New Zealand. *Australian Journal of Zoology 21*: 75–90.

Efford, M. G. 1991: The ecology of an uninfected forest possum population. *In:* Jackson, R. *convenor* Symposium on tuberculosis. *Veterinary Continuing Education Publication 132*. Massey University, Palmerston North, New Zealand. Pp. 41–51.

Efford, M. 1998: Demographic consequences of sex-biased dispersal in a population of brushtail possums. *Journal of Animal Ecology 67*: 503–517.

Efford, M. G.; Warburton, B.; Spencer, N. J. 2000: Home-range changes by brushtail possums in response to control. *Wildlife Research* in press.

Finerty, J. P. 1980: The population ecology of cycles in small mammals : Mathematical theory and biological fact. New Haven, Yale University Press. 234 p.

Green, W. Q. 1984. A review of ecological studies relevant to management of the common brushtail possum. *In*: Smith A. P.; Hume I. D. *ed.* Possums and gliders. Chipping Norton, NSW, Surrey Beatty in assoc. with the Australian Mammal Society. Pp. 483–499.

Hickling, G. J. 1995: Clustering of tuberculosis infection in brushtail possum populations: implications for epidemiological simulation models. *In:* Griffin, F.; de Lisle, G. *ed.* Tuberculosis in wildlife and domestic animals. *Otago Conference Series No.3.* Dunedin, Otago University Press. Pp. 174–177.

Hope, R. M. 1972: Observations on the sex ratio and the position of the lactating mammary gland in the brush-tailed possum, *Trichosurus vulpecula* (Kerr) (Marsupialia). *Australian Journal of Zoology 20*: 131–137.

Jolly, J. N. 1976: Habitat use and movements of the opossum (*Trichosurus vulpecula*) in a pastoral habitat on Banks Peninsula. *Proceedings, New Zealand Ecological Society 23*: 70–78.

Kean, R. I.; Pracy, L. 1953: Effects of the Australian opossum (*Trichosurus vulpecula* Kerr) on indigenous vegetation in New Zealand. *Proceedings of the Seventh Pacific Science Congress 4*: 696–705.

Keber, A. W. 1988: An enquiry into the economic significance of possum damage in an exotic forest near Taupo. Unpublished PhD thesis, 2 vols, University of Auckland, Auckland, New Zealand.

Kerle, J. A. 1984: Variation in the ecology of *Trichosurus*: its adaptive significance. *In:* Smith, A. P.; Hume, I. D. *ed.* Possums and gliders. Chipping Norton, NSW, Surrey Beatty in assoc. with the Australian Mammal Society. Pp. 115–128.

Krebs, C. J. 1995: Two paradigms of population regulation. *Wildlife Research 22*: 1–10.

Link, W. A.; Barker, R. J. 1994: Density estimation using the trapping web design: a geometric analysis. *Biometrics 50*: 733–745.

Little, E. C. S.; Cowan, P. E. 1992: Natural immigration of brushtail possums, *Trichosurus vulpecula*, onto Aroha Island, Kerikeri Inlet, Bay of Islands, New Zealand. *New Zealand Journal of Zoology 19*: 53–59.

Mason, R. 1958: Foods of the Australian opossum (*Trichosurus vulpecula*, Kerr) in New Zealand indigenous forest in the Orongorongo Valley, Wellington. *New Zealand Journal of Science 1*: 590–613.

Meads, M. J. 1976: Effects of opossum browsing on northern rata trees in the Orongorongo Valley, Wellington, New Zealand. *New Zealand Journal of Zoology 3*: 127–139.

Olds, C. B. 1987: Aspects of the ecology of the brush-tail possum (*Trichosurus vulpecula* Kerr, 1792) on Rangitoto Island. Unpublished MSc thesis, University of Auckland, Auckland, New Zealand.

Ostfeld, R. S. 1988: Fluctuations and constancy in populations of small rodents. *American Naturalist 131*: 445–452.

Otis, D. L.; Burnham, K. P.; White, G. C.; Anderson, D. R. 1978: Statistical inference from capture data on closed animal populations. *Wildlife Monographs 62*. 135 p.

Owen, H. J.; Norton, D. A. 1995: The diet of introduced brushtail possums *Trichosurus vulpecula* in a low-diversity New Zealand *Nothofagus* forest and possible implications for conservation management. *Biological Conservation 71*: 339–345.

Paterson, B. M.; Morris, R. S.; Weston, J.; Cowan, P. E. 1995: Foraging and denning patterns of brushtail possums, and their possible relationship to contact with cattle and the transmission of bovine tuberculosis. *New Zealand Veterinary Journal 43*: 281–288.

Pfeiffer, D. U. 1994: The role of a wildlife reservoir in the epidemiology of bovine tuberculosis. Unpublished PhD thesis, Massey University, Palmerston North, New Zealand. 439 p.

Pledger, S. A.; Efford, M. G. 1998: Correction of bias due to heterogeneous capture probability in capture-recapture studies of open populations. *Biometrics 54*: 888–898.

Pollard, E.; Lakhani, K. H.; Rothery, P. 1987: The detection of density dependence from a series of annual censuses. *Ecology 68*: 2046–2055.

Pollock, K. H.; Nichols, J. D.; Brownie, C.; Hines, J. E. 1990: Statistical inference for capture-recapture experiments. *Wildlife Monographs 107*. 97 p.

Pracy, L. T. 1974: Introduction and liberation of the opossum (*Trichosurus vulpecula*) into New Zealand. *New Zealand Forest Service Information Series 45*. 2nd ed. 28 p.

Ramsey, D.; Akins-Sellar, S.; Baldwin, W.; Hurst, D. 1997: The effects of sterilisation on wild populations of the brushtail possum. Landcare Research Contract Report LC9697/142 (unpublished) 25 p.

Riney, T. 1964: The impact of introductions of large herbivores on the tropical environment. 9th technical meeting Nairobi. *IUCN publications, New Series No. 4*: 261–273.

Ripley, B. D. 1976: The second-order analysis of stationary point processes. *Journal of Applied Probability 13*: 255–266.

Shenk, T. M.; White, G. C.; Burnham, K. P. 1998: Sampling-variance effects on detecting density dependence from temporal trends in natural populations. *Ecological Monographs 68*: 445–463.

Sinclair, A. R. E. 1997: Fertility control of mammal pests and the conservation of endangered marsupials. *Reproduction, Fertility and Development 9*: 1–16.

Thomas, M. D.; Hickling, G. J.; Coleman, J. D.; Pracy, L. T. 1993: Long-term trends in possum numbers at Pararaki: Evidence of an irruptive fluctuation. *New Zealand Journal of Ecology 17*: 29–34.

Triggs, S. 1982: Comparative ecology of the possum, *Trichosurus vulpecula*, in three pastoral habitats. Unpublished MSc thesis, University of Auckland, Auckland, New Zealand.

Viggers, K. L.; Spratt, D. M. 1995: The parasites recorded from *Trichosurus* species (Marsupialia: Phalangeridae). *Wildlife Research 22*: 311–332.

Warburton, B. 1977: Ecology of the Australian brush-tailed possum (*Trichosurus vulpecula* Kerr) in an exotic forest. Unpublished MSc thesis, University of Canterbury, Christchurch, New Zealand.

Ward, G. D. 1978: Habitat use and home range of radio-tagged opossums *Trichosurus vulpecula* (Kerr) in New Zealand lowland forest. *In:* Montgomery, G. G. *ed.* The ecology of arboreal folivores. *Symposia of the National Zoological Park*. Washington DC, Smithsonian Institution Press. Pp. 267–287.

White, G. C. 1998: Program MARK [Online]. Available: http://www.cnr.colostate.edu/~gwhite/mark/mark.htm [26 August 1998].

Williams, C. K. 1982: Nutritional properties of some fruits eaten by the possum *Trichosurus vulpecula* in a New Zealand broadleaf-podocarp forest. *New Zealand Journal of Ecology 5*: 16–20.

Appendix 5.1
Population densities of brushtail possums: data from intensive studies of undisturbed populations.

Site	Latitude	Alt. (m)	Habitat	Near pasture	Years	Method*	Effective area (ha)	Density no./ha	Reference
Podocarp-broadleaf									
Silverdale, Auckland	36°40'	100–150	Mixed podocarp-broadleaf forest	Y	1981–82	LT	12.6	4.9	Triggs 1982
Coatesville, Auckland	36°44'	50	Mixed broadleaf-podocarp forest and *Kunzea ericoides*	Y	1995–96	LT	8	3.8	Ji Weihong & M.Clout pers. comm.
Huapai, Auckland	36°47'	50	Mixed broadleaf-podocarp forest and *Kunzea ericoides*	Y	1995–96	LT	6	8.2	Ji Weihong & M.Clout pers. comm.
Porewa, Manawatu	40°03'	160	Mixed broadleaf forest	Y	1996–98	LT	14	11.6	Ramsey *et al.* 1997
Turitea, Manawatu	40°26'	200	Mixed podocarp-broadleaf forest	Y	1995–98	LT	24	3.6	Ramsey *et al.* 1997
Turitea, Manawatu 50% sterilisation	40°26'	270	Mixed podocarp-broadleaf forest	N	1995–98	LT	27	3.6	Ramsey *et al.* 1997
Turitea, Manawatu 80% sterilisation	40°26'	300	Mixed podocarp-broadleaf forest	Y	1995–98	LT	24	7.8	Ramsey *et al.* 1997
Orongorongo Valley ISA	41°21'	115	Mixed podocarp-broadleaf forest	N	1967–97	LT	22	9.1	Crawley 1973; Efford 1991, 1998, unpubl. data
Orongorongo Valley Browns Stream	41°21'	120	Mixed podocarp-broadleaf forest	N	1995–98	LT	24	10.5	Ramsey *et al.* 1997
Orongorongo Valley 80% sterilisation 3 km	41°21'	160	Mixed podocarp-broadleaf forest	N	1995–98	LT	27	3.8	Ramsey *et al.* 1997

Appendix 5.1 continued

Site	Latitude	Alt. (m)	Habitat	Near pasture	Years	Method*	Effective area (ha)	Density no./ha	Reference
Orongorongo Valley Wootton grid	41°22'	100	Mixed podocarp-broadleaf forest	N	1995–98	LT	25	8.9	Ramsey *et al.* 1997
Mt Bryan O'Lynn Pasture edge	42°37'	250–300	Lowland cut-over forest, *Weinmannia racemosa*, treefern, emergent *Metrosideros umbellata*, podocarps	Y	1978	RT + C	c. 10	25.4	Coleman *et al.* 1980
Mt Bryan O'Lynn Midslope	42°37'	300–500	Mixed *M. umbellata/ W. racemosa* and *Libocedrus bidwillii* forests	N	1978	RT + C	c. 50	10.8	Coleman *et al.* 1980
Mt Bryan O'Lynn Upper slope	42°37'	500–1000	Mixed podocarp hardwood forest (*Quintinia acutifolia, W. racemosa, Podocarpus hallii, Pseudowintera colorata* and *L. bidwillii*)	N	1978	RT + C	c. 40	6	Efford *et al.* 1999
Pigeon Flat, Dunedin	45°48'	220–290	Mixed podocarp-broadleaf forest	Y	1993–94	LT	13	19.8	Coleman *et al.* 1980
Chew Tobacco Bay, Stewart Is.	47°00'	0–100	Coastal mixed podocarp-broadleaf forest (*Dacrydium cupressinum, Prumnopitys ferruginea, M. umbellata, W. racemosa, Griselinia littoralis* and *Fuchsia excorticata*)	N	1980	RT + C	102	3.8	Coleman & Green 1984; Coleman & Pekelharing unpubl.
Nothofagus *spp.*									
Mt Misery, Nelson Lakes	41°56'	460–650	*Nothofagus* forest	N	June 1981	C	61.8	0.46	Clout & Gaze 1984
Burwood Bush, Te Anau	45°34'	540–600	*Nothofagus fusca, N. solandri* var. *cliffortioides*	Y	1987	RT	c. 24	1.7	R. van Mierlo pers. comm.
Pinus radiata									
Silverdale, Auckland	36°40'	100–150	*Pinus radiata* plantation	Y	1981–82	LT	9.4	2.7	Triggs 1982

Appendix 5.1 continued

Site	Latitude	Alt. (m)	Habitat	Near pasture	Years	Method*	Effective area (ha)	Density no./ha	Reference
Kinleith Forest, Tokoroa	38°13'	460	*Pinus radiata* plantation	N	Dec 1974	LT + C	24	3.0	Clout 1977; Clout & Efford 1984
Ashley Forest, Canterbury	43°13'	160	*Pinus radiata* plantation	N	1975–76	LT	16	0.9	Warburton 1977
Other									
Silverdale, Auckland	36°40'	100–150	Tall scrub, mixed native & exotic	Y	1981–82	LT	12.4	4.6	Triggs 1982
Rangitoto Island, Auckland	36°47'	10–160	*Metrosideros excelsa* – *M. robusta* forest on lava flows	N	1985–86	LT	633	4.4	Olds 1987
Bridge Pa swamp	39°41'	50	*Salix* sp., *Rubus fructicosus*, *Cortaderia* spp., *Phormium tenax*	Y	April 1988	LT + C	11.7	8.8	Brockie *et al.* 1997
Bridge Pa willows	39°41'	50	*Salix* spp.	Y	April 1988	LT + C	11	16.7	Brockie *et al.* 1997
Bridge Pa farmland	39°41'	50	Open mixed farmland, barns, tree rows, *Cortaderia*	Y	1982–88	LT	298	0.15	Brockie *et al.* 1997
Castle Point, Wairarapa	40°51'	60–270	*Leptospermum* scrub, *Ulex* and remnant broadleaf forest interspersed with pasture	Y	1991–94	LT	21	7.2	Pfeiffer 1994; Paterson *et al.* 1995
Birdlings Valley	43°48'	10–150	Mixed farmland and scrub	Y	1972–74	LT	47	1.0–1.4	Jolly 1976
Banks Peninsula	43°48'	–	Native scrub *Coprosma* spp., *Cytisus scoparius*, *Melicytus ramiflorus* etc.	Y	1992	LT	18	3.8	Hickling pers. comm. 28/8/98 Zippin estimate
Black Forest Station, Mackenzie Basin	44°23'	c. 600	Scrub in gully system	N	1990	RT	92	0.5 Zippin	Hickling pers. comm. 28/8/98 estimate
Burwood Reserve, Te Anau	45°34'	490	*Chionochloa rubra* tussock grassland	N	Jan 1992	LT	11	0	M. Efford unpubl. results

* Methods: LT is live trapping, RT is removal trapping, and C is cyanide.

CHAPTER SIX

Possum Reproduction and Development

Terry Fletcher and Lynne Selwood

(NB: A glossary of terms to aid the general reader can be found in Appendix 6.1)

At present it is thought that biological control of the possum population in New Zealand will be achieved by targeting aspects of possum reproduction such as gonadal development and pituitary control, fertilisation, embryonic development, the female reproductive tract, and lactation. The successful use or otherwise of these particular targets can only be judged by a comparison with what happens in normal reproduction in the possum. This chapter reviews what is known of possum reproduction in New Zealand with reference to that of possums in Australia where appropriate.

Life history

In New Zealand and most of Australia, possums are seasonal breeders; most births occur during autumn but have been recorded in all months of the year (Kerle 1984). W. Q. Green (1984) found that 85% of mature females in New Zealand breed and 87% in Australia. In the Orongorongo Valley, near Wellington, Bell (1981) found that breeding success of females increased with age from 8% at 1 year, to 60% at 2 years, and 85% at 3 years, with a high rate of breeding for several years thereafter. On leaving the pouch, mean life expectancy of a possum in the Orongorongo Valley is 6 years (Brockie *et al.* 1981). Thus an average Orongorongo Valley female may produce six offspring in her lifetime, but this number may be higher in regions where a second breeding season occurs (W. Q. Green 1984). Mortality of pouch young and during juvenile dispersal markedly reduces the number of surviving animals (Bell 1981; Hocking 1981), with about 85% of pouch young in New Zealand surviving and 87% in Australia (W. Q. Green 1984). Apart from the study of Brockie *et al.* (1981), life expectancy has been rarely calculated. In Australia, C. H. Tyndale-Biscoe (retired, ex CSIRO Wildlife & Ecology, Canberra, pers. comm.) reports a 15-year-old female from the Canberra region, which was carrying a pouch young. In New Zealand, Crawley (1970) recorded a 12-year-old female and an 11-year-old male in the Orongorongo Valley. Maximum ages of 14 years have been estimated in New Zealand possums (Brockie *et al.* 1981; Coleman & Green 1984) using cementum deposition in molar teeth as an indicator of age (Pekelharing 1970).

During the breeding season, females, which are polyoestrous, undergo successive cycles of c. 26 days until conception occurs, but if a suckling young dies during the breeding season, oestrus follows after about 9 days (Pilton & Sharman 1962). Pregnancy in the possum is short (c. 18 days) and lactation lasts about 6 months (Tyndale-Biscoe & Renfree 1987; Table 6.1), so a female possum can raise a young from conception to independence within a year. Oestrus does not normally occur after birth, but has been recorded (Tyndale-Biscoe 1955; Pilton & Sharman 1962; Selwood *et al.* 1998). Developmental arrest of early embryos (at the unilaminar blastocyst stage) has not been found in possums (Pilton & Sharman 1962; Selwood *et al.* 1998). Generally, possums give birth to a single young and twins are rare (Tyndale-Biscoe 1955; Kean 1975; Hughes & Hall 1984; Gemmell 1995). Female possums mature at 1–2 years of age depending on locality and population demographics, while most males are mature at 2 years old (Tyndale-Biscoe & Renfree 1987). The possum is thus a fecund and long-lived species.

Breeding season

The breeding season in New Zealand starts about March and usually ends by November (Fig. 6.1; Table 6.1), with the start and finish of the season varying by region and year. The median date of births varies from late-April to mid-May in New Zealand and from April to early-May in Australia (W. Q. Green 1984), with 80–100% of adult females giving birth from mid-April to early June. Within a particular region, the time of the onset of breeding

Table 6.1
Summary of breeding information on the possum.

Maturity Age (y) F	Maturity Age (y) M	Maturity Weight (kg) F	Maturity Weight (kg) M	Length of oestrous cycle (days)	Gestation period (days)	Neonate weight (mg)	Length of lactation (days)	Breeding season (month)	Sperm present (month)
1[1]	2[2]	2.33[2]	2.46[2]	25.7[3]	17.5[3]	170–190[4]	220–290[3]	March to	All[9]
	2.41[3]	2.65[3]					190–210[3]	November[2, 4-8]	
	2.99[7]	3.46[7]							

[1] Gilmore 1969; [2] Crawley 1973; [3] Pilton & Sharman 1962; [4] Kean 1975; [5] Brockie *et al.* 1981; [6] Bell 1981; [7] Fraser 1979; [8] Tyndale-Biscoe 1955; [9] Cummins 1981

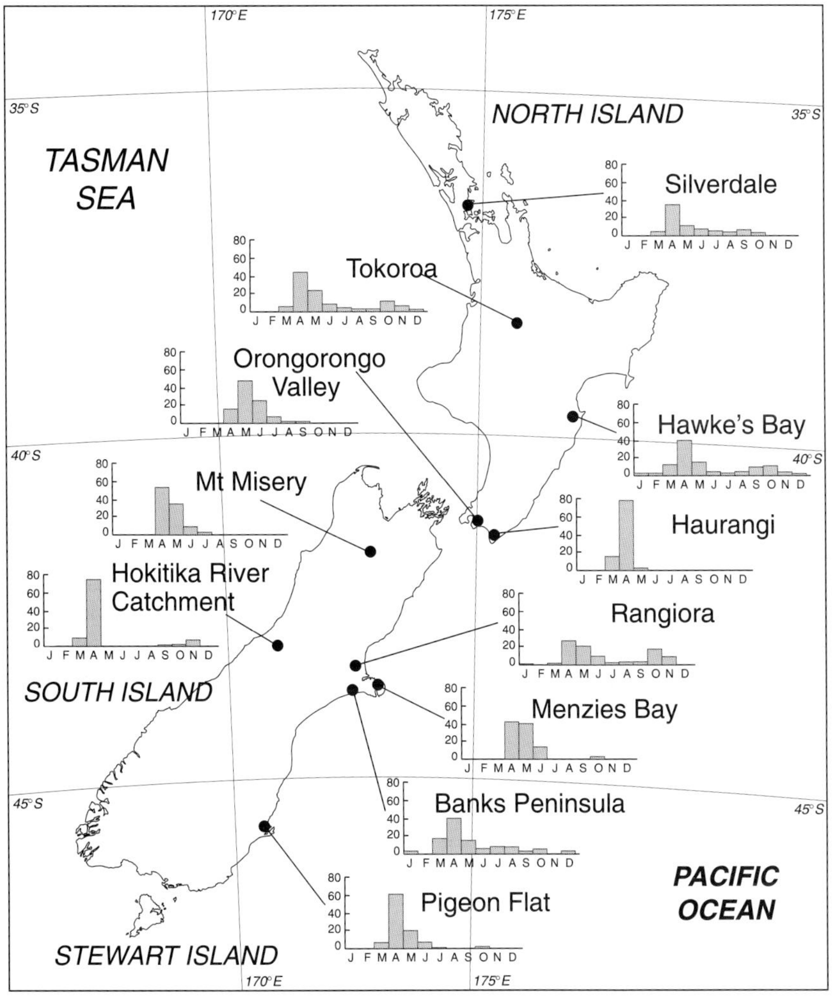

Fig. 6.1
Monthly distribution of births in New Zealand possum populations showing a seasonal pattern of breeding and variation in the incidence and extent of spring breeding. Data from Tyndale-Biscoe 1955; Gilmore 1966; Boersma 1974; Clout 1977; Bell 1981; Triggs 1982; Clout & Gaze 1984; Brockie 1992; Jolly *et al.* 1995; M. Efford, Landcare Research, Dunedin unpubl. data.

does not vary greatly, but the proportion of animals in reproductive condition (i.e., cycling or lactating) at that time can vary greatly. For example, near Palmerston North the proportion of animals in reproductive condition sampled in a 10-day period between 26 March and 18 April was 92% (n=51) in 1995, 45% (n=82) in 1996, 83% (n=60) in 1997, and 16% (n=85) in 1998, although in each year at least one animal had given birth within the 10-day period (Selwood unpubl. data). In some regions a second pulse of breeding occurs between August and November. This second, spring pulse can be slight, such as in the Orongorongo Valley (Bell 1981), or close to autumn breeding levels, such as in the

Wairarapa (Selwood *et al.* 1998). Up to 38% of possums in Australia breed twice a year, and this has been attributed to reproductive failure during autumn or during lactation, or to double breeding by successful females in good habitat (Kerle 1984). A third reason could be earlier breeding in the previous autumn (W. Q. Green 1984).

As with many mammals from temperate regions, onset of the breeding season is controlled by hours of daylight (photoperiod). Changing captive possums from a long- to a short-day photoperiod brings forward the onset (Gemmell *et al.* 1993) while changing possums from a short- to a long-day photoperiod terminates the breeding season (Gemmell & Sernia 1995). Other factors may modulate the onset and length of the breeding season. Bell (1981) showed a clear correlation between earlier breeding seasons and higher mean weights of females, thereby implicating nutrition as a modulator of fecundity in New Zealand. Similarly, Humphreys *et al.* (1984) considered morphological and physiological indices of condition of possum populations in Tasmania and New South Wales, and concluded that food resources were a prime regulator of breeding in Australia.

Anatomy of the male and female reproductive tracts

The reproductive tract of the male possum follows the general marsupial pattern (Temple-Smith 1984: Tyndale-Biscoe & Renfree 1987). The paired testes and epididymides are located in a scrotum behind the penis and there are no ampullae or seminal vesicles. The vasa deferentia lead to a single carrot-shaped prostate gland, which surrounds part of the urethra, posterior to the opening of the bladder. The prostate is divisible into frontal and rear segments on the basis of microscopic appearance of the glandular epithelia and their secretory products (Rodger & Hughes 1973). To the rear of the prostate are two pairs of Cowper's glands joined to the urethra by simple ducts. The penis is withdrawn into an internal, posteriorly directed preputial sac when not erect.

The reproductive tract of the female possum (Fig. 6.2) displays the anatomical features unique to marsupials (Pearson 1945; Tyndale-Biscoe & Renfree 1987). The paired ovaries are surrounded by a membranous fringe (fimbria), which leads into the oviducts where fertilisation occurs. Specialised cells of the oviduct epithelium secrete mucopolysaccharides, which add a mucoid coat to the ovum. The paired uteri open into the vaginal culs-de-sac via separate cervices with a pronounced frontal vaginal expansion, which changes in size and shape with the stage of the oestrous cycle (Pilton & Sharman 1962; Kean 1975). Paired lateral vaginae and a median vaginal canal, which forms a short-lived birth canal (Tyndale-Biscoe 1955, 1966), open into a urogenital sinus, which also receives the urethra.

Reproductive physiology of the male possum

The testis and epididymis weights of adults do not vary significantly with season, but the weight of the prostate gland increases fourfold between mid-February and late March to reach maximum development at the seasonal peaks of conception (Gilmore 1969). The number of Leydig cells in testis interstitial tissue varies with the state of prostate enlargement thereby implicating a hormonal factor in the prostate changes (Gilmore 1969). In the decade that followed Gilmore's work, several workers identified various male sex hormones (androgens) in possum blood. Vinson & Phillips (1969) reported circulating levels up to 50 ng/mL of testosterone in the blood of males and females and Vinson (1974) reported lower levels (up to 10 ng/mL) of testosterone in males. In their review of testicular endocrinology, Carrick & Cox (1977) reported decreasing levels of testosterone in the spermatic vein at different stages of the possum breeding cycle: 114 and 86 ng/mL at the peak, 51 and 47 ng/mL (intermediate), and 36 ng/mL at its lowest level. They also reported that testosterone was the main androgen present in plasma in the breeding season and that a slightly more polar compound was evident in the non-breeding season. Cook *et al.* (1978) measured concentrations of 3.8 ng/mL testosterone, 1.9 ng/mL 5α-dihydrotestosterone (DHT), and 0.4 ng/mL androstanediol in serum collected by cardiac puncture from male possums in December. These workers found that activity of the enzyme 5α-reductase, which reduces testosterone to its metabolites including DHT, was higher in the epididymis than in the prostate. A direct relationship between the prostate and testosterone was suggested when prostate weight was found to be higher in males injected with testosterone enanthate in the non-breeding season than in non-treated males. Cook *et al.* (1978) identified a network of blood vessels surrounding the urethra at the neck of the prostate that offered a local means of transfer of androgens from the testis and epididymis to the prostate.

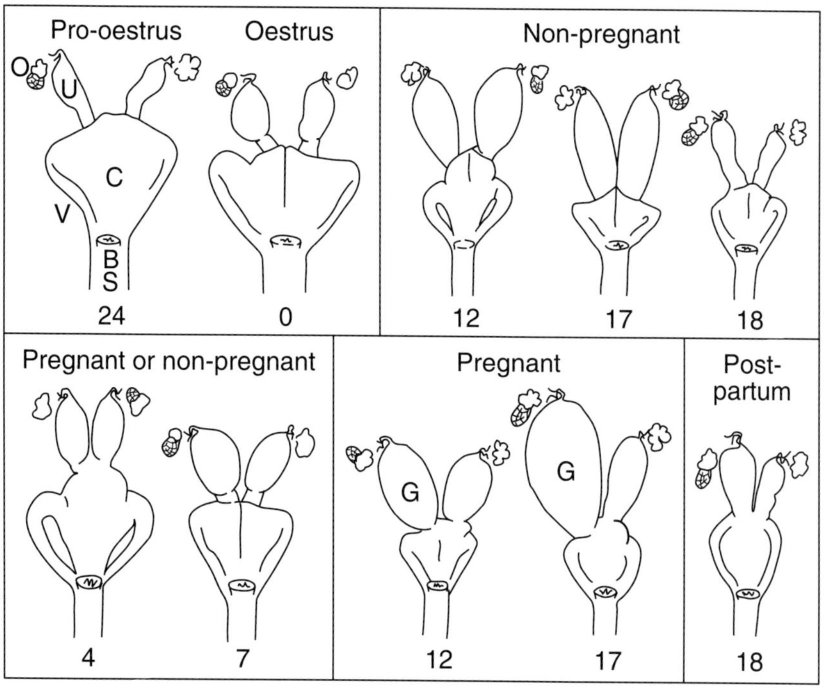

Fig. 6.2
Variation in the relative size of female reproductive organs during the oestrous cycle and pregnancy of the possum. Redrawn from Pilton & Sharman 1962. Numbers below drawings are number of days after oestrus. B, base of bladder; C, median vaginal culs-de-sac; G, gravid uterus; O, ovary; S, urogenital sinus; U, uterus; V, lateral vagina. Most recent corpus luteum shown black.

Allen & Bradshaw (1980) confirmed that testosterone was the major androgen present in peripheral plasma, that DHT was present, and that both varied diurnally. Mean maximum (5.1 ± 1.6 ng/mL) values of testosterone occurred at 0900 hours, and mean minimum (1.2 ± 0.4 ng/mL) values at 1800 hours. (NB All mean values are shown ± SEM). The DHT values were also higher in the morning than at night but the diurnal pattern was less convincing. Furthermore, castrated males and normal females showed diurnal variation in testosterone but not DHT. Testosterone was maximal at 0900 hours (0.74 ± 0.03 ng/mL in castrated males; 0.62 ± 0.04 ng/mL in females) and minimal at 1800 hours (0.18 ± 0.03 ng/mL in castrated males; 0.14 ± 0.02 ng/mL in females). Curlewis & Stone (1985a) also identified testosterone and DHT in plasma from male possums and found testosterone levels were 4–5 times greater than DHT levels. They also measured androgen secretion in blood from females and castrated males, but were unable to demonstrate episodic secretion. Rather, androgen levels in most of the possums declined during the sampling periods, apparently as a response to stress (Curlewis & Stone 1985a). Curlewis & Stone (1985a) also reported that blood taken from possums in the breeding season contained 1.37 ± 0.53 ng testosterone/mL, which was not significantly different to that taken in the non-breeding season (0.22 ± 0.1 ng/ mL). The possums' prostates weighed on average 30.78 ± 3.65 g in the breeding season and fell significantly to 6.74 ± 2.02 g in the non-breeding season, but their epididymis weights were not significantly different (breeding season 1.86 ± 0.07 g; non-breeding 1.57 ± 0.16 g). From the poor correlation between testosterone and DHT levels and prostate weight, they concluded there was no evidence to suggest prostate weight changes were due to seasonal changes in androgen levels (Curlewis & Stone 1985a). In a related study, Curlewis & Stone (1985b) described some of the mechanisms responsible for the prostatic changes. The increase in prostate weight was associated with a decrease in the ratio of DNA to tissue (g), and an increase in the ratios of protein to DNA and RNA to DNA, which indicate an increase in cell numbers and/or accumulation of secretory product. There were no significant changes in these ratios in the epididymis. Androgen receptor levels were significantly elevated in the prostate during the breeding season but not in the epididymis. The epididymis, however, showed a very high level of 5α-reductase activity compared to the prostate regardless of the breeding season (also shown by Cook *et al.* 1978). Prostatic tissue showed a low level of 5α-reductase in both breeding and non-breeding seasons. Both prostate and epididymis tissues showed high levels of DHT, and in the prostate the levels were higher in the breeding season. Gemmell *et al.* (1986) identified an annual cycle of secretion of testosterone in male possums, with peak levels occurring in March when breeding commenced and lowest levels in September. The observed changes in blood testosterone concentrations correlated best with the rate of change of day length thereby implicating photoperiod as a factor in seasonal changes in the reproductive tract of male possums.

Thus increased prostatic growth in the breeding season can be attributed to raised concentrations of testosterone circulating in the blood (Gemmell *et al.* 1986) and elevated androgen receptor levels in the prostate coupled with elevated androgen concentration in tissue (Curlewis & Stone 1985b). The testosterone-binding capacity of possum plasma appears to be low (Sernia *et al.* 1979), which indicates most of the circulating androgen is metabolically active. Significant amounts of DHT occur in the peripheral blood (Cook *et al.* 1978; Allen & Bradshaw 1980; Curlewis & Stone 1985a) but DHT is apparently absent from the testicular vein blood (Carrick & Cox 1977), which implies an extra-testicular source for DHT. The very high levels of 5α-reductase found in the epididymis (Cook *et al.* 1978; Curlewis & Stone 1985b) suggest it is a major site of conversion of testosterone to DHT. Higher concentrations of androgens from the testis and epididymis could be conveyed directly to the prostate by the vas deferens (Cook *et al.* 1978) rather than via the peripheral circulation, which would account for the prostatic enlargement. However, the positive correlation between serum testosterone and prostatic weight, the response of the prostate to testosterone injection (Cook *et al.* 1978), and the biological availability of circulating testosterone (Sernia *et al.* 1979), suggest that peripheral androgens exert a direct control over prostate function in the possum.

The seminiferous tubules and interstitial tissue of the testis have the same basic structure as in eutherian mammals (Temple-Smith 1984; Tyndale-Biscoe & Renfree 1987). The seminiferous tubules are supported in a loose matrix of connective tissue, blood vessels, and lymphatic ducts, with aggregations of Leydig cells found in the spaces between adjacent tubules (Green 1963; Gilmore 1969). Within the tubules, germ cells at various stages of the spermatogenic cycle are found interspersed with Sertoli cells (Setchell 1977). Setchell & Carrick (1973) identified eight stages of spermatogenesis in the male possum, similar to the stages found in eutherian mammals, and found the duration of one spermatogenic cycle is 15 days. The time from division of the spermatocyte to liberation of the spermatozoon is 56 days. The process of spermiogenesis in the possum is generally similar to the process in eutherians, but the final stages of condensation of the nucleus, development of the acrosome, and the changing orientation of the sperm head to the flagellum, appear to be unique marsupial features (reviewed by Temple-Smith 1984). Three stages in morphological maturation were identified during the 11-day transit through the epididymis (Cummins 1976) culminating in the arrival of motile sperm in the cauda epididymis.

Reproductive physiology of the female possum

The oestrous cycle of the possum and its hormonal control is one of the best described of any marsupial (Tyndale-Biscoe 1984; Tyndale-Biscoe & Renfree 1987). Possums are monotocous and ovulate spontaneously from alternate ovaries at successive ovulations (O'Donoghue 1916; Pilton & Sharman 1962). The detection of oestrus and mating from daily vaginal smears and the cycle of changes in vaginal epithelial cells during the oestrous cycle have been well characterised (Pilton & Sharman 1962; Tyndale-Biscoe 1984). More recently, observation of the cycle of vaginal epithelial cells exfoliated into urine and the presence of sperm in urine of the female have been used to detect oestrus and mating (Duckworth *et al.* 1999).

Proliferative phase

The proliferative phase of the oestrous cycle includes the pro-oestrous period of follicular growth, oestrus, and the first few days after ovulation (Pilton & Sharman 1962). During this period, the vaginal complex and urogenital sinus begin to enlarge (Fig. 6.3) and reach maximum development at oestrus (Pilton & Sharman 1962; Kean 1975). The length of the pro-oestrus is about 8–10 days and can be induced by removing pouch young (RPY) to initiate a return to oestrus, or will occur if lactation is not established after birth (Pilton & Sharman 1962; Curlewis & Stone 1986a). The length of the pro-oestrus is somewhat variable between different studies and may reflect differences in how it is measured, with estimates of 4–10 days (Lyne *et al.* 1959), 8 ± 0.2 days (Pilton & Sharman 1962), 9.1 ± 0.6 days (Horn 1981), 9.4 ± 0.4 days (Curlewis & Stone 1986a), 9.6 ± 0.5 days (Gemmell *et al.* 1987), and 10.8 ± 0.5 days (Duckworth *et al.* 1999). In the late pro-oestrus, the epithelial cells of the vaginal culs-de-sac secrete a mucus rich in a carbohydrate-

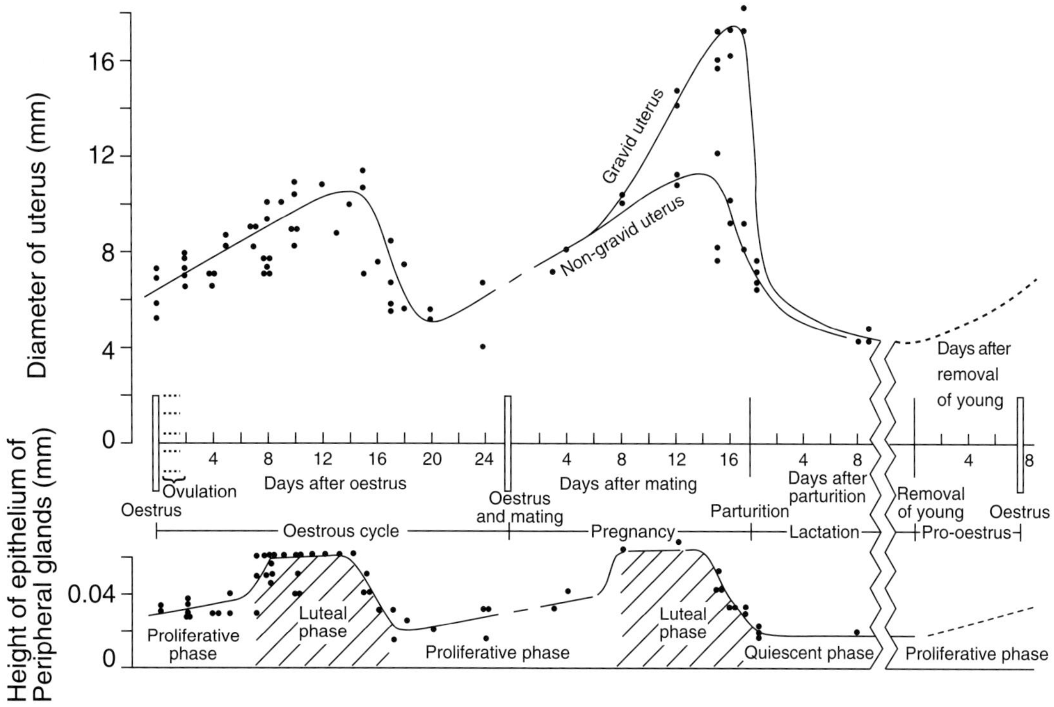

Fig. 6.3

Changes in the possum uterus during the oestrous cycle, pregnancy, and lactation, and after removal of the suckling pouch young. The proliferative phase of the oestrous cycle includes the pro-oestrous period of follicular growth, oestrus, and the first few days after ovulation. The luteal phase of the oestrous cycle is when the uterine epithelium develops. The quiescent phase is when a suckling young is present in the pouch and the uteri regress. Redrawn from Pilton & Sharman 1962.

protein complex (Hughes & Rodger 1971). The mucus may reach 3.5 mL in volume at ovulation after which it rapidly disappears. Treatment of anoestrous possums with oestradiol benzoate stimulated the secretion of similar quantities of mucus to those found at oestrus (Curlewis & Stone 1987), so presumably mucus secretion is under hormonal control. Legge *et al.* (1996) reported increasing concentrations of glycosaminoglycans (GAGs) in the epithelial tissue lining the vaginal cul-de-sac and uterus after stimulation with pregnant mare's serum gonadotrophin (PMSG).

Oestrus and ovulation

At oestrus, the ovary usually contains one large Graafian follicle 4.1–4.9 mm in diameter (Pilton & Sharman 1962; Hughes & Rodger 1971), although Crawford *et al.* (1998) reported a pre-ovulatory follicle 6.5 mm in diameter. There is a pre-ovulatory surge in FSH (follicle-stimulating hormone) secretion (Eckery *et al.* 1998) but its precise relationship to oestrus is uncertain. Oestradiol-17β is the major oestrogenic steroid produced by the growing follicle (Curlewis *et al.* 1985) and is only present in the blood in significant levels (14.5 ± 5.7 pg/mL) on the day of oestrus. Following oestrus, luteinizing hormone (LH) is detected in plasma. The pre-ovulatory LH surge occurs after oestrus, is about 25–30 ng/mL at maximum level, and lasts less than 24 hours (Horn 1981; Eckery *et al.* 1998). Ovulation occurs within 24 hours of oestrus (Shorey & Hughes 1973b), following which the ruptured Graafian follicle undergoes luteinization and transforms into a corpus luteum (CL) over 2–3 days (Shorey & Hughes 1973b).

Luteal phase

The luteal phase of the oestrous cycle is the period when the uterine epithelium develops under the influence of progesterone secreted by the CL (Fig. 6.3). The vaginal complex and urogenital sinus become smaller (Fig. 6.2) while the uteri continue to enlarge (Pilton & Sharman 1962; Shorey & Hughes 1975). The CL reaches its maximum diameter (4.1–4.5 mm) 7–10 days after oestrus, and from day 16 after oestrus, begins to regress (Pilton & Sharman 1962; Shorey & Hughes 1973b). Progesterone is the major steroid secreted by the CL (Curlewis *et al.* 1985), and CL growth is reflected in changing levels of progesterone measured in peripheral blood (Thorburn *et al.* 1971; Shorey & Hughes 1973a; Hinds 1983; Curlewis *et al.* 1985; Gemmell *et al.* 1987). The glandular cells of the uterine epithelium reach maximum height on day 8 (Pilton & Sharman 1962) and the endometrium reaches its maximum width 11–13 days after oestrus and then declines from day 14 (Pilton & Sharman 1962; Shorey & Hughes 1973a) (Fig. 6.3). If either the CL or the ovary is removed before day 7, the uterine epithelium does not proliferate and oestrus follows 8–9 days later (Shorey & Hughes 1975). Complete uterine regression does not occur if the CL or ovary are removed on day 8 and an early return to oestrus does not occur (Shorey & Hughes 1975), but the CL is still essential for maintenance of pregnancy up to day 10 (Sharman 1965).

A one-sided effect of the CL on the size of the uterus on the same side of the body has been described in several studies. Pilton & Sharman (1962) noted that in non-pregnant possums the uterus on the same side as the CL was often larger than the other. In a study of 12 animals Von Der Borch (1963) showed that where the CL was on the left ovary, the left uterus was 19% heavier than the right, but where the CL was on the right ovary, the right uterus was only 1% heavier than the left. The left uterus was found to be significantly heavier than the right and the CL caused a significant increase in the weight of the uterus closest to it. Ligation of the Fallopian tube on the side of the developing follicle did not alter the CL effect, suggesting an ovarian hormone must effect the difference via the circulation. This hypothesis was confirmed by Lee & O'Shea (1977) who described extensive cross-connections and close juxtaposition of branches of the ovarian and uterine blood vessels. No such connections crossed the mid-line above where the uteri or vaginae meet, so a mechanism for the one-sided effect of ovarian hormones is present. The left-side effect remains unexplained but Von Der Borch (1963) noted the CL was on the left ovary in 8 of 12 possums and Curlewis & Stone (1986a) found the CL was on the left side in 20 of 25 possums.

Blood progesterone levels rise steeply on day 8 after oestrus, reaching maximum levels on days 12–14 (Shorey & Hughes 1973a; Hinds 1983; Curlewis *et al.* 1985; Gemmell *et al.* 1987) when the uterine epithelium reaches maximum development and the uteri of non-pregnant possums reach their maximum diameter (Pilton & Sharman 1962). Progesterone levels begin to decline on day 16, reaching basal levels by day 19–20 (Hinds 1983; Curlewis *et al.* 1985; Gemmell *et al.* 1987). The CL may persist as a visible corpus albicans for more than 80 days (Pilton & Sharman 1962).

Steroid hormones act on their target tissues by binding to cellular receptors. Characterisation of possum steroid hormone receptors has given further insight into the mechanisms by which progesterone and oestradiol effect changes in the reproductive tract (Fig. 6.2). Receptors with low capacity / high affinity for oestradiol have been characterised from cytosol and nuclear fractions of the endometrial and vaginal epithelium (Young & McDonald 1982). Oestradiol receptor concentrations were low in anoestrus, lactation, and ovariectomised females, and high at oestrus. Injection of oestradiol into ovariectomised females induced high levels of the oestradiol receptor. The progesterone receptor has been characterised in cytosol preparations of the uterus, vagina, and urogenital sinus of the possum. Concentrations of uterine progesterone receptor were maximal at oestrus, lower at day 5, and minimal at day 13 after mating; in ovariectomised females progesterone receptors were induced by oestradiol and suppressed by progesterone (Curlewis & Stone 1986b). The high levels of oestradiol at oestrus induce progesterone receptor synthesis in the endometrium, while high levels of progesterone in the mid-luteal phase suppress further synthesis. Thus the endometrial progesterone receptors disappear after sufficient progesterone stimulation early in the luteal phase, and the CL is no longer necessary for support of the luteal endometrium after day 7.

Pregnancy

Pregnancy occupies the last few days of the proliferative phase and all of the luteal phase of the oestrous cycle, and birth coincides with the regression of the corpus luteum (Pilton & Sharman 1962). The size, histology, and rate of growth of the CL and the changes observed in the uterine epithelium in the 17 days after oestrus are the same in pregnant and non-pregnant possums (Pilton & Sharman 1962; Shorey & Hughes 1973a). The patterns of secretion of progesterone are also not significantly different in pregnant or non-pregnant animals (Hinds 1983;

Curlewis *et al.* 1985; Gemmell *et al.* 1987). The changes noted in the ovary and vaginal and uterine tissues are similar in pregnant and non-pregnant possums (Pilton & Sharman 1962; Shorey & Hughes 1973a,b). In pregnant possums 8 days after oestrus, the gravid uterus was slightly larger than the non-gravid uterus and at day 12 there was a pronounced difference, which persisted to the day of birth (Fig. 6.2). The difference in diameter of the uteri is presumed to be a mechanical response to intrauterine growth of the foetus (Pilton & Sharman 1962; Kean *et al.* 1964) and is the only known indicator of pregnancy in the possum.

Role of the corpus luteum

The CL has a central role in reproduction of the possum. Once formed from the remnants of the ruptured follicle, the CL of the possum appears to act independently. Despite having receptors for prolactin (milk-stimulating hormone) and LH during pregnancy (Stewart & Tyndale-Biscoe 1982), it is not dependent upon pituitary secretions. When the pituitary gland is removed even 1 day after oestrus, CL development and function is unaltered and pregnancy continues to term, but the foetus dies in the uterus or vaginal cul-de-sac (Hinds 1983). While the CL is essential for the preparation of the secretory epithelium of the uterus, the uterus has no effect on the CL (Clark & Sharman 1965). After day 10, the CL is not essential for maintenance of pregnancy (Sharman 1965), but if it is removed after this time, parturition (birthing) fails. These results point to a role of the pituitary and CL in parturition. Tyndale-Biscoe (1969) demonstrated relaxin-like activity in the CL, and showed that while tissue levels rose in late pregnancy, they were undetectable in early pregnancy and immediately following birth. Injection of porcine relaxin, however, had no detectable effect on the possum reproductive tract. The oxytocic peptide, mesotocin, has been located in the CL and rear pituitary of the possum (Bathgate *et al.* 1995) and oxytocin receptors have been demonstrated in the uterus, mammary gland, and median and lateral vaginae (Sernia *et al.* 1990, 1991). Evidence from other marsupials suggest a role for mesotocin in the birth process and lactation, but this has not been confirmed in the possum.

Development in the uterus

Possum developmental stages and their major features are shown in Table 6.2 and the timetable of embryonic development in Figure 6.4. Improved methods of induced ovulation (Glazier 1998) and artificial insemination (Molinia *et al.* 1998a,b) allow a more detailed account of fertilisation. Fertilisation of oocytes, which are ovulated at meiosis 2, occurs in the ampulla. Zygotes have been obtained from the oviduct and uterus, and transport through the oviduct is rapid. As in other marsupials, the egg, which is free of cumulus cells when shed, receives a mucoid coat as it descends the oviduct and a shell coat at the utero/tubal junction and in the upper uterus. Staining of oviduct and uterus material with a polyclonal antibody from the fat-tailed dunnart (*Sminthopsis orassicaudata*) (Roberts *et al.* 1997) and various histochemical stains show that mucoid coat deposition commences during the early follicular phase of the oestrous cycle and shell deposition begins at the late follicular stage. Removal of the outer egg coats interferes with development *in vitro* (Selwood *et al.* 1998). The zona pellucida and mucoid coat disappear as the embryo expands and the shell is shed around the time of implantation on day 14 (Table 6.2). Research is being done on proteins in the zona pellucida, mucoid coat, and shell as targets for possum immunocontraception.

The rate of intrauterine development of the possum is broadly similar to that in many marsupials. Cleavage and the unilaminar blastocyst stage occupy more than one-third of the gestation period (Fig. 6.4). Formation of the bilaminar and trilaminar stages and embryogenesis to the flat embryo stage is relatively rapid. Shell shedding occurs when the embryo has about 20 somites and the yolk-sac becomes closely applied to the uterine wall on day 13 or 14. The yolk-sac develops to up to 15 mm in diameter, suggesting that the exchange between the developing foetus and the uterus is dependent on the enormous increase in surface area of the yolk-sac placenta.

Embryonic development and cleavage have recently been examined in some detail (Hughes & Hall 1984; Frankenberg & Selwood 1998). The first three cleavages are equal and result in eight blastomeres clustering at one pole of the conceptus, the future embryonic pole. The faint polarity shown by the mature oocyte is more marked in the zygote, with the pronuclei and a mitochondrial-rich cytoplasm accumulating at the future embryonic pole, and electron-lucent vesicles lying within a narrow band of cortical cytoplasm accumulating at the other pole. Much of this electron-lucent material is discharged in a polarised fashion into the cleavage cavity as an extracellular matrix (ECM). Initially it separates the cells from each other and from the zona pellucida,

Table 6.2
Stages of intrauterine development of the possum. Based on information in Hughes & Hall 1984 and Frankenberg & Selwood 1998.

Stage	Description	Dimensions (mm)
Oestrus	Antral follicle development Oocyte maturation	2.0–5.0 (follicle)
Ovulation	Oocyte cumulus-free at meiosis 2	–
Fertilisation	In ampullary region Zygote covered by: mucoid in oviduct, shell in utero/tubal junction and uterus	0.19 (zygote- coats)
Cleavage	Zygote to 32-cell in uterus Pluriblast/trophoblast allocation at 8- to 16-cell stage Enclosed in all coats	0.36–0.36 (+ coats)
Unilaminar blastocyst	Expansion with zona and mucoid compression Pluriblast and trophoblast distinct Gradual loss of zona and mucoid	0.36–0.5 0.5–1.0
Bilaminar blastocyst	Hypoblast allocation from pluriblast to form separate epiblast and hypoblast layers Incomplete bilaminar blastocyst Complete bilaminar blastocyst Epiblast oval	– 1.8 3.2
Trilaminar blastocyst	Slightly ovoid shape Pear-shaped epiblast, with primitive streak and node Mesoderm reaches trophectoderm	4.5–7.5
Embryo	Neurulation, somitogenesis, cardiac formation, proamnion and amnion formation Shell shed	Vesicle diameter: 6.0–10.0
Foetus	Organogenesis, Trilaminar and bilaminar yolk-sac	Vesicle diameter: 10.0–15.0 + HL: 2.0–6.5

but cell-zona adhesion has begun by the early 4-cell stage and cell-cell adhesion at the late 4-cell stage. Cell-zona adhesion begins at the embryonic pole and the developing unilaminar blastocyst epithelium spreads out from this pole. Two cell types, pluriblast and trophoblast, are detectable at the 8-cell stage. The epithelium is completed by a combination of cell division and cell extension. Because the polarised intercellular location and extracellular secretion of ECM is intimately associated with blastocyst formation and the allocation of pluriblast and trophoblast, the ECM has also been targeted as a molecule suitable for immunocontraception. This mode of blastocyst formation is unique to marsupials.

Developmental failure during gestation

Gestation failed in 31% of cases during the luteal phase (Selwood *et al.* 1998), with the majority (47%) because the eggs were not fertilised. Such failures are common in marsupials where insemination needs to occur in sufficient time for the sperm to undergo capacitation and to reach the ampullary region before excessive deposition of the mucoid coat creates a barrier to sperm binding to the zona pellucida. Because of the timing of the sample, the animals examined in this study were undergoing their first or second cycle of the season (Table 6.3). Bell (1981) showed in the Orongorongo Valley study that the first cycle was frequently infertile, so perhaps this is a feature common to many possum populations in New Zealand. For 31% of the failed pregnancies, the appearance of the corpus luteum did not correlate with the size and colour of the uterus. The remainder of failures (22%) were due to embryonic failure and occurred mainly at the cleavage and blastocyst stages, and showed abnormal cleavage patterns or collapse of the blastocysts. A failure of around 25% during cleavage and early blastocyst formation has also been shown in other marsupials (Hartman 1928; Selwood 1983).

Development in the pouch

Marsupials have a short gestation followed by a long lactation period. Lactation is divided into three phases (Tyndale-Biscoe & Janssens 1988; Hinds 1988). The first phase (Fig. 6.5) extends through pregnancy and involves initial development of the mammary gland and the release of the pituitary hormone prolactin just before birth, which

Table 6.3
Breeding success and failure during gestation and pouch life in a sample of 471 wild-caught females examined over 10–14 days in late March and early April at Palmerston North and Christchurch, and in late September and early October at Masterton, between 1994 and 1996. Proliferative and luteal refer to phases of the oestrous cycle. Data from Selwood *et al.* 1998.

Season	No. of animals	Anoestrus	Proliferative phase		Luteal phase		Lactation
			Normal	Fail	Normal	Fail	
Autumn	223	27.4	14.8	20.2	8.1	29.1	0.4
Spring	248	19.4	6.0	14.1	7.3	50.8	2.4
Year	471	23.1	10.2	17.0	7.6	40.6	1.5

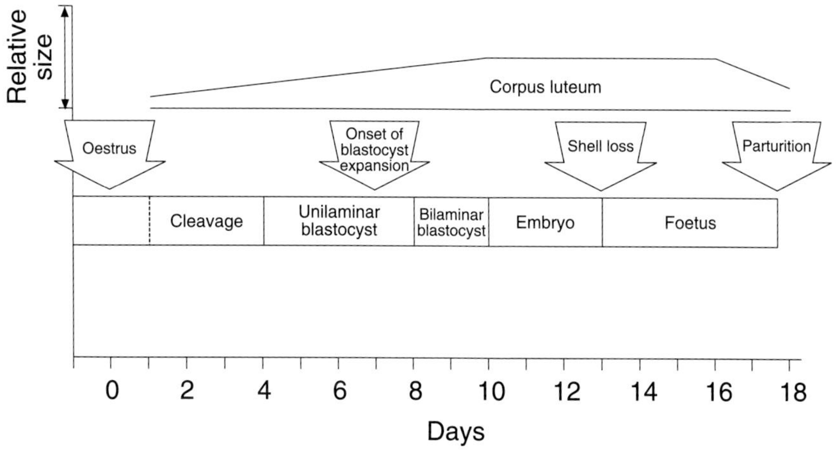

Fig. 6.4
A timetable of possum embryonic development, showing the probable extent of each stage in days. The period of growth, maximum size, and decline in size and function of the corpus luteum (CL) is also shown. The embryo stage is from early primitive streak to C-shaped embryo. Fertilisation occurs at the oestrus peak but the interval between copulation and ovulation appears to vary. Hence the beginning of cleavage is indicated by a dotted line. Based on information from Pilton & Sharman 1962; Shorey & Hughes 1973b; Hughes & Hall 1984.

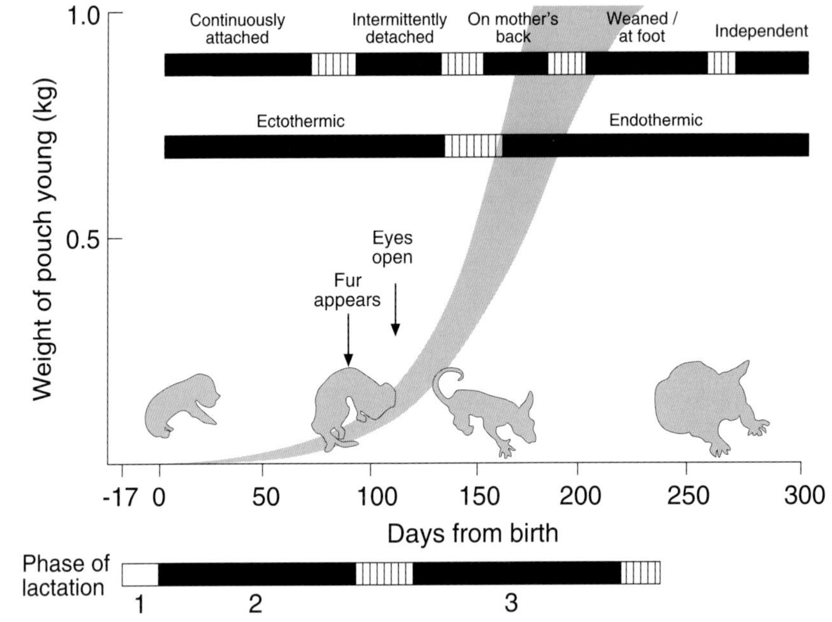

Fig. 6.5
Development of the young possum from parturition to independence relative to change in the lactation cycle. Phases of lactation are (1) preparing for birth; (2) low milk production; and (3) increased milk production and rapid growth of pouch young. Data from Lyne & Verhagen 1957, Gilmore 1969.

presumably acts on the mammary gland to prepare it for lactation. The second phase is characterised by low blood prolactin levels, low milk production, slow growth of the pouch young, and total dependence of the young on milk. The third phase of lactation begins when the pouch young reaches a critical size such that its suckling leads to increased prolactin release from the pituitary, which in turn stimulates the mammary gland to produce more milk, resulting in rapid growth and development of the pouch young and ending in weaning.

Growth and development of the pouch young

The growth of possums for 200 days after birth has been identified in a nomogram for ageing pouch young (Lyne & Verhagen 1957). A number of measurements have been used to age pouch young, of which the most useful and least prone to error is head length (HL) (Pilton & Sharman 1962). Tyndale-Biscoe (1955) and Bell (1981) provide regressions of pouch young HL against age for two populations of possums in New Zealand, but Cowan (1990) warns that growth rates differ between and within populations in different years, which affects the accuracy of age estimations. The single neonate weighs about 200 mg at birth and climbs unaided from the urogenital opening to the pouch where it attaches to one of the two nipples (Lyne *et al.* 1959). It has well-developed lungs, mouth, digestive tract, olfactory epithelium, and fore limbs, and rudimentary hind limbs (Hughes & Hall 1984). Sexual differentiation of the male and female external genitalia and the testis begins before birth, and the ovary after birth (Ullman 1993). For the first 70 days of life the young is naked, permanently attached to the nipple, and growing slowly (Fig. 6.5). During this period the mesonephric kidney matures, sexual differentiation is completed, testicular descent occurs, and oocytes and primordial follicles appear in the ovary (Buchanan & Fraser 1918; Fraser 1919; Turnbull *et al.* 1981; Ullman 1996; Eckery *et al.* 1996). After some 70 days the young may voluntarily release the teat (Lyne & Verhagen 1957). Around 90–110 days, pouch young weight increases more rapidly in response to increases in the size of milk production by the mammary gland (Smith *et al.* 1969; Cowan 1989). Over a short period the eyes open, the fur appears, and homeothermy is established by 140 days (Gilmore 1966). The young exits the pouch for the first time at about 120–140 days and permanently leaves by 170 days. Weaning is usually completed by 240 days after birth when the young become independent juveniles.

Mammary gland

Possum mammary glands follow a cycle of development and regression correlated with the oestrous cycle (Sharman 1962) irrespective of whether the female is pregnant or not (Fig. 6.6a). After ovulation the tubules of the mammary gland divide and the cells lining the ducts increase in size while the

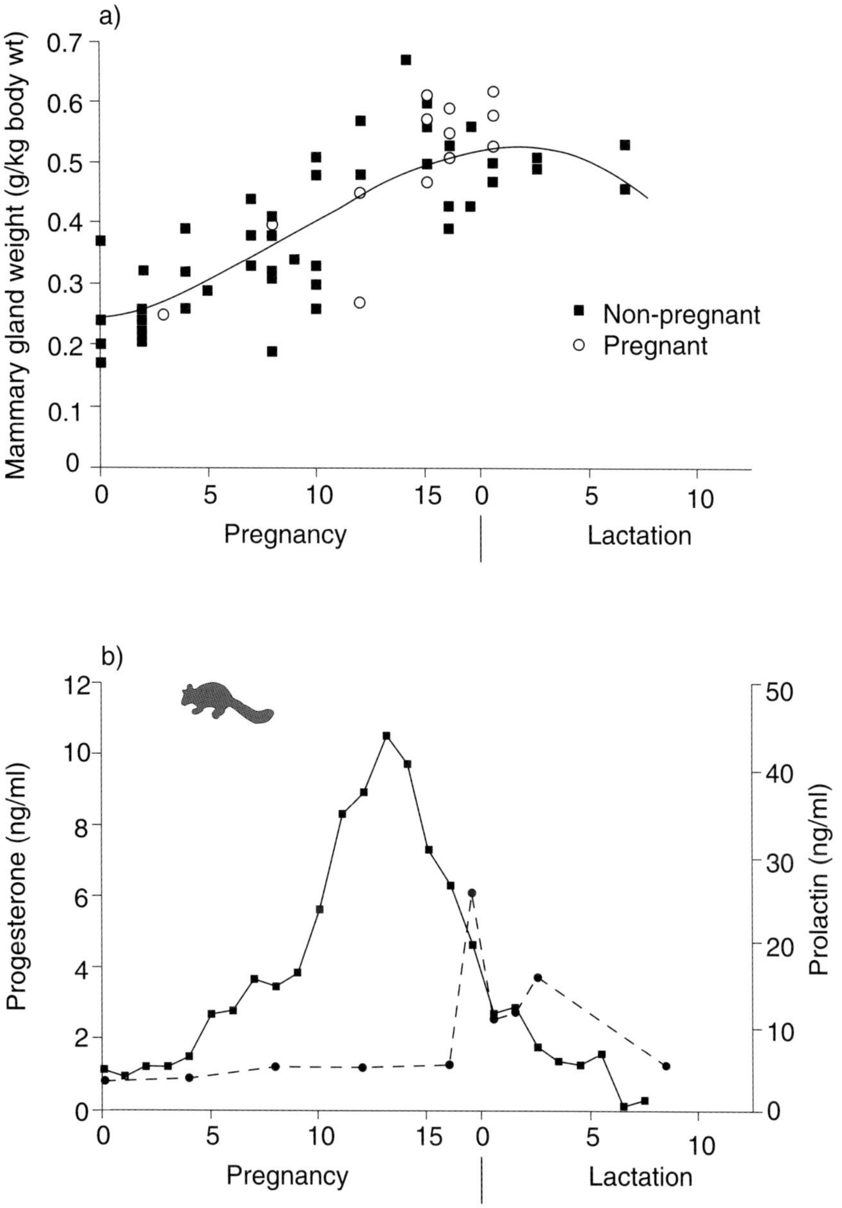

Fig. 6.6
Changes in (a) growth of the mammary gland during the oestrous cycle (■) and pregnancy (o), and (b) blood progesterone (■) and prolactin (●) levels in the possum. Redrawn from data in Gemmell *et al.* 1987, Hinds 1983, and Sharman 1962.

alveoli are formed and begin to accumulate milk-like secretions. The functional equivalence of mammary gland development was fully demonstrated by the normal growth and development of newborn young transferred to the pouches of non-mated females at the same stage after oestrus. It is unlikely that the ovary has any role in the subsequent lactation because in lactating possums the ovaries regress and take on the appearance of those in anoestrous females (Sharman 1962). Like the tammar wallaby (Fletcher *et al.* 1990), a peak of prolactin secretion is associated with birth in the possum (Hinds 1983) (Fig. 6.6b) and has been shown to be essential for initiation of lactation (Fletcher unpubl. observations). After birth only that gland to which the neonate attaches, enlarges and lactates (Fig. 6.7) while the non-suckled gland of a lactating possum regresses, more slowly than in non-lactating females (Sharman 1962). After attachment of the neonate, the suckled gland undergoes a phase of slow growth up to about 85 days (Smith *et al.* 1969) and it is this second phase of lactation that has no equivalent in eutherians. During this period blood prolactin levels (Hinds & Janssens 1986) and milk production are low (P. E. Cowan, Landcare Research, Palmerston North, pers. comm.). After this period there is a rapid increase in mammary gland size associated with increased prolactin levels and milk production. These changes mark the transition to the

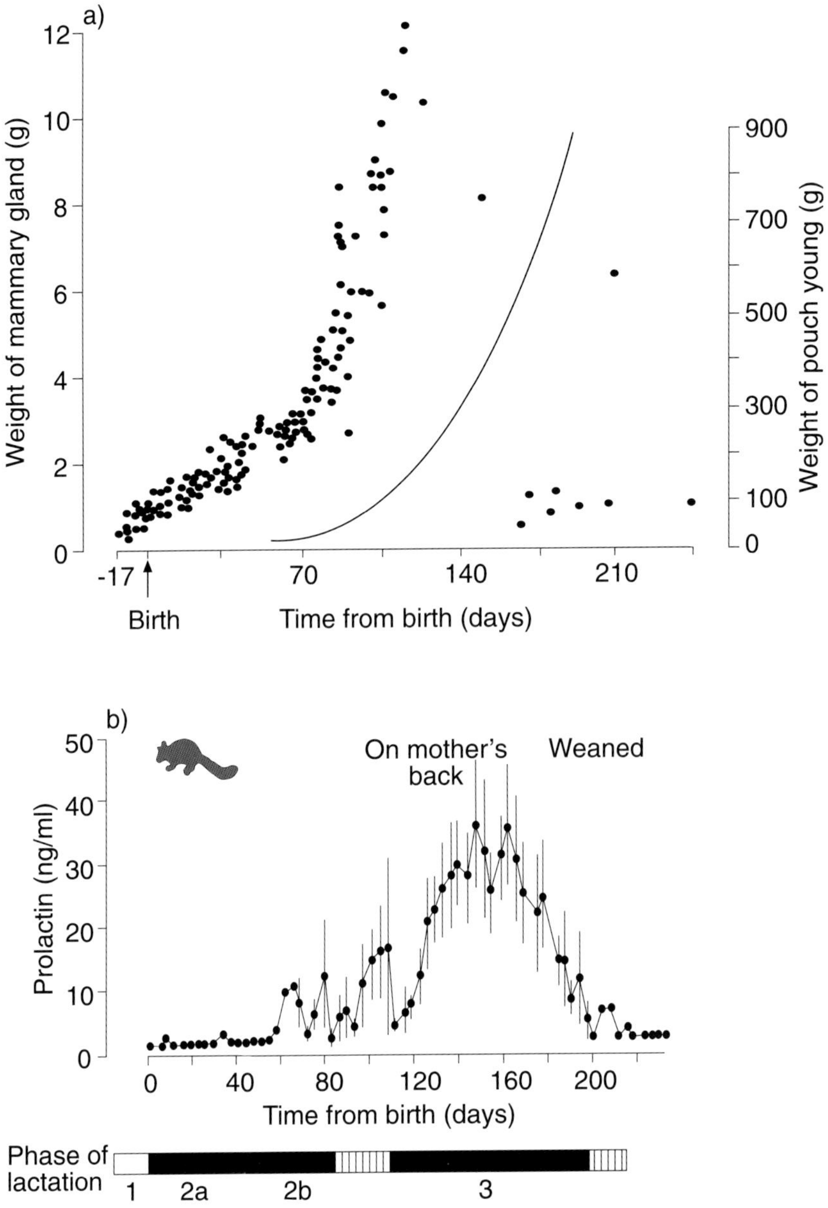

Fig. 6.7
Changes in (a) mammary gland weight (•) and mean pouch young weight (—); and (b) concentration of prolactin (mean/± SEM) during lactation in the possum. Redrawn from data in Hinds & Janssens 1986, Smith *et al.* 1969, and Sharman 1962.

third stage of lactation when prolactin secretion, milk production (P. E. Cowan pers. comm.), and growth of the pouch young is at a maximum. Weaning of the young is associated with declining prolactin levels and rapid regression of the gland (P. E. Cowan pers. comm.; Fig. 6.7).

Milk

Possums give birth to small immature young so all of the nutrients required for their initial growth and development must be supplied through the milk. Early work on the composition of possum milk showed it was very different from that of eutherians and that it changed during lactation (Gross & Bolliger 1959). A more detailed study (Cowan 1989) revealed that, like other marsupials in early lactation, possum milk was dilute, comprising mostly carbohydrate and protein. Late-lactation milk was much more concentrated, comprising mostly lipid and protein (Fig. 6.8). The crossover from early- to late-lactation milk occurred at about 140 days after birth when growth rates were maximal and developmental milestones such as fur formation, establishment of homeothermy, and opening of the eyes occurred. Demmer *et al.* (1998) report the identification and characterisation of 10 major

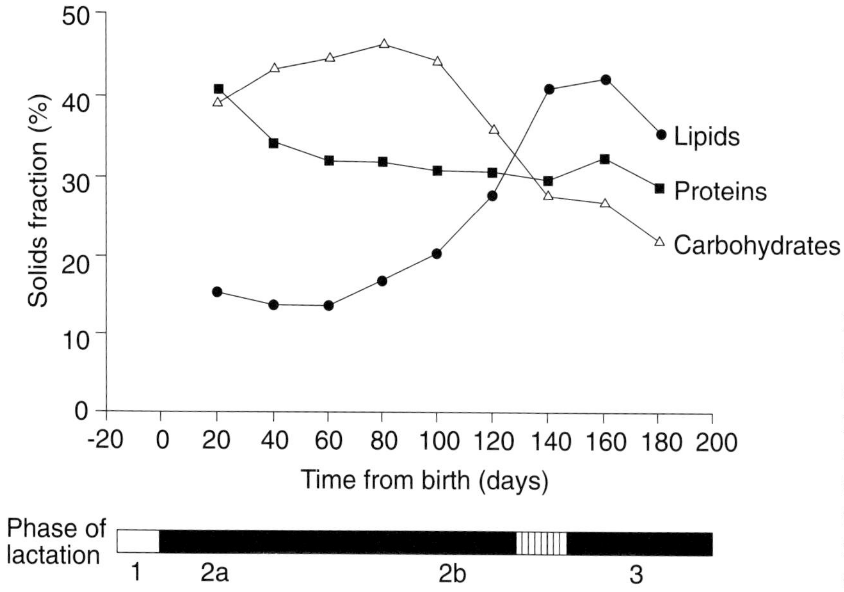

Fig. 6.8
Changes in the relative concentrations of lipids, proteins, and carbohydrates in the solids fraction of possum milk through lactation. Data from Cowan 1989 and Grigor *et al.* 1991.

proteins found in possum milk, but their role is unknown. Three of the genes encoding these proteins are unique to possums and two are expressed in early and late lactation respectively. There are marked changes in concentration of the electrolytes sodium and potassium about 10 days prior to the changes in lipid and carbohydrate, which have the effect of maintaining equal osmotic pressure between milk and maternal plasma (Green & Merchant 1988). Analysis of elements in possum milk (Jolly *et al.* 1996) revealed levels of calcium, phosphorous, copper, iron, zinc, strontium, and manganese that were higher than levels reported in eutherian milk; this is probably due to the need for the pouch young to acquire all its needs from milk. The levels of calcium, phosphorous, sulphur, and potassium increased from early- to late-lactation milk, while levels of sodium, iron, copper, and manganese decreased, and zinc levels were similar throughout. The fatty-acid composition of late-lactation marsupial milk is different to eutherians as short- and long-chain fatty acids are absent (Green & Merchant 1988). Major differences in fatty-acid composition of late-lactation milk of possums and other marsupials may be a consequence of the folivorous diet of the possum (Grigor 1980).

The changes in milk composition probably reflect changing requirements for developmental processes such as bone growth, development of fur, and maturation of digestive organs prior to ingestion of solid food. Milk is also a rich source of immunoglobulins. Marsupials are immunologically incompetent at birth (Yadav 1971) so maternal immunoglobulins passed to the neonate via the milk give protection against some of the micro-organisms that inhabit the pouch (Deakin & Cooper 1998).

Developmental failure in the pouch

Developmental failure during pouch life appears to arise because the pouch young dies rather than through lactation ceasing (Selwood *et al.* 1998). Estimates of the percentage of young lost during pouch life varies from 0–18% (Crawley 1973) to 42–58% (Bell 1981). These field studies of marked animals noted the pouch young as being dead or absent, but did not state cause. Bell (1981) estimated the mean time of death of pouch young, in the podocarp forests studied, to range between 95.4 ± 3.5 (SEM) and 97.7 ± 7.7 days. This is approximately the time when milk composition changes (B. Green 1984) and early-lactation protein levels decline (Demmer *et al.* 1998). Mid-lactation is also when loss of pouch young is most likely to occur in polytocous marsupials (i.e., those presenting many eggs or young at a time; the opposite to possums).

Reproductive technologies

In recent years the urgent need to develop reproductive technologies for conservation and management of possums has seen an upsurge in research and the development of new reagents and technologies. Reliable superovulation techniques have been developed (Rodger & Mate 1988; Glazier & Molinia 1998), which have led to better understanding of the timing of ovulation (Glazier 1998). Considerable effort on cryopreservation of sperm (Mate & Rodger 1991; Rodger *et al.* 1991) has resulted in development of techniques for artificial insemination in the possum, which might be applicable to many other marsupial species currently at risk (Molinia *et al.* 1998a,b). The pituitary hormones LH and FSH have been purified and specific assays developed for their measurement (Moore *et al.* 1997a,b).

The genes for possum LH (Lawrence *et al.* 1997) and FSH (Harrison *et al.* 1998), have been sequenced and significant homologies with non-primate eutherian genes revealed. Development and validation of reagents to factors such as stem cell factor, c-kit and c-kit ligand, GnRH receptor, β_B-Inhibin cDNA (Eckery *et al.* 1998), and androgen receptor have opened up new avenues for future research in possum reproduction. Likewise, the studies of Deakin *et al.* (1998) on the androgen receptor gene and Marshall Graves *et al.*'s (1998) identification of essential genes in the genetic pathways essential for sperm production and sex determination in possums, and possibly identification of a third pathway in marsupials for a switch between scrotum and mammary gland, have opened up new avenues for research into mammalian development and evolution.

Conclusions

Investigations for this chapter have revealed that reproduction of the possum is known in considerable detail. Some obvious gaps in our knowledge should be filled so that the effectiveness of biological control methods can be tested. What causes variations from year to year, in the same location, in breeding success and the proportion of females that are reproductive at a particular time? Why is there a second peak of breeding in some areas and which females do this? A complete table of embryonic and pouch young development should be available. What are the characteristics of possum semen and how can sperm be evaluated? The development of reagents for measurement of pituitary hormones should lead to an understanding of the control of the hypothalamo-pituitary-gonadal axis. How does the endocrine system interact with social behaviour and communication? What factors regulate the interaction of nutrition and reproduction? What regulates the mammary gland and the changes in milk composition during lactation? These questions are some of those for which we can seek answers given the extent of knowledge of possum reproduction available at this time.

Summary

- Most female possums give birth in the autumn (March, April, May) and some again in spring (September, October, November), and the onset of breeding is controlled by photoperiod.
- At any one location there is little annual variation in the onset of breeding, yet the number of possums in breeding condition at the start of the breeding season can vary significantly.
- The anatomy of the reproductive tract of male and female possums is similar to that of other marsupials.
- During the breeding season the prostate gland in males increases in size in response to elevated levels of circulating testosterone and androgen receptors in the prostate.
- While the process of spermatogenisis in male possums is similar to that of other mammals, the final stages are unique to marsupials.
- The oestrous cycle of the possum is the best described of any marsupial. Pro-oestrus is about 8–10 days, pregnancy lasts about 18 days, development in the pouch takes 120–140 days, and young are weaned at around 240 days. Hormonal and anatomical changes associated with the oestrus and development of the young are reviewed.
- At birth, young weigh 200 mg and once in the pouch feed on milk that contains fatty acids, immunoglobulins, and up to 10 major proteins.
- The percentage of young that die in the pouch varies from 0 to 58%.
- Advances in the study of the reproductive biology of possums are supporting the development of new tools for their control.

References

Allen, N. T.; Bradshaw, S. D. 1980: Diurnal variation in plasma concentrations of testosterone, 5α-dihydrotestosterone, and corticosteroids in the Australian brush-tailed possums, *Trichosurus vulpecula* (Kerr). *General and Comparative Endocrinology 40*: 455–458.

Bathgate, R. A.; Parry, L. A.; Fletcher, T. P.; Shaw, G.; Renfree, M. B.; Gemmell, R. T.; Sernia, C. 1995: Comparative aspects of oxytocin-like hormones in marsupials. *Advances in Experimental Medicine and Biology 395*: 639–655.

Bell, B. D. 1981: Breeding and condition of possums *Trichosurus vulpecula* in the Orongorongo Valley, near Wellington, New Zealand, 1966–1975. *In:* Bell, B. D. *ed.* Proceedings of the first symposium on marsupials in New Zealand. *Zoology Publications from the Victoria University of Wellington, No. 74*. Pp. 87–138.

Boersma, A. 1974: Opossums in the Hokitika River catchment. *New Zealand Journal of Forestry Science 4*: 64–75.

Brockie, R. E., 1992: A living New Zealand forest. Auckland, David Bateman. Pp. 65–87.

Brockie, R. E.; Bell, B. D.; White, A. J. 1981: Age structure and mortality of possum *Trichosurus vulpecula* populations from New Zealand. *In:* Bell, B. D. *ed.* Proceedings of the first symposium on marsupials in New Zealand. *Zoology Publications from the Victoria University of Wellington, No. 74*. Pp. 63–83.

Buchanan, G.; Fraser, E. A. 1918: The development of the urogenital system in the Marsupialia, with special reference to *Trichosurus vulpecula*. Part I. *Journal of Anatomy 53*: 35–93.

Carrick, F. N.; Cox, R. I. 1977: Testicular endocrinology of marsupials and monotremes. *In:* Calaby, J. H.; Tyndale-Biscoe, C. H. *ed.* Reproduction and evolution. Canberra, Australian Academy of Science. Pp. 137–141.

Clark, M. J.; Sharman, G. B. 1965: Failure of hysterectomy to affect the ovarian cycle of the marsupial *Trichosurus vulpecula*. *Journal of Reproduction and Fertility 10*: 459–461.

Clout, M. N. 1977: The ecology of the possum (*Trichosurus vulpecula* Kerr) in *Pinus radiata* plantations. Unpublished PhD thesis, University of Auckland, Auckland, New Zealand.

Clout, M. N.; Gaze, P. D. 1984: Brushtail possums (*Trichosurus vulpecula* Kerr) in a New Zealand beech (*Nothofagus*) forest. *New Zealand Journal of Ecology 7*:147–155.

Coleman, J. D.; Green, W. Q. 1984: Variations in the sex and age distributions of brush-tailed possum populations. *New Zealand Journal of Zoology 11*: 313–318.

Cook, B.; McDonald, I. R.; Gibson, W. R. 1978: Prostatic function in the brush-tailed possum, *Trichosurus vulpecula*. *Journal of Reproduction and Fertility 53*: 369–375.

Cowan, P. E. 1989: Changes in milk composition during lactation in the common brushtail possum, *Trichosurus vulpecula* (Marsupialia: Phalangeridae). *Reproduction, Fertility and Development 1*: 325–335.

Cowan, P. E. 1990: Brushtail possum. *In:* King, C. M. *ed.* The handbook of New Zealand mammals. Auckland, Oxford University Press. Pp. 68–98.

Crawford, J. L.; Shackell, G. H.; Thompson, E. G.; McLeod, B. J.; Hurst, P. R. 1998: Preovulatory follicle development and ovulation in the brushtail possum (*Trichosurus vulpecula*) monitored by repeated laparoscopy. *Journal of Reproduction and Fertility 110*: 361–370.

Crawley, M. C. 1970: Longevity of Australian brush-tailed opossums (*Trichosurus vulpecula*), in indigenous forest in New Zealand. *New Zealand Journal of Science 13*: 348–351.

Crawley, M. C. 1973: A live trapping study of Australian brush-tailed possums, *Trichosurus vulpecula* (Kerr), in the Orongorongo Valley, Wellington, New Zealand. *Australian Journal of Zoology 21*: 75–90.

Cummins, J. M. 1976: Epididymal maturation of spermatozoa in the marsupial *Trichosurus vulpecula*: changes in motility and gross morphology. *Australian Journal of Zoology 24*: 499–511.

Cummins, J. M. 1981: Sperm maturation in the possum *Trichosurus vulpecula* : a model for comparison with eutherian mammals. *In:* Bell, B.D. *ed.* Proceedings of the first symposium on marsupials in New Zealand. *Zoology Publications from the Victoria University of Wellington, No. 74*. Pp. 23–37.

Curlewis, J. D.; Stone, G. M. 1985a: Peripheral androgen levels in the male brush-tail possum (*Trichosurus vulpecula*). *Journal of Endocrinology 105*: 63–70.

Curlewis, J. D.; Stone, G. M. 1985b: Some effects of breeding season and castration on the prostate and epididymis of the brushtail possum, *Trichosurus vulpecula*. *Australian Journal of Biological Sciences 38*: 313–326.

Curlewis, J. D.; Stone, G. M. 1986a: Reproduction in captive female brushtail possums, *Trichosurus vulpecula*. *Australian Journal of Zoology 34*: 47–52.

Curlewis, J. D.; Stone, G. M. 1986b: Effects of oestradiol, the oestrous cycle and pregnancy on weight, metabolism and cytosol receptors in the uterus of the brush-tail possum (*Trichosurus vulpecula*). *Journal of Endocrinology 108*: 201–210.

Curlewis, J. D.; Stone, G. M. 1987: Effects of oestradiol, the oestrous cycle and pregnancy on weight, metabolism and cytosol receptors in the oviduct and vaginal complex of the brushtail possum (*Trichosurus vulpecula*). *Australian Journal of Biological Sciences 40*: 315–322.

Curlewis, J. D.; Axelson, M.; Stone, G. M. 1985: Identification of the major steroids in ovarian and adrenal venous plasma of the brush-tail possum (*Trichosurus vulpecula*) and changes in the peripheral plasma levels of estradiol and progesterone during the reproductive cycle. *Journal of Endocrinology 105*: 53–62.

Deakin, J. E.; Cooper, D. W. 1998: Pouch microflora in the brushtail possum and maternal immunity transfer to the pouch young. *In:* Biological control of possums. *The Royal Society of New Zealand Miscellaneous Series 45*: 46–48.

Deakin, J. E.; Harrison, G. A.; Cooper, D. W. 1998. Androgen receptor as a potential target for immunosterilisation in the brushtail possum. *In:* Biological control of possums. *The Royal Society of New Zealand Miscellaneous series 45*: 44–45.

Demmer, J.; Ginger, M. R.; Ross, I. K.; Piotte, C. P.; Grigor, M. R. 1998: Targets in lactation for biocontrol of the common brushtail possum (*Trichosurus vulpecula*). *In:* Biological control of possums. *The Royal Society of New Zealand Miscellaneous Series 45*: 53–58.

Duckworth, J. A.; Scobie, S.; Jones, D. E.; Selwood, L. 1999: Determination of oestrus and mating in captive female brushtail possums, *Trichosurus vulpecula* (Marsupialia: Phalangeridae), from urine samples. *Australian Journal of Zoology 46*: 547–555.

Eckery, D. C.; Tisdall, D. J.; Heath, D. A.; McNatty, K. P. 1996: Morphology and function of the ovary during fetal and early neonatal life: A comparison between the sheep and brushtail possum (*Trichosurus vulpecula*). *Animal Reproduction Science 42*: 551–561.

Eckery, D.; Lawrence, S.; Greenwood, P.; Stent, V.; Ng Chie, W.; Heath, D.; Lun, S.; Vanmontfort, D.; Fidler, A.; Tisdall, D.; Moore, L.; McNatty, K. P. 1998: The isolation of genes, novel proteins, and hormones and the regulation of gonadal development and pituitary function in possums. *In:* Biological control of possums. *The Royal Society of New Zealand Miscellaneous Series 45*: 100–110.

Fletcher, T. P.; Shaw, G.; Renfree, M. B. 1990: Effects of bromocriptine at parturition in the tammar wallaby, *Macropus eugenii*. *Reproduction, Fertility and Development 2*: 79–88.

Frankenberg, S.; Selwood, L. 1998: An ultrastructural study of the role of an extracellular matrix during normal cleavage in a marsupial, the brushtail possum. *Molecular Reproduction & Development 50*: 420–433.

Fraser, E. A. 1919: The development of the urogenital system in the Marsupialia, with special reference to *Trichosurus vulpecula*. Part II. *Journal of Anatomy 53*: 97–129.

Fraser, K. W. 1979: Dynamics and condition of opossum (*Trichosurus vulpecula* Kerr) populations in the Copland Valley, Westland, New Zealand. *Mauri Ora 7*: 117–137.

Gemmell, R. T. 1995: Breeding biology of brushtail possums *Trichosurus vulpecula* (Marsupialia, Phalangeridae) in captivity. *Australian Mammalogy 18*: 1–7.

Gemmell, R. T.; Sernia, C. 1995: Effect of changing from a short-day to long-day photoperiod on the breeding season of the brushtail possum (*Trichosurus vulpecula*). *Journal of Experimental Zoology 273*: 242–246.

Gemmell, R. T.; Cepon, G.; Barnes, A. 1986: Weekly variations in body weight and plasma testosterone concentrations in the captive male possum *Trichosurus vulpecula*. *General and Comparative Endocrinology 62*: 1–7.

Gemmell, R. T.; Hughes, R. L.; Jenkin, G. 1987: Comparative studies on the hormone profiles of progesterone and prostaglandin F metabolite in the possum *Trichosurus vulpecula*. *In:* Archer, M. *ed.* Possums and opossums: Studies in evolution. Sydney, Surrey Beatty in assoc. with the Royal Zoological Society of New South Wales. Pp. 279–291.

Gemmell, R. T.; Cepon, G.; Sernia, C. 1993: Effect of photoperiod on the breeding season of the marsupial possum *Trichosurus vulpecula*. *Journal of Reproduction and Fertility 98*: 515–520.

Gilmore, D. P. 1966: Studies on the biology of *Trichosurus vulpecula* Kerr. Unpublished PhD thesis, University of Canterbury, Christchurch, New Zealand.

Gilmore, D. P. 1969: Seasonal reproductive periodicity in the male Australian brush-tailed possum (*Trichosurus vulpecula*). *Journal of Zoology, London 157*: 75–98.

Glazier, A. M. 1998: Induced ovulation in the brushtail possum (*Trichosurus vulpecula*). *In:* Biological control of possums. *The Royal Society of New Zealand Miscellaneous Series 45*: 98–99.

Glazier, A. M.; Molinia F. C. 1998: Improved method of superovulation in monovulatory brushtail possums (*Trichosurus vulpecula*) using pregnant mare's serum gonadotrophin-luteinizing hormone. *Journal of Reproduction and Fertility 113*: 191–195.

Green, B. 1984: Composition of milk and energetics of growth in marsupials. *In:* Peaker, M.; Vernon, R. G.; Knight, C. H. *ed.* Physiological strategies in lactation . *Symposia of the Zoological Society of London 51*: 369–387.

Green, B.; Merchant, J. C. 1988: The composition of marsupial milk . *In:* Tyndale-Biscoe, C. H.; Janssens, P. A. *ed.* The developing marsupial: Models for biomedical research. Berlin, Springer-Verlag. Pp. 41–54.

Green, L. M. 1963: Interstitial cells in the testes of an Australia phalanger (*Trichosurus vulpecula*). *Australian Journal of Experimental Biology 41*: 99–104.

Green, W. Q. 1984: A review of ecological studies relevant to management of the common brushtail possum. *In:* Smith, A. P.; Hume, I. D. *ed.* Possums and gliders. Chipping Norton, NSW, Surrey Beatty in assoc. with the Australian Mammal Society. Pp. 483–499.

Grigor, M. R. 1980: Structure of milk triglycerols of five marsupials and one monotreme: evidence for an unusual pattern common to marsupials and eutherians but not found in the echidna, monotreme. *Comparative Biochemistry and Physiology 65b*: 427–430.

Grigor, M. R.; Bennett, B. L.; Carne, A.; Cowan, P. E. 1991: Whey protein of the common brushtail possum (*Trichosurus vulpecula*): isolation, characterization and changes in concentration in milk during lactation of transferrin, alpha- lactalbumin and serum albumin. *Comparative Biochemistry and Physiology 98*: 451–459.

Gross, R.; Bolliger, A. 1959: The occurrence of carbohydrates other than lactose in the milk of a marsupial (*Trichosurus vulpecula*). *American Medical Association Journal of Diseases of Children 98*: 758–775.

Harrison, G. A.; Deane, E. M.; Cooper, D. W. 1998: cDNA cloning of luteinizing hormone subunits from brushtail possum and red kangaroo. *Mammalian Genome 9*: 638–642.

Hartman, C. G. 1928: The breeding season of the opossum (*Didelphis virginiana*) and the rate of intrauterine and post natal development. *Journal of Morphology and Physiology 46*: 143–215.

Hinds, L. A. 1983: Progesterone and prolactin in marsupial reproduction. Unpublished PhD thesis, Australian National University, Canberra, Australia.

Hinds, L. A. 1988: Hormonal control of lactation. *In:* Tyndale-Biscoe, C. H.; Janssens, P. A. *ed.* The developing marsupial: Models for biomedical research. Berlin, Springer-Verlag. Pp. 55–67.

Hinds, L. A.; Janssens, P. A. 1986: Changes in prolactin in peripheral plasma during lactation in the brushtail possum *Trichosurus vulpecula*. *Australian Journal of Biological Sciences 39*: 171-178.

Hocking, G. J. 1981: The population ecology of the brush-tailed possum, *Trichosurus vulpecula* (Kerr), in Tasmania. Unpublished MSc thesis, University of Tasmania, Hobart, Australia.

Horn, C. A. 1981: Luteinising hormone in the brushtail possum. Proceedings of the Australian Society for Reproductive Biology Thirteenth Annual Conference, Christchurch, New Zealand. P. 48 abstract

Hughes, R. L., Hall, L. S. 1984: Embryonic development in the common brushtail possum *Trichosurus vulpecula*. *In:* Smith, A. P.; Hume, I. D. *ed.* Possums and gliders. Chipping Norton, NSW, Surrey Beatty in assoc. with the Australian Mammal Society. Pp. 97–212.

Hughes, R. L.; Rodger, J. C. 1971: Studies on the vaginal mucus of the marsupial *Trichosurus vulpecula*. *Australian Journal of Zoology 19*: 19–33.

Humphreys, W. F.; Bradley, A. J.; How, R. A.; Barnett, J. L. 1984: Indices of condition of phalanger populations: a review. *In:* Smith, A. P.; Hume, I. D. *ed.* Possums and gliders. Chipping Norton, NSW, Surrey Beatty in assoc. with the Australian Mammal Society. Pp. 59–77.

Jolly, S. E.; Scobie, S.; Coleman, M. C. 1995: Breeding capacity of female brushtail possums *Trichosurus vulpecula* in captivity. *New Zealand Journal of Zoology 22*: 325–330.

Jolly, S. E.; Morriss, G. A.; Scobie, S.; Cowan, P. E. 1996: Composition of milk of the common brushtail possum, *Trichosurus vulpecula* (Marsupialia: Phalangeridae): concentrations of elements. *Australian Journal of Zoology 44*: 479–486.

Kean, R. I. 1975: Growth of opossums (*Trichosurus vulpecula*) in the Orongorongo Valley, Wellington, New Zealand. 1953–61. *New Zealand Journal of Zoology 2*: 435–444.

Kean, R. I.; Marryatt, R. G.; Carroll, A. L. K. 1964: The female urogenital system of *Trichosurus vulpecula* (Marsupialia). *Australian Journal of Zoology 12*: 18–41.

Kerle, J. A. 1984: Variation in the ecology of *Trichosurus*: its adaptive significance. *In:* Smith, A. P.; Hume, I. D. *ed.* Possums and gliders. Chipping Norton, NSW, Surrey Beatty in assoc. with the Australian Mammal Society. Pp. 115–128.

Lawrence, S. B.; Vanmontfort, D. M.; Tisdall, D. J.; McNatty, K. P.; Fidler, A. E. 1997: The follicle-stimulating hormone β-subunit gene of the common brushtail possum (*Trichosurus vulpecula*): analysis of cDNA sequence and expression. *Reproduction, Fertility and Development 9*: 795–801.

Lee, C. S.; O'Shea, J. D. 1977: Observations on the vasculature of the reproductive tract in some Australian marsupials. *Journal of Morphology 154*: 95–114.

Legge, M.; Hill, B. L.; Shackell, G. H.; McLeod, B. J. 1996: Glycosaminoglycans of the uterine and vaginal cul-de-sac tissue in the brushtail possum (*Trichosurus vulpecula*). *Reproduction, Fertility and Development 8*: 819–823.

Lyne, A. G., Verhagen, A. M. V. 1957: Growth of the marsupial *Trichosurus vulpecula* and a comparison with some higher mammals. *Growth 21*: 167–195.

Lyne, A. G.; Pilton, P. E.; Sharman, G. B. 1959: Oestrous cycle, gestation period and parturition in the marsupial *Trichosurus vulpecula. Nature 183*: 622–623.

Marshall Graves, J. A.; Cook, S. I.; Glas, R.; O'Neill, R. W. J. O.; Pask, A.; Delbridge, M. L. 1998: Genes involved in male reproduction as targets for biocontrol of the brushtailed possum. *In:* Biological control of possums. *The Royal Society of New Zealand Miscellaneous Series 45*: 62–72.

Mate, K. E.; Rodger, J. C. 1991: Stability of the acrosome of the brush-tailed possum (*Trichosurus vulpecula*) and tammar wallaby (*Macropus eugenii*) *in vitro* and after exposure to conditions and agents known to cause capacitation or acrosome reaction of eutherian spermatozoa. *Journal of Reproduction and Development 91*: 41–48.

Molinia, F.; Nickel, M.; Glazier, A.; Rodger, J. 1998a: Artificial insemination and sperm transport in possums. *In:* Biological control of possums. *The Royal Society of New Zealand Miscellaneous Series 45*: 76–78.

Molinia, F. C.; Gibson, R. J.; Brown, A. M.; Glazier, A. M.; Rodger, J. C. 1998b: Successful fertilization after superovulation and laparoscopic intrauterine insemination of the brushtail possum, *Trichosurus vulpecula*, and tammar wallaby, *Macropus eugenii. Journal of Reproduction and Fertility 113*: 9–17.

Moore, L. G.; Ng Chie, W.; Lun, S.; Lawrence, S. B.; Young, W.; McNatty, K. P. 1997a: Follicle-stimulating hormone in the brushtail possum (*Trichosurus vulpecula*): purification, characterization and radioimmunoassay. *General and Comparative Endocrinology 106*: 30–38.

Moore, L. G.; Chie, W. N.; Lun, S.; Lawrence, S. B.; Heath, D. A. McNatty, K. P. 1997b: Isolation, characterization and radioimmunoassay of luteinizing hormone in the brushtail possum. *Reproduction, Fertility and Development 9*: 419–425.

O'Donoghue, C. H. 1916: On the corpora lutea and interstitial tissue of the ovary in the Marsupialia. *Quarterly Journal of Microscopical Science 61*: 433–473.

Pearson, J. 1945: The female urogenital system of the Marsupialia with special reference to the vaginal complex. *Papers and Proceedings of the Royal Society of Tasmania 1944*: 71–97.

Pekelharing, C. J. 1970: Cementum deposition as an age indicator in the brush-tailed possum, *Trichosurus vulpecula* Kerr (Marsupialia). *Australian Journal of Zoology 18*: 71–76.

Pilton, P. E.; Sharman, G. B. 1962: Reproduction in the marsupial *Trichosurus vulpecula. Journal of Endocrinology 25*: 119–136.

Roberts, C. T.; Selwood , L.; Leigh, C. M.; Breed, W. G. 1997: Antiserum to the egg coats of the fat-tailed dunnart (Marsupialia, Dasyuridae) cross-reacts with egg coats of other marsupial and eutherian species. *Journal of Experimental Zoology 278(3)*: 133–139.

Rodger, J. C.; Hughes, R. L. 1973: Studies of the accessory glands of male marsupials. *Australian Journal of Zoology 21*: 303–320.

Rodger, J. C.; Mate, K. E. 1988: A PMSG/GnRH method for the superovulation of the monovulatory brush-tailed possum (*Trichosurus vulpecula*). *Journal of Reproduction and Physiology 83*: 885–891.

Rodger, J. C.; Cousins, S. J.; Mate, K. E. 1991: A simple glycerol-based freezing protocol for the semen of a marsupial *Trichosurus vulpecula*, the common brushtail possum. *Reproduction, Fertility and Development 3*: 119–125.

Selwood, L. 1983: Factors influencing pre-natal fertility in the brown marsupial mouse *Antechinus stuartii. Journal of Reproduction Fertility 68*: 317–324.

Selwood, L.; Frankenberg, S.; Casey, N. 1998: An overview of the importance of the egg coats for embryonic survival in the common brushtail possum: normal development and targets for contraception. *In:* Biological control of possums. *The Royal Society of New Zealand Miscellaneous Series 45*: 89–95.

Sernia, C; Bradley, A. J.; McDonald, I. R. 1979: High affinity binding of adrenocortical and gonadal steroids by plasma proteins of Australian marsupials. *General and Comparative Endocrinology 38*: 496–503.

Sernia, C; Garcia-Aragon, J; Thomas, W. G.; Gemmell, R. T. 1990: Uterine oxytocin receptors in an Australian marsupial, the brushtail possum, *Trichosurus vulpecula. Comparative Biochemistry and Physiology 95*: 135–138.

Sernia, C.; Thomas, W. G.; Gemmell, R. T. 1991: Oxytocin receptors in the mammary gland and reproductive tract of a marsupial, the brushtail possum (*Trichosurus vulpecula*). *Biology of Reproduction 45*: 673–679.

Setchell, B. P. 1977: Reproduction in male marsupials. *In:* Stonehouse, B.; Gilmore, D. P. *ed.* The biology of marsupials. London, Macmillan. Pp. 411–457.

Setchell, B. P.; Carrick, F. N. 1973: Spermatogenesis in some Australian marsupials. *Australian Journal of Zoology 21*: 491–499.

Sharman, G. B. 1962: The initiation and maintenance of lactation in the marsupial, *Trichosurus vulpecula. Journal of Endocrinology 25*: 375–385.

Sharman, G. B. 1965: The effects of the suckling stimulus and oxytocin injection on the corpus luteum of delayed implantation in the red kangaroo. Proceedings of the Second International Congress of Endocrinology. Amsterdam, Exerpta Medica Foundation. Pp. 669–674.

Shorey, C. D.; Hughes, R. L. 1973a: Cyclical changes in the uterine endometrium and peripheral plasma concentrations of progesterone in the marsupial *Trichosurus vulpecula. Australian Journal of Zoology 21*: 1–19.

Shorey, C. D.; Hughes, R. L. 1973b: Development, function, and regression of the corpus luteum in the marsupial *Trichosurus vulpecula. Australian Journal of Zoology 21*: 477–489.

Shorey, C. D.; Hughes, R. L. 1975: Uterine response to ovariectomy during the proliferative and luteal phases in the marsupial, *Trichosurus vulpecula. Journal of Reproduction and Fertility 42*: 221–228.

Smith, M. J.; Brown, B. K.; Frith, H. J. 1969: Breeding of the brush-tailed possum, *Trichosurus vulpecula* (Kerr), in New South Wales. *CSIRO Wildlife Research 14*: 181–193.

Stewart, F.; Tyndale-Biscoe, C. H. 1982: Prolactin and luteinizing hormone receptors in marsupial corpora lutea: relationship to control of luteal function. *Journal of Endocrinology 92*: 63–72.

Temple-Smith, P. D. 1984:Reproductive structures and strategies in male possums and gliders. *In:* Smith, A. P.; Hume, I. D. *ed.* Possums and gliders. Chipping Norton, NSW, Surrey Beatty in assoc. with the Australian Mammal Society. Pp. 89–106.

Thorburn, G. D.; Cox, R. I.; Shorey, C. D. 1971: Ovarian steroid secretion rates in the marsupial *Trichosurus vulpecula. Journal of Reproduction and Fertility 24*: 139.

Triggs, S. J. 1982: Comparative ecology of the possum, *Trichosurus vulpecula*, in three pastoral habitats. Unpublished MSc thesis, University of Auckland, Auckland, New Zealand.

Turnbull, K. E.; Mattner, P. E.; Hughes, R. L. 1981: Testicular descent in the marsupial *Trichosurus vulpecula* (Kerr). *Australian Journal of Zoology 29*: 189–198.

Tyndale-Biscoe, C. H. 1955: Observations on the reproduction and ecology of the brush-tailed possum, *Trichosurus vulpecula* Kerr (Marsupialia), in New Zealand. *Australian Journal of Zoology 3*: 162–184.

Tyndale-Biscoe, C. H. 1966: The marsupial birth canal. *In:* Rowlands, I. W. *ed.* Comparative biology of reproduction in mammals. *Symposia of the Zoological Society of London 15*: 233–250.

Tyndale-Biscoe, C. H. 1969: Relaxin activity during the oestrous cycle of the marsupial, *Trichosurus vulpecula* (Kerr). *Journal of Reproduction and Fertility 19*: 191–193.

Tyndale-Biscoe, C. H. 1984: Reproductive physiology of possums and gliders. *In:* Smith, A. P.; Hume, I. D. *ed.* Possums and gliders. Chipping Norton, NSW, Surrey Beatty in assoc. with the Australian Mammal Society. Pp. 79–87.

Tyndale-Biscoe, C. H.; Janssens, P. A. *ed.* 1988: The developing marsupial: Models for biomedical research. Berlin, Springer-Verlag. 245 p.

Tyndale-Biscoe, C. H.; Renfree, M. B. 1987: Reproductive physiology of marsupials. Cambridge, Cambridge University Press. 476 p.

Ullman, S. L. 1993: Differentiation of the gonads and initiation of mammary gland and scrotum development in the brushtail possum *Trichosurus vulpecula* (Marsupialia). *Journal of Anatomy 187*: 475–484.

Ullman, S. L. 1996: Development of the ovary of the brushtail possum *Trichosurus vulpecula* (Marsupialia). *Journal of Anatomy 189*: 651–665.

Vinson, G. 1974: The control of the adrenocortical secretion in the brush-tailed possum, *Trichosurus vulpecula. General and Comparative Endocrinology 22*: 268–276.

Vinson, G.; Phillips, 1969: Formation of testosterone by a special zone in the adrenal cortex of the brush possum (*Trichosurus vulpecula*). *General and Comparative Endocrinology 13*: 538–539.

Von Der Borch, S. M. 1963: Unilateral hormone effect in the marsupial *Trichosurus vulpecula. Journal of Reproduction and Fertility 5*: 447–449.

Yadav, M. 1971: The transmission of antibodies across the gut of pouch young marsupials. *Immunology 21*: 839–851.

Young, C. E.; McDonald, I. R. 1982: Oestrogen receptors in the genital tract of the Australian marsupial *Trichosurus vulpecula. General and Comparative Endocrinology 46*: 417–427.

Appendix 6.1 Glossary of Terms

acrosome	region of the head of sperm cell that contains hydrolytic enzymes used to digest protective coating of the egg
amnion	membranous fluid-filled sac surrounding the embryo
ampulla(e)	body of oviduct where fertilisation occurs
antral follicle	ovarian follicle with fluid-filled cavity around oocyte
blastocyst	early embryo consisting of blastomeres joined at margins
blastomere	one of the cells formed by the cleavage of a fertilised egg
capacitation	the act of making spermatozoa capable of fertilisation, final stage of sperm maturation
cauda epididymis	tail section of epididymis
conceptus	product of conception
corpus albicans	white fibrous scar on ovary after resorption of corpus luteum
cortical	belonging to the outer part of something
cryopreservation	storing at low temperatures
cumulus cells	cells around oocyte in ovarian follicle
cytoplasm	contents of a cell (protoplasm), excluding its nucleus
ectothermic	dependent on mother for heat
endometrium	mucous membrane lining the uterus
endothermic	capable of internal generation of heat
epiblast	the ectoderm (outermost layer) of an embryo
epididymis(ides)	elongated mass of tubes at back of testis with sperm maturation and storage functions
epithelium	cell sheet formed of one or more layers of cells covering an external surface or lining a cavity
flagellum	tail of sperm
GAGs	glycosaminoglycans
Graafian follicle	dominant follicle in mammalian ovary, in which an ovum matures, has potential to ovulate
gravid	carrying young
homeothermy	self-regulation of an even body temperature
homologies	similarities in structure of an organ or molecule reflecting a common evolutionary origin
hypoblast	endoderm (innermost layer) of an embryo
immunoglobulins	proteins produced in response to a foreign molecule or invading organism. Often bind to a foreign molecule or cell, thereby inactivating it or marking it for destruction by the immune system. Also called antibodies
interstitial	in intervening spaces
Leydig cells	interstitial cells of testis, source of androgen

luteinization	process driven by luteinizing hormone by which ovulated follicle forms corpus luteum
meiosis	special type of cell division by which egg and sperm are produced, resulting in cells with half the chromosome number of parent
mesoderm	middle layer of an embryo
mesonephric kidney	functional in the embryo and for first week after birth
mitochondria	organelles found in most eukaryolic cells, contain enzymes for respiration
monotocous	presenting one egg or offspring at a time
neurulation	stage at which basic vertebra and nervous system are beginning to emerge
oestrus	being "on heat", recurrent cycle of sexual excitability (adjective: oestrous)
oocyte	immature ovum in ovary
oxytocic peptide	hormone with oxytocin-like activity, role in birth process and lactation
pluriblast	multi-layered blastocyst
pole/polar	extremity of the main axis of any spherical body <u>or</u> configuration of electrical charge causing electric dipole within molecule
pre-putial	of the prepuce, foreskin
primordial follicles	ovarian follicles in early stage of development
proamnion	gives rise to the headfold of the amnion
pronuclei	gamete nucleus which after fertilisation unite and form a diploid nucleus
Sertoli cells	cells of testis, apparent source of nourishment to spermatids
somite	one of a series of paired blocks of mesoderm that form during early development and lie on either side of the notocord in the embryo
somitogenesis	beginning of division into body segments
spermatocyte	a cell which may divide into spermatids(immature male sex cells)
spermatogenic cycle	process of transformation of diploid spermatogonia to haploid spermatids
spermatozoon	mature, motile, male sex cell in animals
trophectoderm	(=trophoblast) outer layer of an embryo
trophoblast cell type	layer of ectodermal tissue that forms outer surface of blastocyst, secures egg to wall of uterus, supplies nutrients to embryo
unilaminar blastocyst stage	early embryo with a single layer of cells
yolk-sac	membrane enclosing yolk of egg, source of nutrients to developing embryo
zona pellucida	transparent outer layer of an ovum, secondary egg membrane
zygote	diploid cell formed by union of two gametes (mature germ cells), a fertilised egg

Glossary supplied by Janine Duckworth, Marsupial CRC, Landcare Research, Lincoln.

CHAPTER SEVEN

Predators, Parasites, and Diseases of Possums

Phil Cowan, John Clark, David Heath, Mirek Stankiewicz, and Joanne Meers

Predation, parasitism and diseases (other than bovine tuberculosis (Tb)) have generally been discounted as important influences on possum populations in New Zealand. Recent predator control trials in Australia, however, have shown that the impacts of predation, even on common species like the brushtail possum, may be severe (Anon. 1997). After successful control of foxes using 1080 poison, numbers of brushtail possums increased up to fourfold. Information about parasites and diseases was scarce and patchy (Cowan 1990), but the search for potential biological control agents (Chapter 24) provided the impetus for the first detailed systematic examination of parasites and diseases of possums in New Zealand (Heath *et al.* 1998). This chapter reviews what is known about possum predators, parasites, diseases, and viruses but stops short of linking possum health status with possum population dynamics.

Predators

In New Zealand, live possums are preyed on most frequently by feral cats. In Orongorongo Valley, possum remains were found in up to 38% of cat scats in winter and spring, and newly independent young were often eaten in spring (Fitzgerald & Karl 1979). Farm dogs also kill many possums, accounting for about 20% of the known deaths in one study (Brockie *et al.* 1987). Stoats, weasels, and ferrets prey on young possums and commonly scavenge possum carcasses (King 1990; Smith *et al.* 1995), especially in fur-trapping areas (King & Moody 1982) and after control operations (Alterio 1996). Young possums may be taken occasionally by moreporks, *Ninox novaeseelandiae*, or Australasian harriers, *Circus approximans.* In Australia, possums are preyed on by a wider range of native and introduced predators (Cowan 1990), and predation, especially by foxes, has a major impact on possum numbers (Anon. 1997). The significance of natural predation in the population dynamics of possums in New Zealand is unknown, but is probably much less than that resulting from hunting for furs, control for Tb or conservation management, or road kills.

More than 2 million possum skins were exported each year in 1978–82 (Cowan 1990), leading in many accessible areas to a scarcity of possums and a marked reduction in the size of skins offered for sale (Clout & Barlow 1982). However, the impact of hunting on possums has reduced in recent years with a decline in the export market for possum skins to around 500 000–800 000 annually (Parkes *et al.* 1996). Possum control for Tb or conservation management has been applied over about 4 million hectares in recent years. Assuming an initial average density of 4 possums/ha and an average sustained reduction of 80%, control has reduced possum numbers over that area by about 13 million. In two studies of the fate of dispersing possums on farmland, roadkills and trapping or shooting by farmers were the major causes of mortality (Cowan *et al.* 1996, 1997). A study of urban possums in Australia also found that most possum deaths and injuries were caused by vehicles (Statham & Statham 1997).

Ectoparasites

Mites dominate the ectoparasite fauna of New Zealand possums, both in terms of the number of parasites per possum and the number of possums carrying mites at different geographic locations. *Atellana papilio* was found on all of 59 possums from Banks Peninsula (Sweatman 1962), and on 46% of 125 possums from Central Otago, 42% of which also harboured *Trichosurolaelaps crassipes* and 50% had *Petrogalochirus dycei* (Bowie & Bennett 1983). Clark *et al.* (1997), who surveyed the ectoparasites of possums at 15 sites covering North, South and Stewart islands of New Zealand (Table 7.1), found *A. papilio* on 98% of pelts, *P. dycei* on 93%, *Murichirus anabiotus* on 93% and *T. crassipes* on 99% of pelts examined. One or more of these species were present on all skins sampled. *M. anabiotus* was absent from Chatham Island skins and *T. crassipes*

Table 7.1
Ectoparasites of possums in New Zealand and Australia (from Viggers & Spratt 1995; Clark *et al.* 1997).

	New Zealand	Australia
Acari – Mesostigmata (mites)		
Trichosurolaelaps crassipes	*	*
Haemolaelaps sisyphus		*
Acari – Sarcoptiformes (mites)		
Atellana papilio	*	*
Petrogalochirus dycei	*	*
Murichirus anabiotus	*	
Marsupiopus trichosuri (follicle mite)	*	*
Acari – Trombidiformes (mites)		
Chiggers (10 species)		*
Acari – Ixodoidea (ticks)		
Haemaphysalis spp. (4 species)		*
Ixodes spp. (6 species)		*
Siphonaptera (fleas)		
Fleas (7 species in 2 families)		*
Accidental ectoparasites		
Listrophorus gibbus (rabbit mite)	*	
Notoedres muris (rodent mange mite)		*
Laelapsella humi (mutton-bird mite)		*
Ctenocephalides felis (cat flea)	*	*
Echinophaga gallinacea (stickfast flea)		*
Pulex irritans (human flea)		*
Ornithodorus macmillani (galah tick)		*
Haemaphysalis longicornis (cattle tick)	*	

from Kawau Island skins. The follicle mite, *Marsupiopus trichosuri*, was found only in skins from Kawau Island and Orongorongo Valley, near Wellington. Clark *et al.* (1997) also found nymphs and larvae of the cattle tick, *Haemaphysalis longicornis*, engorged on some pelts from Northland. Recently, an unidentified lung mite was found in lung washings from possums (Heath & F. Aldwell, unpubl. data).

Possums infested with fur mites often show skin irritation and fur loss from the lower back and rump. In Australia, lumbro-sacral dermatitis ("rumpiness") has been attributed to mites (Presidente 1978), although Clark *et al.* (1997) could not confirm this in New Zealand possums. Heavy mite burdens are generally found on possums in poor condition or in high-density populations where sharing of nest sites is common (Pracy 1975). Mites occasionally cause allergic reactions in possum trappers (Clarke 1986).

The atopolemid fur mites are specialised for clasping the fur, using their two front legs to hold hairs into a gutter on their sternum (Clark 1995a). These fur mites are probably best regarded as epizoic rather than parasitic (Clark *et al.* 1997), although *A. papilio* may, by eating surface fur surface scales, cause fur weakness and breakage. By contrast, the follicle mite, *M. trichosuri*, penetrates into the hair follicles and surrounding tissues. The mesostigmatid mite, *T. crassipes*, appears to take blood from its host and to be responsible for damage to the skin and fur loss (Clark 1995a, b).

There are no records of lice from brushtail possums, or the closely related mountain brushtail possum (*T. caninus*), in either Australia or New Zealand (Viggers & Spratt 1995; Clark *et al.* 1997).

Overall, possums in New Zealand have a much smaller ectoparasite fauna than they do in Australia (Table 7.1) because New Zealand lacks more of the species of mites that spend part of their life cycle off the host than are found on Australian possums. Other notable absences from possums in New Zealand, fleas, ticks (other than those from livestock) and chiggers (Viggers & Spratt 1995; Clark *et al.* 1997), are also species with such life cycles.

Endoparasites

Possums in New Zealand also have a very limited endoparasite fauna compared to Australia (Heath *et al.* 1998; Obendorf *et al.* 1998), comprising only a specific tapeworm, *Bertiella trichosuri*, two specific nematodes that dwell in the small intestine, *Parastrongyloides trichosuri* and *Paraustrostrongylus trichosuri*, and a specific coccidian, *Eimeria* sp. (Tables 7.2, 7.3). They may also be infected with the protozoans *Giardia intestinalis, Cryptosporidium parvum*, and *Toxoplasma gondii*, liver fluke (*Fasciola hepatica*), and a range of intestinal, trichostrongyloid nematodes, *Trichostrongylus colubriformis, T. axei, T. vitrinus* and *T. retortaeformis* from livestock and rabbits (Table 7.2). *Bertiella trichosuri* is the only helminth endoparasite with an intermediate host as part of its life cycle that occurs in possums in New Zealand.

The coccidian *Eimeria* sp. is the only parasite reported from all surveys, and has been found to be more common in the North Island (Table 7.3). A species of *Eimeria* occasionally causes diarrhoea in captive possums, but is generally not regarded as being pathogenic (Presidente 1984). Infection occurs at an early age, often in the pouch (Presidente 1982). *Cryptosporidium parvum* infection has been reported from wild (Chilvers *et al.* 1998) and captive possums (Cooke 1998), with clinical disease and diarrhoea recorded in captivity. Possums are thus a potential source of human infection. *Giardia intestinalis* is also widespread among possums in New Zealand (Marino *et al.* 1992; Chilvers *et al.* 1998), but it is unclear at present whether the *Giardia* species carried by possums poses a risk to people or other animals. Infection with *T. gondii* has occasionally been

Table 7.2
Endoparasites and diseases occurring naturally in free-living possums in New Zealand, excluding accidental occurrences.

Endoparasites

Protozoa:	Eimeridae:	*Eimeria* sp.
		Toxoplasma gondii
		Giardia intestinalis
		Cryptosporidium parvum
Cestoda:	Anoplocephalidae:	*Bertiella trichosuri*
Trematoda:	Fasciolidae:	*Fasciola hepatica*
Nematoda:	Strongyloididae:	*Parastrongyloides trichosuri*
	Nicollinidae:	*Paraustrostrongylus trichosuri*
	Trichostrongylidae:	*Trichostrongylus axei, T. colubriformis, T. retortaeformis, T. vitrinus*

Bacterial infections

Leptospira interrogans serovars *balcanica, ballum, copenhageni, pomona, tarassovi, Mycobacterium bovis, M. vaccae, M. avium , M. fortuitum*

Fungal infections

Emmonsia (*Chrysosporium*) *crescens*
Trichophyton mentagrophytes , *T. ajelloi, T. terrestre*
Microsporum cookeri

Viral infections

Whataroa virus
Wobbly possum disease virus
Unidentified viral enteritis
Possum papillomavirus
Uncharacterised possum adenovirus, herpesvirus, coronavirus, coronavirus-like particles, retrovirus

Sources: Presidente 1984; O'Keefe *et al.* 1997; Wickstrom & O'Keefe 1998; Chilvers *et al.* 1998; Cooke 1998; Heath *et al.* 1998; Meers *et al.* 1998; W. A. G. Charleston, Massey University, Palmerston North, unpubl. data.

Table 7.3
Summary of results of possum parasite and leptospirosis surveys 1993–1997.

		Prevalence (%) of infection					
Location	Sample size	*Leptospira balcanica**	*Eimeria* sp.	*Bertiella trichosuri*	*Trichostrongylus* spp.	*Parastrongyloides trichosuri*	*Paraustrostrongylus trichosuri*
North Island							
Northland	206	3.3	21.8	4.4	30.0	60.2	0
Hawke's Bay	158	43.3	10.1	2.9	32.4	15.8	12.8
Wanganui	209	46.7	23.0	0.5	58.9	55.0	2.4
Palmerston North	176	66.7	4.0	2.4	24.0	54.9	70.3
Paraparaumu	53	33.3	20.0	0	23.1	51.9	53.8
Castlepoint	269	70.0	21.4	5.7	33.0	63.2	21.2
Orongorongo Valley	179	46.7	5.2	3.0	37.9	60.4	51.1
South Island							
Nelson	211	0	9.5	24.3	19.4	0	0
Hokitika	159	0	6.3	0.6	26.6	0	0
Banks Peninsula	155	0	6.1	27.5	55.3	0	0
Dunedin	58	0	10.3	0	35.7	0	0
nr Invercargill	236	0	3.8	0.6	34.3	14.0	1.4
Offshore Islands							
Kawau	158	0	16.7	5.2	0	15.5	0
Stewart	194	0	4.6	0	0	0	0
Chatham	214	0	10.9	3.6	16.1	0	0

Sources: Horner *et al.* 1996; Stankiewicz *et al.* 1996a, 1997a, b, 1998; Heath *et al.* 1998.
*n = 30 for *Leptospira* analysis.

reported, and may result in mild to severe encephalitis (Cooke 1998). Since it is capable of infecting virtually all species of mammals (Charleston 1994), the few records of infection in possums in New Zealand are surprising. It may be that possums only rarely contact buried cat faeces, and there are few data on parasites and diseases of possums in urban areas where contact with domestic animals may be more common.

Previous surveys found that the tapeworm *Bertiella trichosuri* infested 19–31% of possums in the North Island, and 0–41% in the South Island, with little difference between forest and pasture feeding populations, contrary to Pracy & Kean's (1968) original suggestion (Khalil 1970; Presidente 1984). It was found in possums throughout the western North Island from Wellington and the Wairarapa to Lake Taupo, but not in those in the forests of the Urewera, Aorangi, or northern Rimutaka ranges (Pracy 1975). In the South Island, it was found in the Golden Bay area, near Westport, and north Canterbury, but not in central Westland or Stewart Island (Warburton 1983). More recent surveys (Table 7.3) have confirmed the widespread but patchy distribution and generally low prevalence of the possum tapeworm, including its occurrence on two of three offshore islands. Its apparent absence from some samples may reflect its marked seasonal occurrence, with prevalence highest in winter (Heath *et al.* 1998 and unpubl. data). Infection with the tapeworm may be mildly pathogenic. Infected female possums from Claverly, North Canterbury, had lower mesenteric fat reserves and slightly reduced fecundity compared with non-infested females (Warburton 1983), although this effect was not observed in possums from Nelson (Heath *et al.* 1998). The life cycle of *B. trichosuri* has not been fully determined. Although oribatid mites have been suggested as its intermediate host, experimental attempts to demonstrate this have been unsuccessful (Clark & Heath, unpubl. data), although Spratt (in Viggers & Spratt 1995) observed development from ingested eggs to the cysticercoid stage in artificially infected soil-dwelling oribatids.

Liver fluke, *Fasciola hepatica*, is found in sheep and cattle in both the North and South islands. Infested possums have been recorded in Taranaki, where up to 70% may carry flukes (Presidente 1984) and in Westland (Cox 1985). However, no flukes were found in the recent survey at Wanganui, Taranaki (Stankiewicz *et al.* 1996a). Fasciolosis is probably the most pathogenic of parasitic infections encountered by possums, causing death within 5–6 months (Spratt & Presidente 1981; Presidente 1984).

The possum-specific nematode, *Parastrongyloides trichosuri*, appears to be common throughout the North Island, being found in all surveys there including Kawau Island (Table 7.3). Burtton (1975) also found it to be common in possums from around Taupo, central North Island. However, in the South Island it has been found only in possums from the Longwood Range, near Invercargill, and Pebbly Hill, Southland (Table 7.3 and unpubl. data). It was not found in possums on either Stewart Island or Chatham Island. In Australia it is found commonly in possums from both mainland Australia and Tasmania (Viggers & Spratt 1995). *Parastrongyloides trichosuri* has the ability to exist in either parasitic or free-living forms (Stankiewicz 1996; Stankiewicz *et al.* 1996a).

Paraustrostrongylus trichosuri, the other possum-specific nematode, although also common throughout the North Island, appears to be more patchily distributed and generally less prevalent than *Parastrongyloides trichosuri* (Table 7.3). It too has been found only in the Longwood Range in the South Island, and it was not present in possums on any of the offshore islands sampled (Table 7.3). Presidente (1984) reported it as being common in possums in Australia.

Despite the very limited distribution of the two possum-specific nematodes in the South Island, experimental infection of possums from the South Island with each parasite showed that they were highly susceptible to such infection (Heath *et al.* 1998). The reasons for the limited distribution of these nematodes in the South Island are currently being investigated. Seasonal and age-related prevalence data (Heath *et al.* 1998) show that in an area where parasites were common, most possums were exposed to infection with *Parastrongyloides trichosuri* and *Paraustronglyus trichosuri* before the age of one year. Thereafter, at any one time, about 30% of possums had no worms, although most had serum antibodies to the parasites, indicating that they had been infected and had become immune. Development of immunity to infection has also been demonstrated in laboratory studies (Heath *et al.* 1998).

Trichostrongylus colubriformis is a sheep intestinal nematode often found in possums in both forest and farmland habitats, but more commonly in the latter (Stankiewicz *et al.* 1996a,b). Unlike the two possum-specific nematodes, it is found commonly in both North and South island possums, although it was absent from two of the three offshore islands (Table 7.3), probably because neither Kawau nor Stewart islands have had sheep on them recently

(Stankiewicz *et al.* 1997a). Other nematodes recorded include *T. axei*, *T. retortaeformis*, and *T. vitrinus*, and in some areas they may be more common than *T. colubriformis* (Stankiewicz *et al.* 1996b,1997a,b).

Possums with higher total numbers of nematodes had lower indices of body fat and condition (Stankiewicz *et al.* 1996a; Heath *et al.* 1998), but it was not clear whether parasites contribute to poor body condition or whether poor body condition promotes parasitism.

A number of possum parasites are notably absent from New Zealand. *Sarcocystis* spp. (Protozoa) were not recorded in any of the recent surveys (Stankiewicz *et al* 1996a; 1997 a,b; 1998) or earlier studies (Smith & Munday 1965; Munday 1969; Charleston 1994). *Adelonema trichosuri* (Nematoda), although widely distributed and common in possums in Australia (Obendorf *et al.* 1998), has never been reported in New Zealand. No microfilariae (larval filarioid nematodes) were seen in the peripheral blood, blood smears, or histological sections of tissues from possums in New Zealand although *Breinlia trichosuri* and *Spratia venacavincola* have been recorded from possums in Australia (Viggers & Spratt 1995). Their life cycle requires a blood-feeding intermediate host, and a suitable one may not occur in New Zealand. *Toxocara canis* (Nematoda) has also never been recorded from possums in New Zealand, although they are known to be susceptible to this infection (Sweatman 1962). *Filostrongylus tridendriticus* (Nematoda) occurs in the lungs of possums from Tasmania (Spratt 1984). Many of the original possum introductions were from Tasmania, but this parasite was either not introduced or has not persisted. *Marsupostrongylus minesi, M. pseudominesi* and *M. longilarvatus* (Nematoda) are present in the lungs of brushtail possums in Australia (Spratt 1979, 1984; Presidente 1984) and *M. minesi* infection may cause clinical illness (Bellamy 1993). *M. longilarvatus* has a wide range of marsupial host species, but its lack of host specificity does not seem to have helped it to establish in New Zealand possums. This may be because *Marsupostrongylus* spp. and *F. tridendriticus* probably have indirect life cycles requiring an obligatory period of development in gastropod intermediate hosts (D. Spratt, CSIRO Wildlife & Ecology, Canberra, pers. comm.).

Bacterial and fungal diseases

Periodontal disease, similar to that of sheep and kangaroos, affects possums in both forest and farmland. The disease usually affects the lower jaw. In the worst cases, the bone may be eroded until only pillars of bone support the teeth, or teeth work loose and fall out. Generally, fewer than 10% of skulls are affected (Cowan & A. J. White unpubl. data).

Possums carry leptospirosis and bovine tuberculosis (Table 7.2), both infective to man and livestock. Bovine tuberculosis is dealt with in detail in Chapter 8. Possums are considered to be the maintenance host for *Leptospira interrogans* serovar *balcanica* (Hathaway *et al.* 1978). Infection is common and widespread in possums in the central and southern North Island (up to 80% of adults), but appears to be absent from possums in the South Island, perhaps because most of the original releases there came from *balcanica*-free Tasmania (Cowan 1990; Horner *et al.* 1996; Heath *et al.* 1998). Sexually immature possums are rarely infected and the prevalence of infection does not appear to differ between the sexes, but increases with age (Cowan *et al.* 1991). Transmission is thought to be primarily by social contact rather than through contact with environmental contamination (Day *et al.* 1997, 1998). The effects of *balcanica* infection on possums appear slight — minor histological changes in the kidneys and a transient loss of body weight (Presidente 1984; Cowan *et al.* 1991; Day *et al.* 1997). Infection of possums with other serovars (*ballum, copenhageni, pomona, tarassovi*) is rare, and *balcanica* apparently is not readily transmitted to humans, even high-risk groups such as possum trappers (D. K. Blackmore, ex Massey University, Palmerston North, unpubl. data). This may be because the New Zealand isolates have been shown, by restriction enzyme analysis, to differ in genotype from the original European isolates (Robinson *et al.* 1982).

Possums also carry fungal infections caused by *Emmonsia (Chrysosporium) crescens, Trichophyton mentagrophytes, T. ajelloi, T. terrestre*, and *Microsporum cookeri.* Pulmonary infection with *E. crescens* is widespread, but does not appear to be of clinical significance (Cooke 1998). Cooke (1998) suggests that records of *Microsporum* spp. are probably from soil or other contamination rather than associated with skin lesions.

A number of other bacterial infections have been reported from ill or dead captive and hand-reared possums, including *Fusobacterium necrophorum, Candida* spp., *Salmonella typhimurium, Clostridium piliforme, Pseudomonas aeruginosa*, and *Yersinia pseudotuberculosis* (Presidente 1984; Cooke 1998).

Viral infections

Until recent surveys related to the search for possible biological control agents (Chapter 24), little was known about viruses of possums. For many years the only virus known to infect possums in New Zealand was Whataroa virus, an arbovirus from the Sinbis group isolated from mosquitoes (Dempster 1964). Twenty-seven percent of possums near Okarito Lagoon in Westland had antibodies to the virus. Experimentally infected possums developed neutralising antibodies, but no virus was found in blood samples from them nor in those from naturally infected possums. Possums are therefore unlikely to be the main hosts for the virus, which circulates in the wild bird populations of Westland, where two species of native mosquitoes serve as vectors (Miles 1973).

More recently, adenovirus precipitating antibodies were detected in possum sera (Rice *et al.* 1991), and subsequent electron microscopy of possum intestinal contents revealed a range of herpesviruses, adenoviruses, coronaviruses and coronavirus-like particles (Rice & Wilks 1996). Studies on *in vitro* propagation and molecular characterisation of possum adenovirus are currently underway. With the use of a range of molecular techniques, retroviral infection of possums has also been detected (Baillie & Wilkins 1998; Meers *et al.* 1998). A papillomavirus was identified from a lesion on the tail of a possum and was subsequently shown by DNA sequencing to be a new species (Meers *et al.* 1998).

A new neurological disease of possums, wobbly possum disease, first detected in a captive colony (Mackintosh *et al.* 1995) was shown to have a viral aetiology (O'Keefe *et al.* 1997). A similar disease has subsequently been found in wild populations (Meers *et al.* 1998). The virus is an RNA, enveloped, heat-sensitive virus about 60 nm in size (O'Keefe & Wickstrom 1998). Sera from infected possums cross-react with antibodies to Borna virus, but there are differences in conditions for propagation (J. O'Keefe, National Centre for Disease Investigation, Upper Hutt, pers. comm.). Affected possums become docile and lethargic, with gradual incoordination, difficulty in climbing, a rolling gait, and blindness. Histology revealed widespread damage to the liver, kidneys, and central nervous system and less severe damage to a variety of other tissues, including skeletal and cardiac muscle (O'Keefe *et al.* 1997). Serum globulin and albumin concentrations changed significantly (O'Keefe *et al.* 1997). In-contact animals, but not animals housed adjacent to infected possums, became infected, suggesting natural transmission does not occur by aerosol (Meers *et al.* 1998). Possums with acute or subacute meningo-encephalitis have been recorded in Australia but the condition appears to differ from wobbly possum syndrome (Hartley 1993; Obendorf *et al.* 1998).

An outbreak of enteritis in another captive possum colony that caused 100% morbidity and 20–65% mortality also appears to have a viral aetiology (Wickstrom & O'Keefe 1997). Histological examination of gut sections revealed an enteritis with similar features to that seen with coronavirus or rotavirus infection in neonates of other species. A cytopathic virus has been isolated and is under investigation (O'Keefe & Wickstrom 1998).

Experimental infections and other pathological conditions

Presidente (1984) lists a range of viral, rickettsial, and chlamydial infections to which possums have been exposed experimentally or for which free-living possums have been screened for antibodies. Of potential concern was the recovery of live virus from possums infected with human influenza virus types A and B, and foot and mouth disease.

Cooke (1993) lists a wide range of naturally occurring lesions encountered during routine autopsies of wild possums.

Conclusions

Possums in New Zealand have few predators of note other than people and motor vehicles, unlike in their native Australia where foxes significantly reduce possum numbers. Possums in New Zealand are infected with a smaller range of parasites and diseases than those in Australia, perhaps reflecting some combination of the small number of animals originally brought from Australia, founder effects, absence of other terrestrial mammals, absence of suitable intermediate hosts, climate and habitat differences. Bovine Tb is the only disease of possums of current economic significance (see Chapter 8). Risks to people from parasites and diseases carried by possums are slight, except for bovine Tb, although the potential of wobbly possum disease virus, *Giardia* sp. and *Cryptosporidium* sp. to infect people is still to be resolved. There is little information about how parasitism and disease might influence the population dynamics of possums in either New Zealand or Australia. It has proved extremely difficult to eliminate natural parasitic infection to allow such an evaluation (Ralston *et al.* 1998; Viggers *et al.* 1998).

Summary

- Possums have few predators of significance in New Zealand other than people and feral cats.
- Possums in New Zealand have many fewer species of ectoparasites and endoparasites than in their native Australia.
- The parasites not present in New Zealand appear to be those that do not complete their life cycle on one individual host, those with indirect life cycles, or those for which possums are not the primary host.
- The effects of parasite infection on the well-being of free-living possums are not well known, but in most cases, they appear to be slight.
- Possums in New Zealand are infected with a range of bacterial, fungal, and viral diseases.
- Bovine tuberculosis is the only disease carried by possums, of current economic significance.
- Several new possum viruses have been discovered recently, and are currently being characterised.
- Studies are needed to determine what contribution predation, parasitism, and disease make to the population dynamics of possums.

References

Alterio, N. 1996: Secondary poisoning of stoats (*Mustela erminea*), feral ferrets (*Mustela furo*), and feral house cats (*Felis catus*) by the anticoagulant poison, brodifacoum. *New Zealand Journal of Zoology 23*: 331–338.

Anonymous 1997: Brushtail possum. *In:* Western Shield progress report, November 1997. Department of Conservation and Land Management, WA, Australia. P.6.

Baillie, G. J.; Wilkins, R. J. 1998: Retroviruses in the common brushtail possum (*Trichosurus vulpecula*). *In:* Biological control of possums. *The Royal Society of New Zealand Miscellaneous Series 45*: 20–22.

Bellamy, T. 1993: Clinical case reports — brushtail possums. *In:* Proceedings of the Wildlife Disease Association Australasian Section Annual General Meeting, North Stradbroke Island. P.38.

Bowie, J.; Bennett, W. 1983: Parasites from possums in Waitahuna forest. *New Zealand Veterinary Journal* 31: 163.

Brockie, R. E.; Fairweather, A. A. C.; Ward, G. D.; Porter, R. E. R. 1987: Field biology of Hawke's Bay farmland possums *Trichosurus vulpecula.* Lower Hutt, Department of Scientific and Industrial Research *Ecology Division Report 10* (unpublished). 40 p.

Burtton, R. A. 1975: An initial study on the gastrointestinal parasites of the brush-tailed opossum *Trichosurus vulpecula* (Kerr 1792). Unpublished BSc (Hons) thesis, Victoria University of Wellington, Wellington, New Zealand.

Charleston, W. A. G. 1994: *Toxoplasma* and other protozoan infections of economic importance in New Zealand. *New Zealand Journal of Zoology 21*: 67–81.

Chilvers, B. L.; Cowan, P. E.; Waddington, D. C.; Kelly, P. J.; Brown, T. J. 1998: The prevalence of infection of *Giardia* spp. and *Cryptosporidium* spp. in wild animals on farmland, southeastern North Island, New Zealand. *International Journal of Environmental Health Research 8*: 59–64.

Clark, J. M. 1995a: Morphological and life-cycle aspects of the parasitic mite, *Trichosurolaelaps crassipes* Womersley, 1956 of trichosurid possums. *New Zealand Veterinary Journal 43*: 209–214.

Clark, J. M. 1995b: The diet of the parasitic mite *Trichosurolaelaps crassipes* Womersley, 1956 and its potential to act as a disease vector. *New Zealand Veterinary Journal 43*: 215–218.

Clark, J. M.; Heath, D. D.; Stankiewicz, M. 1997: The ectoparasites of brushtail possum *Trichosurus vulpecula* in New Zealand. *New Zealand Journal of Zoology 24*: 199–204.

Clarke, P. S. 1986: Allergy to brush-tailed possums. *Medical Journal of Australia 145*: 658–659.

Clout, M. N.; Barlow, N. D. 1982: Exploitation of brushtail possum populations in theory and practice. *New Zealand Journal of Ecology 5*: 29–35.

Cooke, M. M. 1993: A selection of naturally occurring pathological phenomena in the brush-tail possum, *Trichosurus vulpecula. New Zealand Veterinary Journal 41*: 44.

Cooke, M. 1998: Infectious diseases of possums in New Zealand. *Surveillance 25(2)*: 10–12.

Cowan, P. E. 1990: Brushtail possum. *In:* King, C. M. *ed.* The handbook of New Zealand mammals. Auckland, Oxford University Press. Pp. 68–98.

Cowan, P. E.; Blackmore, D. K.; Marshall, R. B. 1991: Leptospiral infection in common brushtail possums (*Trichosurus vulpecula*) from lowland podocarp/mixed hardwood forest in New Zealand. *Wildlife Research 18*: 719–727.

Cowan, P. E.; Brockie, R. E.; Ward, G. D.; Efford, M. G. 1996: Long-distance movements of juvenile brushtail possums (*Trichosurus vulpecula*) on farmland, Hawke's Bay, New Zealand. *Wildlife Research 23*: 237–244.

Cowan, P. E.; Brockie, R. E.; Smith, R. N.; Hearfield, M. E. 1997: Dispersal of juvenile brushtail possums, *Trichosurus vulpecula*, after a control operation. *Wildlife Research 24*: 279–288.

Cox, B. T. 1985: Liver fluke in possums. *Surveillance 12(2)*: 16.

Day, T. D.; Waas, J. R.; O'Connor, C. E. 1997: Effects of experimental infection with *Leptospira interrogans* serovar *balcanica* on the health of brushtail possums (*Trichosurus vulpecula*). *New Zealand Veterinary Journal 45*: 4–7.

Day, T. D.; O'Connor, C. E.; Waas, J. R.; Pearson, A. J.; Matthews, L. R. 1998: Transmission of *Leptospira interrogans* serovar *balcanica* infection among socially housed brushtail possums in New Zealand. *Journal of Wildlife Diseases 34*: 576–581.

Dempster, A. G. 1964: *Trichosurus vulpecula* and *Rattus norvegicus* in the epidemiology of two arboviruses. Unpublished BMedSc thesis, University of Otago, Dunedin, New Zealand.

Fitzgerald, B. M.; Karl, B. J. 1979: Foods of feral house cats (*Felis catus* L.) in forest of the Orongorongo Valley, Wellington. *New Zealand Journal of Zoology 6*: 107–126.

Hartley, W. J. 1993: Central nervous system disorders in the brush-tail possum in Eastern Australia. *New Zealand Veterinary Journal 41*: 44–45.

Hathaway, S. C.; Blackmore, D. K.; Marshall, R. B. 1978: The serologic and cultural prevalence of *Leptospira interrogans* serovar *balcanica* in possums (*Trichosurus vulpecula*) in New Zealand. *Journal of Wildlife Diseases 14*: 345–350.

Heath, D.; Cowan, P.; Stankiewicz, M.; Clark, J.; Horner, G.; Tempero, J.; Jowett, G.; Flanagan, J.; Shubber, A.; Street, L.; McElrea, G.; Chilvers, L.; Newton-Howse, J.; Jowett, J.; Morrison, L. 1998: Possum biological control - parasites and bacteria. *In:* Biological control of possums. *The Royal Society of New Zealand Miscellaneous Series 45:* 13–19.

Horner, G. W.; Heath, D. D.; Cowan, P. E. 1996: Distribution of leptospirosis in possums from New Zealand and its offshore islands. *New Zealand Veterinary Journal 44*: 161.

Khalil, L. F. 1970: *Bertiella trichosuri* n. sp. from the brush-tail opossum, *Trichosurus vulpecula* (Kerr) from New Zealand. *Zoologischer Anzeiger 185*: 442–450.

King, C. M. *ed.* 1990: The handbook of New Zealand mammals. Auckland, Oxford University Press. 598 p.

King, C. M.; Moody, J. E. 1982: The biology of the stoat (*Mustela erminea*) in the National Parks of New Zealand. I–VI. *New Zealand Journal of Zoology 9*: 49–144.

Mackintosh, C. G.; Crawford, J. L.; Thomson, E. G.; McLeod, B. J.; Gill, J. M.; O'Keefe, J. S. 1995: A newly discovered disease of the brushtail possum: wobbly possum syndrome. *New Zealand Veterinary Journal 43*: 126.

Marino, M. R.; Brown, T. J.; Waddington, D. C.; Brockie, R. E.; Kelly, P. J. 1992: *Giardia intestinalis* in North Island possums, house mice and ship rats. *New Zealand Veterinary Journal 40*: 24–27.

Meers, J.; Perrott, M; Rice, M.; Wilks, C. 1998: The detection and isolation of viruses from possums in New Zealand. *In:* Biological control of possums. *The Royal Society of New Zealand Miscellaneous Series 45*: 25–28.

Miles, J. A. R. 1973: The ecology of Whataroa virus, an alpha virus, in South Westland, New Zealand. *Journal of Hygiene (London) 71*: 701–713.

Munday B. L. 1969: The epidemiology of toxoplasmosis with particular reference to the Tasmanian environment. Unpublished MVSc thesis, University of Melbourne, Melbourne, Australia.

Obendorf, D.; Spratt, D.; Beveridge, I.; Presidente, P.; Coman, B. 1998: Parasites and diseases in Australian brushtail possums, *Trichosurus* spp. *In:* Biological control of possums. *The Royal Society of New Zealand Miscellaneous Series 45*: 6–9.

O'Keefe, J.; Wickstrom, M. 1998: Viruses of the brushtail possum in New Zealand. *In:* Biological control of possums. *The Royal Society of New Zealand Miscellaneous Series 45*: 23–24.

O'Keefe, J. S.; Stanislawek, W. L.; Heath, D. D. 1997: Pathological studies of wobbly possum disease in New Zealand brushtail possums (*Trichosurus vulpecula*). *Veterinary Record 141*: 226–229.

Parkes, J. P.; Nugent, G.; Warburton, B. 1996: Commercial exploitation as a pest control tool for introduced mammals in New Zealand. *Wildlife Biology 2*: 171–177.

Pracy, L. T. 1975: Opossums (1). *New Zealand's Nature Heritage 3*: 873–882.

Pracy, L. T.; Kean, R. I. 1968: Tapeworms found in opossums. *New Zealand Journal of Agriculture 117(5)*: 21–22.

Presidente, P. J. A. 1978: Diseases seen in free-ranging marsupials and those held in captivity. *In:* Fauna Part B. *Post-graduate Committee on Veterinary Science Proceedings No. 36.* University of Sydney, Australia. Pp. 457–471.

Presidente, P. J. A. 1982: Common brushtail possum *Trichosurus vulpecula*: maintenance in captivity, blood values, diseases and parasites. *In:* Evans, D.D. *ed.* The management of Australian mammals in captivity. Melbourne, Zoological Board of Victoria. Pp. 55–66.

Presidente, P. J. A. 1984: Parasites and diseases of brushtail possums (*Trichosurus* spp.): occurrence and significance. *In:* Smith, A. P.; Hume, I. D. *ed.* Possums and gliders. Chipping Norton, NSW, Surrey Beatty in assoc. with the Australian Mammal Society. Pp. 171–90.

Ralston, M. J.; Cowan, P. E.; Stankiewicz, M.; Heath, D. D. 1998: A field trial on the effect of parasites on possum fecundity. New Zealand Society for Parasitology Annual Meeting No. 27, Lincoln University, 3–4 September 1998. P. 22 (abstract).

Rice, M.; Wilks, C. R. 1996: Virus and virus-like particles observed in the intestinal contents of the possum, *Trichosurus vulpecula. Archives of Virology 141*: 945–950.

Rice, M.; Wilks, C. R.; Pfeiffer, D.; Jackson, R. 1991: Adenovirus precipitating antibodies in the sera of brush-tailed possums in New Zealand. *New Zealand Veterinary Journal 39*: 58–60.

Robinson, A. J.; Ramadass, P.; Lee, A.; Marshall, R. B. 1982: Differentiation of subtypes within *Leptospira interrogans* serovars *hardjo, balcanica* and *tarassovi*, by bacterial restriction endonuclease DNA analysis (BRENDA). *Journal of Medical Microbiology 15*: 331–338.

Smith, G. P.; Ragg, J. R.; Moller, H.; Waldrup, K. A. 1995: Diet of feral ferrets (*Mustela furo*) from pastoral habitats in Otago and Southland, New Zealand. *New Zealand Journal of Zoology 22*: 363–369.

Smith I. D.; Munday B. L. 1965: Observations of the incidence of *Toxoplasma gondii* in native and introduced feral fauna in eastern Australia. *Australian Veterinary Journal 41*: 285–286.

Spratt, D. M. 1979: A taxonomic revision of the lungworms (Nematoda: Metastrongyloidea) from Australian marsupials. *Australian Journal of Zoology Supplementary Series 67*: 1–45.

Spratt, D. M. 1984. Further studies of lung parasites (Nematoda) from Australian marsupials. *Australian Journal of Zoology 32*: 283–310.

Spratt, D. M.: Presidente, P. J. A. 1981: Prevalence of *Fasciola hepatica* infection in native mammals in southeastern Australia. *Australian Journal of Experimental Biology and Medical Science 59*: 713–721.

Stankiewicz, M. 1996: Observations on the biology of free-living stages of *Parastrongyloides trichosuri* (Nematoda, Rhabditoidea). *Acta Parasitologica 41*: 38–42.

Stankiewicz, M.; Cowan, P.; Jowett, G. H.; Clark, J. M.; Jowett, J.; Roberts, M. G.; Charleston, W. A. G.; Heath, D. D. 1996a: Internal and external parasites of possums *(Trichosurus vulpecula)* from forest and farmland, Wanganui, New Zealand. *New Zealand Journal of Zoology 23*: 345–353.

Stankiewicz, M.; McMurtry, L. W.; Hadas, E.; Heath, D. D.; Cowan, P. E. 1996b: *Trichostrongylus colubriformis, T. vitrinus* and *T. retortaeformis* infection in New Zealand possums. *New Zealand Veterinary Journal 44*: 201–202.

Stankiewicz, M.; Heath, D. D.; Cowan, P. E. 1997a: Internal parasites of possums (*Trichosurus vulpecula*) from Kawau Island, Chatham Island and Stewart Island. *New Zealand Veterinary Journal 45*: 247–250.

Stankiewicz, M.; Cowan, P. E.; Heath, D. D. 1997b: Endoparasites of brushtail possums (*Trichosurus vulpecula*) from the South Island, New Zealand. *New Zealand Veterinary Journal 45*: 257–260.

Stankiewicz, M.; Cowan, P. E.; Heath, D. D. 1998: Endoparasites of possums from selected areas of North Island, New Zealand. *New Zealand Journal of Zoology 25*: 91–97.

Statham, M.; Statham, H. L. 1997: Movements and habits of brushtail possums (*Trichosurus vulpecula* Kerr) in an urban area. *Wildlife Research 24*: 715–726.

Sweatman, G. K. 1962: Parasitic mites of non-domesticated animals in New Zealand. *New Zealand Entomologist 3*: 15–23.

Viggers, K. L.; Spratt, D. M. 1995: The parasites recorded from *Trichosurus* species (Marsupialia: Phalangeridae). *Wildlife Research 22*: 311–332.

Viggers, K. L.; Lindenmayer, D. B.; Cunningham, R. B.; Donnelly, C. F. 1998: The effects of parasites on a wild population of the mountain brushtail possum (*Trichosurus caninus*) in south-eastern Australia. *International Journal for Parasitology 28*: 747–755.

Warburton, B. 1983. Infestations of *Bertiella trichosuri* (Cestoda: Anoplocephalidae) in possums from Claverly, North Canterbury. *New Zealand Journal of Zoology 10*: 221–224.

Wickstrom M. L.; O'Keefe J. S. 1997: Investigation into the cause of an infectious disease outbreak in captive brushtail possums: confirmation of viral etiology. Landcare Research Contract Report LC9798/71 (unpublished) 13 p.

CHAPTER EIGHT

Possums as a Reservoir of Bovine Tb

Jim Coleman and Peter Caley

Bovine tuberculosis (Tb) is a worldwide disease of great antiquity caused by *Mycobacterium bovis.* Cattle (*Bos* spp.) are its natural host. The disease is widespread in temperate areas where, until the early 20th century, it occurred mostly in dairy herds around European cities (Alhaji 1976). *Mycobacterium bovis* belongs to the complex of bacteria that includes *M. tuberculosis* (human Tb).

The host range of *M. bovis* seems unlimited in mammals (O'Reilly & Daborn 1995). In cattle, Tb is a debilitating, chronic, often fatal disease, readily transmitted to other cattle. Humans get infected with bovine Tb mostly by drinking unpasteurised milk from infected cows or by direct contact with infected animals or carcasses. Pasteurisation of milk and inspection of meat at abattoirs are highly effective measures for preventing humans being exposed to infection. However, *M. bovis* infection still causes 100 000 to 200 000 human deaths worldwide annually (Collins *et al.* 1995), with most occurring in developing countries that lack effective disease control programmes in livestock. Human-to-human spread of *M. bovis* has been postulated but it appears likely to be a very rare event (Collins & Grange 1987). Of all Tb infections in humans, the proportion caused by *M. bovis* is typically low (see Chapter 20).

The possum is considered the primary wildlife reservoir of Tb for farmed cattle and deer in New Zealand, and continuing transmission of Tb from tuberculous possums (Fig. 8.1) to livestock is the single greatest barrier to its eradication from New Zealand livestock.

This chapter describes the role of possums as a reservoir host of Tb and their role in infecting other wildlife and livestock.

Tb in New Zealand livestock

There is little doubt that Tb was introduced into New Zealand via cattle in the 1830s–1840s, as happened during European colonisation in many countries throughout the world (Francis 1947). Tuberculosis in deer was not confirmed until 1970, though there is a report of what appears to be a tubercular wild deer in 1954 (Beatson 1985). Tuberculosis was formally recognised in the 1940s as a serious problem in New Zealand dairy herds that reduced productivity and put humans at risk (Jamieson 1960). In 1945 a voluntary testing scheme was introduced for "town milk" supply herds. This involved a diagnostic caudal-fold skin test using tuberculin (a purified protein derivative of *M. bovis*) to screen cattle for the disease. This scheme became compulsory for all town supply herds in 1956, and for all dairy herds in 1961 (O'Neil & Pharo 1995).

During the 1960s, the cattle industry and the government recognised the growing need to meet minimum animal health standards if access to export markets for meat and dairy products was to be maintained. To this end, a national Tb eradication scheme was started on a voluntary basis in 1968 and became compulsory in 1970. In contrast, a voluntary eradication scheme for Tb in deer was introduced in 1985, and became compulsory in 1990. The initial Tb eradication campaign was a traditional one. Herd testing and the slaughter of animals reacting to the test ("reactors") rapidly eradicated tuberculosis from cattle herds provided that reinfection from other sources was negligible. Initial progress in the national Tb scheme for cattle was promising for most dairy and beef herds. It was, therefore, confidently expected that within a short period the disease would be eradicated from cattle. However, in the late 1960s and early 1970s, it had become apparent that on the West Coast of the South Island, central North Island, and south-east North Island in particular, the scheme was failing to eradicate Tb from cattle. Gradually an association between "problem" herds and tuberculous possums unfolded (see Chapter 20), which persists to the present day. Tuberculosis testing remains the primary check on the health status of herds and the means of identifying and removing within-herd sources of infection (see Chapter 20).

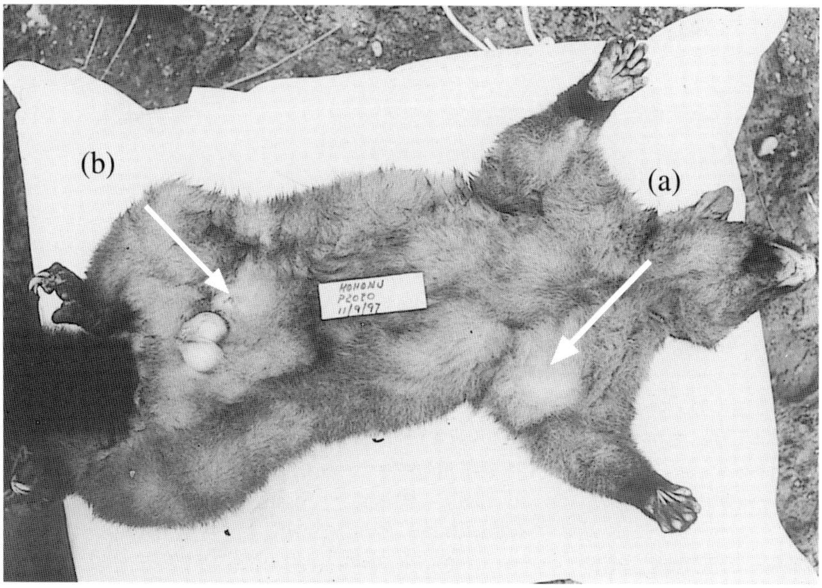

Fig. 8.1
Tuberculous possum with grossly enlarged superficial axillary lymph node (a) and inguinal lymph node (b).
Photo: P. Caley

Tb in possum populations

In New Zealand, Tb was first recorded from free-ranging possums in 1967 (Ekdahl *et al.* 1970), though it appears likely that possum populations were infected as early as 1955–1960 (Morris & Pfeiffer 1995). A seemingly intractable Tb "problem area" for cattle at Waimangaroa, north of Westport in Buller County, had been identified from unexplained local upsurges in the number of tuberculous cattle. The search for a likely reservoir in wild animals (those able to maintain the disease and transmit it to other host species) identified tuberculous possums coexisting with several chronically infected cattle herds. Soon after this discovery, infected possums were identified elsewhere in Buller county, and then in rapid succession in the late 1960s and early 1970s in other widely dispersed areas of New Zealand such as Inangahua and Grey counties, in Wairarapa, the Western Bays of Lake Taupo, North Canterbury, and Otago. In each instance, infected possums coexisted with tuberculous cattle.

Beginning in 1978, similar possum-related disease outbreaks were identified in farmed deer (Carter 1995), and the transmission of Tb from them to coexisting possum populations has been documented several times (Mackereth 1993; de Lisle *et al.* 1995). Indeed, tuberculous wild deer may have been the principal initiator of infection in possum populations, though the evidence from geographical patterns of possum infection suggests that establishment of infection in uninfected possum populations from another species is a rare event (Morris & Pfeiffer 1995).

Mycobacterium bovis is a notifiable organism under the Biosecurity (Notifiable Organisms) Order 1993 and, as such, the detection of its presence in possums, or in other wildlife in areas of New Zealand where it has not been identified before, must be immediately reported to the Chief Technical Officer of the Ministry of Agriculture. Control of Tb in cattle and deer herds in previously disease-free areas is reliant on such notifications.

Current geographic distribution of infected possums

Tb-infected possum populations now occupy about 23.6% or 6.24 million hectares of New Zealand (Animal Health Board 1998). They occur in five major areas, namely, the central North Island, Wairarapa, Westland, North Canterbury, and Otago (see Fig. 20.1, Chapter 20), and in at least 15 discrete areas in Southland, Banks Peninsula, Horowhenua, Manawatu, Hawke's Bay, and Auckland. When combined, these areas (termed Vector Risk Areas (VRA); Animal Health Board 1995, p. 45) contain 75% of New Zealand's infected cattle herds and reactor cattle. In 1988, Batcheler & Cowan reported that, after c. 15 years of Tb possum control, areas with infected possums in both islands of New Zealand had increased at approximately exponential rates, from isolated centres (foci) of infection through amalgamation of neighbouring centres into large areas of discontinuous Tb. Such expansion continues to the present day. One of the major reasons for this expansion appears to be the dispersal of infected juvenile possums (Cowan *et al.* 1996), and the movement (and in some cases escape) of tuberculous farmed deer (Mackereth 1993; de Lisle *et al.* 1995).

Within VRAs, tuberculous possums have been found in a wide variety of habitats and locations, often located close to extensive areas of native forest. Infected possum populations identified in the 1960s and 1970s were principally confined to the medium to high rainfall dairying areas of the West Coast and the lower and central North Island. This is no longer true — infected possum populations now occur across extensive areas of dryland beef and deer farming areas such as those found in South Canterbury and Otago, and only possum populations in Northland, Coromandel, Taranaki, East Coast – Bay of Plenty, mid-Canterbury, and Stewart Island are free of the disease. As revealed through restriction endonuclease analysis (REA), tuberculous possums may show distinct strains of *M. bovis* both within and between Tb VRAs (Collins & de Lisle 1995).

Epidemiology of Tb in possums

Pathology

The pathology of Tb in possums shows that they are inherently highly susceptible to *M. bovis* in ways that may predispose them to become sources of infection for other animals. Experimental inoculation of possums with *M. bovis* has invariably resulted in generalised, rapidly progressive, fatal infections (Bolliger & Bolliger 1948; Corner & Presidente 1980, 1981; Buddle *et al.* 1994). However, this pattern does not accurately reflect the disease pathogenesis in naturally infected free-ranging possums, where the disease often runs a much more prolonged course (Lugton 1997).

Naturally occurring Tb in possums is characterised by grossly visible (macroscopic) and sub-clinical (microscopic) lesions (abscesses). Macroscopic lesions in infected possums occur in up to 10 distinct body sites (Coleman 1988; Jackson *et al.* 1995a; Pfeiffer *et al.* 1995), and for possums from the Ahaura Valley averaged 4.6 sites per possum (Table 8.1). Microscopic lesions have been identified in twice as many sites in macroscopically infected animals, showing a high level of generalisation of the disease (Cooke *et al.* 1995). Lesions are most common in the lungs and in the superficial lymph nodes (75% of all macroscopically lesioned possums), closely followed by infections in the liver (Table 8.1). Macroscopic lesions contain exceptionally large numbers of *M. bovis* organisms (Ekdahl *et al.* 1970). Lesions in superficial and deep lymph nodes can be up to 60 mm in diameter, and usually caseous (cheesy). They may intermittently discharge their contents including tubercle bacilli to the exterior via sinuses (= artificial openings; Cooke *et al.* 1995; Jackson *et al.* 1995b). On incision, such

Table 8.1
Percent prevalence of macroscopic lesions in the most commonly infected sites of possums.

	Hohonu (n=115; Coleman 1988)	Ahaura Valley (n=73; Jackson *et al.* 1995a)	Hauhungaroa (n=76; Hickling unpubl. data)
Organs			
Lung	55.4	72.5	59.2**
Liver	17.0	56.2*	51.3***
Kidney	12.5	40.0	51.3***
Spleen	8.9	25.0	51.3***
Nodes			
Axillary	29.5	65.0	53.9
Iliac/Inguinal	15.2	72.5	53.9
Mesenteric	20.5	26.2	–
Hepatic	7.1	–	–
Mediastinal/bronchial	5.4	28.7	–

Note: * includes hepatic nodes, ** includes mediastinal and bronchial nodes, *** includes hepatic and mesenteric nodes

nodes are usually pale, turgid, and occasionally oedematous (filled with fluid), with small discrete green or cream-coloured foci (Fig. 8.2). Lesions in the lungs vary from discrete nodules up to 60 mm in diameter, to numerous (miliary) lesions 1–2 mm in diameter, irregular consolidations, or widely disseminated lesions involving up to 50% of lung tissue in terminally ill possums (Cooke *et al.* 1995) (Fig. 8.3). Lesions in abdominal organs may be numerous and are usually cream to creamy-white and 1–20 mm in diameter. The size and shape of abdominal lesions vary with the organ infected but, typically, most contain fewer *M. bovis* organisms than lesions in the respiratory or lymph systems.

Virulence

Virulence (the ability of the organism to cause disease in a specified host) has not been compared for the different strains (REA types) of *M. bovis* currently infecting wildlife in New Zealand. However, Corner & Presidente (1981) reported that a possum-adapted strain from New Zealand produced a more rapidly progressing, severe disease in captive possums than that induced by an Australian cattle strain. Innate resistance to Tb occurs to varying degrees in a number of mammalian species (O'Reilly & Daborn 1995). Some possums can develop a degree of immunity to *M. bovis* infection, and a range in susceptibility of individuals to the disease is likely (Lugton 1997).

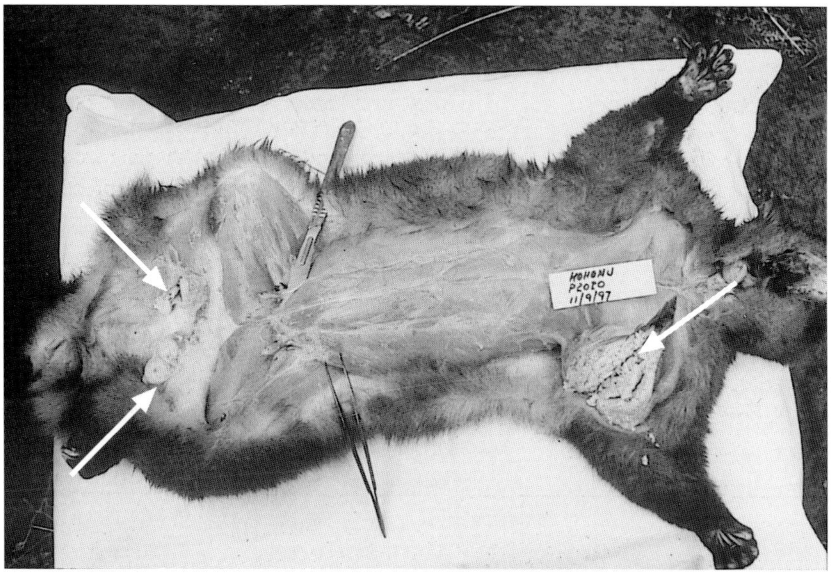

Fig. 8.2
Tuberculous possum showing cheese-like contents of lesions.
Photo: P. Caley

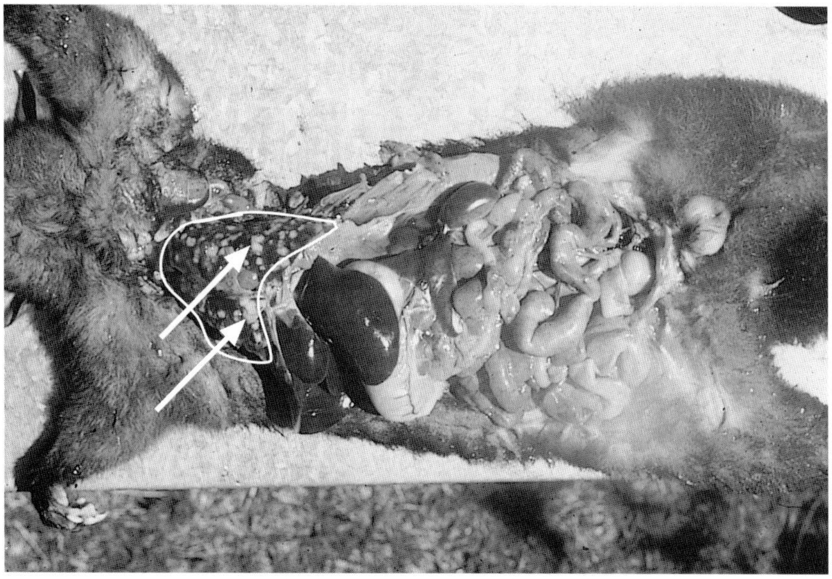

Fig. 8.3
Widely disseminated lesions in the lungs of a tuberculous possum.
Photo: P. Caley

Routes of infection

Transmission between free-living possums can be either pseudo-vertical (mother to young), horizontal (between free-ranging possums during mating, fighting, or simultaneous den sharing), or through indirect environmental contamination of dens or foraging areas. Routes of infection may be indicated by the location of single lesions in possums, representing the site of infection (a "primary complex"). At Hauhungaroa, 38% of all single lesions occurred in the superficial lymph nodes, 31% in the respiratory tract, and 31% in abdominal organs (Pfeiffer *et al.* 1995). Single-site infections in possums in the Hohonu sample (Coleman 1988) were also most common in the lungs and in the superficial lymph nodes. The distribution of single lesions therefore suggests that many possums are infected via the respiratory tract (i.e., from aerosols), and the isolation of *M. bovis* from tracheal washes of infected possums suggests a likely route in possum-to-possum transmission (Jackson *et al.* 1995b).

Macroscopic lesions in superficial lymph nodes could indicate skin infections from bites and scratches. However, this route seems unlikely. No tuberculous bite wounds have ever been recorded in possums. Superficial lymph nodes are commonly lesioned in juvenile possums prior to sexual maturity, but most fighting occurs between adults (Jackson *et al.* 1995a). An alternative explanation for the preponderance of macroscopic lesions in superficial lymph nodes is that they arise from spread from more distant sites via the blood or lymphatic systems, of which microscopic lesions in the lungs are the most likely (Jackson *et al.* 1995b). It is considered that microscopic lesions in the lungs occur more frequently than have been recorded (Jackson *et al.* 1995b), as has been found with Tb infection in badgers (*Meles meles*) (Gallagher *et al.* 1998).

Pseudo-vertical transmission has been proven in pouch and back-riding young (with almost no opportunity of contact with possums other than their mothers), apparently either via aerosol exchange, parental grooming, or the consumption of infected milk (Jackson *et al.* 1995b). Close contact between foraging and feeding possums is normally limited to mother-young pairings and to activities associated with mating. However, Jolly (1976) observed possums congregating and fighting around localised, seasonal food sources such as fruit trees. When feeding on commonly occurring foods, possums keep their distance from each other and generally feed alone (Winter 1976), providing few opportunities for direct disease transmission.

In some situations, simultaneous den sharing could also account for a significant proportion of the adult possum-to-possum (horizontal) disease transmission. Pracy (1975 p. 876) reported that "instances of more than three opossums occupying the one den have been observed frequently where populations are at high or peak levels" and Fairweather *et al.* (1987) recorded three to five animals sharing a single den (a willow stump) at a site near Bridge Pa, Hawke's Bay. Wellington Regional Council pest control staff recorded up to six possums simultaneously sharing one den in the Wairarapa District (M. Hunter, ex Wellington Regional Council, pers. comm.). Lastly, of the possum denning events recorded by Caley *et al.* (1998) at a site near Dunedin, 7% involved simultaneous sharing by up to four possums.

Environmental contamination appears to be less important than animal-to-animal contact in disease transfer (Morris & Pfeiffer 1995). This is because the survival of *M. bovis* outside a host is brief (see below), and lesion sites generally indicate animal-to-animal transmission. For indirect transmission, den sites may be important as they provide opportunity for greater survival (2–3 weeks) of the bacillus than open environments (2 days). Mature possums used an average of 10–15 dens per year, with sequential use by neighbouring animals common (Green & Coleman 1987; Pfeiffer 1994; Paterson *et al.* 1995; Cowan 1989). Cowan (1989) estimated there was about a 50% chance that a den would be occupied by different possums within the probable survival period of deposited tuberculous bacilli. Indirect transmission through contamination of dens between both related and unrelated animals thus seems possible.

The relative importance of each transmission route has not been accurately determined. However, Morris (1995) crudely estimated c. 50% of infections arise from horizontal transmission (mating and fighting), c. 40% from pseudo-vertical transmission (via milk, aerosols, or grooming) and c. 10% from indirect infections. Although the data are not conclusive, it appears that the respiratory route is the principal route of initiation of infection in adult possums (Jackson *et al.* 1995b).

Routes of excretion

The main routes of excretion of *M. bovis* from tuberculous possums are oral, cloacal (urine, faeces), or via discharging sinuses. Cross-sectional studies of lesioned possums suggest that the excretion of bacilli in urine and faeces is rarer (0 of 38 sampled) than that via aerosols from the respiratory tract (9 of 25 tracheal washes) or from superficial sinuses (22 out of 71)(Jackson *et al.* 1995b). Superficial lesions especially may open, drain, and close, and thus not continuously excrete bacilli (Cooke *et al.* 1995; Lugton 1997). That said, most possums with macroscopic lesions appear likely to be excreting *M. bovis* from one or more routes at any one time. In particular, pulmonary lesions are often centred on blood vessels, either blocking airways or spilling *M. bovis* organisms into them, and providing many opportunities for airborne excretion of bacilli (Cooke *et al.* 1995). Lesions also occur in mammary tissue (Jackson *et al.* 1995b) and such possums are likely to excrete bacilli in their milk.

Demography of tuberculous possums

Macroscopic Tb infection is largely independent of age (Jackson *et al.* 1995a), and is more prevalent in males than in females (Coleman 1988; Jackson *et al.* 1995a; Pfeiffer *et al.* 1995), particularly amongst juveniles (Coleman 1988). The difference in prevalence between males and females appears to be caused by behavioural rather than intrinsic physiological factors (Lugton 1997): males forage more widely and juvenile males are more likely than juvenile females to disperse from natal home ranges. Both activities presumably expose males to a greater risk of infection when fighting and maintaining larger activity areas (Jackson *et al.* 1995b). Males, especially, enter winter in poor condition (Bamford 1970). Macroscopically lesioned possums are significantly lighter than possums without macroscopic lesions (Coleman 1988; Pfeiffer *et al.* 1995), but these data are clearly influenced by the physical condition of terminally ill animals. There is no evidence that condition predisposes possums to infection. Whether the prevalence of disease is influenced by the breeding status of females, due perhaps to the physiological stresses associated with lactation compromising the possums immune system, is unclear — no effect was observed by Coleman (1988) and Jackson *et al.* (1995a), but Pfeiffer *et al.* (1995) reported a weak relationship.

Prevalence

The point prevalence (i.e., the proportion of the population infected at any one time) of possums with macroscopic lesions estimated from cross-sectional surveys averages c. 5%, and typically ranges from 1% to 10% (Coleman 1988; Pfeiffer *et al.* 1995; Jackson 1995; Caley *et al.* 1999). However, the prevalence of disease can fluctuate considerably, and occasionally reach much higher levels. For example, 53% of possums in the Ahaura Valley, Westland, in 1992 possessed macroscopic lesions, and subsequent laboratory examination of a sample of the remaining apparently non-lesioned possums revealed further infections giving a total prevalence of at least 60% (Coleman *et al.* 1994). Prevalence rates at Ahaura Valley since 1992 have progressively fallen to more normal levels (i.e., 11, 10, 4, and 2%) over the subsequent 4 years (Coleman *et al.* 1999). Likewise, at the Castlepoint study site in the Wairarapa, the prevalence over a 5-year period averaged about 6%, but at times reached 20% (Jackson 1995). By comparison with the prevalence from broad-scale samples, the prevalence within clusters of infection (i.e., excluding sub-populations free of the disease) ranged from 9% to 32% (Hickling 1995), which more closely approximated the 1992 Ahaura Valley survey.

The low prevalence of macroscopically lesioned possums identified during most extensive surveys is postulated to be caused by the rapid progression and subsequent death of possums after the disease becomes visually detectable (Morris & Pfeiffer 1995). However, time to death after developing clinical Tb is variable — although Jackson (1995) estimated the subsequent mean time to death to be 6 months, possums with readily identifiable superficial lesions have been known to survive for several years (Lugton 1997). Indeed, Lugton (1997) argued strongly that the disease usually runs a prolonged course, with a median survival time after infection in excess of 3 years.

The period prevalence over 1 year, although difficult to determine, has been roughly estimated to be five times that of the point prevalence revealed in standard cross-sectional surveys (also suggesting a rapid progression to death of infected possums; Morris & Pfeiffer 1995). Even so, the prevalence of macroscopically lesioned possums does not appear to vary seasonally with any predictable pattern. Coleman (1988) reported that the prevalence of macroscopically lesioned possums for part (not all) of his Hohonu Range study site was higher during April, June, and August, while Pfeiffer (1994) found Tb prevalence was higher in July, November, and January. However, season had no effect on the prevalence of macroscopically infected possums in the Hauhungaroa Range (Pfeiffer *et al.* 1995), and, after analysing 6 years of data from the Castlepoint site, Lugton (1997) effectively ruled out season as a predictor of prevalence.

Effect on possum population dynamics

When the incidence of *M. bovis* infection in possum populations is low, it appears that possum abundance is largely unaffected, and there appears to be little or no relationship between the prevalence of disease and possum abundance (Barlow 1991b). Any effect of disease-induced mortality is probably obscured by the effects of compensatory survival of non-infected possums and/or immigration (see Efford 1998). However, the prevalence of the disease may become exceedingly high (e.g., 34%, Coleman 1988; 53%, Coleman *et al.* 1994). Given the relatively short life expectancy after the development of macroscopic lesions, Tb-induced mortality resulting from such high levels of infection is likely to exceed birth and immigration rates. Indeed, at Ahaura Valley the abundance of possums (as ascertained by removal trapping) fell with the outbreak of Tb and later increased as the prevalence of the disease fell (Coleman *et al.* 1999).

Spatial aspects

Historically it appeared that the prevalence of infection was highest in possum populations foraging on pasture adjacent to the bush edge. A survey in the Hohonu Range in Westland in 1973–74 recorded 75% of all macroscopically lesioned possums within 50 m of pasture (Coleman 1988). Only 19% were more than 1 km from pasture, with the farthest 4 km into the forest. This pattern was presumed to have been caused by (a) the number and foraging behaviour of bush-edge possums increasing the per capita rate of disease transmission; (b) possums interacting with tuberculous cattle; or (c) Tb spreading progressively deeper into the forest after it had recently established in the possum population there (Coleman 1988). Of these hypotheses, none now survive. Resurveys of some of the same lines 17 and 25 years later have shown almost identical spatial patterns of infection, despite the removal of cattle as a source of infection, the replacement of some of the pasture by dense scrub, and ample time for disease to spread deeper into the forest (Hickling 1991; Coleman & Caley unpubl. data). It is now apparent that distance from the pasture margin is, however, generally correlated with decreasing possum abundance, changing vegetation, increasing altitude, and microclimate, all of which could influence the transmission patterns of Tb in possums. It now seems unlikely that foraging on pasture per se leads to increased per capita rates of Tb transmission. For example, at a time of exceptional prevalence (53%) of macroscopic Tb infection amongst a population of bush-edge possums, there was no decrease in the prevalence of Tb in possums with increasing distance (over 1.5 km) from the forest margin in the Ahaura Valley (Coleman *et al.* 1993). Possum density has been proposed as a factor influencing the prevalence of Tb in possums. Although analysis of broad-scale data based on the possum carrying capacity of the habitat does not support this, perhaps because of the patchiness of the spatial pattern of Tb (Barlow 1991b), crowding of possums on a small scale (a measure of local density) was found to correlate with a higher prevalence of disease (Hickling 1995). This observation follows O'Reilly & Daborn's (1995) thesis that systems that facilitate close contact between animals favour the spread of respiratory diseases including Tb.

Infected possums are typically highly aggregated in Tb foci or "hot spots" (Coleman 1988; Coleman *et al.* 1994; Pfeiffer *et al.* 1995; Caley 1996). For example, 54% of 105 possums with visible lesions taken from the Hohonu Range (Coleman 1988) came from a trap site that captured other infected possums, or from an immediately adjacent (within 50 m) trap site, suggesting possum-to-possum transmission of the disease. Comparative data from four other surveys of tuberculous possum populations showed similar patterns, with infected possums 4–16 times more likely to be trapped close together than if distributed randomly (Hickling 1995). By contrast, annual surveys in the Ahaura Valley have recorded tuberculous possums occurring along c. 5 km of pasture edge (Coleman *et al.* 1994), but even these infected possums are significantly clustered in time and space (Coleman *et al.* 1999).

Tuberculosis foci in possum populations may be long-lasting. Ongoing studies in Westland have recorded infection at the same or nearby trap sites on bush/pasture margins apparently persisting for 16 (Cooke *et al.* 1995) and 25 years (Coleman & Caley unpubl. data). Such infections frequently persist despite possum control, presumably either through more frequent horizontal transmission in favoured longstanding den sites or through their contamination, although neither is proven. Macroscopically lesioned possums continued to be caught from the one small area in the Hohotaka site near Taumarunui, during the first 5 years of ongoing possum control, though not in the subsequent 5 years (Caley 1996; Caley *et al.* 1999).

Persistence of Tb foci in space and time has led to speculation about the possible underlying cause of clustering. Pfeiffer (1994) hypothesised that spatial clustering arises from a combination of pseudo-vertical transmission from mother to daughter, aggregated mating patterns, and environmentally stressful areas. Of these three factors, only pseudo-vertical transmission has been confirmed (Pfeiffer 1994). Environmental stress may play an important role in the development of clinical Tb and therefore its transmission in possums; stressful life events induce the reactivation of infection in humans (O'Reilly & Daborn 1995). Factors contributing to such environmental stress are postulated to include aspect, exposure of habitats to prevailing winds, quality of dens, slope, and elevation (Pfeiffer 1994). Low temperature and poor nutrition appear to compromise the possum's (and other animals) immune system, and this is postulated to lead to infection and the temporal clustering of tuberculosis in possums (van den Oord *et al.* 1995), though evidence to support this from analysis of meteorological data is weak (Lugton 1997). No difference was detected between the degree of exposure of dens containing tuberculous possums and dens in a nearby area free from macroscopically Tb-lesioned possums (Caley 1996). This suggests that den quality (as indicated by exposure) is unimportant in the epidemiology of Tb in possums, or that the differences in den quality are not large enough or are too spatially localised to be easily detected. New foci presumably arise from the dispersal of infected young possums (Cowan *et al.* 1996), or from infections in other species.

Survival outside the host

Although *M. bovis* is an obligate pathogen, it has the ability to survive outside its host for extended periods of time under favourable conditions (Morris *et al.* 1994). Of particular importance for understanding the mechanisms of possum-to-possum and possum-to-cattle transmission of *M. bovis* is the survival time of the organism on pasture and inside possum dens under New Zealand conditions. New Zealand studies have shown Tb organisms sprayed onto damp cloth placed in pasture survived for less than 4 days. Similar material placed on the forest floor survived for about 7 days, with survival increasing as temperatures decreased. Bacilli placed in possum dens survived for up to 28 days in cool months (Jackson *et al.* 1995c). Similar studies in Australia (Duffield & Young 1985) and Britain (Anon. 1979) show low survival of *M. bovis* on grass, soil, or faeces after exposure of 1–2 weeks. It appears that sunlight and natural organisms quickly destroy tubercle bacilli. Therefore contamination of pasture, particularly in summer, may be relatively unimportant in the transmission of the disease either to other possums or to livestock, especially considering that the minimum infective dose for oral infection is high (Morris *et al.* 1994). Survival of *M. bovis* in the carcasses of dead possums may be considerably longer than on pasture, as *M. bovis* has been successfully isolated from the carcasses of dead possums 1 month after death (Pfeiffer 1994).

Interactions with other hosts

Transmission to livestock

Tuberculosis in cattle is primarily a respiratory disease and Francis (1947) estimated that 80–90% of all cattle are infected by inhalation. The mode of transmission between possums and cattle is less well understood and difficult to study, as healthy possums avoid stock wherever possible and provide few situations for the close contact seemingly necessary for Tb transmission. However, direct observations of grazing stock and possums suggest such avoidance may not always be possible. Dominant cattle and deer have been observed to approach semi-sedated ("sick") possums from up to 50 m away, and sniffing and mouthing them (Paterson & Morris 1995; Sauter & Morris 1995). A parallel study of captive deer deliberately exposed to infected possums led to the dominant deer becoming infected (Lugton *et al.* 1997b), and it seems clear that exploratory behaviour of possums by both deer and cattle is likely to expose them to infection. In contrast, sheep showed much less interest in investigating sick possums (Sauter & Morris 1995). Although Tb infection in sheep is a rare disease, it is known to occur from properties with a history of Tb infection in possums (Cordes *et al.* 1981; Davidson *et al.* 1981; Allen 1988; Lugton 1997), as it is for goats (Cousins *et al.* 1993). Such inquisitiveness in stock did not extend to their inspection of dead possums, indicating that possum carcasses are unlikely to play a major role in direct disease transfer to livestock (Paterson & Morris 1995). The mode of transmission from livestock (particularly farmed deer) back to possums is unknown.

Tb in other wildlife

Both wild red deer (*Cervus elaphus*) and sika deer (*Cervus nippon*) have been found infected with Tb (de Lisle & Havill 1985). The prevalence of infection from wild deer living in VRAs can be as high as 37% (Lugton *et al.* 1998). Likewise Tb is commonly identified in feral goats inhabiting VRA's (Allen 1987; Sanson 1988; Lugton 1997). Despite tuberculous possum carcasses posing little risk of infection to livestock, they may be an important source of infection for scavenging species such as feral ferrets (*Mustela furo*), feral cats (*Felis catus*), stoats (*Mustela erminea*), hedgehogs (*Erinaceus europaeus*), and feral pigs (*Sus scrofa*). Tuberculosis infection occurs in all these species at varying levels (Ekdahl *et al.* 1970; Wakelin & Churchman 1991; de Lisle *et al.* 1993; de Lisle 1994; Lugton *et al.* 1995; Ragg *et al.* 1995; Lugton 1997; Caley 1998). Other terrestrial wildlife species found to be infected with bovine tuberculosis in areas occupied by tuberculous possums in New Zealand include rabbits (*Oryctolagus cuniculus*)(Gill & Jackson 1993) and brown hares (*Lepus europaeus occidentalis*)(Cooke *et al.* 1993).

However, not all the above species play a significant role in the wildlife disease cycle. Species are of much more importance to the cycle of Tb if they are reservoir hosts of the disease — as opposed to being spillover or dead-end hosts. In the case of spillover and dead-end hosts, Tb infection will disappear progressively if the disease is eliminated from the species that is acting as the main source. Morris & Pfeiffer (1995) considered that feral cats, feral goats, stoats, hedgehogs, rabbits, and hares are definitely either spillover or dead-end hosts for Tb. Of the remainder, they considered it likely that feral deer are reservoir hosts, despite tuberculous possums being the source of most of the Tb infection in wild deer (Lugton *et al.* 1998). Indeed, deer may be the principal initiator of new areas of wildlife infection, which would make them, along with brushtail possums, the underlying reservoir of Tb in New Zealand (Morris & Pfeiffer 1995).

Whether feral ferrets are reservoir or spillover hosts has been subject to considerable debate (Walker *et al.* 1993; Morris *et al.* 1994; Ragg *et al.* 1995; Ragg 1997; Caley 1998), though lack of data has prevented a definitive judgement of their host status (Morris & Pfeiffer 1995). It appears that tuberculous possums are a major source of Tb infection in ferrets (Lugton *et al.* 1997a; Caley 1998), though this does not preclude them from being reservoir hosts (Caley 1998). Acquisition of infection in feral pigs also seems to occur principally from the ingestion of tuberculous possum carcasses, and other tuberculous carrion, and as many as 96% of feral pigs inhabiting VRAs are infected with Tb (Lugton 1997). There appears to be considerable potential for pig-to-pig transmission to occur under New Zealand conditions. However, based on the evidence from Australia (Corner *et al.* 1981; McInerney *et al.* 1995), it appears likely that feral pigs are spillover hosts in New Zealand also (Morris & Pfeiffer 1995; Lugton 1997).

With so many wildlife species potentially infected with Tb, it is often difficult to determine the relative importance of the various mechanisms of transmission between the various species. In some cases, DNA fingerprinting has assisted in determining the geographical origin of new infections in wildlife, but it is less helpful in revealing in which direction infection spreads (de Lisle *et al.* 1995). Within geographic areas, stock and wildlife often have the same REA type cycling among them, and such strains are not generally specific to one species (Collins & de Lisle 1995).

Effect of possum control on Tb in possum populations

Control of tuberculous possum populations (and of other wildlife hosts) is one part of a "National Pest Management Strategy for Bovine Tuberculosis (Tb)" (see Chapter 20), produced to meet the requirements of the Biosecurity Act 1993. In the absence of external non-possum sources of infection, the eradication of Tb from possum populations requires the basic disease reproductive ratio (the average number of additional infections that each infected possum causes) to be reduced to less than 1.0 (Anderson & May 1991). This ratio has been estimated to be about 1.8 for Tb infection in uncontrolled possum populations (Barlow 1991a). In the absence of immigration, eradication of Tb from a possum population from a defined area may be achieved in several ways. First, all the possums within the area may be killed. This is not considered feasible with current funding levels except for small areas. Second, all tuberculous possums within the area may be killed by culling. This is unlikely to be achieved by a single control operation, but may be achieved by continuous intensive control. Last, the possum population may be driven below a threshold where the combined effects of mortality due to occasional culling and the reduced rate of disease transmission results in disease eradication.

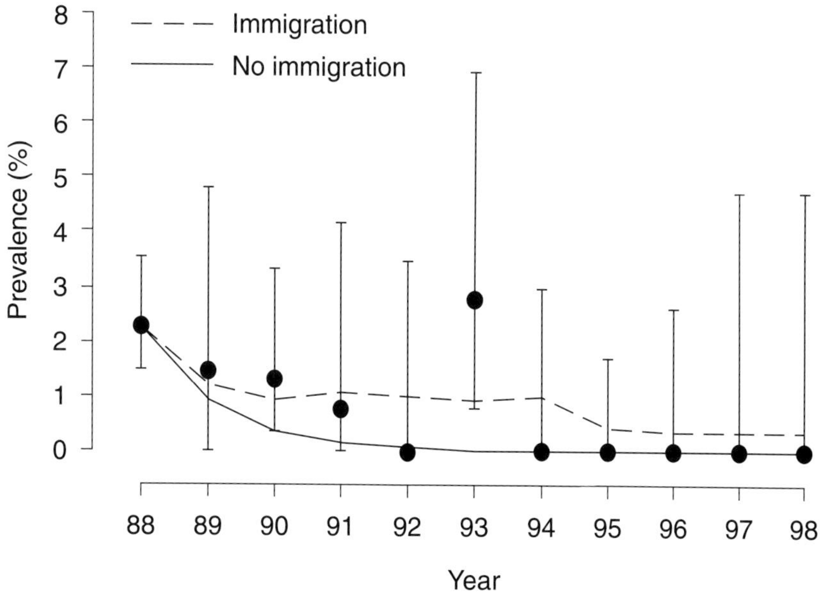

Fig. 8.4
Annual estimates of prevalence of macroscopic *M. bovis* infection in possums at Hohotaka (•) before control (1988) and with regular maintenance control of possums (1989–1996), which has held the possum density at 23% of its pre-control level. Bars represent binomial confidence intervals (not corrected for finite population size). The observed decrease since maintenance control began is significant ($P < 0.05$). Solid and dashed lines represent prevalence predicted by Barlow's possum/Tb model with and without immigration (From Caley *et al.* 1999).

To aid in the management of Tb in possum populations, a number of predictive possum/Tb models have been developed (e.g., Barlow 1991a,b, 1993, 1996; Pfeiffer 1994; Roberts 1996). The Barlow models have played a significant role in the development of the Animal Health Board (AHB) and Ministry of Agriculture and Forestry (MAF) control policies and strategies for management of Tb in New Zealand (Barlow 1994). They predict that, on average, Tb will not persist (or establish) in possum populations held below about 40% of a habitat's carrying capacity, and this level can be achieved by culling at a rate of about 18% per year. However, some assumptions in these models are based on limited data. The models are supported by data from the Hohotaka site near Taumarunui (Caley *et al.* 1995; Caley *et al.* 1999), where ongoing control of possums has been associated with a significant decline in the prevalence of macroscopically infected possums, broadly fitting the predictions of the models (Fig. 8.4).

Tuberculosis persisted for 5 years at Hohotaka in a possum population held at less than 30% of its pre-control level, suggesting transmission can continue at low possum density with immigration (Caley *et al.* 1999). Analysis of these data suggests that the disease contact rate is nonlinear on density (convex up), in contrast with the early assumptions of it being directly proportional to possum density (Barlow 1991a,b). Subsequent models have considered the consequences of varying this assumption (Barlow 1996; Roberts 1996). It appears that larger reductions in density than first thought are needed to effect significant reductions in the per capita transmission rate between possums. So, much higher levels of control will be needed than previously thought to eradicate disease from possum populations.

Conclusions

Direct and indirect evidence of the role of possums as a reservoir host for Tb includes:

(a) the persistence of Tb in possum populations in the absence of readily explainable external sources of the infection (Pfeiffer 1994; Hickling 1991);
(b) the shedding of *M. bovis* organisms into the environment by tuberculous possums enabling the infection of other species (Jackson *et al.* 1995b);
(c) the nearly universal concurrence of infection in possums and in coexisting cattle or deer herds;

(d) the reduction of the incidence of Tb infection in livestock following the effective control of tuberculous possum populations (Tweedle & Livingstone 1994; Caley *et al.* 1995; Pannett 1995);

(e) DNA fingerprinting, which implicates possums as the source of Tb for cattle (Collins *et al.* 1994); and

(f) the apparent role of possums in infecting wildlife such as feral ferrets and deer (Lugton *et al.* 1997a; Lugton *et al.* 1997b; Caley 1998; Nugent *et al.* 1998).

Summary

- The possum is considered the primary wildlife reservoir of Tb for farmed cattle and deer in New Zealand and the single greatest barrier to eradication of Tb from New Zealand livestock.
- The first report of a free-ranging possum with Tb in New Zealand was in 1967 and 4 years later epidemiological investigations reported an association between Tb in possums and the persistence of Tb in adjacent cattle herds.
- Possum populations infected with Tb now occupy about 23.6% of New Zealand, which is also where 75% of the reactor herds and Tb-infected cattle are found.
- Possums with Tb often exhibit macroscopic and microscopic lesions (abscesses) especially in the lungs and lymph nodes.
- Transmission of Tb in possums is either from mother to young, adult to adult (during fighting, mating, and den sharing), and from environmental contamination. Infected individuals are thought to survive for about 6 months.
- Typically 1% to 10% of possums in a Tb-infected population show macroscopic lesions, though infection can vary over time and range up to as high as 60%. The prevalence of Tb in possum populations appears to be unrelated to season.
- Incidence of Tb in possums is typically aggregated into "hot spots" that may endure for several years despite possum control.
- The mode of transmission of Tb from possums to stock is unknown, but is thought to occur when inquisitive stock encounter recently dead or dying possums.
- Although possums have been implicated as the source of Tb in cattle, several other scavenging animals such as ferrets, stoats, feral cats, hedgehogs, and feral pigs are also known to host Tb as spillover hosts. Deer and ferrets are also likely reservoir hosts.
- It appears that higher levels of control will be needed to bring possum densities below a level where disease eradication becomes a possibility.

References

Alhaji, I. 1976: Bovine tuberculosis: a general review with special reference to Nigeria. *The Veterinary Bulletin, Weybridge 46*: 829–941.

Allen, G. M. 1987: Tuberculosis in feral goats. *Surveillance 14(1)*: 13.

Allen, G. M. 1988: Tuberculosis in sheep — a very rare disease. *Surveillance 15(5)*: 8–9.

Anderson, R. M.; May, R. M. 1991: Infectious diseases of humans. Oxford, Oxford University Press. 757 p.

Animal Health Board 1995: National Tb strategy: Proposed national pest management strategy for bovine tuberculosis. Wellington, New Zealand, Animal Health Board. 116 p.

Animal Health Board 1998: Annual report for the year ending 30 June 1998. Wellington, New Zealand, Animal Health Board. 43 p.

Anonymous 1979: Bovine tuberculosis in badgers. Report by the Ministry of Agriculture, Fisheries and Food. London, MAFF.

Bamford, J. 1970: Estimating fat reserves in the brush-tailed possum, *Trichosurus vulpecula* Kerr (Marsupialia: Phalangeridae). *Australian Journal of Zoology 18*: 415–425.

Batcheler, C. L.; Cowan, P. L. 1988: Review of the status of the possum (*Trichosurus vulpecula*) in New Zealand. Unpublished FRI contract report to the Technical Advisory Committee (Animal pests) for the Agriculture Pests Destruction Council, Department of Conservation, and MAFQual Ministry of Agriculture and Fisheries. Wellington. 129 p.

Barlow, N. D. 1991a: Control of endemic bovine Tb in New Zealand possum populations: results from a simple model. *Journal of Applied Ecology 28*: 794–809.

Barlow, N. D. 1991b: A spatially aggregated disease/host model for bovine Tb in New Zealand possum populations. *Journal of Applied Ecology 28*: 777–793.

Barlow, N. D. 1993: A model for the spread of bovine Tb in New Zealand possum populations. *Journal of Applied Ecology 30*: 156–164.

Barlow, N. D. 1994: Bovine tuberculosis in New Zealand: epidemiology and models. *Trends in Microbiology 2*: 119–124.

Barlow, N. D. 1996: The ecology of wildlife disease control: simple models revisited. *Journal of Applied Ecology 33*: 303–314.

Beatson, N. S. 1985. Tuberculosis in red deer in New Zealand. *In*: Fennessy, P. F.; Drew, K. R. *ed.* Biology of deer production. *Bulletin, Royal Society of New Zealand No. 22.* Pp. 147–150.

Bolliger, A.; Bolliger, W. 1948: Experimental transmission of tuberculosis to *Trichosurus vulpecula*. *Australian Journal of Science 10*: 182–183.

Buddle, B. M.; Aldwell, F. E.; Pfeffer, A.; de Lisle, G. W. 1994: Experimental *Mycobacterium bovis* infection in the brushtail possum (*Trichosurus vulpecula*): pathology, haematology and lymphocyte stimulation responses. *Veterinary Microbiology 38*: 241–254.

Caley, P. 1996: Is the spatial distribution of tuberculous possums influenced by den "quality"? *New Zealand Veterinary Journal 44*: 175–178.

Caley, P. 1998: Broad-scale possum and ferret correlates of macroscopic *Mycobacterium bovis* infection in feral ferret populations. *New Zealand Veterinary Journal 46*: 157–162.

Caley, P.; Hickling, G. J.; Cowan, P. E. 1995: Sustained control of possums to reduce bovine tuberculosis in cattle and possum populations in New Zealand. Proceedings: 10th Australian Vertebrate Pest Control Conference, Hobart, Tasmania, 29 May–2 June 1995. Tasmania, Department of Primary Industry and Fisheries. Pp. 276–181.

Caley, P.; Spencer, N. J.; Cole, R. A.; Efford, M. G. 1998: The effect of manipulating population density on the probability of den-sharing among common brushtail possums and the implications for transmission of bovine tuberculosis. *Wildlife Research 25*: 383–392.

Caley, P.; Hickling, G. J.; Cowan, P. E.; Pfeiffer, D. U. 1999: Effects of sustained control of brushtail possums on levels of *Mycobacterium bovis* infection in cattle and brushtail possum populations from Hohotaka, New Zealand. *New Zealand Veterinary Journal 47*: 133–142.

Carter, C. 1995: The eradication of bovine tuberculosis from New Zealand farmed deer herds. *In*: Griffin, F.; de Lisle, G. *ed.* Tuberculosis in wildlife and domestic animals. *Otago Conference Series No. 3.* Dunedin, University of Otago Press. Pp. 354–356.

Coleman, J. D. 1988: Distribution, prevalence, and epidemiology of bovine tuberculosis in brushtail possums, *Trichosurus vulpecula*, in the Hohonu Range, New Zealand. *Australian Wildlife Research 15*: 651–653.

Coleman, J. D.; Drew, K.; Coleman, M. C. 1993: Prevalence and spatial distribution of bovine tuberculosis in a possum population, Ahaura Valley, Westland. Within-forest patterns: December 1992. Landcare Research Contract Report LC9293/90 (unpublished) 10 p.

Coleman, J. D.; Jackson, R.; Cooke, M. M.; Grueber, L. 1994: Prevalence and spatial distribution of bovine tuberculosis in brushtail possums on a forest-scrub margin. *New Zealand Veterinary Journal 42*: 128–132.

Coleman, J. D.; Cooke, M. M.; Jackson, R.; Webster, R. 1999: Temporal patterns in bovine tuberculosis in a brushtail possum population contiguous with infected cattle in the Ahaura Valley, Westland. *New Zealand Veterinary Journal 47*: 119–124.

Collins, C. H.; Grange, J. M. 1987: Zoonotic implications of *Mycobacterium bovis* infection. *Irish Veterinary Journal 41*: 363–366.

Collins, D.; de Lisle G. 1995: *Mycobacterium bovis* strains. *In:* Griffin, F.; de Lisle, G. *ed.* Tuberculosis in wildlife and domestic animals. *Otago Conference Series No. 3.* Dunedin, University of Otago Press. Pp. 202–204.

Collins, D. M.; Radford, A. J.; de Lisle, G. W.; Billman-Jacobe, H. 1994: Diagnosis and epidemiology of bovine tuberculosis using molecular biological approaches. *Veterinary Microbiology 40*: 83–94.

Collins, D.; Kawakami, P.;Wilson, T.; de Lisle, G. 1995: Virulence factors of *Mycobacterium bovis. In:* Griffin, F.; de Lisle, G. *ed.* Tuberculosis in wildlife and domestic animals. *Otago Conference Series No. 3.* Dunedin, University of Otago Press. Pp. 1–4.

Cooke, M. M.; Jackson, R.; Coleman, J. D. 1993: Tuberculosis in a free-living brown hare (*Lepus europaeus occidentalis*). *New Zealand Veterinary Journal 41*: 144–146.

Cooke, M. M.; Jackson, R.; Coleman, J. D.; Alley, M. R. 1995: Naturally occurring tuberculosis caused by *Mycobacterium bovis* in brushtail possums (*Trichosurus vulpecula*): II. Pathology. *New Zealand Veterinary Journal 43*: 315–321.

Cordes, D. O.; Bullians, J. A.; Lake, D. E.; Carter, M. E. 1981: Observations on tuberculosis caused by *Mycobacterium bovis* in sheep. *New Zealand Veterinary Journal 29*: 60–62.

Corner, L. A.; Presidente, P. J. A. 1980: *Mycobacterium bovis* infection in the brush-tailed possum (*Trichosurus vulpecula*): I. Preliminary observations on experimental infection. *Veterinary Microbiology 5*: 309–321.

Corner, L. A.; Presidente, P. J. A. 1981: *Mycobacterium bovis* infection in the brush-tailed possum (*Trichosurus vulpecula*): II. Comparison of experimental infections with an Australian cattle strain and a New Zealand possum strain. *Veterinary Microbiology 6*: 351–366.

Corner, L. A.; Barrett, R. H.; Lepper, A. W. D.; Lewis, V.; Pearson, C. W. 1981: A survey of mycobacteriosis of feral pigs in the Northern Territory. *Australian Veterinary Journal 57*: 537–542.

Cousins, D. V.; Francis, B. R.; Casey, R.; Mayberry, C. 1993: *Mycobacterium bovis* infection in a goat. *Australian Veterinary Journal 70*: 262–263.

Cowan, P. E. 1989: Denning habits of common brushtail possums, *Trichosurus vulpecula*, in New Zealand lowland forest. *Australian Wildlife Research 16*: 63–78.

Cowan, P. E.; Brockie, R. E.; Ward, G. D.; Efford, M. G. 1996: Long-distance movements of juvenile brushtail possums (*Trichosurus vulpecula*) on farmland, Hawke's Bay, New Zealand. *Wildlife Research 23*: 237–244.

Davidson, R. M.; Alley, M. R.; Beatson, N. S. 1981: Tuberculosis in a flock of sheep. *New Zealand Veterinary Journal 29*: 1–2.

de Lisle, G. W. 1994: Mycobacterial infections in pigs. *Surveillance 21(4)*: 23–25.

de Lisle, G. W.; Havill, P. F. 1985: Mycobacteria isolated from deer in New Zealand from 1970–1983. *New Zealand Veterinary Journal 33(8)*: 138–140.

de Lisle, G. W.; Crews, K.; de Zwart, J.; Jackson, R.; Knowles, G. J. E.; Paterson, K. D.; MacKenzie, R. W.; Waldrup, K. A.; Walker, R. 1993: *Mycobacterium bovis* infections in wild ferrets. *New Zealand Veterinary Journal 41*: 148–149.

de Lisle, G. W.; Yates, G. F.; Collins, D. M.; MacKenzie, R. W.; Crews, K. B.; Walker, R. 1995: A study of bovine tuberculosis in domestic animals and wildlife in the MacKenzie Basin and surrounding areas using DNA fingerprinting. *New Zealand Veterinary Journal 43*: 266–271.

Duffield, B. J.; Young, D. A. 1985: Survival of *Mycobacterium bovis* in defined environmental conditions. *Veterinary Microbiology 10*: 193–197.

Efford, M. 1998: Demographic consequences of sex-biased dispersal in a population of brushtail possums. *Journal of Animal Ecology 67*: 503–517.

Ekdahl, M. O.; Smith, B. L.; Money, D. F. L. 1970: Tuberculosis in some wild and feral animals in New Zealand. *New Zealand Veterinary Journal 18*: 44–45.

Fairweather, A. A. C.; Brockie, R. E.; Ward, G. D. 1987. Possums (*Trichosurus vulpecula*) sharing dens: a potential infection route for bovine tuberculosis. *New Zealand Veterinary Journal 35(1/2)*: 15–16.

Francis, J. 1947: Bovine tuberculosis including a contrast with human tuberculosis. London, Staple Press.

Gill, J. W.; Jackson, R. 1993: Tuberculosis in a rabbit: A case revisited. *New Zealand Veterinary Journal 41*: 147.

Gallagher, J.; Monies, J.; Gavier-Widen, M.; Rule, B. 1998: Role of infected, non-diseased badgers in the pathogenesis of tuberculosis in the badger. *The Veterinary Record 142*: 710–714.

Green, W. Q.; Coleman, J. D. 1987: Den sites of possums, *Trichosurus vulpecula*, and frequency of use in mixed hardwood forest in Westland, New Zealand. *Australian Wildlife Research 14*: 285–292.

Hickling, G. 1991: Ecological aspects of endemic bovine tuberculosis infection in brushtail possum populations in New Zealand. *In*: Proceedings: 9th Australian Vertebrate Pest Control Conference, Adelaide. Pp. 332–335.

Hickling, G. 1995: Clustering of tuberculosis infection in brushtail possum populations: implications for epidemiological simulation models.

In: Griffin, F.; de Lisle, G. *ed.* Tuberculosis in wildlife and domestic animals. *Otago Conference Series No. 3.* Dunedin, University of Otago Press. Pp. 174–177.

Jackson, R. 1995: Transmission of tuberculosis (*Mycobacterium bovis*) by possums. Unpublished PhD thesis, Massey University, Palmerston North, New Zealand.

Jackson, R.; Cooke, M. M.; Coleman, J. D.; Morris, R. S. 1995a: Naturally occurring tuberculosis caused by *Mycobacterium bovis* in brushtail possums (*Trichosurus vulpecula*): I. An epidemiological analysis of lesion distribution. *New Zealand Veterinary Journal 43*: 306–314.

Jackson, R.; Cooke, M. M.; Coleman, J. D.; Morris, R. S.; de Lisle, G. W.; Yates, G. F. 1995b: Naturally occurring tuberculosis caused by *Mycobacterium bovis* in brushtail possums (*Trichosurus vulpecula*): III. Routes of infection and excretion. *New Zealand Veterinary Journal 43*: 322–327.

Jackson, R.; de Lisle G. W.; Morris, R. S. 1995c: A study of the environmental survival of *Mycobacterium bovis* on a farm in New Zealand. *New Zealand Veterinary Journal 43*: 346–352.

Jamieson, S. 1960: Bovine tuberculosis: problems and prospects of eradication. *New Zealand Journal of Agriculture 100(4)*: 314–319.

Jolly, J. N. 1976: Habitat use and movements of the opossum (*Trichosurus vulpecula*) in a pastoral habitat on Banks Peninsula. *Proceedings, New Zealand Ecological Society 23*: 70–78.

Lugton, I. W. 1997: The contribution of wild mammals to the epidemiology of tuberculosis (*Mycobacterium bovis*) in New Zealand. Unpublished PhD thesis, Massey University, Palmerston North, New Zealand.

Lugton, I. W.; Johnstone, A. C.; Morris, R. S. 1995: *Mycobacterium bovis* infection in New Zealand hedgehogs (*Erinaceus europaeus*). *New Zealand Veterinary Journal 43*: 342–345.

Lugton, I. W.; Wobeser, G.; Morris, R. S.; Caley, P. 1997a: Epidemiology of *Mycobacterium bovis* infection in feral ferrets (*Mustela furo*) in New Zealand: II. Routes of infection and excretion. *New Zealand Veterinary Journal 45*: 151–157.

Lugton, I. W.; Wilson, P. R.; Morris, R. S.; Griffin, J. F. T.; de Lisle, G. W. 1997b: Natural infection of red deer with bovine tuberculosis. *New Zealand Veterinary Journal 45*: 19–26.

Lugton, I. W.; Wilson, P. R.; Morris, R. S.; Nugent, G. 1998: Epidemiology and pathogenesis of *Mycobacterium bovis* infection of red deer (*Cervus elaphus*) in New *Zealand. New Zealand Veterinary Journal 46*: 147–156.

Mackereth, G. 1993: The Waipawa endemic area: the epidemiological picture. *In:* Proceedings of a deer course for veterinarians. Palmerston North, Deer Branch of the New Zealand Veterinary Association. Pp. 222–228.

McInerney, J.; Small, K. J.; Caley, P. 1995: Prevalence of *Mycobacterium bovis* infection in feral pigs in the Northern Territory. *Australian Veterinary Journal 72*: 448–451.

Morris, R. 1995: Epidemiological principles for tuberculosis control. *In*: Griffin, F.; de Lisle, G. *ed.* Tuberculosis in wildlife and domestic animals. *Otago Conference Series No. 3.* Dunedin, University of Otago Press. Pp. 210–213.

Morris, R. S.; Pfeiffer, D. U. 1995: Directions and issues in bovine tuberculosis epidemiology and control in New Zealand. *New Zealand Veterinary Journal 43*: 256–265.

Morris, R. S.; Pfeiffer, D. U.; Jackson, R. 1994: The epidemiology of *Mycobacterium bovis* infections. *Veterinary Microbiology 40*: 153–177.

Nugent, G.; Whitford, J.; Coleman, J. D.; Fraser, K. W. 1998: Effect of possum (*Trichosurus vulpecula*) control on the prevalence of bovine tuberculosis in wild deer. *In:* Proceedings of the OIE International Congress with WHO-co-sponsorship on anthrax, brucellosis, CBPP, clostridial, and mycobacterial diseases. Pretoria, South Africa, Sigma Press. Pp. 462–466.

O'Neil, B. D.; Pharo, H. J. 1995. The control of bovine tuberculosis in New Zealand. *New Zealand Veterinary Journal 43*: 249–255.

O'Reilly, L. M.; Daborn, C. J. 1995: The epidemiology of *Mycobacterium bovis* infections in animals and man: a review: *Tubercle and Lung Disease 76 (Supplement 1)*: 1–46.

Pannett, G. 1995: Possum control in the Wellington region: how successful has it been? *In*: Griffin, F.; de Lisle, G. *ed.* Tuberculosis in wildlife and domestic animals. *Otago Conference Series No. 3.* Dunedin, University of Otago Press. Pp. 294–296.

Paterson, B. M.; Morris, R. S. 1995: Interactions between beef cattle and simulated tuberculous possums on pasture. *New Zealand Veterinary Journal 43*: 289–293.

Paterson, B. M.; Morris, R. S.; Weston, J.; Cowan, P. E. 1995. Foraging and denning patterns of brushtail possums, and their possible relationship to contact with cattle and the transmission of bovine tuberculosis. *New Zealand Veterinary Journal 43*: 281–288.

Pfeiffer, D. U. 1994: The role of a wildlife reservoir in the epidemiology of bovine tuberculosis. Unpublished PhD thesis. Massey University, Palmerston North, New Zealand. 439 p.

Pfeiffer, D. U.; Hickling, G. J.; Morris, R. S.; Patterson, K. P; Ryan, T. J.; Crews, K. B. 1995: The epidemiology of *Mycobacterium bovis* infection in brushtail possums (*Trichosurus vulpecula* Kerr) in the Hauhungaroa Ranges, New Zealand. *New Zealand Veterinary Journal 43*: 272–280.

Pracy, L. T. 1975: Opossums (1). *New Zealand's Nature Heritage 3*: 873–882.

Ragg, J. R. 1997: Tuberculosis (*Mycobacterium bovis*) epidemiology and the ecology of ferrets (*Mustela furo*) on New Zealand farmland. Unpublished PhD thesis, University of Otago, Dunedin, New Zealand.

Ragg, J.R.; Moller, H.; Waldrup, K.A. 1995: The prevalence of bovine tuberculosis (*Mycobacterium bovis*) infections in feral populations of cats (*Felis catus*), ferrets (*Mustela furo*) and stoats (*Mustela erminea*) in Otago and Southland, New Zealand. *New Zealand Veterinary Journal 43*: 333–337.

Roberts, M. G. 1996: The dynamics of bovine tuberculosis in possum populations, and its eradication or control by culling or vaccination. *Journal of Animal Ecology 65*: 451–464.

Sanson, R. L. 1988: Tuberculosis in goats. *Surveillance 15(2)*: 7–8.

Sauter, C. M.; Morris, R. S. 1995: Dominance hierarchies in cattle and red deer (*Cervus elaphus*): Their possible relationship to the transmission of bovine tuberculosis. *New Zealand Veterinary Journal 43*: 301–305.

Tweedle, N. E.; Livingstone, P. 1994: Bovine tuberculosis control and eradication programs in Australia and New Zealand. *Veterinary Microbiology 40*: 23–39.

van den Oord, Q. G. W.; van Wijk, E. J. A.; Lugton, I. W.; Morris, R. S.; Holmes, C. W. 1995: Effects of air temperature, air movement and artificial rain on the heat production of brushtail possums (*Trichosurus vulpecula*): an exploratory study. *New Zealand Veterinary Journal 43*: 328–332.

Wakelin, C. A.; Churchman, O. T. 1991: Prevalence of bovine tuberculosis in feral pigs in Central Otago. *Surveillance 18(5)*: 19–20.

Walker, R.; Reid, B.; Crews, K. 1993: Bovine tuberculosis in predators in the Mackenzie Basin. *Surveillance 20(2)*: 11–14.

Winter, J. W. 1976: The behaviour and social organisation of the brush-tail possum (*Trichosurus vulpecula*: Kerr). Unpublished PhD thesis, University of Queensland, Brisbane, Australia.

CHAPTER NINE

Impact of Possums on Primary Production

Stephen Butcher

Possums are of interest and concern to primary producers because they are a vector for bovine tuberculosis (bovine Tb, see Chapters 8 and 20) and because they eat or damage crops and pasture. Information on the timing, extent, and economics of losses caused by possums in rural areas is at best limited, and largely qualitative, often based on anecdotal comment and casual observation rather than measurement (Keber 1988). This chapter will examine what is known about possum effects on primary production (pastoral farming, horticulture, forestry) other than the transmission of Tb from possums to livestock.

Pastoral farming

Information on the impact of possums on pasture is conflicting. Hackwell & Bertram (1999) assert that possums consume NZ$12 million worth of pasture annually, yet 20 years previously the Agricultural Pests Destruction Council (APDC) considered the impact to be low (APDC 1980). Damage can be locally severe, especially on newly sown pasture or in areas within 40 m of adjacent forest (APDC 1980), as possums will travel up to 1200 m through forest to feed along the pasture edge (Green & Coleman 1981; Coleman *et al.* 1985).

Estimates of pasture loss have been derived in two ways. Firstly, by comparing differences in production in areas inside and outside of exclusion fences (Spurr & Jolly 1981) and, secondly, by analysing diet and estimating the amount of pasture eaten by possums (Gilmore 1965a,b, 1967; Harvie 1973; Warburton 1978; Coleman *et al.* 1985) and multiplying the quantity eaten by an estimate of the possum density to calculate the foregone stock equivalents (Harvie 1973).

Spurr & Jolly (1981) reported localised losses of up to 26% in a choumoellier/swede fodder crop, and estimated that the lost yield from this 0.4-ha block over 3 months would have supported 336 ewe equivalents for 1 day. Harvie (1973) estimated foregone production due to possum browsing of 2.5 sheep per hectare, with pasture comprising about 30% of the possums' diet. In Taranaki, the regional council pest control officers currently advise farmers that, for every 159 possums that graze on their farms, they forego production of an additional dairy cow (G. Railton, Taranaki Regional Council, pers. comm.). This is based on pasture consumption per possum per night of 0.11 kg (Harvey 1973) and consumption per cow per day of 17.5 kg (Maynard *et al.* 1979). On this basis a typical Taranaki dairy farm of 100 ha, with a possum population of 0.5–4.5 possums per hectare, would be forgoing production from an additional three milking cows. For 1999, the forgone production probably represents a loss of approximately NZ$2800 per farm. This estimate assumes that three milking cows produce approximately 900 kg of milk per year valued at NZ$3.15/kg (R. Leech, Wrightsons, Christchurch, pers. comm.).

Pastoral production lost due to possum browsing can be illustrated by comparing grazing potential before and after possum control. Loss of grazing potential is most obviously noticed by those farmers who practise intensive grazing and feed budgeting. Following possum control, there can be a noticeable increase in grazing capacity in pasture adjacent to bush/pasture margins. Such instances were observed following a Taranaki possum control operation in 1996 (M. Kennedy, National Possum Control Agencies, pers. comm.), but they are infrequently documented or measured.

Regrettably it appears that little progress has been made in quantifying pastoral farming losses due to possum browsing even though the need has been repeatedly recognised for more than 30 years (Harvie 1973; Keber 1988; Batcheler & Cowan 1988), and there are still few data to support or refute the assertion that possums are a significant pest to pastoral farms.

Horticulture

The effect of possums on horticultural production is even less well documented than damage to pasture sites. Possums feed on a variety of horticultural crops, such as pip and stone fruit, citrus, kiwifruit, avocados, macadamia nuts, walnuts, vegetables, and flowers (Table 9.1). They damage horticultural crops by eating the bark of fruit trees and vines; breaking branches and eating the buds, fruit, flowers, and

Table 9.1
Commercial crops consumed by possums in New Zealand.

Crop	Reference	Crop	Reference
Fruit & Vegetables		***Orchard trees***	
Beans	2	Pip fruit	2
Broccoli	1	Apples	1
Brussels sprouts	1	Nashi pears	1
Cabbages	1	Citrus fruit	2
Carrots	2	Stone fruit	2
Cauliflower	1	Peaches	1
Celery	1	Plums	1
Corn	2	Avocados	2
Courgettes	1	Cherimoyas	2
Kumara	1	Feijoas	1
Lettuces	1	Macadamias	2
Parsley	1	Persimmons	11
Parsnips	2	Tamarillos	1
Peas	2	Walnuts	9
Pumpkins	2		
Silverbeet	2	***Stock feed***	
Tomatoes	1	Pasture grasses	3
Grapes	8	Pasture weeds	7
Kiwifruit	2	Clover	4
Passion fruit	1	Choumoellier	4
Pepinos	1	Lucerne	4
Rock melons	1	Oats	10
		Swedes	4
Berry fruit		Turnips	4
Blackberries	3		
Blueberries	1	***Shelterbelt / erosion control tree species***	
Boysenberries	1	Poplar (*Populus* spp.)	5
Strawberries	1	Willow (*Salix* spp.)	5
Flowers		***Plantation trees***	
Carnations	2	Macrocarpa (*Cupressus* sp.)	3
Cyclamen	2	Douglas fir (*Pseudotsuga menziesii*)	6
Gladioli	2	Eucalypt (*Eucalyptus* spp.)	2
Godetias	2	Pine (at least 12 *Pinus* spp. including *P. radiata*)	1
Polyanthuses	2		
Roses	2		

1. Batcheler & Cowan 1988; 2. Nelson 1983; 3. Harvie 1973; 4. Spurr & Jolly 1981; 5. Jolly & Spurr 1981; 6. Warburton 1978; 7. Gilmore 1967; 8. Cowan 1991; 9. Jolly 1976; 10. Statham & Rayner 1995; 11. G. Hoskins, Agriculture New Zealand, Kumeu. pers. comm.

leaves; and by spoiling fruit (Nelson 1983; Cowan 1991). Damage tends to be highly seasonal and occurs when crops are maturing (Jolly 1976; Batcheler & Cowan 1988). Possum damage can also be influenced by whether plants are already protected by guards and nets against other vertebrate pests such as rabbits, hares, and birds (G. Hoskins, Agriculture New Zealand, Kumeu, pers. comm.).

While most damage is to maturing crops, some, such as branch breakage, can affect future crops by adversely affecting tree or vine growth and form years after the damage has occurred (C. Cook, Agriculture New Zealand, Nelson, pers. comm.). Even when the extent of damage is small, losses can be high especially where crops have a high unit value, such as nashi pears and avocados (Batcheler & Cowan 1988).

Damage to favoured crops tends to be patchy, and losses can be high when possums move in to an area to feed. For example, 1–2 ha of carrots were lost over 3 nights in a Pukekohe block that yielded 50 possums shot on the first night of a subsequent control operation (B. Hart, Agriculture New Zealand, Pukekohe, pers. comm.).

In addition to their direct effects on horticultural production, possums also seriously damage willows and poplars planted as shelter belts around orchards and market gardens. Damage to newly planted shelter belts often means a second planting is needed, increasing costs to the grower, and leading to slower establishment. This in turn leads to delays in the establishment and growth of crops, reducing yield and profit. Also, shelter belt plantings are a suitable habitat for possums (Batcheler & Cowan 1988), allowing populations to establish amongst crops or orchards.

The extent to which possums damage horticultural crops has been poorly quantified although Batcheler & Cowan (1988) noted that their effects are not significant enough to warrant much interest except at the local, farm level. It is clear, however, that possum damage to horticulture is seasonal, patchy, and only prevalent in areas adjacent to trees and native forest (APDC 1980).

Forestry

Although 72 tree species are used in commercial forestry in New Zealand, most commercial production comes from plantations, of which 90% are *Pinus radiata* (Ministry of Agriculture 1996). Consequently most possum damage to forestry affects *P. radiata*, though possum damage has been recorded on 11 other species of pine, eucalypts, poplars, and willows (Batcheler & Cowan 1988).

Possum damage in *P. radiata* forests is largely confined to young forests (Cowan 1991) and characteristically involves the browsing of terminal shoots, breakage of leaders and upper laterals, bark stripping, and staminate cone loss (Clout 1977; Keber 1988). Such damage was first reported in pine plantations in several areas in New Zealand in 1920 (Kirk 1920) and again in the North Island by the then New Zealand State Forest Service in 1932 when plantation areas were being established (Clout 1977). In the 1930s, possums were not considered a serious problem as there were few possums in the central North Island (Wodzicki 1950). By the 1960s and early 1970s, however, when the original 1930s plantings were harvested and replanted and new plantations were being established (Ministry of Agriculture 1996), possums had spread into most plantation areas and their damage had reached significant levels. In surveys of damage in exotic plantings between 1970 and 1975, up to 90% of young *P. radiata* were found to be browsed at some sites and up to 50% were thought to have been killed by damage caused by possums (Clout 1977). By the 1980s the amount of damage in pine plantations had abated, as trees had grown beyond the stage at which they were most vulnerable (0–14 years, Martin 1973) and the level of new plantings had declined (Ministry of Agriculture 1996).

As the 1980s progressed, the impact of possums on exotic forestry came to be considered only minor, although there were still areas of locally severe damage (Cowan 1991) where the damage was largely confined to the upper whorls of trees (K. Wright, Wellington Regional Council, pers. comm.). New plantings increased again in 1995, when about 95 000 ha were established, so it can reasonably be expected that losses of trees due to possums have also increased where possum densities are high.

Although possum damage has often been reported and losses assessed in some forest compartments, the cost of the damage has rarely been modelled. Griffiths (1985) sought to estimate whether control operations could be economically justified based on the value of the wood lost through damage, as estimated from the predicted loss of final crop stems, the stumpage and tree volume at harvesting. If losses due to possum browse were around 15%, he estimated the cost of these at harvesting (in 1985 dollars) as being about NZ$317 per hectare. Such losses are significant considering that the cost of reducing possum numbers, and hence risk of browse damage, by aerial poisoning in 1996 was approximately NZ$25 per hectare. At today's stumpage rates, which range from NZ$50 to $150 per cubic metre (Agri-fax 1998), even a 5% loss could be expected to cost between $282 and $840 per hectare at harvesting. This is based on final stocking rates of 333 stems per hectare, a log volume at harvest of 2.25 m^3, a rotation length of 25 years, and a discount rate of 7%.

Other impacts

Possums can lower productivity or add to costs in other areas of primary production. The honey industry, for example, has been particularly hard hit by possum damage. Production of bush honey has declined considerably, particularly in areas where possum damage to native bush has been high (Parliamentary Commissioner for the Environment 1994). The decline in flowering of native tree species such as rātā *(Metrosideros* spp.) and kāmahi (*Weinmannia racemosa*) has been identified as the main factor affecting bush honey production (Wodzicki 1950). An indication of the likely size of the decline in honey production caused by possums comes from data on the increase in [pōhutukawa] honey production during a possum eradication programme on Rangitoto Island (which is about 2300 ha in size). In 1993, the year after 90% of the possums were killed, honey production increased nearly sevenfold from 2.5 tonnes to 17 tonnes (Livingstone 1994). As possums eat clover flowers and pasture weeds (Harvie 1973), it is likely they also depress clover honey production, but this needs to be quantified.

Possums also damage newly planted willows (*Salix* spp.) and poplars *(Populus* spp.) commonly used for catchment protection and erosion control. They browse the buds and foliage, bite the bark, and break branches and leaders of most of the willow and poplar varieties planted (Jolly & Spurr 1981), and even clones specifically bred and planted for their resistance to possum browse are damaged or killed (Batcheler & Cowan 1988). In some areas possum damage to protection and erosion control plantings has been as high as 20% (Cowan 1991) and led to the cessation of planting programmes (Jolly 1980). In 1980, the annual cost of losses to catchment plantings was estimated to be NZ$800,000 dollars (Thomas *et al.* 1984), which in 1999 dollars represents losses of about NZ$3 million (adjusted for inflation using a conservative rate of 7% p.a.). Levels of catchment plantings have substantially declined since the early 1980s (L. Savage, Gisborne District Council, pers. comm.) and some farmers have even removed willows from waterways in an attempt to limit the number of possums on their farms (C. Arnesen, Agriculture New Zealand, Masterton, pers. comm.). In recent years possum damage to poplars and willows has declined due to improved possum control, continued development and planting of tree stock that are less palatable to possums (Wilkinson 1999), and the use of Dynex solid PVC sleeves and Netlon plastic mesh sleeves (D. Cameron, Wellington Regional Council, pers. comm.) to exclude possums from plantings. At NZ$8–10 per planted pole, the cost of replacing poles killed by possums can be considerable, and risks of erosion extended. Current plantings in the North Island are about 150 000 poles per annum (D. Knowles, Taranaki Regional Council, pers. comm.). Consequently if 10–20% of poles are lost due to possum damage, then annual losses are likely to be between NZ$150,000 and $300,000.

Other as yet unquantified but likely impacts of possums on primary production are the direct and indirect losses due to the transmission of diseases and parasites. Possums carry and transmit a range of parasites and diseases, such as liver fluke *Fasciola hepatica,* tape worm *Bertiella trichosuri,* coccidia *Eimeria* sp.; protozoans such as *Giardia* sp., *Cryptosporidium* sp., and *Leptospira* spp.; and nematodes such as *Trichostrongylus* spp. (see Chapter 7). While transmission of these diseases cannot be exclusively attributed to possums, as other vectors exist, the role of possums in the transmission and maintenance of parasites and diseases may be important and significantly reduce farm production. For instance, outbreaks of *Leptospira* spp. have killed up to 18% of sheep on some New Zealand farms (R. Marshall, Palmerston North, pers. comm.). Similarly, the high infection rate of *Leptospira* in possums (up to 66%, see Chapter 7) may well contribute to the comparatively high number of people infected with *Leptospira* spp. in New Zealand (10 : 100 000), which is noted as being one of the highest in the world (Hathaway 1981).

Summary

- Although it is recognised that possums have an impact on primary production, the extent of this is generally anecdotal and unquantified.
- Possum impacts on pastoral farming have been estimated in terms of grass eaten by possums rather than in dollars terms. Competition between stock and possums is only likely to be significant when the shared resource, grass, is in limited supply.
- Possum damage to horticulture is widespread. Damage has been reported to at least 46 varieties of fruit and vegetables. Damage is seasonal, patchy, and only prevalent in areas adjacent to trees and native forest.
- Most damage to forestry plantations occurs in stands of *P. radiata* up to 14 years of age. The prevalence of possum damage in New Zealand plantations peaked in the 1930s, 1960s, and the 1990s, corresponding with peaks in new plantings and replanting of harvested areas. The value of a 5% loss at planting in a *P. radiata* plantation represents losses at harvesting of between NZ$282 and $840 per hectare at current prices.
- Possums probably depress the honey harvest, but the extent of foregone production is unknown.
- Even when damage on a national scale is small, losses to individual farmers can be considerable.
- Possums can cause considerable damage to willows and poplars planted for catchment protection and erosion control, but the extent of losses varies from year to year and is not currently recorded.
- Possums carry a range of diseases and parasites that can infect humans and domestic stock.

References

Agri-Fax 1998: Pine log prices — Week beginning June 16th 1998. *New Zealand Forest Industries Magazine 29(8)*: 4.

Agricultural Pests Destruction Council (APDC) 1980: Opossum Survey Report 1980. 39 p.

Batcheler, C. L.; Cowan, P. E. 1988: Review of the status of the *possum (Trichosurus vulpecula)* in New Zealand. Unpublished FRI contract report to the Technical Advisory Committee (Animal Pests) for the Agriculture Pests Destruction Council, Department of Conservation, and Ministry of Agriculture and Fisheries, Wellington, 129 p.

Clout, M. N. 1977: The ecology of the possum (*Trichosurus vulpecula* Kerr) in *Pinus radiata* plantations. Unpublished PhD thesis, University of Auckland, Auckland, New Zealand.

Coleman, J. D.; Green, W. Q.; Polson, J. G. 1985: Diet of brushtail possums over a pasture-alpine gradient in Westland, New Zealand. *New Zealand Journal of Ecology 8*: 21–35.

Cowan, P. E. 1991: The ecological effects of possums on the New Zealand environment. *In:* Jackson, R. *convenor.* Symposium on tuberculosis. *Publication No. 132, Veterinary Continuing Education.* Palmerston North, New Zealand, Massey University. Pp. 73–88.

Gilmore, D. P. 1965a: Food of the opossum, *Trichosurus vulpecula*, in pastoral areas of Banks Peninsula, Canterbury. *Proceedings, New Zealand Ecological Society 12*: 10–13.

Gilmore, D. P. 1965b: Opossums eat pasture. *New Zealand Journal of Agriculture 110*: 284–286.

Gilmore D. P. 1967: Foods of the Australian opossum (*Trichosurus vulpecular* Kerr) on Banks Peninsula, Canterbury, and a comparison with other selected areas. *New Zealand Journal of Science 10*: 235–279.

Green, W. Q.; Coleman, J. D. 1981: A progress report on the movements of possums between native forest and pasture. *In:* Bell, B. D. *ed.* Proceedings of the first symposium on marsupials in New Zealand. *Zoology Publications from the Victoria University of Wellington No. 74.* Pp. 51–62.

Griffiths, A. 1985: Economics of wild animal control/management in production forests. *In:* Newton, A. E. *ed.* Proceedings of the 58th course in wild animal management: Changing the emphasis towards the 1990s. Rotorua, Forest Research Institute. Pp. 79–83.

Hackwell, K.; Bertram, K. 1999: Pests and weeds. The cost of restoring an indigenous dawn chorus. Wellington, New Zealand Conservation Authority. 71 p.

Hathaway, S. C. 1981: Leptospirosis in the possum *Trichosurus vulpecula. In:* Bell, B. D. *ed.* Proceedings of the first symposium on marsupials in New Zealand. *Zoology Publications from the Victoria University of Wellington No. 74.* Pp. 157–162.

Harvie, A. E. 1973: Diet of the opossum (*Trichosurus vulpecula* Kerr) on farmland northeast of Waverley, New Zealand. *Proceedings, New Zealand Ecological Society 20*: 48–52.

Jolly, J. N. 1976: Habitat use and movements of the opossum (*Trichosurus vulpecula*) in a pastoral habitat on Banks Peninsula. *Proceedings, New Zealand Ecological Society 23*: 70–78.

Jolly, J. N. 1980: Assessment of sleeve barriers for protection of erosion control plantings from damage by possums, *Trichosurus vulpecula. Protection Forestry Report No. 169.* Christchurch, Protection Forestry Division, Forest Research Institute. 18 p.

Jolly, J. N.; Spurr, E. B. 1981: Damage by possums *Trichosurus vulpecula* to erosion-control plantings. *In:* Bell, B. D. *ed.* Proceedings of the first symposium on marsupials in New Zealand. *Zoology Publications from the Victoria University of Wellington No. 74.* Pp. 205–210.

Keber, A. W. 1988: An enquiry into the economic significance of possum damage in an exotic forest near Taupo. Unpublished PhD thesis, 2 vols, University of Auckland, Auckland, New Zealand.

Kirk, H. B. 1920: Opossums in New Zealand: report on Australian opossums in New Zealand. *Appendix to the Journals of the House of Representatives of New Zealand H–28*: 1–12.

Livingstone, P. 1994: Infra red camera trail to identify possum and other feral/wild animal density/location. *In:* O'Donnell, C. F. J. *comp.*

Possums as conservation pests. Proceedings of an NSSC Workshop ... 29–30 November 1994. Wellington, Department of Conservation. P. 54.

Martin, J. T. 1973: Report on opossum damage in New Zealand Forest Products exotic forests. Unpublished report, New Zealand Forest Products.

Maynard, L. A.; Loosli, J. K.; Hintz, H. F.; Warner, R. G. 1979: Animal nutrition. New York, McGraw-Hill.

Ministry of Agriculture 1996: New Zealand forestry statistics 1995. Wellington, Ministry of Agriculture. 158 p.

Nelson, P. C. 1983: Possum control - Horticultural crops and orchards. Poisons, repellents and protective measures. Aglink HPP 273, 1st revision. Wellington, Ministry of Agriculture and Fisheries. 4 p.

Parliamentary Commissioner for the Environment 1994: Possum management in New Zealand. Wellington, New Zealand, Office of the Parliamentary Commissioner for the Environment. 196 p.

Spurr, E. B.; Jolly, J. N. 1981: Damage by possums *Trichourus vulpecula* to farm crops and pasture. *In:* Bell, B. D. *ed.* Proceedings of the first symposium on marsupials in New Zealand. *Zoology Publications from the Victoria University of Wellington No. 74.* Pp. 197–203.

Statham, M.; Rayner, P. J. 1995: Loss of pasture and crop to native animals in Tasmania. Proceedings: 10th Australian Vertebrate Pest Control Conference, Hobart, Tasmania, 29 May–2 June 1995. Tasmania, Department of Primary Industry and Fisheries. Pp. 171–176.

Thomas, M. D.; Warburton, B.; Coleman, J. D. 1984: Brush-tailed possum *(Trichosurus vulpecula)* movements about an erosion-control planting of poplars. *New Zealand Journal of Zoology 11*: 429–436.

Warburton, B. 1978: Foods of the Australian brush-tailed opossum (*Trichosurus vulpecula*) in an exotic forest. *New Zealand Journal of Ecology 1*: 126-131.

Wilkinson, A. G. 1999: Introduced forest trees in New Zealand: recognition, role and seed source. 18. The Poplars – *Populus* spp. *FRI Bulletin No. 124.* Rotorua, Forest Research Institute (NZ).

Wodzicki, K. A. 1950: Introduced mammals of New Zealand: an ecological and economic survey. *DSIR Bulletin No. 98.* Wellington, Department of Scientific and Industrial Research.

CHAPTER TEN

Damage to Native Forests

Ian Payton

New Zealand's early separation from the Gondwanan landmass resulted in a flora that evolved in the absence of mammalian herbivory (Fleming 1979). Although birds such as kererū (*Hemiphaga novaeseelandiae*) and kōkako (*Callaeas cinerea*) may have exerted localised browsing pressure on the canopies of at least some tree and shrub species (Clout & Hay 1989), there are no reports of native vertebrates having a significant impact on forest canopies. Brushtail possums (*Trichosurus vulpecula*) thus took up a largely unoccupied niche in New Zealand's native forest ecosystems (Caughley 1989). This chapter examines changing attitudes to the presence of this nocturnal arboreal foliovore, reviews the literature on patterns of possum damage in New Zealand forests and the rate and extent of these changes, and outlines options for monitoring possum damage to forest communities.

Changing attitudes

Despite the presence of possums in some New Zealand forests as early as 1840, their detrimental impacts were not recognised until much later and were the subject of considerable controversy (Pracy 1974; Cowan 1992). E. Phillips Turner (in McKinnon & Coughlan (1960)) summed up the debate in the annual report of the State Forestry Department for 1918 when he noted that "where these animals [possums] are abundant they would have an appreciably deterrent effect on natural vegetation", but concluded that "the damage they do may be compensated for by the return from the skins. A more serious pest are imported deer."

In 1919, as a result of continuing disagreement between pro- and anti-possum lobbies, the government commissioned Professor H. B. Kirk to investigate the advisability of further possum liberations. Kirk (1920) reported that he had "found no native tree that has, in my opinion, been killed by an opossum", and concluded that he was "convinced that opossums do no serious damage to the New Zealand bush." On these grounds he recommended that "opossums may, in my opinion, with advantage be liberated in all forest districts except where the forest is fringed by orchards or has plantations of imported tree species in the neighbourhood." Perham (1924) similarly failed to attribute forest damage in Westland to browsing by brushtail possums. He observed that while many southern rātā (*Metrosideros umbellata*) and fuchsia (*Fuchsia excorticata*) trees on Mt Tuhua were dead or dying others remained quite healthy, but concluded that the damage could not be the result of possum browsing because "if the opossum is to be blamed, all the specimens of a [plant] species in a forest where opossums are numerous, should be in approximately the same state." We now recognise this pattern of damage as typical of possum browsing. The perception that deer and other ungulate species were the primary damaging agents in native forests and that possum browsing had few consequences for the native vegetation was also held by Cockayne (1928), and strongly influenced public opinion and government policies toward the possum until the mid-1940s. This was despite growing concerns (e.g., Skipworth 1928; Zotov *et al.* 1938) that the selective feeding habits of possums might pose a threat to native forest communities.

By the early 1940s there was increasing evidence that expanding possum populations were seriously damaging at least some native forest communities. Kehoe (1946) lamented the demise of southern-rātā-dominated forests in Westland and the resultant reduction in rātā honey production, and both Pracy & Kean (1949) and Zotov (1947) documented damage by possums and deer to native forests in the lower North Island ranges. In 1955 escalating tree mortality in Westland hill-country forests prompted the Conservator of Forests to report that "damage by this animal [possum] in protection forests is becoming more widespread and serious yearly" (McKinnon & Caughlan 1960 p. 45). That year the Government commissioned an interdepartmental taskforce to determine causes of canopy tree mortality in the southern rātā–kāmahi (*Weinmannia racemosa*) forests of Westland. Their report (Chavasse 1955 summary p. 2) concluded that "opossums are primarily responsible for major defoliation of the canopy trees, that ground browsing animals are responsible for destruction of shrub and ground layers, for trampling and destroying the duff layer, and for the complete inhibition of regeneration. Both are responsible for the alteration of the microclimate, possibly leading to damage by drought, and for the destruction of optimum conditions for insectivorous birds. These activities lead to insect damage to weakened trees, (and possibly to fungal attack at the tree roots) causing death. Insects may also have been encouraged, particularly those feeding on stems and branches, by the reduction in species of birds brought about by introduced disease, cats, rats and mustelids."

Since that time numerous studies have described possum damage to both native plant species and forest communities. Despite the conclusions of Chavasse (1955) that the dieback process was multifaceted, most of these studies have related observed changes in vegetation composition or structure solely to the impacts of introduced mammalian herbivore populations.

The idea that possums have been the primary cause of the increased tree mortality and widespread forest dieback that has been observed in many New Zealand forests over the last 50 years has not been without its critics. For the Ruahine Ranges, Mosley (1978) and Grant (1989) have argued that the large-scale forest collapse and accelerated rates of erosion reported in the 1950s were initiated during a period of increased storminess during the late 1930s, at a time when the forests were already in decline, and before introduced animal (possum and deer) populations could have become a major factor. In support of his contention that deterioration of the forest cover is fundamentally natural but perhaps exacerbated by introduced animals Mosley (1978) cites observations by Elder (1958) on the old and even-aged nature of much of the dead and dying kāmahi forest, and the early and complete death of kāmahi on exposed ridges and steep faces during periods of drought. Rather than initiating periods of instability Mosley (1978 p. 21) concludes that introduced animal populations "have probably been responsible primarily for prolonging the period of above-average erosion by retarding recovery of the forest cover."

In the Kaimai Ranges the role of browsing animals in the widespread mortality and increased landslide frequency of mid to upper elevation forests (Dale & James 1977) was challenged by Jane & Green (1983, 1986). They argued that soil and climatic factors predisposed the forests to periodic drought damage, and used tree-ring data to link forest mortality episodes to periods of severe drought in 1914 and 1946. As in the Ruahine Ranges, introduced animal (possum, goat) populations were seen as delaying forest recovery rather than initiating forest dieback.

Veblen & Stewart (1982a) similarly questioned the role of possums in the extensive dieback and collapse of many Westland southern rātā–kāmahi forests. They noted that canopy dieback in these forests had been reported before possum populations were present (Holloway 1957; Morgan 1908), and proposed that much of the apparently excessive tree mortality observed since the 1940s could be explained by natural stand dynamics. In support of their hypothesis they argued that the apparently even-aged structure of many southern rātā–kāmahi stands, which is a feature of Westland hill country forests, would eventually result in many trees in a given area becoming senescent, and therefore more susceptible to a range of damaging agents, at approximately the same time. If, as Wardle (1980) suggested, the abundance of even-aged southern rātā–kāmahi stands along the Alpine Fault (a crustal plate boundary that traverses Westland) resulted from a major earthquake that triggered massive landslides, the synchronous senescence of these stands could be expected to result in widespread forest dieback. In this explanation possum browsing is seen as one of a range of deleterious factors triggering the demise of already overmature or senescent forest stands.

Subsequent studies have reinforced the conclusions reached by Chavasse (1955) that browsing by brushtail possums has resulted in increased levels of canopy dieback in New Zealand forests over the last 50 years, and that this has had flow-on effects throughout the forest ecosystem. Not all damage to native forests, however, is attributable to possums and/or other introduced mammalian browsers. Additional factors known to trigger or accelerate dieback in New Zealand forest communities include drought, earthquakes, flooding, frost, hailstorms, insects and fungal pathogens, snow, and wind (summarised in Wardle 1984; Stewart 1989; Wardle 1991).

Patterns of damage

Despite their presence throughout most forested areas on the main islands of New Zealand, the impact of brushtail possums has been far from uniform. This variability occurs within and between plant populations, communities, and ecosystems. It is influenced by a range of biotic and abiotic factors, which may predispose plant communities to possum damage, trigger damage episodes, or accelerate the rate of vegetation change (Manion 1981; Stewart & Rose 1988).

Variable impacts within populations of the same plant species

Within forest stands possum browsing is frequently concentrated on only a few trees, which may be heavily defoliated or killed (termed salt-and-pepper dieback) while neighbouring individuals of the same species remain largely unaffected (Fig. 10.1; Kean & Pracy 1953; Elder 1965). The link between possums and the death of individual trees was demonstrated by Meads (1976), who monitored possum browsing of northern rātā (*Metrosideros robusta*) trees in the Orongorongo Valley near Wellington between 1969 and 1974. Over this period average levels of possum browsing of northern rātā increased, 3 of 24 trees in the study died as a result of continued heavy browsing, and 5 recovered when fitted with metal sheaths that prevented possums gaining access to the canopy. This led Meads to predict that ongoing possum browsing would result in accelerated mortality of northern rātā in the Orongorongo Valley, as existing trees died and browsing pressure on remaining trees increased. In 1990, all 21 trees surviving in 1974 were still alive (Cowan *et al.* 1997). Annual reassessment of 15 of these trees between 1990 and 1994 showed that while average levels of possum browsing were increasing, this varied markedly between years and from tree to tree. Based on these observations and long-term datasets on possum density and climate in the Orongorongo Valley, Cowan *et al.* (1997) suggest that the high level of tree mortality in 1970–1974 and the increasing damage observed between 1990 and 1994 are likely to have resulted from a combination of above-average possum numbers and stress factors such as drought.

A link between drought stress and susceptibility to defoliation has also been suggested for insect-related decline (Hosking & Hutcheson 1986) and mortality (Hosking & Kershaw 1985) in beech (*Nothofagus*) forests, and is consistent with observed patterns of southern rātā mortality in Westland rātā-kāmahi forests, where trees on steep ridges with shallow soils appear less tolerant of possum browsing than those on nearby sites with deeper soils.

The extent to which selective browsing is the result of animal behaviour or is related to differences in the foliar chemistry of individual plants is not clear. For southern rātā heavy defoliation has been shown to increase nitrogen, phosphorus, and potassium concentrations in the regrowth foliage (Payton 1989a), which may provide a rationale for the repeated browsing of individual trees.

Variable impacts between populations of the same plant species

Regional differences in the possum-related decline of native plant species have been most clearly documented for fuchsia (Batcheler 1983; Sweetapple & Nugent 1999) and the loranthaceous mistletoes (de Lange & Norton 1997). Possum browsing of fuchsia, a small tree species characteristic of regenerating forest and forest margins, has all but eliminated this species from many forested areas (Coleman *et al.* 1980; Green 1984). For south Westland forests Pekelharing *et al.* (1998) have shown a strong relationship between possum density and fuchsia mortality, with most trees dying within a few years of possum numbers exceeding a trap-catch density index of 25%. Reduction of the possum population reversed this trend, even where trees were heavily defoliated. Not all fuchsia populations are similarly affected, and a reddish-leaved form remains common in many eastern localities despite being browsed by long-established possum populations (Batcheler 1983). Possum browsing rarely kills all fuchsia trees and, even where there are high levels of mortality, a few healthy and apparently unbrowsed individuals can usually be found (Sweetapple & Nugent 1999; I. J. Payton pers. obs.).

A parallel situation exists with the loranthaceous mistletoes. All five extant New Zealand species (*Alepis flavida, Peraxilla colensoi, P. tetrapetala, Ileostylus micranthus, Tupeia antarctica*) were formerly more widespread and abundant, and there is a large body of circumstantial evidence that links their decline and local extinction to the spread of brushtail possums (summarised in Ogle & Wilson 1985; Ogle 1997). But the relationship between possum browsing and mistletoe decline is not as clear as much of the anecdotal evidence suggests. Both *Peraxilla* species and *A. flavida* remain common in central and southern South Island beech forests with long-established possum populations, and *T. antarctica* and *I. micranthus* are locally abundant on a range of native and adventive host species throughout New Zealand (de Lange *et al.* 1997).

Fig. 10.1
Possum-related death of individual southern rātā trees (salt-and-pepper dieback) within an otherwise healthy forest stand. Architect Creek, Westland National Park.
Photo: L.E. Burrows.

Research studies are similarly inconclusive. In Nelson Lakes National Park low densities (<2/ha) of possums had a marked adverse effect on *Peraxilla* and *Alepis* species little more than a decade after possums first colonised the area (Wilson 1984), while in south Westland *P. colensoi* was little affected in forests with similar possum densities (0.5–1/ha) and a longer (30–40 year) history of possum occupation (Owen & Norton 1995). Thus while possum browsing is undoubtedly an important factor in the decline of at least some mistletoe populations, it is only one of several factors affecting the ongoing health of New Zealand mistletoe species. These include habitat and host tree loss, overcollecting, invertebrate herbivory (de Lange 1997), and the loss of pollinating and dispersal agents (Ladley & Kelly 1996; Ladley *et al.* 1997).

Variable impacts between plant communities

While brushtail possums are reported as browsing the foliage, flowers, and fruit of a wide range of plant species (Chapter 2), a few key species normally form the bulk of the diet. These typically fall into two broad groups: canopy dominants (e.g., kāmahi, kohekohe (*Dysoxylum spectabile*), northern and southern rātā) and shrub or small tree species (e.g., fuchsia, pāte (*Schefflera digitata*), wineberry (*Aristotelia serrata*)) that are characteristic of seral or regenerating vegetation. Where these species are major structural components of forests, possum damage is extensive and may lead to complete canopy collapse over large areas (Batcheler 1983; Payton 1987; Rose *et al.* 1992). Conversely, where browsing of the dominant tree species is minimal (e.g., beech species), floristic composition but not forest structure is typically affected (Wardle 1984).

The relationship between possum damage to native forests and the factors that predispose forest stands to dieback, trigger dieback episodes, or accelerate the dieback process, is best understood for the southern rātā–kāmahi forests of central Westland. Large-scale canopy dieback in these forests is common in the schist landscapes east of the Alpine Fault (Fig. 10.2) and rare on granitic substrates west of the faultline. East of the faultline a combination of high rainfall (5000–12 000 mm/annum) and tectonic activity has produced steeply dissected landscapes characterised by high rates of uplift and erosion (O'Loughlin & Pearce 1982), and rapid soil development (Basher *et al.* 1985). In this geologically active landscape southern rātā–kāmahi forests are a mosaic of communities whose composition is in large part dependent on the length of time since the site was last disturbed. They contain a high proportion of seral shrub and small tree species and are prime possum habitat (Wardle 1977; Reif & Allen 1988). Although stable landforms occur in areas where schist is the parent material, they are not a major feature of the landscape (Harrison 1985). West of the Alpine Fault the more indurated and less fractured granitic substrates produce soils that are typically older and less fertile than those derived from schist. On these more stable landscapes the frequency of disturbed sites and therefore seral vegetation decreases. As the time since the last major disturbance increases, soil fertility declines and preferred plant species are progressively replaced by

Fig. 10.2
Canopy collapse of montane forests dominated by southern rātā and kāmahi, Kokatahi Valley, Westland.
Photo: Robert B. Allen.

those characteristic of low fertility sites (Reif & Allen 1988), which are little browsed by brushtail possums.

Stand structure is also a factor that predisposes southern rātā–kāmahi forests to the damaging impacts of brushtail possums. Canopy dieback is greatest in mature stands containing a predominance of large trees (Fig. 10.2), and is unknown from the dense, young, even-aged pole stands that establish after landslides (Fig. 10.3; Wardle 1971; Stewart & Veblen 1982; Allen & Rose 1983). The reason for the greater susceptibility of larger (older) southern rātā is not clear, but may relate to a reduced ability to withstand the effects of defoliation (Payton 1983, 1985). Alternatively, mature stands may provide better possum habitat (Wardle 1971) and therefore be subject to greater browsing pressure. Although young pole stands are browsed by possums, they are characterised by higher tree vigour than older trees (Payton 1983), and a lack of possum-preferred seral species (Reif & Allen 1988). Where they are still undergoing natural thinning, canopy size is small, and the loss of individual trees (whether through competition or possum browsing) does not affect the ability of the stand to maintain an intact canopy cover (Payton 1988). Once the natural thinning process is complete, however, the loss of larger-crowned trees leads to canopy opening and continuing damage by accelerating factors such as wind, insects, and fungal pathogens (Payton 1988, 1989b). Where landscapes are less disturbed, regeneration patterns are characterised by smaller-scale (single tree to small group) replacement (Stewart & Veblen 1982). The survival of these mixed-age and uneven-structured stands after the loss of canopy trees appears to depend on their ability to fill the gaps created and retain a wind-resistant canopy (Payton 1988, 1989b).

The relationship between landscape stability and possum-related damage to forest communities provides a useful basis for determining the relative susceptibility of similar forests elsewhere in New Zealand to browsing by brushtail possums. Northern rātā–kāmahi forests in the Ruahine Ranges, which were subject to widespread canopy collapse from the 1940s (Elder 1965; Cunningham 1979), occur on intensely folded and faulted greywacke substrates, which are highly shattered and erodible (Mosley 1978). As on the schist substrates in Westland, disturbance increases the seral component of the vegetation, and larger-scale replacement patterns promote even-aged forest structures (Elder, in Mosley 1978). Conversely, in the Catlins district of S.E. Otago, where predominantly mixed-age southern rātā–kāmahi forests with a low seral component occur on mature landforms, the presence of possums for over 100 years does not appear to have significantly altered the composition of forest canopies (Fig. 10.4; Halkett & Leitch 1976).

Variable impacts between forest ecosystems

Conifer-broadleaved forests in the northern North Island are floristically diverse (Wardle 1991), contain a wide range of possum-palatable species (Cowan 1992; Chapter 2), and as such are vulnerable to the damaging impacts of brushtail possums (Payton *et al.* 1997a). The most conspicuous damage to forest canopies is the severe defoliation and death of

Fig. 10.3
Dense, young, even-aged pole stand of southern rātā and kāmahi surrounded by mature forest of similar composition affected by possum-related dieback. Kokatahi Valley, Westland.
Photo: Robert B. Allen.

Fig. 10.4
Southern rātā-kāmahi forests on mature landforms in the Catlins district, south-east Otago. Despite the presence of possums for over 100 years, forest canopies are little altered.
Photo: Ralph B. Allen.

northern rātā and kohekohe. Shrub and small tree species such as five-finger (*Pseudopanax arboreus*), fuchsia, māhoe (*Melicytus ramiflorus*), patē and raukawa (*Raukaua edgerleyi*), and the mamaku tree fern (*Cyathea medullaris*) are also defoliated and killed. Not all canopy tree species are palatable to possums, however, and where species such as kauri (*Agathis australis*) and taraire (*Beilschmiedia tarairi*) predominate, forest canopies are less modified. In Northland and on the Coromandel Peninsula damage is greatest in the south reflecting the relatively recent spread of possums into northern areas (Chapter 1). Damage can be rapid and affect both current and future plant generations. On the Coromandel Peninsula Ogden & Buddenhagen (1995) measured a > 50% loss in the basal area of kohekohe over 10 years, and concluded that, if this trend continued, the formerly dominant canopy species would be almost entirely lost as a canopy tree within 10–15 years. Similarly Esler (1983 p. 115) noted that kohekohe would almost certainly become a prominent species in logged forest areas in the Waitakere Ranges "but the plants are seriously damaged by opossums before the saplings reach the canopy."

Pōhutukawa (*Metrosideros excelsa*) forests are a feature of the Northland and Coromandel Peninsula coastlines, extending south towards Gisborne and New Plymouth on the east and west coasts, respectively. Possum damage to these forests occurs throughout their range but is most conspicuous on the east coast of Northland between Auckland and the Bay of Islands (Forest Research Institute 1989). Damage to individual pōhutukawa trees appears greatest in areas that are farmed or otherwise developed (Pracy 1978). Where coastal forests are less disturbed, possum damage is most severe within the salt spray zone, and in addition to pōhutukawa, whārangi (*Melicope ternata*), *Pittosporum umbellatum*, karamū (*Coprosma lucida*), and taupata (*C. repens*) are heavily browsed and killed. The damage does not appear to be caused by salt spray as forest cover on offshore rock stacks has remained healthy (N. Clunie, in Batcheler & Cowan 1988). On Rangitoto Island, a recent volcanic cone in the Hauraki Gulf from which brushtail possums and rock wallabies (*Petrogale penicillata*) have now been eliminated, pōhutukawa is the dominant and often sole canopy species on the lava flows (Atkinson 1959). In these forests most possum damage occurred in plant communities growing on blocky ('a'ā) lava flows (I. J. Payton unpubl. data), which in addition to pōhutukawa contain a range of possum-palatable understorey species. Possum browsing, canopy dieback, and tree mortality was greatest in the vicinity of black-backed gull (*Larus dominicanus*) colonies where the additional nutrient input (Francis 1970) is reflected in higher foliar concentrations of nitrogen and phosphorus (I. J. Payton unpubl. data). By contrast, canopy dieback is rare or absent in stunted pōhutukawa forests growing on sheet (pahoehoe) lava, which have a floristic composition characteristic of infertile sites and which are little browsed by brushtail possums.

Tawa (*Beilschmiedia tawa*) is the dominant canopy species in conifer/broadleaved forests that extend from the Bay of Plenty lowlands and Raukumara Ranges in the east, to the western King Country and inland Wanganui west of the Volcanic Plateau. These forests are regarded as good habitat for possums. Apart from the defoliation and death of scattered emergent northern rātā, possums have not greatly modified tawa forest canopies on ridge and hillslope sites. Most change occurs in the gully communities where possum-preferred shrub and small tree species are heavily browsed, and in some areas are replaced by *Uncinia*-dominated clearings, with ungulate (usually deer or goat) browsing preventing forest regeneration. In areas such as the Matemateonga and Ruahine ranges, where kāmahi replaces tawa as the dominant canopy species, canopy dieback is more evident and forest structures are more extensively modified (Batcheler 1983; I. J. Payton in Batcheler & Cowan 1988; Rogers & Leathwick 1997).

The pattern of possum damage typically observed in lower altitude conifer-broadleaved forests at southern North Island and South Island localities has been described from the Orongorongo Valley near Wellington by Fitzgerald (1976) and Campbell (1984, 1990). The forests are floristically diverse with a wide range of possum-palatable tree species. Possum-related defoliation and death of emergent northern rātā has been recorded from forests in the Orongorongo Valley since the 1930s (R. I. Kean, in Campbell 1984), and elsewhere in the region has resulted in the local extinction of this species (Druce 1971; Rogers & Leathwick 1997). Over the 60 years that vegetation records have been kept, possum-palatable species formerly common in the Orongorongo Valley have become uncommon and/or restricted in their distribution (e.g., fuchsia, five-finger, kāmahi, tītoki (*Alectryon excelsus*), toro (*Myrsine salicina*), tutu (*Coriaria arborea*)); rare and threatened by possum browsing (*Paratrophis banksii*, *Plagianthus regius*); or are presumed to be extinct (*Peraxilla tetrapetala*) (Fitzgerald 1976; Campbell

1984). If present trends continue, kāmahi is predicted to become extinct locally and northern rātā increasingly uncommon (Allen *et al.* 1997). The resulting forests will have a greater proportion of tree ferns, and more short-lived, small-diameter tree species such as pigeonwood (*Hedycarya arborea*), which are not favoured by possums (Campbell 1990).

A shift towards less palatable or unpalatable species was also noted in lowland forest and shrubland communities on Kapiti Island during the 1970s (Atkinson 1992). Before brushtail possums were exterminated in 1986 (Cowan 1992), they were reported as defoliating and killing emergent (northern rātā), canopy (e.g., kohekohe, tawa, hīnau (*Elaeocarpus dentatus*)), shrub (e.g., fuchsia, taupata, toru) and tree fern (mamaku) species. Kāmahi, however, which is locally abundant, was generally only lightly browsed. As in other areas, browsing damage and tree mortality varied greatly between individual plants of the same species, and from year to year.

Montane conifer-broadleaved forests in which Hall's tōtara (*Podocarpus hallii*) and pāhautea (*Libocedrus bidwillii*) predominate are a feature of the wetter mountains of central and southern New Zealand. In the North Island, canopy dieback of Hall's tōtara–pāhautea stands is widespread in Egmont National Park, Hihitahi Forest Sanctuary and the Ruahine Ranges, with 50–80% of emergent canopies showing severe dieback or collapse (Rogers 1997). Dieback patterns parallel those of southern rātā–kāmahi forests, with young even-aged stands, those on stable, low fertility sites, and those in wetter and colder climates least affected (Rogers 1997). Dead spars of both Hall's tōtara and pāhautea are also a feature of montane southern rātā–kāmahi forests in central Westland (Holloway 1959; Wardle 1978). Although possums are known to browse pāhautea and Hall's tōtara (Coleman *et al.* 1985; Nugent *et al.* 1997; Rogers 1997), their role in the dieback process is not well understood. Both species have slow to very slow growth rates (Veblen & Stewart 1982b) and an apparently low ratio of foliage to total biomass especially in mature trees. This suggests a tight energy balance and the ability of even low levels of possum browsing to trigger dieback (Payton 1983, 1987).

Beech and beech-broadleaved canopies are a feature of hill-country forests over large areas of the axial ranges in both the North and South Islands. Possums feed on the leaves and bark of beech trees, but they are not preferred foods, and damage to individual trees is usually minimal. Monotypic beech forests characteristic of drier and cooler climates contain few possum-preferred species, and are little damaged. As climates become wetter and warmer, the added broadleaved canopy (e.g., northern or southern rātā, kāmahi), and subcanopy (e.g., fuchsia, *Pseudopanax* spp.) component increases the susceptibility of beech forest communities to modification by brushtail possums (Wardle 1984; Rogers & Leathwick 1997). Where these species are an important component of regenerating beech forest communities, possum and ungulate browsing can lead to their replacement by turf communities or unpalatable woody species such as horopito (*Pseudowintera colorata*) (Wardle 1967; James 1974).

A recent survey of possum damage to beech-broadleaved forests in the northern South Island (Rose *et al.* 1995) found southern rātā, Hall's tōtara and kāmahi the canopy species most heavily browsed, with 10% of trees estimated to have died over the last 15–30 years. Shrub and small tree species affected included fuchsia, five-finger, māhoe, mountain ribbonwood (*Hoheria lyallii*) and wineberry, and damage appeared to be progressive rather than synchronous as observed in some rātā-kāmahi forest communities. Similar patterns of possum damage have been reported from beech-broadleaved forests in northern (Wardle 1974) and southern Westland (Rose *et al.* 1993; Owen & Norton 1995), and western Otago (Smale *et al.* 1996; Burrows *et al.* 1997). Canopy species most affected are southern rātā and Hall's tōtara. Damage is most extensive and conspicuous in seral forest communities and the understorey of mature forests, which often contain high frequencies of fuchsia, patē, and wineberry (Rose *et al.* 1993). Reports of possum damage to forest communities in the southern South Island and on Stewart Island are sparse, despite the sometimes lengthy presence of possum populations. On Stewart Island, where coastal conifer-broadleaved forests are subject to periodic dieback from wind-blown salt spray (Stewart & Burrows 1989), possums are reported as killing canopy trees at Lord's River (Fig. 10.5; L. E. Burrows, Landcare Research, Lincoln, pers. comm.), and decimating fuchsia-dominated gully communities (Coleman & Pekelharing 1991).

Rate and extent of change

Brushtail possum populations have now modified most New Zealand forests. The rate and extent of these changes varies widely. Colonising possum

Fig. 10.5
Possum-induced dieback of southern rātā–kāmahi forests, Lord's River, Stewart Island.
Photo: L. E. Burrows

populations grow slowly at first, go through a relatively short-lived irruptive phase to reach peak density, then decline sharply to lower and more stable levels (Pracy 1975). Damage to individual trees (termed salt-and-pepper dieback) is typically observed within 10–25 years of colonisation (Leutert 1988; Payton *et al.* 1997a). Where forest communities are predisposed to possum damage, change may be rapid, especially during the irruptive phase. In the southern Ruahine Ranges conspicuous defoliation of kāmahi was evident in mid-altitude forests within 10 years of possums colonising the area, and complete canopy collapse occurred within 15–20 years (Batcheler 1983). Comparison of aerial photographs taken in 1946 and 1978 showed that tall forest in this area had been reduced by 68%, and that shrublands and regenerating forests formerly dominated by fuchsia and wineberry were now chiefly composed of horopito and māhoe. For Westland southern rātā–kāmahi forests Rose *et al.* (1992) estimated that 20% of canopy trees (predominantly southern rātā and Hall's tōtara) had died over the previous 40 years, and Pekelharing & Batcheler (1990) report 40% of canopy trees defoliated between 1960 and 1973 in forest communities in the Taramakau Valley. On the east bank of the Adams River (a tributary of the Wanganui River, Westland) an irruptive possum population severely defoliated or killed 50% of southern rātā trees over a 2-year period in the late 1980s (Rose *et al.*1992), and 6 years later had effected almost 100% canopy mortality in some areas (I. J. Payton pers. obs.). Where possum populations are post-peak and/or forest communities less vulnerable to modification, vegetation change is more gradual and is characterised by the progressive decline of possum-preferred species (e.g., Fitzgerald 1976; Thomas *et al.* 1993).

Options for monitoring possum damage to forest communities

Although the influence of introduced animal pests on New Zealand's native forest ecosystems has been extensively monitored for over 50 years (Stewart *et al.* 1989), attempts to quantify possum damage to forest canopies are more recent (Meads 1976; Leutert 1988; Payton 1988; Pekelharing & Batcheler 1990; Atkinson 1992; Payton *et al.* 1997a). Techniques that have been used to assess changes to New Zealand forest canopies include:

Descriptive accounts: These range from broad overviews (e.g., Kean & Pracy 1953; Batcheler & Cowan 1988) to detailed descriptions of possum damage to specific areas or forest communities (e.g., Zotov 1947; Pekelharing & Reynolds 1983). They are largely subjective and strongly influenced by the author's observational skills and perceptions (e.g., Kirk 1920). The reconnaissance (Recce) plot method, a standardised non-area method of vegetation description (Allen 1992) that records animal (deer, goat, possum etc.) damage to plant species, has been used to describe the impact of ungulates on forest understorey vegetation (e.g., Wardle 1974) and could be used to assess the extent of possum damage to forest canopies.

Permanent forest plots: These use repeated measures of tagged trees, saplings, and seedlings to monitor change (Allen 1993). They provide some information about changes taking place in the forest canopy in that they identify turnover in the canopy tier. Although not sensitive short-term indicators of canopy damage, permanent (20 m × 20 m) forest plots provide detailed information on longer-term changes in both the composition and structure of the vegetation that are not readily obtainable using other methods.

Point-height intercept: This technique involves projecting a point or cylinder up (or down) through the vegetation and recording what it intercepts. It provides information on forest composition and structure (Park 1973; Leathwick *et al.* 1983), and with repeat measurement can be used to detect changes to forest canopies. In forests, the multi-layered nature of the vegetation, the accurate measurement of the height of intercepts, and the need to determine whether foliar intercepts are browsed, pose major problems. Data collection is also time-consuming. This technique has proved effective for monitoring non-forest communities (Dickinson *et al.* 1992), but has little to recommend it as a method for monitoring canopy vegetation in tall forest.

Photopoints: These are oblique photographs taken from a marked point. They can provide a good visual record of changes in the foliage cover and extent of dieback in tree canopies, but are not readily amenable to quantitative analysis (Beadel 1987), and cannot be used to assess the extent or severity of browsing. Their use is most appropriate where tree canopies are clearly visible, and where there have been no changes in the understorey that would obscure the field of vision.

Hemispherical (fish-eye) photography: The technique uses a wide-angle (fish-eye) lens to take photographs looking up through the forest canopy. In New Zealand it has been used to quantify understorey light environments in conifer-broadleaved (Veblen & Stewart 1982b) and mountain beech (Hunt & Hollinger 1988) forests. Fish-eye photography can provide an accurate measure of canopy cover and light penetration (Lasko 1980). It does not allow individual plant species to be identified or browsing to be quantified. The development of computerised analysis systems (Chan *et al.* 1986; Rich 1989) has largely eliminated the lengthy time required for image analysis, making fish-eye photography a practical means of assessing changes in forest canopy cover. To monitor change (i.e., repeat measurement) the ability to relocate the exact position from which the original photo was taken is critical. As with other photographic techniques, it is easy to amass a much larger number of photographs than are needed or can be readily processed.

Direct observation: This approach has been used to assess the impact of insect populations (e.g., Wickman 1979; Fox & Morrow 1983), nutrient deficiency (e.g., Hunter *et al.* 1991), and pollutants (e.g., Innes 1988) on forest health. Most studies are of one, or a few, dominant timber-producing species. Tree canopies are assigned to one of a series of predetermined classes, often defined using diagrams or photographs (Innes 1990). Because the approach relies on subjective assessments rather than counts or measurements, careful attention needs to be paid to questions of reliability and repeatability (Innes 1988; Innes & Boswell 1990). In New Zealand several studies have quantified possum damage to forest canopies by direct observation of either individual leaf bunches (Meads 1976; Payton 1988) or whole tree canopies (Leutert 1988; Atkinson 1992; Payton *et al.* 1997a). The Foliar Browse Index method (Payton *et al.* 1999) combines direct observation with the use of indicator species to provide an assessment of possum damage to forest communities.

Aerial photography: Visual aerial photographic interpretation is commonly used for large-scale mapping of forest vegetation and damage to forest canopies (e.g., Avery 1966). In New Zealand, Rose *et al.* (1988) have shown that the amount of dieback visible on aerial photographs is strongly correlated with ground-based measures of defoliation for major canopy tree species such as southern rātā and Hall's tōtara. Dieback is mapped in spatial units defined by topographic features (e.g., creeks, ridges, or bluff systems) using a predetermined damage scale. Recent studies include mapping possum damage to forests in Nelson and Marlborough (Rose 1994; Rose *et al.* 1995), central and southern Westland (Rose *et al.* 1988, 1992, 1993), and western Otago (Smale *et al.* 1996; Burrows *et al.* 1997).

Airborne video: This technology uses a video camera fitted to a fixed-wing plane or helicopter and a map and image-processing system, to obtain and analyse images of the forest canopy (Pywell & Myhre 1990; Hosking *et al.* 1992; Hosking 1995). Compared to aerial photography, video images are low cost, can be obtained in weather conditions not suitable for aerial photography, and are immediately available.

Disadvantages include a lower resolution than for photographic film, and the potentially less permanent nature of videotape as a storage medium. As with other remote-sensing techniques, the usefulness of airborne video hinges on the ability of the image-processing system to resolve and quantify the relevant forest canopy characteristics.

Remote sensing of aerial photographic or satellite images: This approach has the potential to provide large-scale, low-cost assessment of changes in forest canopy condition, once smaller areas are able to be resolved and the relationships between spectral signature, vegetation type, and degree of damage have been established. Trials using colour-infrared aerial photographs of pōhutukawa forest on Rangitoto Island in the Hauraki Gulf have shown a good relationship between remotely sensed measures of total green leaf biomass and field scores of percentage leaf cover (Trotter 1992). In common with other remote-sensing techniques, image analysis must be accompanied by adequate ground survey to confirm the validity of the analyses and to determine the cause of the changes or damage observed.

Summary

- Despite their presence in some New Zealand forests as early as 1840, the detrimental impacts of brushtail possums were not recognised for nearly a century and were the subject of considerable controversy. While possum browsing has increased levels of canopy dieback in New Zealand forests, not all forest damage can be attributed to possums or other introduced mammalian browsers. Additional factors reported as damaging New Zealand forest communities include drought, earthquakes, flooding, frost, hailstorms, insects and fungal pathogens, snow, and wind.
- Brushtail possum damage to New Zealand forest communities varies widely. Where preferred foods are dominant, canopy species damage can be extensive and may lead to complete canopy collapse. Conversely, where browsing of the dominant tree species is minimal, floristic composition but not forest structure is typically affected.
- Possum damage varies within and between plant populations, communities, and ecosystems. It is influenced by a range of biotic and abiotic factors, which may predispose plant communities to possum damage, trigger damage episodes, or accelerate the rate of vegetation change.
- Within forest communities possum browsing is frequently concentrated on a few trees, which may be heavily defoliated or killed while neighbouring individuals of the same species remain largely unaffected. At a regional scale, plant species such as mistletoe or fuchsia may be either threatened with extinction or co-exist with long-established possum populations. In neither case do we understand the reasons for these patterns of damage.
- Floristic composition and stand structure predispose forest communities and ecosystems to damage by brushtail possums. Both are largely determined by the stability of the landscape. Forests most affected are characterised by abundant seral shrub and small tree species and large-scale replacement patterns, typical of landscapes prone to natural geologic disturbance. As landforms mature and the frequency of disturbed sites and seral vegetation declines, preferred plant species are progressively replaced by those characteristic of low-fertility habitats, which are little browsed by brushtail possums.
- Where forest communities are highly predisposed to possum damage, change may be rapid, especially where colonising populations have not yet reached peak density. Conspicuous defoliation of forest canopies may occur within 10 years of colonisation, and complete canopy collapse within 15–20 years. Where possum populations are post-peak and/or forest communities less vulnerable to modification, vegetation change is more gradual and is characterised by the progressive decline of possum-preferred plant species.
- Techniques that have been used to assess changes to New Zealand forest canopies include descriptive accounts, permanent (20 m × 20 m) forest plots, point-height intercept, photopoints, hemispherical (fish-eye) photography, direct observation, and the visual or remotely sensed interpretation of aerial photographs, videotape, or satellite images. Foliar Browse Index, a recently developed method that uses a direct observation approach, is most commonly used for monitoring possum damage to forests.

References

Allen, R. B. 1992: RECCE: an inventory method for describing New Zealand vegetation. *FRI Bulletin No. 181*. Christchurch, Ministry of Forestry. 25 p.

Allen, R. B. 1993: A permanent plot method for monitoring changes in indigenous forests. Christchurch, Manaaki Whenua – Landcare Research. 35 p.

Allen, R. B.; Rose, A. B. 1983: Regeneration of southern rata (*Metrosideros umbellata*) and kamahi (*Weinmannia racemosa*) in areas of dieback *Pacific Science 37*: 433–442.

Allen, R. B.; Fitzgerald, A. E.; Efford, M. G. 1997: Long-term changes and seasonal patterns in possum (*Trichosurus vulpecula*) leaf diet, Orongorongo Valley, Wellington, New Zealand. *New Zealand Journal of Ecology 21*: 181–186.

Atkinson, I. A. E. 1959: Forest vegetation of the inner islands of the Hauraki Gulf. *Proceedings, New Zealand Ecological Society 7*: 29–33.

Atkinson, I. A. E. 1992: Effects of possums on the vegetation of Kapiti Island and changes following possum eradication. DSIR Land Resources Contract Report 92/52 (unpublished) 68 p.

Avery, T. E. 1966: Foresters guide to aerial photo interpretation. USDA *Forest Service, Agriculture Handbook No. 304*. Washington DC, USA.

Basher, L. R.; Tonkin, P. J.; Daly, G. T. 1985: Pedogenesis, erosion and revegetation in a mountainous, high rainfall area – Cropp River, central Westland. *In:* Campbell, I. B. *ed.* Proceedings of the soil dynamics and land use seminar, Blenheim, May 1985. New Zealand Society of Soil Science, Lower Hutt, and New Zealand Soil Conservators Association. Pp. 49–64.

Batcheler, C. L. 1983: The possum and rata-kamahi dieback in New Zealand: a review. *Pacific Science 37*: 415–426.

Batcheler, C. L.; Cowan, P. E. 1988: Review of the status of the possum (*Trichosurus vulpecula*) in New Zealand. Unpublished FRI contract report to the Technical Advisory Committee (Animal Pests) for the Agriculture Pests Destruction Council, Department of Conservation, and Ministry of Agriculture and Fisheries. 129 p.

Beadel, S. M. 1987: Assessment of vegetation condition using permanent photographic points, Horomanga Catchment, Urewera National Park. *Regional Report Series No.2*. Rotorua, Department of Conservation (unpublished) 48 p.

Burrows, L. E.; Pekelharing, C. J.; Savage, T. J. 1997: Canopy dieback and the impact of possums in the conservation forests of west Otago. Landcare Research Contract Report LC9697/137 (unpublished) 24 p.

Campbell, D. J. 1984: The vascular flora of the DSIR study area lower Orongorongo Valley, Wellington, New Zealand. *New Zealand Journal of Botany 22*: 223–270.

Campbell, D. J. 1990: Changes in structure and composition of a New Zealand lowland forest inhabited by brushtail possums. *Pacific Science 44*: 277–296.

Caughley, G. 1989: New Zealand plant-herbivore systems: past and present. *New Zealand Journal of Ecology 12*: 3–10.

Chan, S. S.; McCreight, R. W.; Walstad, J. D.; Spies, T. A. 1986: Evaluating forest vegetative cover with computerized analysis of fisheye photographs. *Forest Science 32*: 1085–1091.

Chavasse, C. G. R. 1955: Mortality in rata/kamahi protection forests, Westland. Unpublished report. Wellington, New Zealand Forest Service.

Clout, M. N.; Hay, J. R. 1989: The importance of birds as browsers, pollinators and seed dispersers in New Zealand forests. *New Zealand Journal of Ecology 12*: 27–33.

Cockayne, L. 1928: Monograph on the New Zealand beech forests. Part 2. The forests from the practical and economic standpoints. *Bulletin No.4. New Zealand Forest Service*. Wellington. 59 p.

Coleman, J. D.; Pekelharing, C. J. 1991: The role of brushtail possums in coastal forest dieback on Stewart Island. Unpublished manuscript held at Landcare Research, Lincoln.

Coleman, J. D.; Gillman, A.; Green, W. Q. 1980: Forest patterns and possum densities within podocarp/mixed hardwood forests on Mt Bryan O'Lynn, Westland. *New Zealand Journal of Ecology 3*: 69–84.

Coleman, J. D.; Green, W. Q.; Polson, J. G. 1985: Diet of brushtail possums over a pasture-alpine gradient in Westland, New Zealand. *New Zealand Journal of Ecology 8*: 21–35.

Cowan, P. E. 1992: The eradication of introduced Australian brushtail possums, *Trichosurus vulpecula*, from Kapiti Island, a New Zealand nature reserve. *Biological Conservation 61*: 217–226.

Cowan, P. E.; Chilvers, B. L.; Efford, M. G.; McElrea, G. J. 1997: Effects of possum browsing on northern rata, Orongorongo Valley, Wellington, New Zealand. *Journal of the Royal Society of New Zealand 27*: 173–179.

Cunningham, A. 1979: A century of change in the forests of the Ruahine Range, North Island, New Zealand: 1870-1970. *New Zealand Journal of Ecology 2*: 11–21.

Dale, R. W.; James, I. L. 1977: Forest and environment in the Kaimai Ranges. *Technical Paper No. 65, Forest Research Institute*. Wellington, New Zealand Forest Service. 34 p.

de Lange, P. J. 1997: Decline of New Zealand loranthaceous mistletoes – a review of non-possum threats. *In*: de Lange, P. J.; Norton, D. A. *ed.* New Zealand's loranthaceous mistletoes. Proceedings of a workshop hosted by Threatened Species Unit, Department of Conservation, Cass, 17–20 July 1995. Pp. 155–163.

de Lange, P. J.; Norton, D. A. ed. 1997: New Zealand's loranthaceous mistletoes. Proceedings of a workshop hosted by Threatened Species Unit, Department of Conservation, Cass, 17–20 July 1995. Wellington, Department of Conservation. 220 p.

de Lange, P. J.; Norton, D. A.; Molloy, B. P. J. 1997: Conservation status of New Zealand loranthaceous mistletoes: a comment on the application of IUCN Threatened Plant Committee Red Data Book categories. In: de Lange, P. J.; Norton, D. A. ed. New Zealand's loranthaceous mistletoes. Proceedings of a workshop hosted by Threatened Species Unit, Department of Conservation, Cass, 17–20 July 1995. Pp. 171–177.

Dickinson, K. J. M.; Mark, A. F.; Lee, W. G. 1992: Long-term monitoring of non-forest communities for biological conservation. *New Zealand Journal of Botany 30*: 163–179.

Druce, A. P. 1971: The flora of the Aorangi Range, southern Wairarapa with notes on the vegetation. *Bulletin, Wellington Botanical Society 37*: 4–29.

Elder, N. L. 1958: Southern Ruahine Range: ecological report. Unpublished New Zealand Forest Service report. Wellington, New Zealand.

Elder, N. L. 1965: Vegetation of the Ruahine Range: an introduction. *Transactions of the Royal Society of New Zealand (Botany) 3*: 13–66.

Esler, A. E. 1983: Forest and scrubland zones of the Waitakere Range, Auckland. *Tane 29*: 109–117.

Fitzgerald, A. E. 1976: Diet of the opossum *Trichosurus vulpecula* (Kerr) in the Orongorongo Valley, Wellington, New Zealand, in relation to food-plant availability. *New Zealand Journal of Zoology 3*: 399–419.

Fleming, C. A. 1979: The geological history of New Zealand and its life. Auckland, Auckland University Press/Oxford University Press. 141 p.

Fox, L. R.; Morrow, P. A. 1983: Estimates of damage by herbivorous insects on Eucalyptus trees. *Australian Journal of Ecology 8*: 139–147.

Francis, D. A.1970: Studies of the Dominican gull (*Laras dominicanus*) on Rangitoto Island. *Tane 16*: 91–95.

Grant, P. J. 1989: A hydrologist's contribution to the debate on wild animal management. *New Zealand Journal of Ecology 12*: 165–169.

Green, W. Q. 1984: A review of ecological studies relevant to management of the common brushtail possum. In: Smith, A. P.; Hume, I. D. ed. Possums and gliders. Chipping Norton, NSW, Surrey Beatty in assoc. with the Australian Mammal Society. Pp. 483–499.

Halkett, L. M.; Leitch, J. V. 1976: Catlins State Forest Park management plan. Invercargill, New Zealand Forest Service, Southland Conservancy. 91 p.

Harrison, J. B. J. 1985: Soil distribution and landscape dynamics, Camp Creek, Westland. In: Campbell, I. B. ed. Proceedings of the soil dynamics and land use seminar, Blenheim, May 1985. New Zealand Society of Soil Science, Lower Hutt, and New Zealand Soil Conservators Association. Pp. 65–77.

Holloway, J. T. 1957: Charles Douglas – observer extraordinary. *New Zealand Journal of Forestry 7*: 35–40.

Holloway, J. T. 1959: Noxious-animal problems of the South Island alpine watersheds. *New Zealand Science Review 17*: 21–28.

Hosking, G. P. 1995: Evaluation of forest canopy damage using airborne videography. *New Zealand Forestry 40*: 24–27.

Hosking, G. P.; Hutcheson, J. A. 1986: Hard beech (Nothofagus truncata) decline on the Mamaku Plateau, North Island, New Zealand. *New Zealand Journal of Botany 24*: 263–269.

Hosking, G. P.; Kershaw, D. J. 1985: Red beech death in the Maruia Valley, South Island, New Zealand. *New Zealand Journal of Botany 23*: 201–211.

Hosking, G. P.; Herbert, J.; Dick, M. A.; Hutcheson, J. A. 1989: Tackling the pohutukawa health problem. *What's New in Forest Research No. 178.* Rotorua, Forest Research Institute. 4 p.

Hosking, G. P.; Firth, J. G.; Brownlie, R. K.; Shaw, W. B. 1992: Airborne videophotography evaluation in New Zealand: opening a present from Uncle Sam. Proceedings Resource Technology 92 Conference 5: 137–144. Washington DC, USA.

Hunt, J.; Hollinger, D. 1988: Understorey light environment in a mountain beech forest. Unpublished Forest Research Institute contract report. Christchurch, Ministry of Forestry. 9 p.

Hunter, I. R.; Rodgers, B. E.; Dunningham, A.; Prince, J. M.; Thorn, A. J. 1991: An atlas of radiata pine nutrition in New Zealand. *FRI Bulletin No.165.* Rotorua, Ministry of Forestry. 24 p.

Innes, J. L. 1988: Forest health surveys: problems in assessing observer objectivity. *Canadian Journal of Forest Research 18*: 560–565.

Innes, J. L. 1990: Assessment of tree condition. *Forestry Commission Field Book 12.* London, HMSO. 96 p.

Innes, J. L.; Boswell, R. C. 1990: Reliability, presentation, and relationships among data from inventories of forest condition. *Canadian Journal of Forest Research 20*: 790–799.

James, I. L. 1974: Mammals and beech (*Nothofagus*) forests. *Proceedings, New Zealand Ecological Society 21*: 41–44.

Jane, G. T.; Green, T. G. A. 1983: Episodic forest mortality in the Kaimai Ranges, North Island, New Zealand. *New Zealand Journal of Botany 21*: 21–31.

Jane, G. T.; Green, T. G. A. 1986: Etiology of forest dieback areas within the Kaimai Range, North Island, New Zealand. *New Zealand Journal of Botany 24*: 513–527.

Kean, R. I.; Pracy, L. 1953: Effects of the Australian opossum (*Trichosurus vulpecula* Kerr) on indigenous vegetation in New Zealand. *Proceedings of the seventh Pacific Science Congress 4*: 696–705.

Kehoe, E. L. 1946: Behold your native forests. Catholic Writers' Movement pamphlet. Wellington. 24 p.

Kirk, H. B. 1920: Opossums in New Zealand: report on Australian opossums in New Zealand. *Appendix to the Journals of the House of Representatives of New Zealand H-28*: 1–12.

Ladley, J. J.; Kelly, D. 1996: Dispersal, germination and survival of New Zealand mistletoes (Loranthaceae): dependence on birds. *New Zealand Journal of Ecology 20*: 69–79.

Ladley, J. J.; Kelly, D. A.; Robertson, A. E. 1997: Pollination biology of New Zealand Loranthaceae mistletoes: pollinators, nectar production, and explosive flowering. *New Zealand Journal of Botany 35*: 345–360.

Lasko, A. N.1980: Correlations of fisheye photography to canopy structure, light climate, and biological responses to light in apple trees. *Journal of the American Society of Horticultural Science 105*: 43–46.

Leathwick, J. R.; Hay, J. R.; Fitzgerald, A. E. 1983: The influence of browsing by introduced mammals on the decline of North Island kokako. *New Zealand Journal of Ecology 6*: 55–70.

Leutert, A. 1988: Mortality, foliage loss, and possum browsing in southern rata (*Metrosideros umbellata*) in Westland, New Zealand. *New Zealand Journal of Botany 26*: 7–20.

Manion, P. D. 1981: Tree disease concepts. New Jersey, US, Prentice Hall.

McKinnon, A. D.; Coughlan, L. 1960: Data on the establishment of some introduced animals in New Zealand forests. Vol. II. Unpublished report. Wellington, New Zealand Forest Service.

Meads, M. J. 1976: Effects of opossum browsing on northern rata trees in the Orongorongo Valley, Wellington, New Zealand. *New Zealand Journal of Zoology 3*: 127–139.

Morgan, P. G. 1908: The geology of the Mikonui subdivision, north Westland. *New Zealand Geological Survey Bulletin (New Series) No.6.* Wellington, Government Printer.

Mosley, M. P. 1978: Erosion in the south-eastern Ruahine Range: its implications for downstream river control. *New Zealand Journal of Forestry 23*: 21–48.

Nugent, G.; Fraser, K. W.; Sweetapple, P. J. 1997: Comparison of red deer and possum diets and impacts in podocarp-hardwood forest, Waihaha catchment, Pureora Conservation Park. *Science for Conservation 50.* Wellington, Department of Conservation. 61 p.

Ogden, J.; Buddenhagen, C. 1995: Long term forest dynamics and the influence of possums and goats on kohekohe (*Dysoxylum spectabile*) forest in the Kauaeranga Valley, Coromandel Peninsula – some preliminary results. *In:* O'Donnell, C. F. J. *comp.* Possums as conservation pests. Proceedings of an NSSC Workshop . . . 29–30 November 1994. Wellington, Department of Conservation. Pp. 17–23.

Ogle, C. C. 1997: Evidence for the impacts of possums on mistletoes. *In:* de Lange, P. J.; Norton, D. A. *ed.* New Zealand's loranthaceous mistletoes. Proceedings of a workshop hosted by Threatened Species Unit, Department of Conservation, Cass, 17–20 July 1995. Wellington, Department of Conservation. Pp. 141–147.

Ogle, C.; Wilson, P. 1985: Where have all the mistletoes gone? *Forest and Bird 16(3)*: 10–13.

O'Loughlin, C. L.; Pearce, A. J. 1982: Erosion processes in the mountains. *In:* Soons, J. M.; Selby, M. J. *ed.* Landforms of New Zealand. Auckland, Longman Paul. Pp. 67–79.

Owen, H. J.; Norton, D. A. 1995: The diet of introduced brushtail possums *Trichosurus vulpecula* in a low-diversity New Zealand *Nothofagus* forest and possible implications for conservation management. *Biological Conservation 71*: 339–345.

Park, G. N. 1973: Point height intercept analysis: a refinement of point analysis for structural quantification of low arboreal vegetation. *New Zealand Journal of Botany 11*: 103–114.

Payton, I. J. 1983: Defoliation as a means of assessing browsing tolerance in southern rata (*Metrosideros umbellata* Cav.). *Pacific Science 37*: 443–452.

Payton, I. J. 1985: Southern rata (*Metrosideros umbellata* Cav.) mortality in Westland, New Zealand. *In:* Turner, H.; Tranquillini, W. *ed.* Establishment and tending of subalpine forest: research and management. Proceedings of the 3rd IUFRO workshop P1.07-00, 1984. *Swiss Federal Institute of Forestry Research, Report 270.* Pp. 207–214.

Payton, I. J. 1987: Canopy dieback in the rata (*Metrosideros*) - kamahi (*Weinmannia*) forests of Westland, New Zealand. *In:* Fujimori, T.; Kimura, M. *ed.* Human impacts and management of mountain forests. Proceedings of the 4th IUFRO Workshop P1.07-00, 1987. Forestry and Forest Products Research Institute, Ibaraki, Japan. Pp. 123–136.

Payton, I. J. 1988: Canopy closure, a factor in rata (*Metrosideros*)-kamahi (*Weinmannia*) forest dieback in Westland, New Zealand. *New Zealand Journal of Ecology 11*: 39–50.

Payton, I. J. 1989a: Impact of defoliation on Westland rata-kamahi forests. 4. Quality of rata foliage: a possible explanation for the browsing habits of brush-tail possums. Unpublished Forest Research Institute report, Christchurch. 9 p.

Payton, I. J. 1989b: Fungal (*Sporothrix*) induced mortality of kamahi (*Weinmannia racemosa*) after attack by pinhole borer (*Platypus* spp.). *New Zealand Journal of Botany 27*: 359–368.

Payton, I. J.; Forester, L.; Frampton, C. M.;Thomas, M. D. 1997a: Response of selected tree species to culling of introduced Australian brushtail possums *Trichosurus vulpecula* at Waipoua Forest, Northland, New Zealand. *Biological Conservation 81*: 247–255.

Payton, I. J.; Pekelharing, C. J.; Frampton, C. M. 1999: Foliar browse index: a method for monitoring possum (*Trichosurus vulpecula*) damage to plant species and forest communities. Lincoln, New Zealand, Manaaki Whenua - Landcare Research. 62 p.

Pekelharing, C. J.; Batcheler, C. L. 1990: The effect of control of brushtail possums (*Trichosurus vulpecula*) on condition of a southern rata/kamahi (*Metrosideros umbellata/Weinmannia racemosa*) forest canopy in Westland, New Zealand. *New Zealand Journal of Ecology 13*: 73–82.

Pekelharing, C. J.; Reynolds, R. N. 1983: Distribution and abundance of browsing mammals in Westland National Park in 1978, and some observations on their impact on the vegetation. *New Zealand Journal of Forestry Science 13*: 247–265.

Pekelharing, C. J.; Parkes, J. P.; Barker, R. J. 1998: Possum (*Trichosurus vulpecula*) densities and impacts on fuchsia (*Fuchsia excorticata*) in south Westland, New Zealand. *New Zealand Journal of Ecology 22*: 197–203.

Perham, A. N. 1924: Progress report of investigation of the opossum-genus *Trichosurus* in New Zealand. Appendix D *in:* The opossum industry in New Zealand. Unpublished New Zealand Forest Service report, Wellington. 10 p.

Pracy, L. T. 1974: Introduction and liberation of the opossum (*Trichosurus vulpecula*) into New Zealand. *New Zealand Forest Service Information Series No. 45.* 2nd ed. 28 p.

Pracy, L. T. 1975: Opossums (1). *New Zealand's Nature Heritage 3*: 873–882.

Pracy, L. T. 1978: Opossum survey, North Auckland Region County Pest Destruction Boards. Unpublished New Zealand Forest Service report, file 90/7/11A. 9 p.

Pracy, L. T.; Kean, R. I. 1949: Control of opossums an urgent problem. *New Zealand Journal of Agriculture 78*: 353–358.

Pywell, H. R.; Myhre, R. J. 1990: Monitoring forest health with airborne videophotography. Proceedings Resource Technology 90 Conference. Washington DC, USA.

Reif, A.; Allen, R. B. 1988: Plant communities of the steepland conifer-broadleaved hardwood forests of central Westland, South Island, New Zealand. *Phytocoenologia 16*: 145–224.

Rich, P. M. 1989: A manual for analysis of hemispherical canopy. Manual LA-11733-M. New Mexico, USA, Los Alamos National Laboratory.

Rogers, G. M. 1997: Trends in health of pahautea and Hall's totara in relation to possum control in central North Island. *Science for Conservation 52.* Wellington, Department of Conservation. 49 p.

Rogers, G. M.; Leathwick, J. R. 1997: Factors predisposing forests to canopy collapse in the southern Ruahine Range, New Zealand. *Biological Conservation 80*: 325–338.

Rose, A. B.; Pekelharing, C. J.; Hall, G. M. 1988: Forest dieback and the impact of brushtail possums in the Otira, Deception, and Taramakau catchments, Westland. Unpublished Forest Research Institute contract report. Christchurch, Ministry of Forestry. 27 p.

Rose, A. B.; Pekelharing, C. J.; Platt, K. H. 1992: Magnitude of canopy dieback and implications for conservation of southern rata-kamahi (*Metrosideros umbellata-Weinmannia racemosa*) forests, central Westland, New Zealand. *New Zealand Journal of Ecology 16*: 23–32.

Rose, A. B.; Pekelharing, C. J.; Platt, K. H.; Woolmore, C. B. 1993: Impact of invading brushtail possum populations on mixed beech-broadleaved forests, south Westland, New Zealand. *New Zealand Journal of Ecology 17*: 19–28.

Rose, A. B. 1994: A review of possums and possum-vulnerable species in Nelson/ Marlborough Conservancy. Landcare Research Contract Report LC9394/119 (unpublished) 79 p.

Rose, A. B.; Platt, K. H.; Pekelharing, C. J.; Moore, T. J.; Suisted, P.; Savage, T. J. 1995 : Forest dieback and the impact of possums in Nelson/ Marlborough Conservancy. Landcare Research Contract Report LC9596/12 (unpublished) 42 p.

Skipworth, M. R. 1928: Opossums in our forests. *Te Kura Ngahere 2*: 13–15.

Smale, M. C.; Pekelharing, C. J.; Savage, T. J. 1996: Canopy dieback and the impact of possums in the eastern forests of Mount Aspiring National Park. Landcare Research Contract Report LC9596/138 (unpublished) 8 p.

Stewart, G. H. 1989: Ecological considerations of dieback in New Zealand's indigenous forests. *New Zealand Journal of Forestry Science 19*: 243–249.

Stewart, G. H.; Burrows, L. E. 1989: The impact of white-tailed deer *Odocoileus virginianus* on regeneration in the coastal forests of Stewart Island, New Zealand. *Biological Conservation 49*: 275–293.

Stewart, G. H.; Rose, A. B. 1988: Factors predisposing rata-kamahi (*Metrosideros umbellata-Weinmannia racemosa*) forests to canopy dieback Westland, New Zealand. *Geojournal 17*: 217–223.

Stewart, G. H.; Veblen, T. T. 1982: Regeneration patterns in southern rata (*Metrosideros umbellata*) - kamahi (*Weinmannia racemosa*) forest in central Westland, New Zealand. *New Zealand Journal of Botany 20*: 55–72.

Stewart, G. H.; Johnson, P. N.; Mark, A. F. 1989: Monitoring terrestrial vegetation for biological conservation. *In:* Craig, B. *ed.* Proceedings of a symposium on environmental monitoring in New Zealand with emphasis on protected natural areas. Wellington, Department of Conservation. Pp. 199–208.

Sweetapple, P.; Nugent, G. 1999: Provenance variation in fuchsia (*Fuchsia excorticata*) in relation to palatability to possums. *New Zealand Journal of Ecology 23*: 1–10.

Thomas, M. D.; Hickling, G. J.; Coleman, J. D.; Pracy, L. T. 1993: Long-term trends in possum numbers at Pararaki: Evidence of an irruptive fluctuation. *New Zealand Journal of Ecology 17*: 29–34.

Trotter, C. M. 1992: Estimating possum browse damage to lowland indigenous forest using remote sensing – a feasibility study. *Proceedings of the 6th Australasian Conference on Remote Sensing 1*: 400–405.

Veblen, T. T.; Stewart, G. H. 1982a: The effects of introduced wild animals on New Zealand forests. *Annals of the Association of American Geographers 72*: 372–397.

Veblen, T. T.; Stewart, G. H. 1982b: On the conifer regeneration gap in New Zealand: the dynamics of *Libocedrus bidwillii* stands on South Island. *Journal of Ecology 70*: 413–436.

Wardle, J. 1967: Vegetation of the Aorangi Range, southern Wairarapa. *New Zealand Journal of Botany 5*: 22–48.

Wardle, J. 1974: Influence of introduced mammals on the forest and shrublands of the Grey River headwaters. *New Zealand Journal of Forestry Science 4*: 459–486.

Wardle, J. A. 1984: The New Zealand beeches: ecology, utilisation and management. Wellington, New Zealand Forest Service. 447 p.

Wardle, P. *Comp.* 1971: Biological flora of New Zealand. 6. *Metrosideros umbellata* Cav. [Syn. *M. lucida* (Forst. f.) A. Rich.] (Myrtaceae) southern rata. *New Zealand Journal of Botany 9*: 645–671.

Wardle, P. 1977: Plant communities of Westland National Park (New Zealand) and neighbouring lowland and coastal areas. *New Zealand Journal of Botany 15*: 323–398.

Wardle, P. 1978: Regeneration status of some New Zealand conifers, with particular reference to *Libocedrus bidwillii* in Westland National Park. *New Zealand Journal of Botany 16*: 471–477.

Wardle, P. 1980: Primary succession in Westland National Park and its vicinity, New Zealand. *New Zealand Journal of Botany 18*: 221–232.

Wardle, P. 1991: Vegetation of New Zealand. Cambridge, Cambridge University Press. 672 p.

Wickman, B. E. 1979: Douglas-fir tussock moth handbook: How to estimate defoliation and predict tree damage. USDA Forest Service, *Agriculture Handbook No. 550.* Washington DC, USA. 15 p.

Wilson, P. R. 1984: The effects of possums on mistletoe on Mt Misery, Nelson Lakes National Park. *In:* Dingwall, P. R. *comp.* Protection and parks. Essays in the preservation of natural values in protected areas. Proceedings of Section A4e, 15th Pacific Science Congress, Dunedin, February 1983. *Information Series No. 12.* Wellington, Department of Lands and Survey. Pp. 53–60.

Zotov, V. D. 1947: Forest deterioration in the Tararuas due to deer and opossum. *Transactions and Proceedings of the Royal Society of New Zealand 77*: 162–165.

Zotov, V. D.; Elder, N. L.; Beddie, A. D.; Sainsbury, G. O. K.; Hodgson, E. A. 1938: An outline of the vegetation and flora of the Tararua Mountains. *Transactions and Proceedings of the Royal Society of New Zealand 68*: 259–324.

CHAPTER ELEVEN

Evidence of Possums as Predators of Native Animals

Richard Sadleir

New Zealanders have debated the impact possums have on native flora since 1918, yet their impact as predators of native fauna remained virtually unrecognised until about two decades ago (Hay 1981). Recent work in Australia has shown that possums are predators of the glossy black-cockatoo *Calyptorhynchus lathami* (see Chapter 22) and, in New Zealand, researchers over the past 6 years have discovered that the predation of eggs and nestlings of birds is widespread (Brown *et al.* 1993; Innes *et al.* 1994). This chapter reviews the evidence that possums eat native birds, snails, and insects. Some comments are made on dietary competition between possums and native birds.

Initial evidence

Evidence that possums eat native animals came initially from remains left by possums after eating eggs, chicks, adult birds, or native snails. In 1979 the tattered remains of a kōkako (*Callaeas cinerea wilsoni*) nestling were found at Mapara in the King Country (Hay 1981). The head, breast, and wings had obviously been eaten, and for the first time it was suggested that possums eat birds as oral pellets containing chewed feathers were found in the nest (Fig. 11.1). Brown *et al.* (1993) drew a similar conclusion about the predatory nature of possums when he discovered an adult female kōkako that had been eaten, in Rotoehu Forest, Bay of Plenty. Subsequent direct observations at night using infrared video cameras (see below) and experiments where possums were fed non-native bird species (Brown *et al.* 1996) confirmed that these signs are characteristic of possum predation. Not only did the presence of oral pellets containing feathers suggest possums ate birds, but possums were further implicated as predators of birds by the presence of their fur at bird's nests. McLennan (1988) found possum fur on the remains of eaten eggs of North Island brown kiwi (*Apteryx australis mantelli*) in Hawke's Bay. Brown *et al.* (1993) found possum fur on the egg or in the nest cavity of the North Island saddleback (*Philesturnus carunculatus rufusater*) and kererū/kūkupa (native pigeon) (*Hemiphaga novaeseelandiae*) (see also Pierce & Graham 1995). More recently, McLennan (1997) noted that a dead adult male kiwi at Lake Waikaremoana had possum fur lodged under its claws. The kiwi was found outside the opening of its nest with clumps of feathers within.

Not long after possums were first found to eat birds, a study by Meads *et al.* (1984) noted that large land snails of the genus *Powelliphanta* were subject to considerable levels of predation in northwest Nelson, which they tentatively attributed to native parrots kākā (*Nestor meridionalis*) or kea (*N. notabilis*). Later observations and experimental feeding of snails to possums revealed that the shell damage attributed to parrots was actually caused by possums (K. Walker, Department of Conservation, Nelson, pers. comm.) (Fig. 11.2). Similarly when Efford & Bokeloh (1991) fed snails to possums during feeding trials, three out of four possums ate the small carnivorous snail *Wainuia urnula* leaving crushed shells. Similar damage was later found on 44% of shells collected during a *W. urnula* shell survey in the Wainuiomata water catchment near Wellington (M. Efford, Landcare Research, Dunedin, pers. comm.).

In subsequent captive possum feeding trials, K. Walker and G. Elliott (Department of Conservation, Nelson, pers. comm.) found four of six possums readily ate the much larger *Powelliphanta* land snails, which also had characteristic holes and scratches on the shell (Fig. 11.2). Since then possum-damaged shells have been found in many snail populations in the Marlborough Sounds, in the Nelson and North Westland areas, and in the Kaimanawa and Ruahine ranges. From the proportion of damaged shells seen, and a decline in the size of the live snail populations, Walker and Elliott concluded that possum predation is having a major detrimental effect on *Powelliphanta superba*, *P. gilliesi*,

P. annectens, *P. hochstetteri*, *P. lignaria*, and *P. marchanti*.

Literature searches and queries to three herpetologists have failed to produce any evidence or accounts of possums eating other vertebrates such as lizards or frogs. Despite the lack of evidence, however, it seems highly likely that possums will eat lizards and frogs, considering their carnivorous nature (i.e., they will eat insects and scavenge meat in ferret traps and from deer carcasses, see below and Chapter 2).

Direct evidence

The presence of native animal remains in stomach contents or faecal pellets is direct evidence of predation by possums. A detailed 5-year study by Cowan & Moeed (1987) described the invertebrates eaten by possums in the Orongorongo Valley. They analysed 2596 faecal pellets, nearly half of which contained invertebrates, and found that stick insects (*Phasmatodea*), cicadas (*Hemiptera*), wētā (*Stenopelmatidae, Rhaphidophorioae*), beetles (*Coleoptera*), fly larvae (*Bibionidae*), and mites (*Acari*) comprised over 80% of the insects identified. Most consumption was in summer (Dec–Feb) and autumn (Mar–May). The authors concluded that "in most circumstances, invertebrates comprised only a small part of the diet of possums. The consumption of invertebrates by possums indicates their opportunistic feeding habits" (Cowan & Moeed 1987 p. 163). However, they suggest that small localised populations of larger sluggish nocturnal species such as giant wētā (*Deinacridia* sp.), large stag beetles (*Geodorcus* sp.), and large weevils (subfamily Cylindrorrhinae) may be at risk from possum predation. Cowan & Moeed (1987) include a review of another 10 papers that report possums eating invertebrates and Nugent *et al.*, (Chapter 2) provide additional evidence that insect larvae (in particular) can at times constitute up to 28% of stomach contents.

There are far fewer published accounts of bird remains in possum stomachs or faecal pellets. An early report (Perham 1924) described nestlings and feathers in possum stomachs, but several other studies of the contents of large numbers of possum stomachs (reported in Morgan 1981, for example) have not found feathers. However Parkes & Thomson (1995 p. 34) report a possum that had eaten a greenfinch *Carduelis chloris*. A. Cox (Department of Conservation, Southland, pers. comm.) examined the stomach contents of over 700 possums that were trapped during an eradication programme on Codfish Island in 1984–1987. One possum stomach was full of the fresh meat from predation on the chicks of sooty shearwaters (*Puffinus griseus*). Similarly, 5 of 23 possums on Stewart Island had also eaten shearwater chicks (see Chapter 2).

The clearest evidence is direct observation of predation by possums. For birds, Brown *et al.* (1993) list the following:

(a) A possum chasing a fledgling Australasian harrier-hawk/kāhu (*Circus approximans*) down a tree, catching it on the ground, and carrying it up another tree (J. Roberts, pers. comm.).
(b) A possum shot while eating the eggs of fantail (*Rhipidura fulginosa*) (G. Priest, pers. comm.).
(c) Possums eating dead chicks of Westland black petrel (*Procellaria westlandica*) (L. van Bijk, pers. comm.).

A captive possum has been observed catching a house sparrow (*Passer domesticus*) and eating parts of its head and breast before other possums in the pen also ate flesh from the carcass (Morgan 1981) Similarly, a wild possum was seen eating a land snail (*P. hochstetteri hochstetteri*) in the Flora Valley, Kahurangi National Park, in 1987 (J. Ryan, New Zealand Forest Service, Nelson, pers. comm.).

The permanent recording of such direct evidence using time-lapse video cameras at nests provides the clearest evidence that possums are predators (Brown *et al.* 1993; Innes *et al.* 1994; James 1995; Innes *et al.* 1996; J. McLennan, ex Landcare Research, Havelock North, pers. comm.). The video camera is used at night with infrared lighting, with the camera located some distance away from the nests. Innes *et al.* (1996) reported that the fledging rates of kōkako from filmed nests (3 of 19) were the same as from unfilmed nests (8 of 46), indicating that camera surveillance had little effect on the nesting birds. The technique has resulted in some remarkable observations of predator-prey behaviour (see Fig. 11.3a–d). Firstly, possums have been recorded attacking and eating adults on nests, chicks on nests, and eggs. Secondly, not all the nests visited by possums resulted in successful predation. Very occasionally, female kōkako defended their nest. J. Innes (Landcare Research, Hamilton, pers. comm.) video-taped a possum hitting a kōkako with a blow, but the bird remained on the nest, and in one of three possum–chick encounters the possum did not touch the chick.

Few papers give data that allow an estimation of what proportion of predation on native animals can

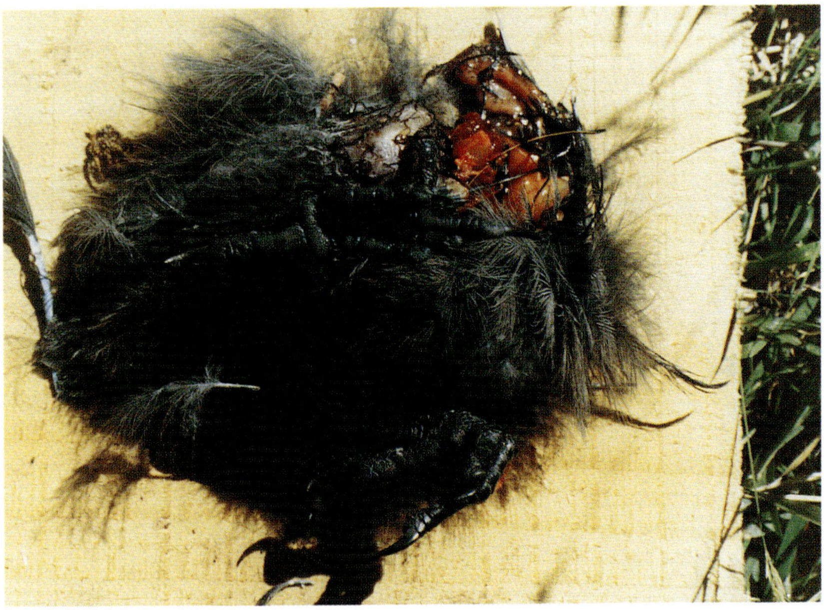

Fig. 11.1 (above)
The head, breast, and wings and chewed feather pellets left after possum predation of a kōkako nesting at Mapara 1979.
Photo by R. Hay

Fig. 11.2 (above)
Shells of land snails eaten by possums (s) *Powelliphanta superba*, (h) *P. hochstetteri*, and (g) *P. gilliesi*.
Photo by K. Walker

definitely be attributed to possums. Those that do are given in Table 11.1. It seems that the more direct the evidence, the higher the rate of possum predation observed, which implies that indirect methods may considerably underestimate the true incidence. Workers on kōkako populations (Innes *et al.* 1996; Innes *et al.* 1999a) are certain that predation by possums (and ship rats *Rattus rattus*) are major causes of low recruitment into these populations. Evidence from Wenderholm, Auckland, indicates that this may also be true for kererū (James & Clout 1996). Strong support for this is provided by pest management outcomes at Motatau, Northland. Prior to possum and rat control at Motatau none of 13 kūkupa nests monitored survived past the egg stage, while after the rat and possum control operations all seven of the nests monitored produced fledglings (Innes *et al.* 1999b).

Dietary competition

There are several studies pointing out dietary overlap or possum impacts on plant species eaten by native animals (Fitzgerald 1984; Powlesland 1987; Cowan 1990; Cowan & Waddington 1990) but none have demonstrated actual competition in a population sense. Indeed, Innes *et al.* (1999a) were convinced that, although there is overlap in diet, the observed correlations between possum spread and kōkako declines (Leathwick *et al.* 1983) were probably due to predation rather than competition. Chapter 22 highlights the weaknesses of the published evidence demonstrating responses of native animal populations after the removal, or drastic reduction in numbers, of possums, so whether there really is an effect of competition on native birds remains unclear.

In a perceptive but apparently overlooked paper, Moller (1989) noted the need for essential, experimental, field manipulation to prove or disprove the importance of competition. In discussing the possible effects of biological control, he correctly emphasised that concentrating on a single species can obscure understanding of the complex interactions in forest ecosystems.

Fig. 11.3a
Possum captured by video photography after preying on kōkako eggs at a nest in a tree fern, Rotoehu Forest, November 1993.

Fig. 11.3b
The first recorded sequence of a possum eating a bird's egg in New Zealand. Recorded on video at Rotoehu Forest, December 1991. Enhanced image shows the possum approaching a kōkako nest.

Fig. 11.3c
The same possum arrives at the kōkako nest.

Fig. 11.3d
The possum proceeds to eat the kōkako egg.

Table 11.1
Incidence of possum predation on bird species.

Bird species	Site	Incidence	Author
Kōkako	Rotoehu	4 of 10 predations (40%) 4 at 19 nests observed (21%)	Innes *et al.* 1994
Kūkupa	Northland	2 of 31 predations (6%)	Pierce & Graham 1995
Kōkako	Rotoehu	11 of 32 predations (34%) 11 of 65 nests observed (17%)	Innes *et al.* 1996
Kiwi	NZ	1 of 14 adults radio-tracked killed by possum	McLennan *et al.* 1996

Summary

- Indirect signs, such as partially consumed eggs and corpses, have incriminated possums as predators of native birds. Holes and scratches on shells, and feeding experiments, have shown that possums are involved in native snail predation.
- Stomach and faecal analysis has shown possums eat birds and many species of insects.
- There are a few accounts of possums seen eating chicks, eggs, or carrion birds.
- Time-lapse video cameras have recently shown possums eating adult birds, their chicks, and eggs, all in the nest.
- Few research papers indicate the incidence of possum predation. Figures given of the low number of predation incidents observed range from 6% to 40%.
- Although invoked by several authors, dietary overlap between several native bird species and possums provides no credible evidence for competition per se.

Acknowledgements

I thank all those who provided me with source material, especially Phil Cowan, Andy Cox, John Innes, John McLennan, Ray Pierce, Hugh Robertson, and Kath Walker. Three referees' comments improved the chapter.

References

Brown, K.; Innes, J.; Shorten, R. 1993: Evidence that possums prey on and scavenge birds' eggs, birds and mammals. *Notornis 40*: 169–177.

Brown, K. P.; Moller, H.; Innes, J. 1996: Sign left by brushtail possums after feeding on bird eggs and chicks. *New Zealand Journal of Ecology 20*: 277–284.

Cowan, P. E. 1990: Fruits, seeds, and flowers in the diet of brushtail possums, *Trichosurus vulpecula*, in lowland podocarp/mixed hardwood forest, Orongorongo Valley, New Zealand. *New Zealand Journal of Zoology 17*: 549–566.

Cowan, P. E.; Moeed, A. 1987: Invertebrates in the diet of brushtail possums, *Trichosurus vulpecula*, in lowland podocarp/broadleaf forest, Orongorongo Valley, Wellington, New Zealand. *New Zealand Journal of Zoology 14*: 163–177.

Cowan, P. E.; Waddington, D. C. 1990: Suppression of fruit production of the endemic forest tree, *Elaeocarpus dentatus*, by introduced marsupial brushtail possums, *Trichosurus vulpecula*. *New Zealand Journal of Botany 28*: 217–224.

Efford, M.; Bokeloh, D. 1991: Some results from a study of the land snail *Wainuia urnula* (Pulmonata: Rhytididae) and their implications for snail conservation. DSIR Land Resources contract report to Department of Conservation 91/1 (unpublished). 17 p.

Fitzgerald, A. E. 1984: Diet overlap between kokako and the common brushtail possum in central North Island, New Zealand. *In:* Smith, A. P.; Hume, I. D. *ed.* Possums and gliders. Chipping Norton, NSW, Surrey Beatty in assoc. with the Australian Mammal Society. Pp. 569–573.

Hay, J. R. 1981: The kokako. Forest Birds Research Group Report. Not seen, as cited in Brown *et al.* (1993, op. cit.).

Innes, J.; Crook, B.; Jansen, P. 1994: A time-lapse video camera system for detecting predators at nests of forest birds : A trial with North Island kōkako. *In:* Bishop, I. *ed.* Proceedings of the Resource Technology '94 Conference, Melbourne. University of Melbourne Press. Pp. 439–448.

Innes, J.; Brown, K.; Jansen, P.; Shorten, R.; Williams, D. 1996: Kokako population studies at Rotoehu Forest and on Little Barrier Island. *Science for Conservation 30*. Wellington, New Zealand, Department of Conservation. 39 p.

Innes, J.; Hay, R.; Flux, I.; Bradfield, P.; Speed, H.; Jansen, P. 1999a: Successful recovery of North Island kokako, *Callaeas cinerea wilsoni* populations, by adaptive management. *Biological Conservation 87*: 201–214.

Innes, J.; Nugent, G.; Prime, K. 1999b: Pigeons versus possum: 7-0 at Motatau. *He Kōrero Paihama – Possum Research News 11*: 1–2.

James, R. E. 1995: Breeding ecology of the New Zealand pigeon at Wenderholm Regional Park. Unpublished MSc thesis, University of Auckland, Auckland, New Zealand. Not seen, as cited in James & Clout (1996) op. cit.

James, R. E.; Clout, M. N. 1996: Nesting success of New Zealand pigeons (*Hemiphaga novaeseelandiae*) in response to a rat (*Rattus rattus*) poisoning programme at Wenderholm Regional Park. *New Zealand Journal of Ecology 20*: 45–51.

Leathwick, J. R.; Hay, J. R.; Fitzgerald, A. E. 1983: The influence of browsing by introduced mammals on the decline of North Island kokako. *New Zealand Journal of Ecology 6*: 55–70.

McLennan, J. A. 1988: Breeding of North Island brown kiwi, *Apteryx australis mantelli,* in Hawke's Bay, New Zealand. *New Zealand Journal of Ecology 11*: 89–97.

McLennan, J. 1997: Ecology of brown kiwi and causes of population decline in Lake Waikaremoana catchment. *Conservation Advisory Science Notes 167*. Wellington, New Zealand, Department of Conservation. 25 p.

McLennan, J. A.; Potter, M. A.; Robertson, H. A.; Wake, G. C.; Colbourne, R.; Dew, L.; Joyce, L.; McCann, A. J.; Miles, J.; Miller, P. J.; Reid, J. 1996: Role of predation in the decline of kiwi, *Apteryx* spp., in New Zealand. *New Zealand Journal of Ecology 20*: 27–35.

Meads, M. J.; Walker, K. J.; Elliott, G. P. 1984: Status, conservation, and management of land snails of the genus *Powelliphanta* (Mollusca: Pulmonata). *New Zealand Journal of Zoology 11*: 277–306.

Moller, H. 1989: Towards constructive ecological engineering; the biological control of pests for the restoration of mainland habitats. *In:* Norton,

D.A. *ed.* Management of New Zealand's natural estate. *New Zealand Ecological Society Occasional Publication 1*: 89–93.

Morgan, D. R. 1981: Predation on a sparrow by a possum. *Notornis 28:* 167–168.

Parkes, J. P.; Thomson, C. 1995: Management of thar Part II: Diet of thar, chamois, and possums. *Science for Conservation* 7. Wellington, New Zealand, Department of Conservation. Pp. 22–42.

Perham, A. N. 1924: Progress report of investigation of the opossum-genus *Trichosurus* in New Zealand. Appendix D *in:* The opossum industry in New Zealand. Unpublished New Zealand Forest Service report, Wellington. 10 p. Not seen, as cited in Brown *et al.* (1993) op. cit.

Pierce, R. J.; Graham, P. J. 1995: Ecology and breeding biology of kukupa (*Hemiphaga novaeseelandiae*) in Northland. *Science & Research Series No. 91.* Wellington, New Zealand, Department of Conservation. 32 p.

Powlesland, R. G. 1987: The foods, foraging behaviour and habitat use of North Island kokako in Puketi State Forest, Northland. *New Zealand Journal of Ecology 10*: 117–128.

CHAPTER TWELVE

Monitoring Possum Populations

Bruce Warburton

To run any business effectively, managers need to have reliable information on the outcomes of their actions and operations. Management of possum populations is no exception, and to be effective, pest managers must carry out monitoring to know what their control operations achieve. This chapter covers issues of sample design, statistical precision and accuracy, and examines the range of population monitoring methods currently used in New Zealand for estimating the relative and/or absolute abundance of possums.

Monitoring – what is it?

In the context of possum management, monitoring can be divided into operational monitoring, which aims to determine the impact that management has on possum numbers, and outcome (performance) monitoring, which aims to measure the effectiveness of the management in terms of the response of the resource being protected from possums. Both forms of monitoring are essential for good possum management. Operational monitoring enables managers to measure the effectiveness of their operational tactics, and outcome monitoring enables managers to determine the benefits of their management.

Operational monitoring has been routinely carried out since possum control started in the mid-1950s (Batcheler 1978), but outcome monitoring has only recently been accepted as a necessary part of any possum management programme (Payton *et al.* 1997). In the past, the success of possum control operations was judged solely on the estimated percent kill achieved (i.e., a reduction in possum numbers of 70% or more was generally acceptable). Implicit with this approach was that if the population could be reduced by 70% or greater, then the resource would be protected. There are ample examples (e.g., the Whitcombe Valley, Westland) where estimated kills greater than 70% were achieved but the resource was not protected, either because the control was carried out too late to halt the decline in forest condition, or the population was not reduced to sufficiently low levels to get a response in the resource.

Monitoring pest populations can impose significant costs on operational budgets (often 10–20% of total budget (Warburton & Cullen 1993), and therefore pest managers need to understand why monitoring is being carried out, what information is required from the monitoring, and how the information can be used for making management decisions. If no useable knowledge is gained from monitoring, or the knowledge is not used to make management decisions, then funds should not be wasted on monitoring.

Before monitoring is undertaken, the following questions should be addressed:

(a) Who needs to know – i.e., who is the end-user of the knowledge?
(b) What do they need to know – i.e., what knowledge is required?
(c) What level of precision is required – i.e., how important is it to get the answer right?
(d) What funds should be allocated to monitoring?

Who needs to know?

Who needs to know the information and knowledge obtained from monitoring will depend on who is funding it, whether the control operation has a high public profile, and what the implications are of operational success or failure. For example, monitoring pest populations can involve routine operational monitoring (determining the percent kill), the outcome of which may be of interest to only the manager who carried out the operation. However, if the control operation has a high public profile, the users of any monitoring results could include politicians, interest groups, the general public, or the media. If the control operation was carried out under a commercial contract, with payment based on performance, the results are important to both the contractor and employer. There can be many potential

Table 12.1
The potential end-users of pest population monitoring information and knowledge at a local, regional, and national level.

Local	Regional	National
Landuser	Local politicians	Animal Health Board
Landcare group	Pest managers	AgriQuality NZ
Pest manager	Planners/Policy staff	Planners/Policy (AgriQuality NZ/MOH/MfE)
Contractors & Subcontractors	Environmental managers	National level politicians
Dept of Conservation (DOC)	Ratepayers	Taxpayers
Agribusiness	Regional Animal Health	Producer boards
Interest groups (Non-government organisations)	Interest Groups (NGOs)	Agribusiness
Iwi	AgriQuality NZ	Interest Groups (NGOs)
Media	Media	Media
Research	DOC	DOC
	Iwi	Iwi
	Research	Research

NGO = non-governmental organisation, MfE = Ministry for the Environment, MOH = Ministry of Health.

end-users of monitoring results (Table 12.1), and for each monitoring operation managers need to identify the most likely end-users "who need to know".

The "who needs to know" can be divided into four main groups:

(a) Beneficiaries/exacerbators (e.g., the landowner/occupier);
(b) Funders (e.g., landowner, ratepayer, and taxpayer);
(c) Technicians (e.g., pest managers and researchers);
(d) Interested parties (e.g., environmental groups, planners, and politicians).

What do they need to know?

Population monitoring is carried out to obtain three main types of information:

(a) Pest population status (i.e., what is happening out there and is there a problem with the pest?);
(b) Funding information (e.g., was the control operation successful and am I getting my money's-worth?); and
(c) Technical information (e.g., why did the control operation fail, if it did?).

Pest population monitoring can provide information on whether the population is stable, declining, or increasing. Information on changes in relative abundance of a pest population, or just what the current pest abundance is, might be needed by managers if they have to determine if compliance with Regional Pest Management Strategy (PMS) requirements is being achieved. The distribution and relative abundance of pests are also needed to determine the relationship between relative pest abundance and resource impacts. Knowledge of these relationships is necessary if realistic and effective compliance levels (levels that eliminate unsustainable resource impacts) are to be set.

Ratepayers, taxpayers, local and national politicians, and buyers of pest control all want to know if their money is being spent efficiently. That is, funders want to know that the kills are maximised while the costs are minimised within the limits of the technical and biological constraints of the operation. To obtain this information, the relative abundance of pests before and after control are used to estimate percent kill.

Pest managers and researchers need to know if a control operation was successful or not. If the operation has failed to achieve a desired control objective (percent kill or residual trap-catch frequency), then the operation's technical methods and materials can be scrutinised for probable reasons for failure. This allows for improvements to future operations and ensures technical standards are adhered to. When testing new toxins, comparison of percent kill results allows the efficiencies of the old and the new to be compared.

What level of precision is required?

Unfortunately, managers can never know how many possums are in a piece of forest or what the true kill is when they carry out a control operation. The only option they have available is to obtain an *estimate* of pest abundance or population change using one of several monitoring methods. The monitoring methods chosen (spotlight counts, capture per unit effort, etc.) are used to sample the population, and estimates obtained from this sampling will have a sampling error. Because the sampling error (precision) depends on the sample effort and the spatial variation in the population being measured, how much sampling and how much money must be spent on monitoring depends on the quality of the information and knowledge the end-user wants. If the end-user needs to be sure that the true population abundance or percent kill is within a small range 9 times out of 10, then much more effort and cost will be required than if the end-user was satisfied with less precision and a lower level of confidence (e.g. only correct 6 times out of 10). Statistics cannot determine what precision or confidence the end-user requires. This must be set by pest managers before any monitoring is undertaken to ensure that the money and effort spent on monitoring will provide the required information.

What funds should be allocated to monitoring?

Because monitoring costs can be high, monitoring is often seen as an unacceptable burden on operational costs, and most managers attempt to reduce monitoring costs to a minimum. If monitoring costs are to be minimised, it is preferable to monitor fewer operations but do these well rather than monitor more operations by reducing the amount of monitoring in any one operation. Possible justifications for electing not to carry out operational monitoring are when the method used is routine, it has previously been shown to be effective, and most importantly, when a high level of quality assurance in control is employed. Thus, if aerial operations can be carried out with a high standard of quality assurance (bait, toxic loading, bait spread, and weather), there may be sufficient justification not to monitor such an operation. The decision on whether to monitor or not will depend on the size of the operation and the need for an estimate of the operator's success or failure.

Another potential strategy for reducing monitoring costs is to monitor only after control to obtain an index of the abundance of the residual population. This approach requires some understanding of the relationship between an index of population abundance (e.g., percent catch in traps) and the level at which the population needs to be maintained for the resource to be protected. Some Department of Conservation conservancies favour this approach as it focuses on the residual population and the resource rather than the percent kill which, even if high (and therefore indicating a successful operation), may not achieve the desired resource outcome.

Population monitoring has costs and benefits. The costs are the sum of the direct costs of carrying out the monitoring plus the value of the increased probability of causing an operational failure because some of the operational funds were diverted into monitoring. The benefits are the value of the pest impacts that are avoided because monitoring allows operational failures to be identified and remedied (Choquenot & Warburton 1998).

Monitoring to determine if a target reduction or residual density has been reached can be set in a hypothesis-testing framework. That is, a null hypothesis might be that the control operation will achieve a 5% residual trap catch, and only if the monitoring proves, with a given level of confidence, that the actual residual percent catch is 5% or greater, will that hypothesis be rejected. However, because monitoring has a sampling error, sometimes the percent catch will indicate that the control operation has failed when it has been successful (Type I error) or that it has been successful when it has failed (Type II error). The probability of correctly concluding that the operation has failed is termed the statistical power (Skalski & Robson 1992; Steidl *et al.* 1997). The statistical power can be improved by increasing survey effort (i.e., spend more funds on monitoring).

So what proportion of control funds should be allocated to monitoring? Choquenot & Warburton (1998), using the hypothesis-testing framework and cost-benefit analysis, contrasted the marginal change in benefits and costs to identify the proportional investment in monitoring that provided the best return. Using current monitoring costs, and benefits and costs of possum impacts based on differential rates of recovery to target levels, they suggested that 15–20% of possum control funds should be allocated to monitoring. This is considerably more than the 2–10% that is routinely allocated by control agencies (Parliamentary Commissioner for the Environment 1994).

Monitoring possum populations: some statistical requirements

There are two statistical criteria that need to be optimised when monitoring possum populations: accuracy and precision. In terms of percent kill, accuracy is how close the estimated percent kill is to the actual, but unknown, percent kill. Precision provides an estimate of the range within which the actual kill will lie with some level of confidence (95% is most commonly used). The challenge for monitoring is to provide unbiased and precise estimates of the kill or relative population abundance. Biases can result from inappropriate sampling or intrinsic biases in the monitoring method being used.

Sampling biases

No matter what monitoring method is used (e.g., traps, bait interference, spotlighting), the method will not provide robust estimates unless the sampling strategy used, samples the control area with no biases. Possums are not distributed randomly through forest or farmland (i.e., each sample point, such as a trap, does not have an equal chance of capturing a possum) and if sampling is not random, then biased estimates of population density or percent kill may result.

Stratification

Possum populations are rarely evenly distributed, and one method for addressing the problem of heterogeneous populations and to increase the precision for a fixed sampling effort is to stratify the monitoring area into sub-areas (strata) of less variable densities (Stuart 1984; Krebs 1999). Unfortunately in most management situations this can only be done at a very coarse level, such as separating major vegetation types or perhaps on the basis of altitude. Although stratifying does not guarantee that within each stratum populations are homogeneous, stratified sampling will always give as precise estimates, if not more precise, as non-stratified sampling.

The main reason why field staff often do not adhere to random sampling is because of cost — it costs more (i.e., takes more person days) to locate and run randomly placed sampling lines than lines that are located within easy access of roads or walking tracks.

When different control treatments are being used in one control area (e.g., aerial 1080 and contract hunters), the area should be stratified at least to separate these treatments. Stratification does not necessarily mean a greater monitoring effort is required. Rather it ensures that a proportion of the sampling effort is allocated to monitoring each stratum, usually on the basis of the proportional contribution each stratum makes to the total area being monitored.

Method biases

Method biases result from problems that are inherent in the monitoring method being used. For example, two potential biases for trap-catch monitoring are trap saturation and trap shyness. If traps become saturated as possum density increases, then the index obtained from the possum catch will always underestimate relative density when possum numbers are high. Conversely, at very low densities traps might not detect the presence of possums and therefore underestimate their abundance. A second bias may result if possums become trap-shy after a control operation in which traps were used. Such a behavioural response would result in the population after control being underestimated. In terms of estimating percent kills, the first bias will result in an underestimated kill and the second an overestimate. Non-target species interference can also produce biases, especially when the interferences are incorrectly attributed to possums.

Monitoring methods

Population monitoring methods can be divided into two broad categories: (a) estimates of population size (i.e., an estimate of the number of possums and possibly their density), and (b) relative indices (i.e., an index that has a proportional but unknown relationship to density). To obtain estimates of absolute numbers or density (to estimate density you need to know the area that the estimate of numbers is associated with and sometimes this is not known), considerable trapping effort is required, and therefore because of cost, methods that provide estimates of absolute density are generally used only for research purposes. Most management operations rely on index methods because they are cheaper to obtain and because, for most wild animal management, index methods are sufficient (Caughley 1977).

Estimates of population size

The most common method used in New Zealand for estimating the number of possums ($\hat{N}$) is mark-recapture or capture-recapture. This method requires possums to be captured, marked (usually with serial numbered ear tags), and released. The simplest form of mark-recapture is the Lincoln–Petersen method, often called the Lincoln index, and it takes the form:

$$\frac{m_2}{n_2} = \frac{n_1}{\hat{N}} \quad \text{or} \quad \hat{N} = \frac{n_1 n_2}{m_2} \qquad (1)$$

where n_1 is the number of possums marked and released, n_2 is the number of possums captured in a second trapping session, and of those n_2 possums, m_2 are marked (Pollock *et al.* 1990).

The method has several assumptions (Seber 1982):

- The population is closed. That is, there can be no births, deaths, immigration, or emigration;
- All individual possums have the same likelihood of being captured;
- Marking does not affect the animals catch-ability;
- Marks (ear tags) are not lost or overlooked by the observer;
- The second sample is a simple random sample.

An extension of the Petersen single-capture-and-release method is Schnabel's census for closed populations (see Otis *et al.* 1978). This method allows for multiple capture periods, and the capture history of individuals from a series of trapping sessions are modelled to determine their capture probabilities and other statistics. Different models have been developed to cope with different sets of assumptions about the sources of variation in the capture probabilities (Otis *et al.* 1978). For example, one such model incorporates behavioural response to trapping by using one capture probability for newly captured animals and another for second or subsequent captures. Closed population mark-recapture data can be analysed using a computer package such as CAPTURE (Otis *et al.* 1978; ([online] available URL: http://www.mbr.nbs.gov/software.html), or MARK ([online] available URL: http://www.cnr.colostate.edu/~gwhite/mark/mark.html).

The assumption of closure for the above models is often difficult to satisfy, and a series of mark-recapture models that cope with open populations have been developed (Nichols 1992). The first of these open population models was the Jolly–Seber model (Otis *et al.* 1978; Seber 1982; Pollock *et al.* 1990). The model allows capture probabilities to vary among sampling periods with both additions to and losses from the population. Estimates of population size, survival rates, and capture probabilities can be obtained for open population data using computer programs such as MARK or POPAN5 ([online] available URL: http:// www.cs.umanitoba.ca/~popan), or JOLLY (URL: http://www.mbr.nbs.gov/software.html) (last seen 19 November 1999). For assistance with setting up the analyses for JOLLY and CAPTURE, the Mark-Recapture Analysis Interface ([online] available URL: http://www.landcare.cri.nz/information/software/ mark_recapture) can be used.

Minimum number alive

Some vertebrate ecologists prefer to use a more basic analysis of capture data that ignores the proportion of unmarked animals in the population. This method is known as the enumeration method (Pollock *et al.* 1990), minimum number alive (Krebs 1999), or known number alive (Caughley & Sinclair 1994). The enumeration estimator is:

$$\hat{N}_i = n_i + z_i \qquad (2)$$

where n_i is the number of possums captured in the *i*th trapping session, and z_i is the number of possums caught before and after session *i* and therefore known to be alive at time *i* (Pollock *et al.* 1990). Although this estimator is simple, it makes the assumption that the unmarked population is zero; therefore, it will always provide an underestimate of population numbers — hence the label, minimum number alive. If the underestimate or negative bias was consistent, this estimator, as an index, could be a useful management tool. However, Pollock *et al.* (1990) showed that over a series of trapping sessions, particularly if a session had low capture probabilities, the negative bias of the enumeration estimator could be substantial. Consequently the estimator might indicate significant changes in population size when in fact population size had not changed but only the capture probabilities.

Estimates of population density

The above mark-recapture methods provide estimates of population size ($\hat{N}$) not density (i.e., possums/ha). To obtain density estimates it is necessary to determine to what area the population size relates. To determine density from population size estimates obtained from a grid of traps, the simple approach is to use the area of the grid. This approach, however, can produce significant overestimation of density because of boundary or edge effects (Otis *et al.* 1978). That is, some of the animals included in the population size estimate have home ranges that extend outside the grid. Dice (1938) developed the concept of a boundary strip based on the size of the animal's home range as a method for resolving the problem of determining the effective area of the trapping grid. Including Dice's (1938) boundary-strip method, there are three main approaches to addressing this problem:

- Use an estimate of home range size to obtain an estimate of the boundary required to be added to the area of the grid;
- Use a series of subgrids within a larger grid to enable the boundary effect to be estimated;
- Use a series of "assessment" lines at increasing distances out from the edge of the grid to assess the distance at which marked animals on the grid are not captured.

Otis *et al.* (1978) and Seber (1982) discuss these various options in detail.

Relative indices

Six monitoring methods have been used to obtain relative abundance indices of possums in New Zealand: (a) Spotlight counts, (b) Bait interference, (c) Bait-take from bait stations, (d) Faecal pellet counts, (e) Trap catch, and (f) Mortality-sensing radio-transmitters.

Spotlight counts

Spotlight counts are a favoured monitoring method because the target animal is seen, and this is an appealing attribute of the method compared to one that uses only sign of the animal, such as faecal pellet counts. The method involves driving along a road or track and counting all the possums seen along a fixed route. The count route can be divided into kilometre sections and the counts totalled for each section. Spotlight counts can also be made at fixed points rather than continuous counts, and this method was recommended by Brockie *et al.* (1989) because the count stations could be independent sampling units from which a valid statistical error could be estimated. For both counting methods it is very important to keep all variables as consistent as possible. Changes in time of night, vehicle speed, direction of travel, spotlight wattage, wind speed, precipitation, moonlight, and observers should be kept to a minimum. Brockie *et al.* (1989) recommended counts should be completed in the first 3 hours after sunset, carried out at at least 25 stations, and repeated for 6 fine nights. This high number of nights was necessary to reduce the variance to reasonable levels. Bamford (1970) attempted to identify the critical factors that might influence possum's nightly visibility. Although he found maximum temperature during the previous 12 hours and relative humidity at the time of counting to significantly influence counts, these two factors only accounted for 30% of the variation. Thus, the majority of spotlight count variation cannot be explained.

Because the method generally requires the use of a vehicle, and spotlights can only be used in habitats with open vegetation, the places that are sampled are restricted to roads, 4WD tracks, and farmland. Consequently, the application of accepted sampling principles is usually ignored (i.e., random selection of sampling routes), and this could lead to considerable biases in the abundance estimates. Also, the method is not suitable for estimating possum abundance in forest.

Baddeley (1985) provides a method for analysing spotlight-count data using logarithmic transformation and counts from a non-treatment block. Because of the high variability of possum spotlight counts and the restricted sampling often required when using the method, spotlight counts should only be used for obtaining crude estimates of possum abundance.

Bait interference

Monitoring possum abundance using bait interference relies on possums eating a piece of bait, such as a piece of apple or citrus, cereal pellet, wax block, a small pile of flour and icing sugar, or flour-paste baits. The proportion of baits interfered with provides an index of relative abundance. Although the method has been used since the 1970s for monitoring possums in New Zealand, the problem of individual possums being able to take more than one bait (often referred to as contagion) has prevented the method being widely accepted. Bamford (1970) investigated the contagion problem using flour-paste baits in inverted bottle tops held above the ground by wire and found that, as long as baits were at least 40 m apart, there was little evidence of contagion. Jane (1981), however, found that even at 40-m spacing bait-take increased from night to night presumably as possums learned where the bait lines were. To address this problem, Jane (1981) proposed using the number of nights as a correction to account for possums remembering baits from night to night. Spurr (1995), using baits at 40-m spacings, examined the contagion problem further and found the spatial distribution of bait interference along bait lines was random for at least the first night and usually random for up to 5 nights. Spurr (1995) also compared changes in the bait interference index with possums killed on cyanide baits and found a significant correlation between the two.

Because it is impossible to determine how many possums interfere with each station, bait interference is recording only the presence or absence of possums at the stations. However, if the possum interference occurs at random, mean frequency of interferences can be transformed to mean number of interferences per station (Caughley 1977). The proportion of stations with 0, 1, 2 ... interferences follows the poisson distribution for which the first term is $\underline{e}^{-\bar{x}}$, thus the number of stations having no interferences is:

$$1 - f = \underline{e}^{-\bar{x}} \qquad (3)$$

where f is frequency of baits interfered with and $\bar{x}$ the estimated number of interferences per bait; and if , for example, f equals 0.6 then $\bar{x}$ equals 0.92 (i.e., –Ln (1 – 0.6)).

Although this method has the advantage of being very cost-efficient, it has not been adequately validated for it to be adopted as a robust monitoring method.

Bait-take from bait stations

The weight of bait eaten from bait-stations has also been used as an index of possum abundance (Walker & Hickling 1987). Bait-stations that can hold up to 1 kg of baits are established along lines at spacings of 100–300 m. The amount of bait eaten is monitored for 3 nights and this provides an index of possum abundance. A non-treatment block is recommended because of seasonal changes in bait consumption. This method was used extensively by the Ministry of Agriculture and Fisheries (MAF) staff to monitor Tb-related possum control operations in the late 1980s and early 1990s. However, because some monitoring results were disputed with poor kill estimates being attributed to rats and other non-target animals removing baits, the method has fallen out of favour.

Faecal pellet counts

Faecal pellet counts have been used to monitor possum populations in New Zealand since the 1950s (Riney 1957). The basic method involves counting the number of possum faecal pellets on fixed radius plots (usually 80-cm radius). Up to 120 (usually 100) plots are located at intervals of 10–20 m along lines that follow a predetermined compass bearing (Baddeley 1985). There are three variations of faecal pellet counts:

(a) Presence-absence
This method involves searching each plot for the presence of one or more faecal pellets. No attempt is made to count or estimate the total number of pellets on the plots (Baddeley 1985). As with bait interference the presence-absence data can be corrected with the poisson density transformation, and this should be done because frequency of plots with pellets will not have a linear relationship with possum numbers, particularly at high densities. Presence-absence methods should be used only to obtain broad classes of abundance (i.e., high, medium, low).

(b) Mean pellets per plot
This method involves either counting all pellets on each plot (i.e., the total count method) or estimating the mean number of pellets per plot using a point-distance nearest-neighbour method (Batcheler 1971, 1975). The total count method provides the most accurate estimate of pellet abundance and should be used in preference to the other options.

(c) Recruitment rate
Recruitment rate of faecal pellets can be estimated either from clearing plots of all pellets and then counting the number of pellets that have been recruited over a period of days, or by correcting two density estimates (e.g., a month apart) obtained from total counts or the point-distance nearest-neighbour method, using an estimate of pellet decay rate. Pellet decay rate is estimated by marking a number of pellets (usually with a flagged bicycle spoke) and assessing their presence or disappearance after a selected period. This method provides an estimate of the number of pellets recruited per day, and if the number of pellets voided per day is known (for possums it is about 100, Fitzgerald 1977), then the number of possums can be estimated.

Unfortunately, each of the estimates (total pellets, disappearance rate, defecation rate) has an error, so the final estimate of possum numbers must be treated with suspicion.

Because defecation rate and decay rate can change between seasons and even over shorter periods (e.g., defecation rate can double over a 2-week period, D. Morgan, Landcare Research, Lincoln, pers. comm.), all pellet counts should be carried out with a non-treatment block in which the natural changes in faecal deposition can be estimated and used to correct changes in the treatment block.

Trap catch

Trap catch was first evaluated as a method for monitoring possums in the mid-1960s (Batcheler et al. 1967). Although the method was not used extensively during the 1970s and 1980s it is now the main monitoring method used by regional councils, Department of Conservation, and researchers. In 1996 a National Trap-Catch Protocol and training course was developed by the National Possum Control Agencies (NPCA PO Box 11-461, Wellington) to ensure the method and its application was standardised.

The Trap-Catch Protocol standardises the way trap sites should be selected and what lure to use, and how possum escapes, sprung-but-empty traps, and non-target captures are handled in the analysis. The trap-catch method currently used in the National Trap-Catch Protocol uses the total possums caught over 3 consecutive nights as an index of relative density. Although the analysis method does not attempt to correct for trap saturation, the potential bias is unlikely to be significant if percentage catches are below 30%. The method is used for monitoring percent kill and relative residual possum abundance (termed the residual trap catch (RTC)). The success of many possum control operations is determined by comparing the post-control RTC to some predefined target catch. For example, most control operations that are carried out to reduce the prevalence of bovine Tb have RTC targets of 5%, but RTC targets to protect conservation resources can range from zero to 10%.

The trap-catch method is a basic form of a catch-effort model based on the assumption that the number of possums caught is proportional to the effort (i.e., trap-nights) put into catching them (Seber 1982). Several methods are available for calculating an estimate of $\hat{N}$ from catch-effort data, such as Leslie's method, which essentially calculates a regression of the cumulative catch against the catch from each sample (Caughley 1977). As with mark-recapture methods, there are assumptions that need to be satisfied when using Leslie's method.

These are:

(a) all animals have an equal probability of being caught;
(b) the population is not sufficiently dense that one animal interferes with the capture of another;
(c) there are no losses or additions to the population (i.e., the population is closed).

Another method to analyse trap-catch data when sample effort (number of traps set) is held constant is the Zippen removal model (Seber 1982). This method uses the probability of capture ($\hat{p}$) to estimate the total number of possums that could potentially be available for capture from a trap-line.

$$\hat{N} = x_s / (1 - \hat{q}^s) \qquad (4)$$

where x is the number of possums captured on s nights, and $\hat{q}$ *is* $(1 - \hat{p})$. An explicit solution for $\hat{p}$ provided by Seber (1982) is:

$$\hat{p} = \frac{3X - Y - \sqrt{(Y^2 + 6XY - 3X^2)}}{2X} \qquad (5)$$

where X = 2(possums caught on night one) + possums caught on night two, and Y = total possums caught.

Both $\hat{N}$ and $\hat{p}$ can be calculated for each trap-line, but the model fails if the catch on night three is not less than night one. Often this can happen for individual lines, and in this case $\hat{p}$ can be calculated by pooling the trap-catch data from all lines and then applying the mean $\hat{p}$ to the individual lines.

If the catch on night three is less than night one for all trap-lines, then an explicit solution for 3 trap-nights is:

$$\hat{N} = \frac{6X^2 - 3XY - Y^2 + Y\sqrt{(Y^2 + 6XY - 3X^2)}}{18(X - Y)} \qquad (6)$$

where X and Y are the same as for calculating $\hat{p}$ above.

As for mark-recapture estimates, the Zippen removal method only provides estimates of $\hat{N}$ not density. To convert $\hat{N}$ to a density, the area around each line for which $\hat{N}$ applies needs to be known. This area has not been estimated for possums and is likely to be quite variable between habitats.

Research is being carried out to determine the accuracy of the trap-catch method and the potential of using wax blocks for bait interference. Additionally, trap-catch data are being used to assess the options for optimising the mix of trap lines, number of traps per line, and the number of nights so that for any given cost, precision can be maximised, or for any fixed precision, cost can be minimised.

Mortality-sensing radio-transmitters

A direct estimate of percent kill can be obtained by using mortality-sensing radio-transmitters. This method does not provide an estimate of relative abundance before and after control, but only measures the proportional change in the population. Because there are few assumptions underlying this method, it is likely to provide the least biased estimates of all the monitoring methods available. Two assumptions associated with this method are that the possums radio-collared are a random sample of the population (i.e., they have the same probability of being killed during the control operation as the rest of the population) and that none die from other causes over the period of the operation. The precision of the kill estimates will largely depend on the number of radio-transmitters used with no less than 30 recommended. Each radio-transmitter costs about \$250–\$300, so unless all or most of the transmitters are retrieved (this can require considerable effort and cost especially if a low kill is achieved), the method is relatively expensive.

Radio-transmitters are attached to a random sample of possums scattered through the proposed control area, and after the control has been completed each transmitter can be checked to determine if the possums have been killed (the pulse rate of the transmitter doubles if the possum does not move for 12 hours). The number of transmitters that indicate a dead possum expressed as a percentage of the total transmitters put on possums is the estimate of the percent kill.

The precision of the estimate is obtained using a binomial confidence interval:

$$\text{Approx. 95\% CL} = 2\sqrt{\frac{pq}{n}} \qquad (7)$$

where *t* is the students *t* value for sample size *n*, *p* is the proportion of possums with radio-transmitters that are killed, and *q* is 1 – *p*. Thus if 30 possums had radio-transmitters attached, and an 80% kill was achieved, the 95% CIs would be ±14.6%.

Conclusions

Population monitoring is an essential part of possum management, providing managers with information on the effectiveness of individual control operations and on the responses of population over time. Although population monitoring is often perceived as competing for control funds, without monitoring there is no way of ensuring control funds are being spent effectively or efficiently.

Although there are a number of methods available for monitoring possums, whatever method is chosen, basic sampling rules must be adhered to if potential sampling biases are to be avoided. Because an increasing number of possum control operations are now being carried out under contract, it is essential that the monitoring results are as defendable as possible. It is often difficult to satisfy all the assumptions that the various monitoring methods have (e.g., capture probabilities not being affected by weather or season). However, there is no excuse for not following good sampling design.

None of the monitoring methods available are bulletproof. They all have assumptions that to varying degrees are violated, and users must ensure that environmental and observer factors that can influence monitoring results are minimised as much as possible.

Future research aims to identify the extent of some of the environmental factors that could have an impact on trap-catch estimates, such as season and habitat differences, as well as the effect that trap-shyness might have on estimates following control.

Because many possum populations have now been reduced to very low levels, managers are requesting monitoring methods that can statistically differentiate between residual possum densities that might only differ by 3–4% catch. This requirement needs an increase in precision, which can most easily be achieved by increasing the number of sampling units used. However, because additional monitoring effort increases cost, current research on the trap-catch method is assessing the potential of optimising trap-catch lines to ensure the mix of the number of trap- lines and number of traps per line is achieving maximum precision for minimum cost. Alternative monitoring methods, such as that used by Zielinski & Stauffer (1996) for monitoring fishers (*Martes pennanti*) and Brown & Miller (1998) for monitoring stoats (*Mustela erminea*), might be necessary for sampling low-density populations.

Recent research trials have been assessing the potential of using trapping-webs (Link & Barker 1994) for estimating absolute density as an alternative to using the more expensive capture-recapture methods. Such a method might provide us with a cost-effective method to assess whether New Zealand really does have 70 million possums.

Summary

- To run any business effectively, managers need to have reliable information on the outcomes of their actions and operations and the management of possum populations is no exception.
- Monitoring can be divided into operational monitoring, which aims to determine the impact that management has on possum numbers, and outcome (performance) monitoring, which aims to measure the effectiveness of the management in terms of the response of the resource being protected from possums.
- Monitoring pest populations can impose significant costs on operational budgets (often 10–20% of total budget), and therefore pest managers need to clarify why monitoring is being carried out, what information is required from the monitoring, and how the information can be used for making management decisions.
- Monitoring provides an *estimate* of pest abundance or population change, and underlying these estimates are statistical models that have assumptions, biases, and technical limitations and therefore sampling errors.
- About 15–20% of possum control funds should be allocated to monitoring.
- Population monitoring methods can be divided into two broad categories: (a) estimates of population size (i.e., an estimate of the number of possums and possibly their density), and (b) relative indices (i.e., an index that has a proportional but unknown relationship to density).
- Mark-recapture methods can be used to obtain estimates of possum numbers and require possums to be captured, marked (usually with serial-numbered ear tags), and released. The simplest form of mark-recapture for closed populations is the Lincoln–Petersen method, often called the Lincoln index, with more sophisticated methods such as Jolly–Seber models having been developed for open populations.
- There are six monitoring methods that have been used to obtain relative abundance indices of possums in New Zealand: spotlight counts, bait interference, bait-take from bait stations, faecal pellet counts, trap catch, and mortality-sensing radio-transmitters.
- The trap-catch method is the most frequently used monitoring method for which a National Trap-Catch Protocol has been developed.
- Current research is assessing the accuracy of the current trap-catch monitoring method and the potential of using bait interference.

References

Baddeley, C. J. comp. 1985: Assessments of wild animal abundance. Forest Research Institute, Christchurch, *FRI Bulletin No. 106.* 46 p.

Bamford, J. M. 1970: The influence of weather on the nocturnal activity of possums. *Protection Forestry Report No. 64.* Christchurch, Protection Forestry Division, Forest Research Institute. 12 p.

Batcheler, C. L. 1971: Estimation of density from a sample of joint point and nearest-neighbour distances. *Ecology 52*: 703–709.

Batcheler, C. L. 1975: Development of a distance method for deer census from pellet groups. *Journal of Wildlife Management 39*: 641–652.

Batcheler, C. L. comp. 1978: Report to Minister of Forests and Minister of Agriculture and Fisheries on compound 1080, its properties, effectiveness, dangers, and use. Wellington, New Zealand Forest Service (unpublished) 68 p.

Batcheler, C. L.; Darwin, J. H.; Pracy L. T. 1967: Estimation of opossum (*Trichosurus vulpecula*) populations and results of poison trials from trapping data. *New Zealand Journal of Science 10:* 97–114.

Brockie, R. E.; Rhoades, D. A.; Ward, G. D. 1989: Spotlight counts for assessing possum control on farmland. Ecology Division Report 19, Lower Hutt, Department of Scientific and Industrial Research. 29 p.

Brown, J. A.; Miller, C. J. 1998: Monitoring stoat *Mustela erminea* control operations: Power analysis and design. Biomathematics Research Centre, University of Canterbury Report No. 164 (unpublished).

Caughley, G. 1977: Analysis of vertebrate populations. London, John Wiley. 234 p.

Caughley, G.; Sinclair, A. R. E. 1994: Wildlife ecology and management. Oxford, Blackwell Scientific Publications. 334 p.

Choquenot D.; Warburton, B 1998: How much pest monitoring is enough? Allocation of monitoring resources in pest management programmes. Landcare Research Contract Report LC9899/05 (unpublished) 29 p.

Dice, L. R.1938: Some census methods for mammals. *Journal of Wildlife Management 2*: 119–130.

Fitzgerald, A. E. 1977: Number and weight of faecal pellets produced by opossums. *Proceedings, New Zealand Ecological Society 24*: 76–78.

Jane, G. T. 1981: Application of the poisson model to the bait interference method of possum *Trichosurus vulpecula* assessment. *In:* Bell, B. D. *ed.* Proceedings of the first symposium on marsupials in New Zealand. *Zoology Publications from the Victoria University of Wellington No. 74.* Pp. 185–195.

Krebs, C. J. 1999: Ecological methodology. California, Addison Wesley Longman. 620 p.

Link, W. A.; Barker, R. J. 1994: Density estimation using the trapping web design: a geometric analysis. *Biometrics 50*: 733–745.

Nichols, J. D. 1992: Capture-recapture models — using marked animals to study population dynamics. *BioScience 42*: 94–102.

Parliamentary Commissioner for the Environment 1994: Possum management in New Zealand. Wellington, New Zealand, Office of the Parliamentary Commissioner for the Environment. 196 p.

Payton, I. J.; Pekelharing, C. J.; Frampton, C. M. 1997: Foliar browse index: a method for monitoring possum damage to forests and rare or endangered plant species. Landcare Research Contract Report LC9697/60 (unpublished) 64 p.

Pollock, K. H.; Nichols, J. D.; Brownie, C.; Hines, J. E. 1990: Statistical inference for capture-recapture experiments. *Wildlife Monographs 107.* 97 p.

Riney, T. 1957: The use of faeces counts in studies of several free-ranging mammals in New Zealand. *New Zealand Journal of Science and Technology 38(B)*: 507–532.

Seber, G. A. F. 1982: The estimation of animal abundance and related parameters. London, Charles Griffin. 654 p.

Skalski J. R.; Robson, D. S. 1992: Techniques for wildlife investigations — Design and analysis of capture data. San Diego, Academic Press. 237 p.

Spurr, E. B. 1995: Evaluation of non-toxic bait interference for indexing brushtail possum density. *New Zealand Journal of Ecology.19*: 123–130.

Steidl, R. J.; Hayes, J. P.; Schauber, E. 1997: Statistical power analysis in wildlife research. *Journal of Wildlife Management 61*: 270–279.

Stuart, A. 1984: The ideas of sampling. High Wycombe, UK, Charles Griffin. 91 p.

Otis, D. L.; Burnham, K. P.; White, G. C.; Anderson, D. R. 1978: Statistical inference from capture data on closed animal populations. *Wildlife Monographs 62.* 135 p.

Walker, R.; Hickling, G. J. 1987: Bait stations for assessment and control of possum populations. Forest Research Institute unpublished report. 9 p.

Warburton, B.; Cullen, R. 1993: Cost-effectiveness of different possum control methods. Landcare Research Contract Report LC9293/101 (unpublished) 21 p.

Zielinski, W. J.; Stauffer, H. B. 1996: Monitoring *Martes* populations in California: Survey design and power analysis. *Ecological Applications 6*: 1254–1267.

CHAPTER THIRTEEN

Techniques Used for Poisoning Possums

David Morgan and Graham Hickling

In recent decades poison-baiting has been the main method of large-scale possum control in New Zealand, other methods generally being less cost-effective (see Chapter 15 for a discussion of alternative control methods). In part, this reflects the ease with which aerial sowing of toxic baits can control possums over large areas of inaccessible country. Toxic baits are also used extensively for ground-based control. Together, control using aerial and ground application of toxic baits is presently carried out annually on over 2 million hectares of New Zealand — approximately 12% of the total land area. This widespread use of poisons is probably the most controversial aspect of possum management in New Zealand, and particularly the use of 1080 (sodium monofluoroacetate), which is facing increasing public concern (Parliamentary Commissioner for the Environment 1994; Eason 1995). In particular, aerial application of 1080 baits has been vigorously opposed by various groups demanding that more environmentally, culturally, and socially acceptable methods of possum control be found (Morgan 1998). Furthermore, use of poisons to protect agricultural production is an issue that could influence the purchasing behaviour of overseas consumers of New Zealand's produce (Williams 1994).

Pest managers in the 1990s have increasingly made a distinction between "initial" and "maintenance" control operations (see Chapter 19). Researchers have consequently focused on reducing the cost and impacts of initial aerial control, while expanding the range and effectiveness of ground-based maintenance methods. In this chapter we describe how poisons and baits are applied by air or by ground-based methods, and summarise the refinements that have been made through ongoing research to improve the efficiency and sustainability of the control work.

Aerial poisoning

Large-scale aerial poisoning with 1080 (Fig. 13.1) has been used since 1956 as the primary means of achieving rapid and substantial knockdowns of possum populations. Considerable effort has been made since the mid-1970s to refine the efficiency and environmental safety of the technique, and it is used to control possums on areas of up to 20 000 ha in a single operation for which the field work may be completed within a few days. To sustain the benefits given by initial aerial control, it is usually followed in accessible areas by ground control to maintain the possum population at a low level. However, in less accessible areas, "maintenance" control is achieved by repeated aerial control on a 5- to 10-year cycle, depending on the level of population recovery that can be tolerated (Hickling 1995a).

Advantages and disadvantages

Aerial distribution of 1080 baits is presently the most efficient method of controlling possums over large or inaccessible areas, particularly where it is important to achieve a rapid reduction in the possum population. Typical costs are in the range of $15 to $25 per hectare (Morgan *et al.* 1997a). The main disadvantage of aerial control operations is the strong public perception that broadcast distribution of toxic bait comprises an unacceptable environmental risk. As a consequence there are complex regulatory procedures that must be followed to minimise risk. A considerable body of environmental research data has been gathered in response to these public concerns — much of this information is presented in Chapters 14 and 16. In brief, sampling of stream water within 1080 drop zones has shown that contamination of water supplies is extremely unlikely (Eason *et al.* 1999). Impacts on those populations of common bird species that have been studied are generally negligible (Spurr 1994; Chapter 16), and in many cases removal of possums should benefit bird populations through a reduction in the competition for food (Leathwick *et al.* 1983) or less predation, particularly if rodents are controlled at the same time as possums (Brown *et al.* 1993; Innes *et al.* 1999; Powlesland *et al.* 1999). The 1080 compound breaks down rapidly in all living systems and hence will not accumulate in ecosystems like DDT or other pesticides (Parfitt *et al.* 1994). Nevertheless, 1080 is a highly toxic compound and dogs, livestock, and other non-target animals have been occasionally killed following aerial spreading of 1080 baits. While most dog deaths occur as a result of the dog eating the stomach of poisoned possums, livestock are killed usually after eating baits. In both cases, the risks can be minimised by both precise and minimal application of toxic baits, and by landowners preventing domestic animals from straying into poisoned areas.

The 1080 toxin

Sodium monofluoroacetate is the only toxin currently in use for aerial control of possum populations. This compound, first produced synthetically in Belgium in 1896, is also known as 1080 as this was the laboratory acquisition number given to it when it was tested as a rat poison at the Patuxent Wildlife Research Centre (USA) in 1944. It occurs naturally in some South African, South American, and Australian plants of the *Gastrolobium*, *Acacia*, and *Oxylobium* genera, and it is fully biodegradable (see Chapter 14).

Sodium monofluoroacetate is a fine white powder, which is highly soluble in water and stable under normal storage. Although there is wide variation in the susceptibility of animal species to the toxin, the compound works in all animals by disrupting the "Krebs cycle" and hence the production of energy. Symptoms of poisoning vary between species. Possums stop eating within 30–60 minutes of consuming 1080. They become lethargic and die between 5 and 40 hours later, depending on the dose consumed. While baits were typically loaded with 0.08% 1080 until the mid-1990s, they are now more commonly loaded at a 0.15% concentration. Research has shown that a concentration of 0.15% is more likely to impart a lethal dose to a possum that eats a single bait (Henderson *et al.* 1999a).

The development, properties, toxicology, and environmental fate of 1080 are summarised in Chapter 14. They are also comprehensively reviewed in Haydock & Eason (1997).

Baits for aerial operations

Cereal-based pellets are presently the most commonly used bait type for aerial control of possums (e.g., 74% of managers surveyed use aerially sown pellets). Pellets are easily handled and, provided moisture content does not exceed 14% (Henderson *et al.* 1998), they can be stored for up to 6 months in cool, dry conditions. The pellets are cylindrical in shape, measure 16 mm or 20 mm in diameter and approximately 25–30 mm in length. In addition to 1080, the pellets also contain green dye to deter birds from eating bait (Caithness & Williams 1971) and flavours such as cinnamon or orange, which are much liked by possums. The flavouring is, however, incorporated primarily to mask the odour of 1080 that possums otherwise detect (Morgan 1990).

Carrot bait coated with 1080 is also very effective in controlling possums. Preparation of carrot baits (Fig. 13.2) is usually undertaken at a specially designated airstrip just prior to use, using specialised cutting equipment and appropriately skilled staff. These requirements result in less usage of carrot bait than pellets in possum control (e.g., less than 40% of Department of Conservation and regional council managers surveyed use aerially sown carrot bait). However, there are some advantages in using carrot baits. Carrot bait is even more palatable to possums than cereal pellet bait (Henderson & Frampton 1999). In addition, large quantities of carrots can be bought cheaply, and carrot baits are more resistant to rainfall than pellet baits, withstanding over 200 mm of rain before the 1080 starts to leach out, compared with only 10 mm for pellet baits (Bowen *et al.* 1995).

Carrot bait is most commonly prepared using a "Reliance cutter" fitted with a 20-mm cutting grid (i.e., knives arranged in squares of 20 mm × 20 mm). Large carrots are fed into the machine to produce mainly cube-shaped pieces weighing about 6 g. Chaff (i.e., pieces weighing less than 2 g) is screened out as it is hazardous to non-target species and sub-lethal to possums (Batcheler 1982). Baits are then sprayed, using calibrated equipment, with a solution of 1080 to produce a concentration on baits of 0.08% or 0.15%, as for pellet bait. The solution also contains a green dye as a bird deterrent, and a flavour such as cinnamon oil as a mask for the poison.

Aerial application of toxic bait

Bait is aerially applied by either helicopters with specially designed underslung buckets or by conventional top-dressing aeroplanes. Kills of 80–95% are achieved within 2 days of sowing, as possums generally eat bait as soon as they encounter it (Morgan 1982).

Bait with no poison or green dye is sometimes distributed first (i.e., "pre-feeding") particularly if carrot bait is being used. This practice has been found to improve average kills in aerial control operations

Fig. 13.1
An aerial 1080 control operation in progress. Carrot baits are prepared on the airstrip while pellets are delivered to the airstrip ready for use.

Fig. 13.2
Carrot baits being prepared. The carrot-cutting equipment rejects small pieces of "chaff" (orange material in the picture), which would be sub-lethal to possums, hazardous to non-target animals, and wasteful. Bait of an acceptable size is treated with 1080, a mask for the toxin such as cinnamon oil, and green dye.

from 78% to 90% (Henderson *et al.* 1998), and if the bait is dyed and flavoured in the same way as the subsequent toxic bait, it has been shown to reduce the risk of bait shyness among surviving possums (Moss *et al.* 1998). However, because pre-feeding increases operational costs considerably, it is generally only used with carrot bait as this bait type is less expensive than pellets.

Until recently, a frequent reason for possums surviving aerial operations was the poor bait coverage sometimes achieved (Fig. 13.3, from Morgan 1994). This problem has been largely overcome by the use of differential global positioning system (DGPS) guidance equipment, which has enabled a marked improvement in the precision of aerial sowing. Since fewer gaps in coverage occur, kills of 90% or more are now being achieved more regularly than in previous operations carried out without the use of this technology. Where gaps do occur, they can be readily identified by DGPS and subsequently treated. In practice, DGPS is better suited to use with helicopters rather than with fixed wing aircraft, as the slower flying speed and greater manoeuvrability of helicopters provides more opportunity for the pilot to respond to the accurate guidance signals given by the system.

Application rates of baits have declined markedly since the early 1970s (Morgan *et al.* 1997a; Fig. 13.4) as a consequence of improvements in bait quality and accuracy of aerial distribution. Presently, using

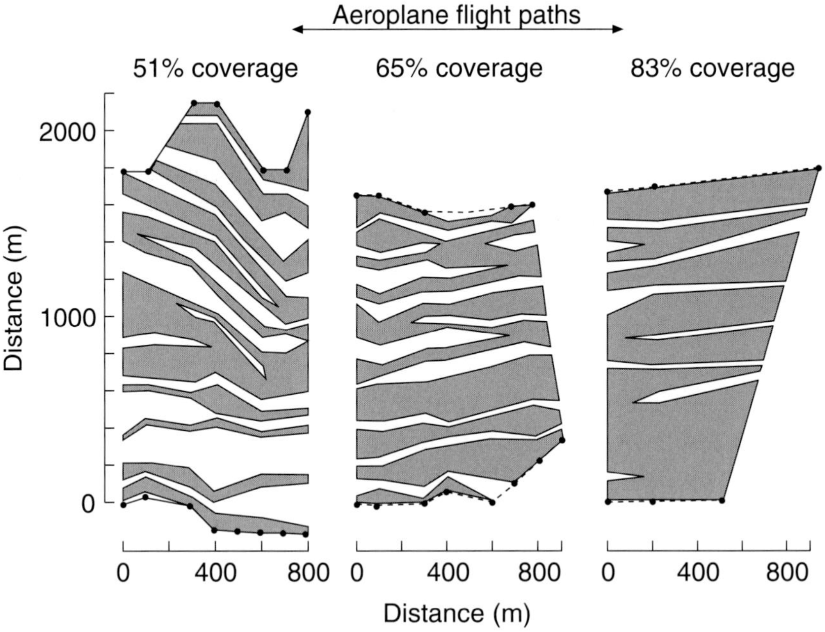

Fig. 13.3
Aerial baiting coverage in three areas of a control operation at Pureora Forest. The operation was conducted in 1992 without the use of DGPS guidance. The distorted shape of the "gaps" in coverage is due to the differing topographic gradients encountered during the field survey of the bait distribution.

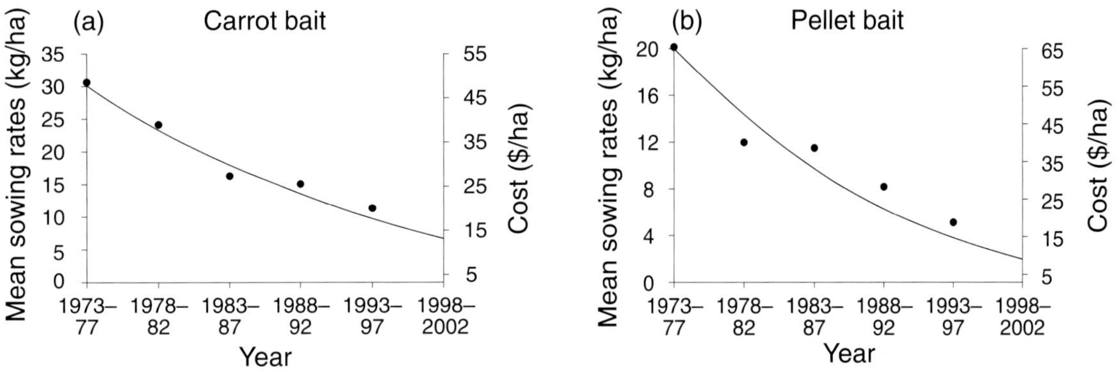

Fig. 13.4
Trends in the application rates of carrot (a) and pellet (b) baits used in aerial control of possums.

sowing rates of 5 kg/ha for pellet and 10 kg/ha for carrot baits is estimated to be saving in the vicinity of NZ$9 million per annum compared with earlier, higher application rates. Since trials have indicated that rates as low as 1.5 kg/ha can be effective, further reductions should be possible as new machinery for application of baits at such low rates becomes available (Morgan *et al.* 1997a). When baits containing 0.15% 1080 are applied at a rate of 5 kg/ha, there will be only 7.5 g of 1080 — about a teaspoonful — distributed per hectare.

Great emphasis is placed on "quality assurance" in all aspects of aerial control operations to ensure that high kills are achieved economically and with a high degree of human and environmental safety. Nevertheless, unexpected failures still occur, and these may result from possums' preference for specific foods at certain times of the year, or from their lower susceptibility to 1080 in warmer temperatures (Oliver & King 1983).

Ground-based poisoning

Ground-based poisoning is used to achieve both initial and maintenance control. In recent years there has been an increase in the use of ground-based poisoning. This has resulted from, firstly, the increasing emphasis being placed on preventing reinfestation of key habitats after initial control (e.g., farm/forest boundaries), and secondly, the preference by the general public for alternative control techniques to aerial poisoning. Complementing this trend is the increase in the amount of research being conducted on improving ground-based poisoning.

Advantages and disadvantages

Compared with aerial poisoning, baits can be located with greater accuracy rendering the method more suitable for use on farmland and land commonly visited by people. Compliance with regulations is therefore less complex than for aerial poisoning. Some of the materials and strategies that are used in ground-based poisoning (e.g., 1080 paste bait in bait stations) provide for relatively inexpensive control, but the cost of other poisons that can be used without a licence tend to make control relatively expensive. Ground-poisoning also has the disadvantage of being time-consuming, necessitating careful planning to avoid conflicts with farming schedules and the requirements of other land users.

1080 baits for ground control

The most commonly used 1080 baits in ground-based control are cereal pellets, as used in aerial control, and apple paste. While the short field-life of pellets is often desirable for use in aerial operations as possums can be expected to encounter baits rapidly, ground-based poisoning strategies often allow up to 5 weeks for possums to find baits. Trials on the rainfall resistance of baits (Bowen *et al.* 1995) showed that cereal pellets break down rapidly due to rainfall and can become less palatable when ambient moisture is absorbed (Henderson & Morriss 1996). Even dampness encourages rapid microbial growth in cereal pellets with a resultant loss of palatability (Henderson & Frampton 1999). Alternatively carrot baits containing 1080 can be used as the 1080 is less affected by moisture in this bait type. Pellets with water-resistant coatings have recently been introduced (e.g., Pestoff®, Animal Control Products) to improve durability under field conditions.

Apple-based 1080 paste is often used for ground control. As with pellets, the paste typically contains cinnamon oil to mask the poison and a green dye to deter birds and identify the bait as toxic.

Alternative poisons and baits

A variety of alternative baits and poisons are available for controlling and harvesting possums, and new materials are under development. Poisons can be categorised as fast-acting ("acute") or slower-acting ("subacute" or "chronic").

Cyanide is a highly acute toxin typically used at a concentration of 60% wt:wt in an oily paste formulation that protects it from being rapidly deactivated by moisture. Cyanide kills through its supression of cellular respiration, which particularly affects the central nervous system resulting in respiratory arrest and death (Haydock & Eason 1997). Unfortunately, in repeatedly poisoned areas many possums exhibit shyness to cyanide baits (Warburton & Drew 1994), due to the hydrocyanic acid gas that is slowly emitted. As a consequence the kills achieved using such baits are often unsatisfactory. Moreover, cyanide-shy populations may become difficult to control with other toxins (Henderson *et al.* 1997a). A new formulation of encapsulated cyanide, Feratox®, is designed to eliminate this gaseous emission.

The tendancy for 1080 and cyanide – both acute toxins – to trigger bait shyness in possum populations has become increasingly apparent in recent years. Possums that eat sub-lethal 1080 baits become shy towards that bait type (Morgan *et al.* 1996) and the shyness may persist for 12 months or more (Morgan & Meikle 1997; O'Connor & Mathews 1997). There is consequently increasing awareness of the need to avoid over-reliance on any single toxin such as 1080. As the use of 1080 and cyanide is restricted for use to licensed operators, there is also a need for alternative poisons to be available to the increasing number of landowners and private possum-control agencies carrying out possum control.

Pindone pellets, containing a first-generation anticoagulant similar to warfarin, are sometimes used for possum control but are not very effective for this purpose. Individual possums need to eat 1–2 kg of pindone pellets to receive a lethal dose (Eason & Jolly 1993). Talon® and Pestoff® baits, containing the second-generation (i.e., considerably more potent than first-generation) anticoagulant brodifacoum, are more effective, but very expensive to use when possums are numerous (Henderson *et al.* 1997b). Furthermore, as brodifacoum is a persistent compound, care must be taken to avoid using it in habitats where it may accumulate in scavengers such as wild pigs (Eason *et al.* 1999). Campaign® is a cereal pellet bait containing the subacute toxin cholecalciferol (vitamin D_3), to which possums are particularly sensitive. The compound causes rapid resorption of calcium from the blood and subsequent formation of calcium deposits in the arteries surrounding the heart leading to death by heart failure within 2–6 days of eating bait (Haydock & Eason 1997). While Campaign® is more costly than 1080 and brodifacoum baits, it causes a loss of appetite, which avoids the overconsumption of bait once a lethal dose has been ingested. With appropriate use of pre-feeding, the cost of poisoning with Campaign® can be realistic even for high-density possums populations (Henderson *et al.* 1997b), allowing it to be used for initial control in areas where 1080 poisoning is inappropriate. Other advantages of Campaign® are that birds are highly resistant to its toxic effects, and the risk of dogs being poisoned by scavenging possum carcasses is lower than for 1080. All three poisons can be used by unlicensed users, thus providing useful alternatives to the restricted 1080 and cyanide.

Trials with bait-shy possums have shown that it is primarily the bait material, rather than the poison, that possums become shy towards (Morgan *et al.* 1996). While the development of bait shyness can be minimised by high standards of bait preparation and application (Henderson & Morriss 1996), the growing reliance on repeated maintenance control increases the potential for generating shyness, so different bait types should be used in successive maintenance operations. For example, where 1080 pellets have been used within the previous 3 years, it would be prudent to switch to using toxic paste, carrot, or gel baits (Morgan 1999) in follow-up maintenance control.

Use of paste baits can be problematic if the fruit paste attracts bees. Recent trials (e.g., Jordan *et al.* 1997; Morgan *et al.* 1997b) have sought additives that discourage bees and native insects feeding on paste baits while retaining the baits' palatability to possums. Gel baits, containing either 1080 or cholecalciferol, have been developed as a ready-to-use item (Morgan & Henderson 1996). The custom-designed bait stations ensure maximum safety to users and non-target animals (Morgan 1999), and because the bait is highly durable it requires only an occasional replacement of the gel cartridge.

The cinnamon and other flavouring agents used may also have a role in luring possums to the baits, but only at close-range (Morgan *et al.* 1995). They may also deter some non-target species such as kākā (Hickling 1997). Attempts to develop pheromone, sound, or visual lures for possums have so far met with little success (e.g., Carey *et al.* 1997). The large amounts (e.g., ~1700 tonnes of 1080 baits per annum) of bait used in possum control operations each year represents a significant cost and efforts are underway to develop cheaper baits using waste material from food-processing industries (Morgan & Henderson 1996).

Ground-based application of toxic bait

Ground-based application of toxic baits involves placement of paste baits on earth "spits" (i.e., clods of earth) or trees, presentation of baits in various designs of bait station, or hand-broadcasting of pellets baits.

In open country the control area is usually destocked, and apple paste bait is then applied using an applicator gun onto upturned earth spits. These are prepared, using a grubber, at regular intervals throughout farmland and particularly along bush edges (Fig. 13.5). Non-toxic pre-feed paste is usually presented for several days preceding the application of 1080 paste bait. The benefit of pre-feeding with paste is, however, not as clear as with aerial operations. An improvement in average kills from 76% to 84% was shown from nine field trials, but statistical analysis suggested that further trials would be required to confirm an improvement due to pre-feeding in most situations (Thomas & Morgan 1998). After 1080 paste has been presented, baits are checked and, where possum numbers are high, they are usually replenished for several days. The spits are then turned back to bury remaining bait, thereby allowing livestock to be brought back into the area. Along bush edges, paste is sometimes laid on branches or pieces of tin or cardboard, which allows uneaten baits to be removed.

Most possums move away from 1080 spit lines before dying, so hunters must use fast-acting cyanide paste (or traps) if they are to retrace their bait lines to retrieve possum carcasses for skinning. It is common to surround cyanide baits, which are usually laid in the approximate size of a pea, with a mixture of flour, icing sugar, and a lure such as cinnamon or aniseed. Encapsulated cyanide (i.e., Feratox®) can be used very effectively in a variety of ways to both control and harvest possums. The capsules can be mixed with non-toxic feed pellets or non-toxic peanut paste in a special bait station that prevents the costly problem of rats removing the capsules. Alternatively the capsules can be presented with the non-toxic feed bait in small, sealable plastic bags that are suspended from a rat-proof wire stake.

Bait stations are typically used where it is important to avoid exposing livestock, people, pets, or native fauna to toxic bait. Possums are good climbers, so raised stations can be placed out of reach of many non-target species (Fig. 13.6), although research has demonstrated that this reduces the level of control if pre-feeding is not done (Henderson *et al.* 1998). Stations targeting possums on the ground should be raised at least 40 cm to minimise interference by birds and rodents.

Small stations (e.g. "KK", "Romark", "Sentry", or horizontally mounted, plastic flower pots) holding individual small amounts of paste or pellet bait will provide good control only if they are sufficiently numerous and are serviced regularly. Larger stations, typically plastic tubes or containers (e.g., "PhilProof", "Kilmore") attached to trees and holding 1 kg or more of pelleted bait, allow numerous possums to feed at each station over periods of days or weeks (Fig. 13.6). In well-populated areas, 10 or more possums may congregate to feed at such stations, with subordinate individuals waiting until dominant possums have fed (Henderson & Hickling 1997). Such stations are particularly effective when placed at 50-m intervals along forest margins because they intercept the natural movement of possum to and from pasture (Hickling *et al.* 1990). Stations can also provide effective control if they are placed in grids at 150-m intervals, requiring pre-feeding, in areas of continuous forest, or without pre-feeding at 100-m intervals in small reserves and plantations (Thomas *et al.* 1997).

Pre-feeding at bait stations increases the amount of toxic bait subsequently eaten by individual possums due to familiarity with it (Henderson *et al.* 1999b). Also, more possums find toxic bait after pre-feed has been presented (Thomas *et al.* 1995). The pre-feeding is generally carried out over 2 or more weeks in order to attract the majority of possums (Hickling *et al.* 1990). Some possums have been recorded travelling a kilometre or more to feed at such stations, but many others do not. While numerous possums may feed at a station, the majority will have travelled only 100 m or less. Since females typically have smaller home ranges than males (see Chapter 3), bait stations that are spaced too far apart will result in a higher survival of females and consequently more rapid repopulation of the area through breeding.

Any possums that survive their initial encounter

Fig. 13.5
Apple paste baits containing 1080 being applied to earth spits. The earth spits are later turned over to render the area safe once again for livestock.

Fig. 13.6
A possum feeding from a bait station placed up a tree. Pre-feeding ensures that bait stations placed in trees are discovered by the majority of possums, resulting in better control.

with toxic bait at a bait station quickly learn to distinguish between, and sort through, toxic and non-toxic baits. For this reason, toxic and non-toxic baits should never be mixed in the same station. Furthermore, these possums will consume prefeed (i.e., non-toxic) baits in subsequent control but avoid toxic baits (Hickling *et al.* 1991). Similar problems have been reported when control is attempted in areas where possums have previously survived rabbit control operations that used low-concentration baits (Hickling 1994; Hickling *et al.* 1998). Bait-shy possums can be targeted by using slow-acting poisons such as brodifacoum or cholecalciferol, preferably in a new bait formulation (Ross *et al.* 1997). Use of brodifacoum is normally only cost-effective for control of residual populations because possums tend to overconsume these expensive baits.

Cereal 1080 pellets are occasionally broadcast by hand in small forest reserves where access by livestock and the public can be curtailed. Further detail on the optimum ways of using different types of possum baits, and of combining them sequentially over time to avoid bait shyness, can be found in reports and manuals prepared for professional pest management personnel (e.g., Henderson *et al.* 1998, 1999b).

Legislation relating to possum poisoning

Some possum control options require compliance with complex legislation, which can significantly increase operational costs. Details of how this legislation impinges on possum control are included in Nelson (1996), a Ministry of Health publication (1995), and a Ministry of Agriculture publication (1998). (See also Appendix 1).

The main statues that relate to the use of poisons for the control of possums are the following: Biosecurity Act 1993; Civil Aviation Act 1990; Hazardous Substances and New Organisms Act 1996 (which will incorporate the Dangerous Goods Act 1974, the Pesticides Act 1979, and the Toxic Substances Act 1979); Health Act 1956; Health and Safety in Employment Act 1992; Pesticides (Vertebrate Pest Control) Regulations 1983; and the Resource Management Act 1991. Four further pieces of legislation relate primarily to toxic residues: Agricultural Compounds and Veterinary Medicines Act 1997; Dairy Industry Act 1952; Food Regulations 1983; and the Meat Act 1981.

The future: better integration of aerial and ground-based control?

Simulation modelling, and field experience, suggest that the most cost-effective way to control a possum population is often to use large-scale aerial control to achieve a high primary kill, which is then maintained by regular ground-based control (see Chapter 19). The maintenance work should not rely on a single control method because overly frequent application of that one method is likely to induce behavioural resistance in the residual possums population (Hickling 1995b). Ongoing research into maintenance control has consequently focused on cost-effective ways of utilising several different poisons within a control programme, and of ways to integrate poison-based control with alternative techniques such as trapping. Field trials have shown that if acute poisons are used ineffectively (due to the use of old or moist bait for example), surviving possums become bait shy and little further success can be expected through the repeated use of acute toxins. By contrast, baits containing a chronic poison such as brodifacoum will be sampled by surviving possums, and because no ill-effects are experienced for some days, amounts consumed gradually increase until effective control is achieved (Henderson *et al.* 1997b).

The present challenge facing pest managers is to maintain and, where possible, enhance the efficacy of poison-based control strategies, while maintaining public and market acceptance of such control. Increasingly, consumers in New Zealand and overseas are demanding that food is produced using humane practices and that it is free of toxic residues. For these reasons, possum control by poisoning is likely to come under increasing scrutiny. While an intensive effort is presently underway to find biological control methods that are more sustainable and possibly more socially acceptable, poisoning is likely to remain the mainstay of the control effort in the short- to medium-term. Even if current efforts to develop biocontrol were to prove successful, biocontrol and toxins would still need to be used together as components of an integrated control strategy.

Given the disastrous ecological and economic consequences of failing to control possums (see Chapters 8–11) it is imperative that new poisoning practices are developed that are affordable (at least in the medium term), effective, and acceptable to the wider community here and overseas. The use of multispecies baiting strategies (Morgan 1993), although potentially problematic (Hickling *et al.* 1998), is a further direction of interest.

Consequently, there is continuing interest in the development of alternative toxicants. What must be recognised, however, is that the costs and time associated in bringing a novel toxicant through the full process of efficacy testing, risk assessment, and regulatory approval is extremely time-consuming and expensive (Lotti 1987). There will thus be a continuing need to refine the use of existing poisons.

Summary

- Control of possum populations over large areas is presently highly dependent on the use of toxic baits.
- The range of bait types and poisons available is limited, with most use being made of 1080 in cereal pellet, paste, and carrot baits; cyanide in paste, and encapsulated cyanide combined with either non-toxic pellet or pastes; brodifacoum in cereal pellets; and cholecalciferol in cereal pellets.
- Aerial control, using 1080 baits, is used over large areas of inaccessible country. The technique is rapid and cost-effective. Public perception of the method is its main drawback.
- Ground-based control is carried out using a variety of bait types on more accessible land where baits must either be retrieved or buried after control.
- Pre-feeding, where non-toxic baits are presented first, improves the effectiveness of most poisoning methods, but should be considered carefully as it increases costs appreciably.
- Long-term control of possums involves the repeated use of toxic baits. Different types of bait and poison are used in such "maintenance" control operations to avoid the build-up of bait-shy populations.
- The use of poisons for possum control is coming under increasing scrutiny both in New Zealand and among overseas markets where consumers are increasingly concerned about both the safety (human and environmental) and the humaneness of food production. Continued future use of poisoning, which is essential to New Zealand's agricultural and environmental needs, is therefore likely to be achieved by development of new products and refinement of existing ones to address these public concerns.

References

Batcheler, C. L. 1982: Quantifying "bait quality" from number of random encounters required to kill a pest. *New Zealand Journal of Ecology 5*: 129–139.

Bowen, L. H.; Morgan, D. R.; Eason, C. T. 1995: Persistence of sodium monofluoroacetate (1080) in baits under simulated rainfall. *New Zealand Journal of Agricultural Research 38*: 529–531.

Brown, K.; Innes, J.; Shorten, R. 1993: Evidence that possums prey on and scavenge birds' eggs, birds and mammals. *Notornis 40*: 169–177.

Caithness, T. A.; Williams, G. R. 1971: Protecting birds from poisoned baits. *New Zealand Journal of Agriculture 122(6)*: 38–43.

Carey, P. W.; O'Connor, C. E.; McDonald, R. M.; Matthews. L. R. 1997: Comparison of the attractiveness of acoustic and visual stimuli for brushtail possums. *New Zealand Journal of Zoology 24*: 273–276.

Eason, C. T. 1995: Sodium monofluoroacetate (1080): an update on recent environmental toxicology data and community concerns. Proceedings: 10th Australian Vertebrate Pest Control Conference, Hobart, Tasmania, 29 May–2 June 1995. Tasmania, Department of Primary Industry and Fisheries. Pp. 39–43.

Eason, C. T.; Jolly, S. E. 1993: Anticoagulant effects of pindone in the rabbit and Australian brushtail possum. *Wildlife Research 20*: 371–374.

Eason, C. T.; Milne, L.; Potts, M.; Morriss, G.; Wright, G. R. G.; Sutherland, O. R. W. 1999: Secondary and tertiary poisoning risks associated with brodifacoum. *New Zealand Journal of Ecology 23*: 219–224.

Haydock, N.; Eason. C. T. 1997: Vertebrate pest control manual. Wellington, New Zealand, Department of Conservation. 88 p.

Henderson, R. J.; Frampton, C. M. 1999: Avoiding bait shyness in possums by improved bait standards. Landcare Research Contract Report LC9899/60 (unpublished) 50 p.

Henderson, R. J.; Hickling, G. J. 1997: Possum behaviour as a factor in sub-lethal poisoning during control operations using cereal baits. Landcare Research Contract Report LC9798/03 (unpublished) 26 p.

Henderson, R. J.; Morriss, G. A. 1996: Sub-lethal poisoning of possums with acute pesticides used in bait stations. *Proceedings of the forty-ninth New Zealand Plant Protection Conference*: 137–142.

Henderson, R.J.; Morriss, G.A.; Morgan, D.R. 1997a: Cyanide-induced shyness to cereal baits. Landcare Research Contract Report LC9798/62 (unpublished) 14 p.

Henderson, R. J.; Morriss, G. A.; Morgan, D. R. 1997b: The use of different types of toxic bait for sustained control of possums. *Proceedings of the fiftieth New Zealand Plant Protection Conference*: 382–390.

Henderson, R. J.; O'Connor, C. E.; Morgan, D. R. 1998: Current practices in sequential use of possum baits. Landcare Research Contract Report LC9899/09 (unpublished) 69 p.

Henderson, R. J.; Frampton, C. F.; Morgan, D. R.; Hickling, G. J. 1999a: Efficacy of baits containing 1080 for the control of captive brushtail possums. *Journal of Wildlife Management 63:* 1138–1151.

Henderson, R. J.; Morgan, D. R.; Eason, C. T. 1999b: Manual of best practice for ground control of possums. Landcare Research Contract Report LC9899/84 (unpublished). 80 p.

Hickling, G. J. 1994: Behavioural resistance by vertebrate pests to 1080 toxin: implications for sustainable pest management in New Zealand. *In:* Seawright, A. A.: Eason, C.T. *ed.* Proceedings of the science workshop on 1080. *The Royal Society of New Zealand Miscellaneous Series 28:* 151–158.

Hickling, G. J. 1995a: Action thresholds and target densities for possum pest management. *In:* O'Donnell, C. F. J. *comp.* Possums as conservation pests. Proceedings of an NSSC Workshop . . . 29–30 November 1994. Wellington, Department of Conservation. Pp. 47–52.

Hickling, G. J. 1995b: Implications of learned and innate behavioural resistance for single tactic control of vertebrate pests. Proceedings: 10th Australian Vertebrate Pest Control Conference, Hobart, Tasmania, 29 May–2 June 1995. Tasmania, Department of Primary Industry and Fisheries. Pp. 303–308.

Hickling, G. J. 1997: Effect of green dye and cinnamon oil on consumption of cereal pest baits by captive North Island kaka (*Nestor meridionalis*). *New Zealand Journal of Zoology 24*: 239–242.

Hickling, G. J.; Thomas, M. D.; Grueber, L. S.; Walker, R. 1990: Possum movements and behaviour in response to self-feeding bait stations. Forest Research Institute Contract Report FWE90/9 (unpublished) 17 p.

Hickling, G. J.; Heyward, R. P.; Thomas, M. D. 1991: An evaluation of possum control using 1080 toxin in bait feeders. Forest Research Institute Contract Report FWE91/61 (unpublished) 18 p.

Hickling, G. J.; Thomas, M. C.; Moss, Z. N.; Ross, J. G. 1998: Rabbit control can induce toxic bait aversion in other species: evidence and potential solutions. Proceedings: 11th Australian Vertebrate Pest Conference, Bunbury. Pp. 325–330.

Innes, J.; Nugent, G.; Prime, K. 1999: Pigeons versus possum: 7-0 at Motatau. *He Kōrero Paihama – Possum Research News 11*: 1–2.

Jordan, B.; Moller, H.; Hickling, G. 1997: Field trial to assess possum preference for a new pest control paste. *Proceedings of the fiftieth New Zealand Plant Protection Conference*: 377–381.

Leathwick, J. R.; Hay, J. R.; Fitzgerald, A. E. 1983: The influence of browsing by introduced mammals on the decline of North Island kokako. *New Zealand Journal of Ecology 6*: 55–70.

Lotti, M. 1987: Production and use of pesticides. *In:* Costa, L. G.; Galli, C. L.; Murphy, S. D. *ed.* Toxicology of pesticides: experimental, clinical and regulatory perspectives. Berlin, Springer-Verlag. Pp. 11–17.

Ministry of Agriculture 1998: Regulatory framework for the control of agricultural compounds under the Agricultural Compounds and Veterinary Medicines Act 1997. Wellington, Ministry of Agriculture. P. 26.

Ministry of Health 1995: Model permit conditions for the use of sodium monofluoroacetate (1080) issued by the Medical Officer of Health (MoH). Wellington, Ministry of Health (unpublished).

Morgan, D. R. 1982: Field acceptance of non-toxic and toxic baits by populations of the brushtail possum (*Trichosurus vulpecula* Kerr). *New Zealand Journal of Ecology 5*: 36–43.

Morgan, D. R. 1990: Behavioural responses of brushtail possums, *Trichosurus vulpecula*, to baits used in pest control. *Australian Wildlife Research 17*: 601–613.

Morgan, D. R. 1993: Multi-species control by aerial baiting: a realistic goal? *New Zealand Journal of Zoology 20*: 367–372.

Morgan, D. R. 1994: Improved cost-effectiveness and safety of sodium monofluoroacetate (1080) possum control operations. *In:* Seawright, A.A.; Eason, C.T. *ed.* Proceedings of the science workshop on 1080. *The Royal Society of New Zealand Miscellaneous Series 28:* 144–150.

Morgan, D. R. 1998: Community participation in research to improve efficiency of ground-based possum control. Proceedings: 11th Australian Vertebrate Pest Control Conference, Bunbury. Pp. 93–98.

Morgan, D. R. 1999: Risks to non-target species from use of a gel bait for possum control. *New Zealand Journal of Ecology 23*: 281–287.

Morgan D. R.; Henderson, R. J. 1996: Development of new types of possum baits. *In:* Improving conventional control of possums. *The Royal Society of New Zealand Miscellaneous Series 35:* 62–64.

Morgan, D. R.; Meikle, L. 1997: Persistence of learned aversions by possums to 1080 baits and "bait-switching" as a solution. Landcare Research Contract Report LC9697/125 (unpublished) 10 p.

Morgan, D. R.; Innes, J.; Frampton, C. M.; Woolhouse, A. D. 1995. Responses of captive and wild possums to lures used in poison baiting. *New Zealand Journal of Zoology 22*: 123–129.

Morgan D. R.; Morriss, G.; Hickling, G. J. 1996: Induced 1080 bait-shyness in captive brushtail possums and implications for managament. *Wildlife Research 23*: 207–211.

Morgan, D. R.; Thomas, M. D.; Meenken, D.; Nelson, P. C. 1997a: Less 1080 bait usage in aerial operations to control possums. *Proceedings of the fiftieth New Zealand Plant Protection Conference*: 391–396.

Morgan, D. R.; Ward-Smith, T.; McQueen, S. 1997b: Responses of native birds and bats to "bee-safe" possum paste. Landcare Research Contract Report LC9697/126 (unpublished) 19 p.

Moss, Z. N.; O'Connor. C. E.; Hickling. G. J. 1998: Implications of prefeeding for the development of bait aversions in brushtail possums (*Trichosurus vulpecula*). *Wildlife Research 25*: 133–138.

Nelson, P. 1996: The safety requirement in pest control. *In:* Improving conventional control of possums. *The Royal Society of New Zealand Miscellaneous Series 35:* 11–14.

O'Connor, C. E.; Mathews, L. R. 1997: Duration of cyanide-induced conditioned food aversions in possums. *Physiology and Behavior 62*: 931–933.

Oliver, A. J.; King, D. R. 1983: The influence of ambient temperatures on the susceptibility of mice, guinea-pigs and possums to compound 1080. *Australian Wildlife Research 10*: 297–301.

Parfitt, R. L.; Eason, C. T.; Morgan, A. J.; Wright, G. R.; Burke, C. M. 1994: The fate of sodium monofluoroacetate (1080) in soil and water. *In:* Seawright, A.A.; Eason, C.T. *ed.* Proceedings of the science workshop on 1080. *The Royal Society of New Zealand Miscellaneous Series 28:* 59–66.

Parliamentary Commissioner for the Environment 1994: Possum management in New Zealand. Wellington, Office of the Parliamentary Commissioner for the Environment. 196 p.

Powlesland, R. G.; Knegtmans, J. J. W.; Marshall, I. S. J. 1999: Costs and benefits of aerial 1080 possum control operations using carrot baits to North Island robins (*Petroica australis longipes*), Pureora Forest Park. *New Zealand Journal of Ecology 23*: 149–159.

Ross, J.; Hickling, G. J.; Morgan, D. R. 1997: Use of subacute and chronic toxicants to control sodium monofluoroacetate (1080) bait shy possums. *Proceedings of the fiftieth New Zealand Plant Protection Conference*: 397–400.

Spurr, E. B. 1994: Review of the impacts on non-target species of sodium monofluoroacetate (1080) in baits used for brushtail possum control in New Zealand. *In:* Seawright, A.A.; Eason, C.T. *ed.* Proceedings of the science workshop on 1080. *The Royal Society of New Zealand Miscellaneous Series 28:* 124–133.

Thomas, M. D.; Morgan, D. R. 1998: Is prefeeding necessary in possum control with 1080 paste? Landcare Research Contract Report LC9899/34 (unpublished) 12 p.

Thomas, M. D.; Frampton, C. M.; Briden, K. W.; Hunt, K. G. 1995: Evaluation of brodifacoum baits for maintenance control of possums in small forest reserves. *Proceedings of the forty-eighth New Zealand Plant Protection Conference*: 256–259.

Thomas, M. D.; Mason, J.; Briden, K. W. 1997: Optimising the use of bait stations for possum control in native forest. Landcare Research Contract Report LC9697/45 (unpublished) 16 p.

Warburton, B.; Drew, K. W. 1994: Extent and nature of cyanide-shyness in some populations of Australian brushtail possums in New Zealand. *Wildlife Research 21*: 599–605.

Williams, J. M. 1994: Food and fibre markets and societal trends: implications for pest management. *In:* Seawright, A.A.: Eason, C.T. *ed.* Proceedings of the science workshop on 1080. *The Royal Society of New Zealand Miscellaneous Series 28:* 20–32.

CHAPTER FOURTEEN

Toxicants Used for Possum Control

Charles Eason, Bruce Warburton, and Ray Henderson

In New Zealand there are six toxicants currently registered for possum control: 1080 (sodium monofluoroacetate), cyanide, cholecalciferol (Vitamin D_3), phosphorus, brodifacoum, and pindone. Their use for killing possums raises a variety of concerns including risks and persistence, which are often different for each toxicant because of the differences in their physical and chemical properties and their toxicological modes of action.

The ideal toxicant for possum control should be inexpensive, usable by farmers, and humane. It should have an antidote and be degradable in soil and water. As far as possible it should be species-specific. It should not leave persistent residues in livestock and should have a low risk of primary or secondary poisoning in non-target species. It should be supported by a comprehensive database of efficacy, toxicology, and risk assessment studies that satisfy local concerns as well as national and international regulatory agencies. Such an ideal pesticide has not yet been identified and may never be. The advantages and disadvantages of the six toxicants currently used for possum control are summarised below and some of the research questions being addressed regarding their safety and efficacy are outlined. Currently the range of toxicants and baits available allows those involved in possum control to select the most appropriate combination for sustained control of possums, provided they are familiar with the advantages and disadvantages of the different toxicants and baits.

Toxicants and their delivery systems

The most extensively used toxicant in New Zealand is 1080, which is incorporated in cereal pellets or applied to carrots for aerial sowing and for use in bait stations. Phosphorus, cyanide, cholecalciferol and 1080 are also used in pastes for ground-based control, and cyanide is currently being manufactured as a encapsulated pellet. Brodifacoum, pindone, and cholecalciferol are incorporated in cereal baits and only used in bait stations. Cholecalciferol and 1080 are also available in a gel-bait formulation. This selection of toxicants with different properties in a range of different bait types provides those involved in possum control with a selection of "tools" to use in different circumstances. For example, high-density populations are reduced most effectively by fast-acting poisons, e.g., 1080, cyanide, or cholecalciferol (Henderson *et al.* 1997; Hickling *et al.* 1999; Ross 1999), whereas slower-acting toxicants are chosen for killing possums present at low densities or when "bait shyness" or aversion to fast-acting toxicants is suspected (Henderson *et al.* 1997). Pre-feeding and the use of different bait types allow practitioners to avoid or mitigate bait shyness (Morgan *et al.* 1996; Moss *et al.* 1998). Safety concerns can also influence the choice of toxicant used. For example, when controlling possums on or adjacent to farmland, the use of baits containing cholecalciferol or cyanide will reduce the risk of secondary poisoning of farm or pet dogs. Current projections suggest that existing and new toxicants and delivery systems will be required to control possums at least until 2017 (National Science Strategy Committee (NSSC) 1998) when new alternative control methods may be available.

Sodium monofluoroacetate (1080)

Sodium monofluoroacetate is both a natural plant toxin and one of the most toxic rodenticides known. Monofluoroacetate occurs naturally in plants in Australia and appears to be one of the secondary plant compounds that can cause death of livestock, having evolved as a defence mechanism against browsing mammals (Twigg *et al.* 1996). It is highly effective at killing possums and has been used in New Zealand since the 1950s. Monofluoroacetate is converted in poisoned animals to fluorocitrate, which inhibits the tricarboxylic acid cycle resulting in energy deprivation and death. The heart, lungs, and brain appear to be the main target organs with death resulting from respiratory depression and cardiac failure. However, it is not selective and can be lethal to all mammals and other animals that eat baits (Table 14.1). Dogs (*Canis familaris*) in particular are susceptible to both primary and secondary poisoning (Eason 1997).

This toxin is readily dissolved and diluted by water, and is degraded by micro-organisms, so long-term environmental contamination is unlikely (Seawright & Eason 1994; Eason 1997; Ogilvie *et al.* 1998; Booth *et al.* 1999; Eason *et al.* 1999d). Under mild wet conditions, 1080 residues will usually disappear from baits and soil within 1–4 weeks (Bowen *et al.* 1995; King *et al.* 1994). However, 1080 may persist in bait, in the soil, or in carcasses for several weeks or months in cold or dry conditions (Parfitt *et al.* 1994; Meenken & Booth 1997). So, while 1080 has many desirable properties in terms of its effectiveness and its lack of persistence in the environment, in contrast, its persistence in possum carcasses for many months puts farm dogs at considerable risk because of the unique susceptibility of this species to even trace amounts of the poison (Meenken & Booth 1997; Eason 1997).

Because 1080 is used more widely in New Zealand than elsewhere in the world the fate of 1080 in water and in non-target species, especially livestock and invertebrates (Eason *et al.* 1992; Eason *et al.* 1993; Rammel 1993; Eason *et al.* 1994a,b; Eason 1997; Eason *et al.* 1998; Booth *et al.* 1999; Booth & Wickstrom 1999), has been comprehensively assessed. This work is complemented by metabolic studies on the breakdown of fluoroacetate and fluorocitrate in the US (Tecle & Casida 1989), Canada (Sykes *et al.* 1987), Australia (King *et al.* 1994), and South Africa (Meyers 1994), and by recent regulatory toxicology studies to meet US EPA (Environmental Protection Agency) and local registration requirements (Fagerstone *et al.* 1994; Eason *et al.* 1999d). Over 70 studies have been completed on the non-target impacts of 1080 on wildlife in New Zealand (reviewed by Spurr 1994a,b). Environmental studies indicate that 1080 causes minimal water and soil contamination, and it appears that there have been no long-term impacts on bird and invertebrate populations. Recent studies have provided additional data on the fate of 1080 and metabolites such as fluorocitrate in the

Table 14.1
Sodium monofluoroacetate (compound 1080): summary of key features.

Advantages	Disadvantages
Highly effective for achieving a rapid reduction in possum numbers.	Controversial, especially aerial operations.
The only poison available for aerial application.	Secondary poison risk (especially dogs) from possum carcasses.
Cheap compared to most other poisons.	No effective antidote.
Biodegradable in the environment.	Can only be used by licensed operators in government agencies.
Kills other pest species. Can kill stock and insects.	Generates bait shyness if target animal gets sub-lethal doses.
Can achieve consistently high kills.	Poor quality bait causes bird deaths.
High-quality efficacy data exist to support both aerial and ground-baiting techniques.	

1080 is the most widely used toxicant for achieving rapid reductions in possum numbers over large areas. It is used in carrot, cereal, paste, and gel baits. Cinnamon is used as an added flavour primarily to mask the taste of 1080. Carrot baits are screened to remove small baits so that the risk to non-target birds is minimised. Cereal baits are also used extensively in bait stations. 1080 paste baits are used extensively for ground-based follow-up maintenance control.

environment (Booth *et. al.* 1999; Potts *et al.* 1999) and the histopathological effects of 1080 at sub-lethal doses in mammals (see Eason *et al.* 1994a,b; Cook *et al.* 1999; Eason *et al.* 1999c,d; O'Connor *et al.* 1999). Underpinning research on the best ways of using 1080 and on safety and toxicology has been intense because 1080 bait still provides the most effective tool for killing possums.

Current and future research continues to focus on three distinct areas: improved efficacy, environmental studies, and toxicology. Improved efficacy studies seek to answer questions such as "Can possums be killed more effectively with 1080, minimising the risk of inducing bait shyness?" (Henderson *et al.* 1999). New initiatives to improve uptake of baits by target species are further linked to reducing non-target impacts and environmental contamination. Environmental studies are addressing questions relating to the impact of 1080 on rarer bird species and invertebrates, as well as more precisely defining the fate of 1080 and its metabolites. Toxicology studies are addressing questions such as "Is 1080 a humane poison and does it have any long-term sub-lethal effects on non-target species, including livestock and humans?" Recent data indicates that 1080 is not a mutagen (an agent causing mutations) but is teratogenic (causes malformations in embryos), and may cause degeneration of the testes, renal failure, cardiotoxicity, or neurological effects, which dictates the need for extreme care when handling 1080 (Eason *et al.* 1999d). Linking new toxicology data with the environmental fate and non-target impact studies produces a clearer picture of the risk and benefits of using 1080 for possum control. Workers in the pest control industry and dogs scavenging carcasses are at greatest risk from unwanted exposure or poisoning, hence considerable care is required in the manufacture and use of baits (Chi *et al.* 1996; Eason *et al.* 1999d; Cook *et al.* 1999). High-quality operational and safety procedures can minimise the risk of contamination of livestock and wildlife, including game, but constant vigilance is required since 1080 is a potent broad-spectrum toxin.

Cyanide (Feratox® and pastes)

Like 1080, cyanide has also been used for several decades in New Zealand for killing possums. Cyanide interferes with oxygen metabolism causing cytotoxic hypoxia, which depresses the central nervous system causing respiratory depression and death. Cyanide is inexpensive and very fast-acting (death usually occurs in 10 to 26 minutes) with few undesirable effects from a welfare perspective (Gregory *et al.* 1998). It does not persist in the environment. As with other bait types the time that baits remain toxic depends on the properties of the formulation, rainfall, and how well the baits are protected from rainfall. However, its instability and associated volatility are a potential risk to pest control operators and hunters, who must handle the material with extreme care. In a number of countries cyanide is considered too hazardous for pest control. Cyanide has in the past not been the toxicant of choice because standard cyanide pastes are only moderately effective, some possums have an innate aversion to their smell (Warburton & Drew 1994), and possums sub-lethally poisoned become bait shy (Hickling *et al.* 1999). Field trials completed in the early 1990s found that the proportion of cyanide-shy possums in four populations ranged from 12% to 54% (Warburton & Drew 1994). The advantages and disadvantages of cyanide are summarised in Table 14.2. Small numbers of birds (e.g., kiwi and weka) have been killed by cyanide paste, but the risk to birds is considered low when compared to 1080 (Spurr 1994a).

Recently completed research has focused on two areas, improved efficacy and humaneness. Cyanide formulations with low emission rates of hydrogen cyanide have been developed in an effort to make the toxicant more effective even in areas where cyanide-shyness has developed (Warburton & Drew 1994). Technical difficulties in producing a formulation with no smell and acceptable environmental breakdown characteristics were partially resolved in 1997. A cyanide pellet (Feratox®) is now available that is safer to use than standard pastes. Collaborative studies between Landcare Research and Massey University conducted in 1996–98 have demonstrated that, from a welfare perspective, cyanide is the most favoured toxin for possum control (Gregory *et al.* 1998). Because of this, and because the risk of secondary poisoning of dogs is very low compared with 1080, phosphorus, or brodifacoum, further efforts to encourage the safe and more effective use of cyanide appears warranted.

Phosphorus paste

Phosphorus, introduced originally for killing rabbits in the 1920s, is still used in New Zealand for possum control, but deregistration on the grounds of inhumaneness has been considered (Table 14.3). Other older non-specific toxicants such as arsenic and strychnine are no longer used in New Zealand for a number of reasons, including inhumaneness,

Table 14.2
Cyanide (paste, encapsulated cyanide Feratox®): summary of key features.

Advantages	Disadvantages
Cheap (1–2 cents per bait).	Hazardous to users.
Its very rapid action minimises welfare impacts and enables possum carcasses to be recovered.	Available only to licence holders.
Low secondary poison risk.	Can induce poison aversion.
Can achieve moderate to high kills (70–90%).	Pastes can result in very poor kills if possums are cyanide-shy, hence not favoured by pest control agencies.
Encapsulated cyanide has no emission of hydrogen cyanide gas, therefore safer to use and supposedly not detectable by possums.	Amyl nitrite is available but its use is controversial*
Encapsulated cyanide not adversely affected by wet weather as cyanide paste is, and pellets can be recovered and reused.	
Not environmentally persistent.	

* Hydroxycobalamin and kalocyanor antidotes are reasonably effective but are not appropriate in a first-aid context.

Cyanide paste is mainly used by commercial skin hunters and contract hunters. Pea-sized pieces of paste are placed with a small handful of flour and icing sugar on a rock or stick. Often the poison lines will be pre-fed with the non-toxic flour/icing sugar mix. Cyanide paste was the sole toxicant used by hunters because its rapid action allowed the recovery of the carcasses. Feratox® (a pea-sized coated cyanide pellet) was developed to overcome the problem of cyanide aversion that many possum populations had developed from intensive use of cyanide paste. The pellets are placed in a bait station with either similar-sized cereal feed pellets or in a peanut-butter paste.

Table 14.3
Phosphorus: summary of key features.

Advantages	Disadvantages
An effective poison with kills > 90% achieved.	Has a greater animal welfare impact than cyanide or 1080, but appears not to be as "bad" as generally thought**.
Less public opposition than with 1080*.	Secondary poisoning risk to dogs and birds.
	Risk of fire.
	Available only to licence holders.
	Antidotes of limited value.

* When farmers oppose the use of 1080 they will often accept phosphorus as a replacement

** Studies show that the symptoms of phosphorus poisoning in possums differ from those earlier reported (Rammell & Fleming 1978)

Phosphorus is used as a paste and generally applied on turf spits on the ground. Phosphorus is favoured for controlling possums in areas where the risk of primary or secondary poisoning of dogs from 1080 is high.

low palatability, and bait shyness. The RSPCA in New Zealand is opposed to the use of any toxicant that causes any animal to suffer and is particularly opposed to the use of arsenic, strychnine, and phosphorus (Loague 1994). Despite its long-term use in New Zealand there are very few published research data on its effectiveness, fate in the environment, or persistence in carcasses. Recent research suggests that phosphorus is less inhumane than previously supposed, and is perhaps less inhumane than anticoagulants (Eason *et al.* 1998). However, it still lacks the advantages of cyanide or cholecalciferol as it is known to cause secondary poisoning of birds (Sparling & Federoff 1997) and dogs, and antidotes are of limited value (Gumbrell & Bentley 1995).

Brodifacoum (Talon® and Pestoff®) and Pindone

Possum bait containing brodifacoum was first registered in New Zealand in 1991 and a pindone cereal bait was registered in 1992 (Nelson & Hickling 1994). Two brodifacoum baits are currently available for possum control under the trade names of Talon® and Pestoff®. Tables 14.4 and 14.5 summarise the advantages and disadvantages of these toxicants. Both compounds are classified as anticoagulant poisons interfering with the synthesis of prothrombin and other clotting factors. Earlier in the 1990s toxicity studies showed pindone to be a less effective toxicant than brodifacoum (Eason *et al.* 1994c). Pindone is used successfully against rabbits in Australia and New Zealand. However, a comparison of the sensitivity of rabbit and possum to pindone found that the possum is relatively tolerant to the substance (Eason & Jolly 1993). Its use has declined as dissatisfied users have noted that very large amounts of bait are used before measurable reductions in possum numbers are achieved.

Efficacy trials on captive possums have shown that the brodifacoum oral LD_{50} value is higher than the published value of 0.17 mg/kg (Godfrey 1985). In the wild, brodifacoum baits need to be applied for 2–4 months to be highly effective (Henderson *et al.* 1997). Because of the comparatively low toxicity of pindone, very large amounts of pindone bait are likely to be eaten before possums are killed.

The toxicology and non-target effects of brodifacoum have been reviewed elsewhere (Eason & Spurr 1995). The risk to dogs and wildlife of secondary poisoning with brodifacoum is high and there are growing concerns regarding its persistence

Table 14.4
Brodifacoum (Talon®, Pestoff®): summary of key features.

Advantages	Disadvantages
Available to general public.	Registered for use only in bait stations.
Is effective against possums that have developed poison/bait aversion.	Expensive compared to 1080 or cyanide.
An antidote (vitamin K) is available, but protracted treatment is needed.	Possums take 2–4 weeks to die (animal welfare issues need to be resolved).
	Possums can eat excessive amounts of baits (increases costs).
	High risk of secondary poisoning of non-target species.
	Persistent (> 1 year) in liver of vertebrates (can enter food chain and put at risk meat for human consumption*).

* Residues are being detected with increasing frequency in wildlife carcasses including game meat.

Cereal baits containing brodifacoum are used in bait stations for maintaining possums at low numbers after the initial population has been reduced with fast-acting toxicants such as cyanide, 1080, or cholecalciferol. The slow action of the toxicant enables it to overcome any toxicant or bait aversion possums might have developed.

Table 14.5
Pindone: summary of key features.

Advantages	Disadvantages
Available to general public.	Not an effective poison for possums. Some possums can eat in excess of 2 kg of bait and are not killed.
Less persistent than brodifacoum.	
An antidote (vitamin K) is available, but protracted treatment is needed.	Risk of secondary poisoning.

Has been used for killing possums but it is generally ineffective unless large quantities of bait are continually supplied.

and transfer through game meat to humans and a number of different wildlife species (Eason *et al.* 1996b,c; Eason *et al.* 1999a).

Brodifacoum and pindone are more effective at controlling rats and rabbits, respectively, than possums. Nevertheless anticoagulants do appear to have a place in current control strategies, particularly in areas with low-density possum populations and following initial reduction of possum populations with fast-acting poisons where residual possums may have bait shyness (Henderson *et al.* 1997).

Current and future environmental research questions for brodifacoum and pindone are similar to those for 1080. The wisdom of large-scale field use of a persistent pesticide has been challenged (Eason *et al.* 1996a, 1999a). For example, feral pigs could feed on poisoned possum carcasses and remain contaminated for months or years after brodifacoum has been used (Eason *et al.* 1999a). Pindone is less persistent, but still results in secondary poisoning (Martin *et al.* 1994). Less research has been conducted on pindone, principally because it is less effective than the other toxicants used for possum control.

Cholecalciferol (Campaign®)

Cholecalciferol (Vitamin D_3) was introduced for possum control in 1995. It acts by elevating plasma calcium, causing heart failure. Calcium homeostasis in possums appears to be poor (Eason 1991), and possums are particularly sensitive to this toxicant. Acute toxicity data for cholecalciferol in cereal bait indicate that the LD_{50} for possums is 18 mg/kg, and the susceptibility of possums is increased by the addition of calcium carbonate to bait (Jolly *et al.* 1993). The advantages and disadvantages of cholecalciferol are summarised in Table 14.6.

Because low toxicity to birds is important when considering toxicants for controlling possums in New Zealand wildlands, 10 mallard ducks were dosed with 2000 mg/kg cholecalciferol. All the birds survived without showing any clinical signs, thus confirming an oral LD_{50} of >2000 mg/kg in the mallard duck (Eason *et al.* 1999b). At present there is insufficient data to be confident that cholecalciferol is not hazardous to native birds and insects. Nevertheless the results with ducks are encouraging and acute toxicity tests have been conducted using common native species, the weka (*Gallirallus australis*) and wētā (*Hemideina crassidens*), as well as canaries and chickens. Weka were not affected and the results from other native and non-native species confirm that cholecalciferol is less toxic to birds than 1080 or brodifacoum (Eason *et al.* 1999b). Regardless of these comparatively low-risk factors it is important that this toxicant is used in carefully positioned bait stations to minimise the risk of killing birds. Secondary poisoning of dogs and wildlife with cholecalciferol is less probable since possums take several days to die and, as they lose their appetite, there should not be an accumulation of toxic baits in the stomach of dead animals (Eason *et al.* 1996a,c; Wickstrom *et al.* 1999). The toxicity of poisoned cholecalciferol possum carcasses to cats and dogs has confirmed a low secondary-poisoning risk when compared with 1080 or brodifacoum (Eason *et al.* 1996a; Wickstrom *et al.* 1999).

As indicated above, current and future research questions for cholecalciferol relate to further investigation of primary and secondary poisoning risks to non-target species, its relative humaneness, and its persistence in the environment and animals. Initial batches of cholecalciferol bait produced in 1995 absorbed water and crumbled, reducing its effectiveness. Considerable effort has been put into reformulating the bait to make it more robust and a

Table 14.6
Cholecalciferol (Campaign®): summary of key features.

Advantages	Disadvantages
Available to general public.	Expensive compared to 1080 and cyanide.
An acute toxin that can achieve rapid knockdown of possum numbers.	Not registered for aerial application.
Low risk of secondary poisoning.	Treatment for accidental poisoning is available, but is complex.
Lower risk to birds than for 1080.	
A useful single-dose alternative to 1080.	

A relatively new toxicant for use in bait stations that poses a low risk of secondary poisoning to dogs and birds.

new gel bait containing cholecalciferol should be available in 1999, which will provide pest control workers with greater flexibility and safer tools.

Discussion

Continued innovations in the use of existing and new toxicants and baits will be required if poisons are to be retained for possum control. It is anticipated that emerging biocontrol technologies will reduce reliance on poisoning. To retain toxicants for the next 15–20 years, the toxicological databases that underpin their use must be robust. Studies need to provide enough toxicological data to satisfy the regulatory authorities in New Zealand and overseas with regard to effectiveness and risk to non-target species. The data packages generated for registration or reassessment must provide sufficient information to allow for human health risks to be comprehensively assessed. Laboratory and field information must be provided to allow the proper consideration of any environmental effects arising from the existing or proposed use of any toxicant. Research efforts in New Zealand have focused on improving the effectiveness of possum control and generating data on the environmental toxicology and non-target impacts of vertebrate pesticides. This experimental research has been important as it addresses both community and scientific questions relating to the use of these toxicants. In the case of 1080 an extensive database on experimental and regulatory toxicology has been generated (Seawright & Eason 1994). Nevertheless an ongoing review of the data requirements and technical information for 1080 and other toxicants is essential. Since 1080 is the most intensively used toxicant for possum control and the only one that is both aerially sown in baits and used in bait stations (Livingstone 1994), it is not surprising that more research questions are being generated with regard to its use (Eason 1997).

A factor influencing the future of the vertebrate pesticides used in New Zealand for pest control will be how they are perceived by the public. Unfortunately the process for involving communities in pest control decisions has not always kept pace with the research on technological improvements. Real community involvement in research and possum control has increased steadily during the 1990s. "Force feeding" scientific information at community meetings to groups opposed to a particular technique (e.g., aerial sowing 1080 baits) understandably has produced a negative response, which is counterproductive and devalues science.

Mechanisms of disseminating information have improved through the efforts of the National Possum Control Agencies (NPCA), but research on the process of technical transfer and community involvement in pest control decisions, including the acceptance of different techniques on the grounds of humaneness or cost-effectiveness, is now required. Complex issues are involved and adverse reactions to pest control, and to 1080 in particular, are inevitable. Adverse public reactions are likely to lead to further restrictions on the use of poisoning and traps unless ongoing dialogue is sustained across the range of interest and pressure groups.

Further development of new vertebrate pesticides should only be considered if candidates demonstrate significant efficacy, welfare, selectivity, persistence, and non-target advantages. The cost of developing a new vertebrate pesticide could be as high as several million dollars. Smarter use of existing products to reduce non-target exposure and smarter products such as Feratox® and Campaign® are a significant step forward. Hence the reformulation of existing toxicants, such as cyanide, and the development of more target-specific bait types and baiting strategies, developed by researchers working with community groups, offers a rational way forward for short- to medium-term improvement in possum control. Quality control standards that minimise exposure of non-target species, including humans, are essential (Eason *et al.* 1997). This review demonstrates there are no ideal toxicants for killing possums; but some have advantages over others (see Table 14.7). An array of different toxicants and bait types allows practitioners to choose the most suitable toxicant or combinations to achieve their goals. In time, greater efforts must be made to use toxicants that score highly from welfare and/or environmental safety perspectives instead of those that are considered inhumane or unacceptable for other reasons. This review document does not cover the symptoms of poisoning or the treatment of antidotes. For further details on these issues readers are referred to the Vertebrate Pest Control Manual (Haydock & Eason 1997).

Table 14.7
Summary of characteristics of poisons used for controlling possums in New Zealand in order of effectiveness.

Cyanide

- rapid action, most humane
- cyanide aversion influences effectiveness
- low environmental persistence
- low secondary poison risk
- effective antidotes lacking

1080

- moderately rapid and humane
- essential for aerial control
- very effective
- low environmental persistence
- secondary poison risks
- no antidote

Cholecalciferol

- effective
- lower toxicity to birds than 1080
- low risk of secondary poisoning
- expensive compared with 1080 or cyanide

Phosphorus

- causes longer periods of pain and sickness compared to cyanide & 1080
- effective
- causes secondary poisoning
- effective antidotes lacking

Brodifacoum

- possums take 2–4 weeks to die
- effective against low density, poison/bait-shy possums
- very persistent
- high secondary poisoning risk
- widespread contamination of wildlife and game possible
- antidote available
- expensive compared with 1080 or cyanide

Pindone

- possums take 2–3 weeks to die
- low effectiveness
- moderate persistence
- low secondary poison risk
- antidote available

Summary

- Six toxicants are currently available for possum control in New Zealand, all of which have advantages and disadvantages.
- Sodium monofluoroacetate is currently the most appropriate for broad-scale use, but care must be taken to reduce non-target exposure, and dogs are at high risk of secondary poisoning.
- Cyanide is effective and the most preferred poison from a welfare perspective.
- Cholecalciferol and cyanide have a low risk of secondary poisoning when compared to 1080.
- Phosphorus has come under considerable animal welfare scrutiny, but in fact appears to be less inhumane than anticoagulants.
- Possums poisoned with brodifacoum take a long time to die. This, coupled with the persistence and accumulation of brodifacoum in the food chain, is grounds for limiting the use of this toxicant.
- Pindone is less persistent than brodifacoum, but not very effective for killing possums.

References

Booth, L. H.; Wickstrom, M. 1999: The toxicity of sodium monofluoroacetate (1080) to *Huberia striata*, a New Zealand native ant. *New Zealand Journal of Ecology 23*: 161–165.

Booth, L. H.; Ogilvie, S. C.; Wright, G. R.; Eason, C. T. 1999: Degradation of sodium monofluoroacetate (1080) and fluorocitrate in water. *Bulletin of Environmental Contamination and Toxicology 62*: 34–39.

Bowen, L. H.; Morgan, D. R.; Eason, C. T. 1995: Persistence of sodium monofluoroacetate (1080) in baits under simulated rainfall. *New Zealand Journal of Agricultural Research 38*: 529–531.

Chi, C. H.; Chen, K. W.; Chan, S. H.; Wu, M. H.; Huang, J. J. 1996: Clinical presentation and prognostic factors in sodium monofluoroacetate intoxication. *Clinical Toxicology 34*: 707–712.

Cook, C.; Eason, C. T.; Wickstrom M. 1999: Changes in brain glutamate and gamma aminobutynic acid (GABA) concentrations induced by sodium monofluoroacetate (1080) exposure. *Journal of Neuroscience* (in press).

Eason, C. T. 1991: Cholecalciferol as an alternative to sodium monofluoroacetate (1080) for poisoning possums. *Proceedings of the forty-fourth New Zealand Weed and Pest Control Conference*: 35–37.

Eason, C. T. 1997: Sodium monofluoroacetate toxicology in relation to its use in New Zealand. *Australasian Journal of Ecotoxicology 31*: 57–64.

Eason, C. T.; Jolly, S. E. 1993: Anticoagulant effects of pindone in the rabbit and Australian brushtail possum. *Wildlife Research 20*: 371–374.

Eason, C. T.; Spurr, E. B. 1995: Review of the toxicity and impacts of brodifacoum on non-target wildlife in New Zealand. *New Zealand Journal of Zoology 22*: 371–379.

Eason, C. T.; Wright, G. R.; Fitzgerald, H. 1992: Sodium monofluoroacetate (1080) water residue analysis after large-scale possum control. *New Zealand Journal of Ecology 16*: 47–49.

Eason, C. T.; Gooneratne, R.; Wright, G. R.; Pierce, R.; Frampton, C. M. 1993: The fate of sodium monofluoroacetate (1080) in water, mammals, and invertebrates. *Proceedings of the forty-sixth New Zealand Plant Protection Conference*: 297–301.

Eason, C. T.; Gooneratne, R.; Fitzgerald, H.; Wright, G.; Frampton, C. 1994a: Persistence of sodium monofluoroacetate in livestock animals and risk to humans. *Human & Experimental Toxicology 13*: 119–122.

Eason, C. T.; Gooneratne, R.; Rammell, C. G. 1994b: A review of the toxicokinetics and toxicodynamics of sodium monofluoroacetate in animals. *In:* Seawright, A.A.; Eason, C.T. *ed.* Proceedings of the science workshop on 1080. *The Royal Society of New Zealand Miscellaneous Series 28*: 82–89.

Eason, C. T.; Henderson, R.; Thomas, M. D.; Frampton, C. M. 1994c: The advantages and disadvantages of sodium monofluoroacetate and alternative toxins for possum control. *In:* Seawright, A. A.; Eason, C. T. *ed.* Proceedings of the science workshop on 1080. *The Royal Society of New Zealand Miscellaneous Series 28*: 159–166.

Eason, C. T.; Meikle, L.; Henderson, R. J. 1996a: Testing cats for secondary poisoning by cholecalciferol. *Vetscript 9*: 26.

Eason, C. T.; Wright, G. R.; Batcheler, D. 1996b: Anticoagulant effects and the persistence of brodifacoum in possums (*Trichosurus vulpecula*). *New Zealand Journal of Agricultural Research 39*: 397–400.

Eason, C. T.; Wright, G. R.; Meikle, L. 1996c: The persistence and secondary poisoning risks of sodium monofluoroacetate (1080), brodifacoum, and cholecalciferol in possums. *In:* Timm, R. M.; Crabb, A. C. *ed.* Proceedings: Seventeenth Vertebrate Pest Conference, California. Davis, University of California. Pp. 54–58.

Eason, C. T.; Wickstrom, M.; Gregory, N. 1997: Product stewardship, animal welfare, and regulatory toxicology constraints on vertebrate pesticides. *Proceedings of the fiftieth New Zealand Plant Protection Conference*: 206–213.

Eason, C.; Wickstrom, M.; Milne, L.; Warburton, B.; Gregory, N. 1998: Implications of animal welfare considerations for pest control research : the possum as a case study. *In:* Proceedings of the joint ANZCCART/NAEAC conference held in Auckland, New Zealand, 19–20 September 1997. Wellington, The Royal Society of New Zealand. Pp. 125–131.

Eason, C. T.; Milne, L.; Potts, M.; Morriss, G.; Wright, G. R. G.; Sutherland, O. R. W. 1999a: Secondary and tertiary poisoning risks associated with brodifacoum. *New Zealand Journal of Ecology 23*: 219–224.

Eason, C. T.; Ogilvie, S.; Milne, L. 1999b: The acute toxicity of cholecalciferol to birds and insects. Poster paper presented at meeting organised by the New Zealand Ecological Society on "Ecological consequences of poisons used for mammalian pest control", Christchurch, 9–10 July 1998: Programme and abstracts. P. 39.

Eason, C. T.; Wickstrom, M.; Milne, L.; Arthur, D. G.; Gooneratne, S. R. 1999c: The acute and long-term effects of exposure to sodium monofluoroacetate (1080) in sheep. Poster paper presented at meeting organised by the New Zealand Ecological Society on "Ecological consequences of poisons used for mammalian pest control", Christchurch, 9–10 July 1998 : Programme and abstracts. P. 36.

Eason, C. T.; Wickstrom, M.; Turck, P.; Wright, G. R. G. 1999d: A review of recent regulatory and environmental toxicology studies on 1080: results and implications. *New Zealand Journal of Ecology 23*: 129–137.

Fagerstone, K. A.; Savarie, P. J.; Elias, D. J.; Schafer Jr., E. W. 1994: Recent regulatory requirements for pesticide registration and the status of

Compound 1080 studies conducted to meet EPA requirements. *In:* Seawright, A. A.; Eason, C. T. *ed.* Proceedings of the science workshop on 1080. *The Royal Society of New Zealand Miscellaneous Series 28:* 33–38.

Godfrey, M. E. R. 1985: Non-target and secondary poisoning hazards of second-generation anticoagulants. *Acta Zoologica Fennica 173*: 209–212.

Gregory, N. G.; Milne, L. M.; Rhodes, A. T.; Litten, K. E.; Wickstrom, M.; Eason, C. T. 1998: Effect of potassium cyanide on behaviour and time to death in possums. *New Zealand Veterinary Journal 46*: 60–64.

Gumbell, R. C.; Bentley, G. R. 1995: Secondary phosphorus poisoning in dogs. *New Zealand Veterinary Journal 43:* 25–26.

Haydock, N.; Eason, C. T. 1997: Vertebrate pest control manual. Wellington, New Zealand, Department of Conservation. 88 p.

Henderson, R. J.; Morriss, G. A.; Morgan, D. R. 1997: The use of different types of toxic bait for sustained control of possums. *Proceedings of the fiftieth Plant Protection Conference*: 382–390.

Henderson R. J.; Frampton, C. M.; Morgan, D. R.; Hickling, G. J. 1999: Efficacy of baits containing 1080 for the control of captive brushtail possums. *Journal of Wildlife Management 63:* 1138–1151.

Hickling, G. J.; Henderson, R. J.; Thomas, M. C. C. 1999: Poisoning mammalian pests can have unintended consequences for future control: two case studies. *Journal of Ecology 23*: 267–273.

Jolly, S. E.; Eason, C. T.; Frampton, C. 1993: Serum calcium levels in response to cholecalciferol and calcium carbonate in the Australian brushtail possum. *Pesticide Biochemistry and Physiology 47*: 159–164.

King, D. R.; Kirkpatrick, W. E.; Wong, D. H.; Kinnear, J. E. 1994: Degradation of 1080 in Australian soils. *In:* Seawright, A. A.; Eason, C. T. *ed.* Proceedings of the science workshop on 1080. *The Royal Society of New Zealand Miscellaneous Series 28:* 45–49.

Livingstone, P. G. 1994: The use of 1080 in New Zealand. *In:* Seawright, A. A.; Eason, C. T. *ed.* Proceedings of the science workshop on 1080. *The Royal Society of New Zealand Miscellaneous Series 28*: 1–9.

Loague, P. 1994: An animal welfare perspective. *In:* Animal welfare in the twenty-first century: ethical, educational, and scientific challenges. Proceedings of the conference held at School of Medicine, Christchurch. Pp. 109–113.

Martin, G. R.; Kirkpatrick, W. E.; King, D. R.; Robertson, I. D.; Hood, P. J.; Sutherland, J. R. 1994: Assessment of the potential toxicity of an anticoagulant, pindone (2-pivalyl-1, 3-indandione), to some Australian birds. *Wildlife Research 21:* 85–93.

Meenken, D.; Booth, L. H. 1997: The risk to dogs of poisoning from sodium monofluoroacetate (1080) residues in possum (*Trichosurus vulpecula*). *New Zealand Journal of Agricultural Research 40:* 573–576.

Meyers, J. J. M. 1994: Fluoroacetate metabolism of *Pseudomonas cepacia. In:* Seawright, A. A.; Eason, C. T. *ed.* Proceedings of the science workshop on 1080. *The Royal Society of New Zealand Miscellaneous Series 28*: 54–58.

Morgan, D. R.; Morriss, G.; Hickling, G. J. 1996: Induced 1080 bait-shyness in captive brushtail possums and implications for management. *Wildlife Research 23:* 207–211.

Moss, Z. N.; O'Connor, C. E.; Hickling, G. J. 1998: Implications of prefeeding for the development of bait aversions in brushtail possums (*Trichosurus vulpecula*). *Wildlife Research 25:* 133–138.

Nelson, P. C.; Hickling, G. J. 1994: Pindone for rabbit control: Efficacy, residues and cost. Proceedings: Sixteenth Vertebrate Pest Conference, Santa Clara, California, 28 February–3 March 1994. Davis, University of California. Pp. 217–222.

National Science Strategy Committee (NSSC) for Possum and Bovine Tuberculosis Control, 1998: Annual report. Wellington, New Zealand, NSSC. 79 p.

O'Connor, C. E.; Milne, L. M.; Arthur, D. G.; Ruscoe, W. A.; Wickstrom, M. 1999: Toxicity effects of 1080 on pregnant ewes. *Proceedings of the New Zealand Society of Animal Production 59:* 250–253.

Ogilvie, S. C.; Booth, L. H.; Eason, C. T. 1998: Uptake and persistence of sodium monofluoroacetate (1080) in plants. *Bulletin of Environmental Contamination and Toxicology 60*: 745–749.

Parfitt, R. L.; Eason, C. T.; Morgan, A. J.; Wright, G. R.; Burke, C. M. 1994: The fate of sodium monofluoroacetate (1080) in soil and water. *In:* Seawright, A. A.; Eason, C. T. *ed.* Proceedings of the science workshop on 1080. *The Royal Society of New Zealand Miscellaneous Series 28*: 59–66.

Potts, M.; Wright, G. R. G.; Eason, C. T. 1999: Dust and 1080 residue contamination after aerial sowing of 1080 baits for possum control. Poster paper presented at meeting organised by the New Zealand Ecological society on "Ecological consequences of poisons used for mammalian pest control", Christchurch, 9–10 July 1998: Programme and abstracts. P. 37.

Rammell, C. G. 1993: Persistence of compound 1080 in sheep muscle and liver. *Surveillance 20(1)*: 20–21.

Rammell, C. G; Fleming, P. A. 1978: Compound 1080: Properties and use of sodium monofluoroacetate in New Zealand. Wellington, New Zealand, Ministry of Agriculture and Fisheries. 137 p.

Ross, J. G. 1999: Cost-effective control of 1080 bait-shy possums (*Trichosurus vulpecula*). Unpublished PhD thesis, Lincoln University, Canterbury, New Zealand. 195 p.

Seawright, A. A.; Eason, C. T. *ed.* 1994: Proceedings of the science workshop on 1080. *The Royal Society of New Zealand Miscellaneous Series 28.* 173 p.

Sparling, D. W.; Federoff, N. E. 1997: Secondary poisoning of kestrels by white phosphorus. *Ecotoxicology 6*: 239–247.

Spurr, E. B. 1994a: Review of the impacts on non-target species of sodium monofluoroacetate (1080) in baits used for brushtail possum control in New Zealand. *In:* Seawright, A. A.; Eason, C. T. *ed.* Proceedings of the science workshop on 1080. *The Royal Society of New Zealand Miscellaneous Series 28*: 124–133.

Spurr, E. B. 1994b: Impacts on non-target invertebrate populations of aerial application of sodium monofluoroacetate (1080) for brushtail possum control. *In:* Seawright, A. A.; Eason, C. T. *ed.* Proceedings of the science workshop on 1080. *The Royal Society of New Zealand Miscellaneous Series 28*: 116–123.

Sykes, T. R.; Quastel, J. H.; Adams, M. J.; Ruth, T. J.; Nonjawa, A. A. 1987: The disposition and metabolism of fluorine-18 fluoroacetate in mice. *Biochemical Archives 3(3):* 317–324.

Tecle, B.; Casida, J. E. 1989: Enzymatic defluorination and metabolism of fluoroacetate, fluoroacetamide, fluoroethanol and (-)-erthro-fluorocitrate in rats and mice examined by ^{19}F and ^{13}C NMR. *Chemical Research in Toxicology 2:* 429–435.

Twigg, L. E.; King, D. R.; Bowen, L. H.; Wright, G. R.; Eason, C. T. 1996: Fluoroacetate content of some species of the toxic Australian plant genus *Gastrolobium*, and its environmental resistance. *Natural Toxins 4:* 122–127.

Warburton, B.; Drew, K. W. 1994: Extent and nature of cyanide-shyness in some populations of Australian brushtail possums in New Zealand. *Wildlife Research 21:* 599–605.

Wickstrom, M.; Henderson, R. J.; Milne, L.; Eason, C. T. 1999: Risk of cholecalciferol secondary poisoning to cats and dogs. Poster paper presented at meeting organised by the New Zealand Ecological society on "Ecological consequences of poisons used for mammalian pest control", Christchurch, 9–10 July 1998: Programme and abstracts. P. 33.

CHAPTER FIFTEEN

Non-toxic Techniques for Possum Control

Thomas Montague and Bruce Warburton

Although poisoning continues to be the most extensively used technique for controlling possums in New Zealand, trapping, shooting, chemical repellents, and physical barriers have been and still are important tools in the management of brushtail possums (*Trichosurus vulpecula*) and the problems they cause. These alternative non-toxic techniques have been developed to satisfy legislative, environmental, and practical constraints, as well as personal preferences for a non-toxic solution to the possum problem. This chapter details these techniques as they are permitted to be used in New Zealand.

Trapping

General

The first traps to be used for capturing possums in New Zealand were leg-hold traps imported from England and Australia (Table 15.1). These traps were most probably first used for trapping rabbits but were equally suitable for trapping brushtail possums. The Lanes-Ace trap (Fig. 15.1) became available in about 1946 (L.T. Pracy, retired, ex-Agriculture Pest Destruction Council, pers. comm.) and was the most popular trap used for four decades, during which time many hundreds of thousands were imported into New Zealand. The Lanes-Ace trap, as well as its long-spring predecessors, were commonly called gin traps, a term derived from "engine", a mechanical device (Bateman 1971). Some US Oneida jump traps and double-coil-spring traps were imported prior to the 1970s, yet US traps did not become popular in New Zealand until the 1980s when one, the Victor No.1 double-coil spring, became the most popular trap for capturing possums (Fig. 15.1). The popularity of the Victor No.1 principally arose from its comparatively light weight, compact design, capture efficiency, and competitive price. Between 1995 and 1998, about 133 000 Victor No.1 traps were imported into New Zealand (M. Woodcraft, Woodcraft Ltd, Mt Maunganui, pers. comm.).

In the late 1970s there was an increase in the importation, manufacture, and use of purportedly humane kill traps (Table 15.1), although many were not accepted by trappers and most are now no longer commercially available. Cage traps, mainly New Zealand made, are also used for trapping possums primarily in urban areas and by researchers who need to capture, tag, and release possums.

In response to increasing public pressure during the 1970s to prohibit leg-hold traps, research was undertaken to compare the capture efficiency and humaneness of the wide variety of traps that were commercially available. Unfortunately, compared to leg-hold traps, the kill traps were relatively ineffective at catching possums (Warburton 1982). However, the research showed that improvements in welfare could be achieved by using the smaller Victor No.1 traps without significantly reducing the capture efficiency below that obtained using Lanes-Ace or Victor No.1$^1/_2$ traps (Warburton 1992). Although further reductions in injuries to captured possums could be achieved by using the Victor No.1 and No.1$^1/_2$ Soft Catch traps (traps with padded jaws, centre-mounted chain, and shock spring on the chain), their unacceptably high escape rates (>20%) precluded these traps from being widely adopted by trappers (Warburton 1998).

The first kill traps tested (Banya, Bigelow 1$^1/_2$, Conibear, and Kaki: Fig. 15.1) proved significantly less efficient than leg-hold traps at capturing possums and, far from being humane, frequently caused gross injuries to possums that were not struck in vital positions (Warburton 1982). Subsequent trials with two more recently available kill traps (BMI 160[1] and LDL 101: Fig. 15.1) showed that these traps have the potential to be both capture-efficient and acceptably humane (Warburton & Orchard 1996). The Timms trap (Fig. 15.1), a kill trap specifically designed for capturing possums, kills very effectively

[1] The BMI 160 and Conibear 160 look almost identical, but the BMI 160 has stronger springs.

Table 15.1
Traps (and country of manufacture) that have been used for capturing possums in New Zealand.

Leg-hold traps	Kill traps
Long-spring traps	Banya (NZ)
	BMI 160 (US)
Defiance (Eng)	Bigelow $1^1/_2$ (US)
Dinkum (Aust)	Clamp (NZ)
Eagle Lanes (Aust)	Conibear 110, 126, 160, 220 (US)
Kangaroo (Aust)	Kaki (NZ)
Kiwi (*)	Timms (NZ)
Lanes-Ace (Aust)	
Side Bothams (Eng)	
Double-coil-spring traps	
BMI No. $1^1/_2$ (US)	
Bridger No. 1 (Taiwan**)	
Duke No. 1 (Korea**)	
Montgomery No. $1^1/_2$ (US)	
Oneida (US)	
Victor No.1, No.$1^1/_2$ (US)	

* Unknown country of manufacture.
** Traps marketed through the US.

(Warburton *et al.* 2000), but compared to leg-hold traps it is less capture-efficient (Miller 1993).

There have been few restrictions on using traps in New Zealand, with the only legislation (The Animals Protection Act 1960) stipulating that all traps must be checked at least once every 24 hours. At the time of writing, new animal welfare legislation has just been adopted (Animal Welfare Act 1999) and this legislation will enable the National Animal Welfare Advisory Committee (NAWAC) to prohibit certain traps if evidence shows they are not acceptably humane. The new legislation will also require leg-hold traps to be checked within 12 hours of sunrise on the day following the traps being set, but kill traps will be able to be left for indefinite periods between checks.

The future approval of traps will most likely be based on how they perform against defined standards. These standards will be similar to those developed by the International Organization for Standardization (ISO) committee (Jotham & Phillips 1994; Warburton 1995). Essentially, kill traps will be assessed for their ability to render target animals unconscious within 3 minutes, and restraining traps, such as leg-hold traps and cage traps, on the frequency and severity of various injuries.

Using traps

Traps, like most tools, should be used correctly, and Pracy & Kean (1969), Moresby (1984), and Swan (1996) provide excellent accounts on how to establish a range of trap sets for different conditions. Because of the more recent arrival of kill traps, these publications unfortunately do not provide much information on how best to use the increasing variety of kill traps becoming available. Only as trappers use and experiment with these new traps will recommended "best practice" be developed. Additional information on trapping techniques can be obtained from Fur Facts, a publication of the Opossum Fur Producers Association, or by attending courses run by some polytechnics.

Cage and box traps

Cage and box traps are less suitable for managing possums except in areas where the capture of domestic pets and stock is of concern. These traps are most often used by researchers from Landcare Research, universities, and the Department of Conservation, who require possums to be captured, tagged, and released. Cage traps are also used in areas where protection of non-target species such as kiwi and weka is of concern. The disadvantage of cage traps is their weight and size, which makes them difficult to transport. A typical wire cage trap for possums produced by Grieve Wire (Christchurch) (Fig. 15.1) weighs 3 kg and measures 60 cm long by 26 cm wide and 28 cm high (collapsed dimension is 0.006 m^3) and in 1999 cost about NZ$77. The model currently produced has a spring-assisted closing door with a pendulum trigger hook. The trigger is most often lured with a slice of apple. The catch rate of cage traps is generally lower than that expected when using leg-hold traps. Woodcraft Ltd (Mt Maunganui) sell Havahart collapsable box traps that retail for about NZ$91, and several other retailers, such as stock and station agents, sell various locally manufactured cage traps.

Kill traps

Kaki

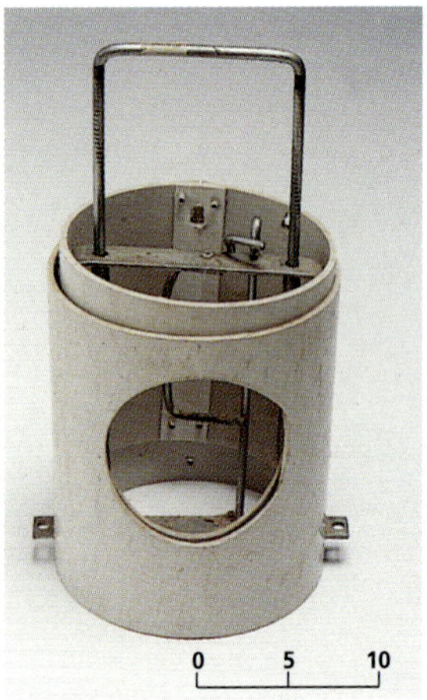

Timms

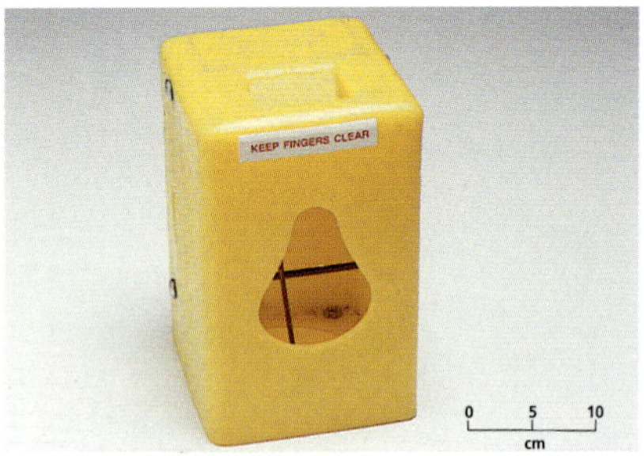

Banya

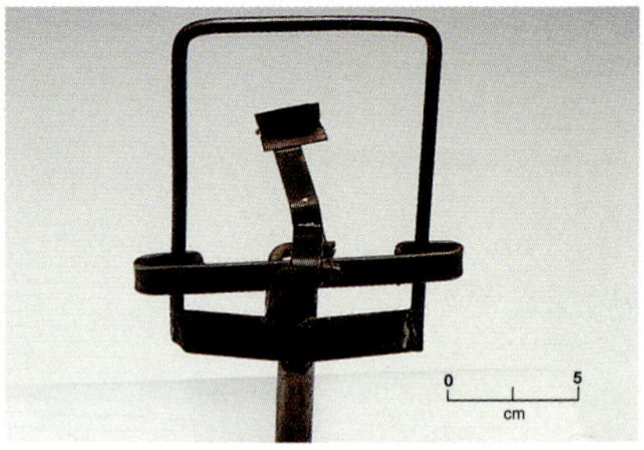

Bigelow 1 1/2

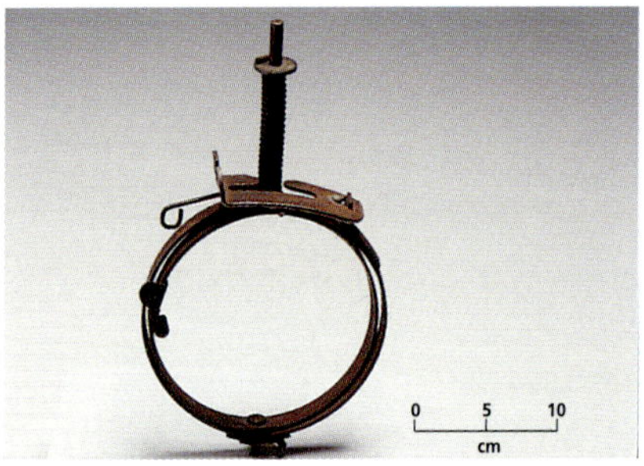

LDL 101

Conibear 160/BMI 160

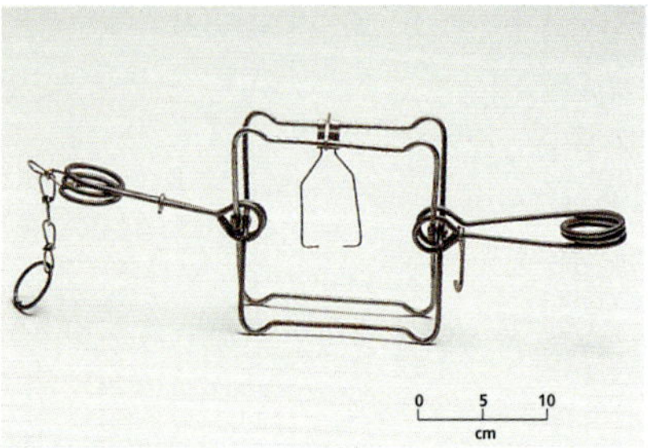

Fenn Mark 5

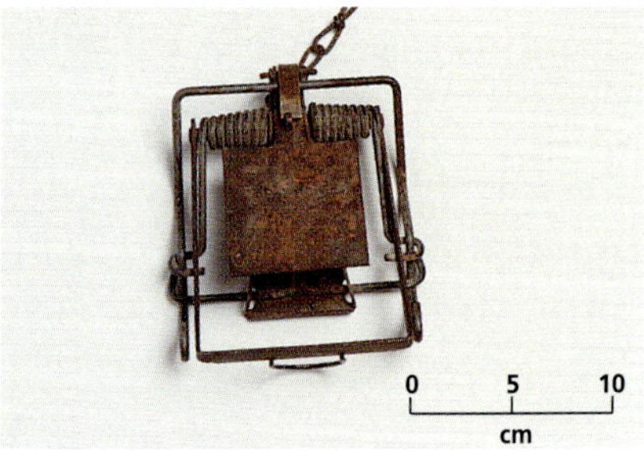

Leg-hold traps

Victor No. 1 1/2 Soft Catch

Victor No. 1 1/2

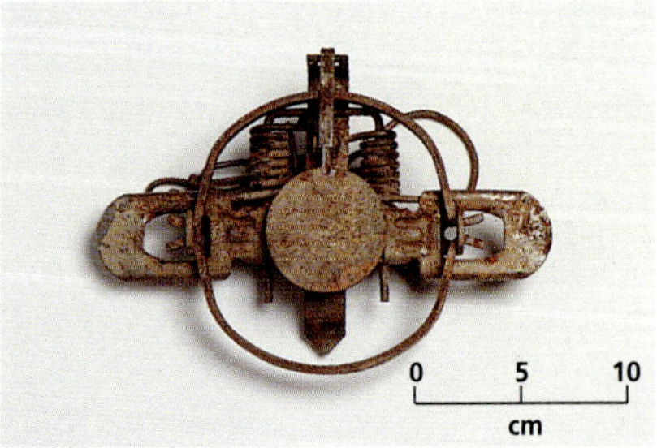

Lanes-Ace

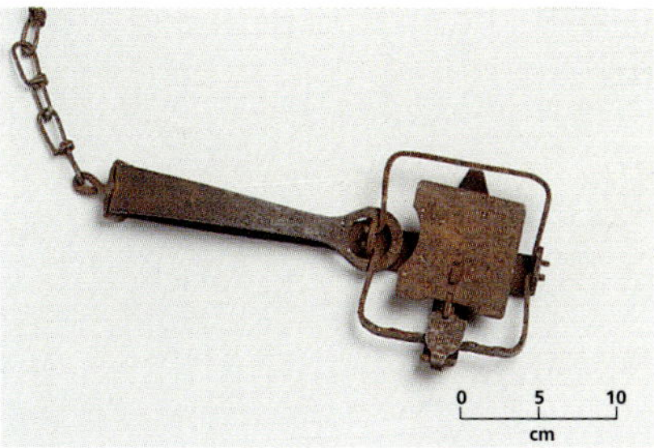

Victor No. 1

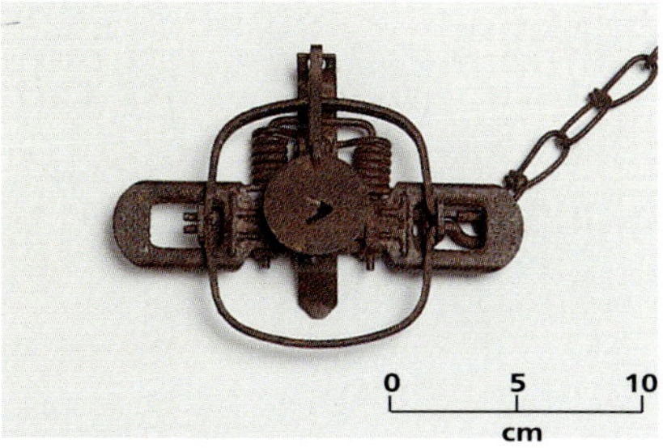

Victor No. 1 Soft Catch

Live capture traps

Grieve wire

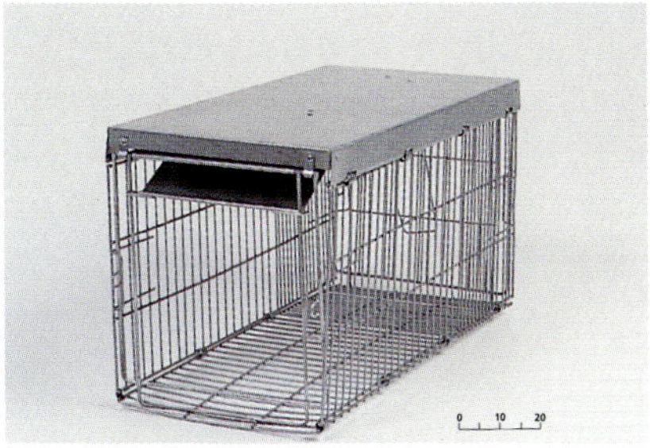

Jamieson

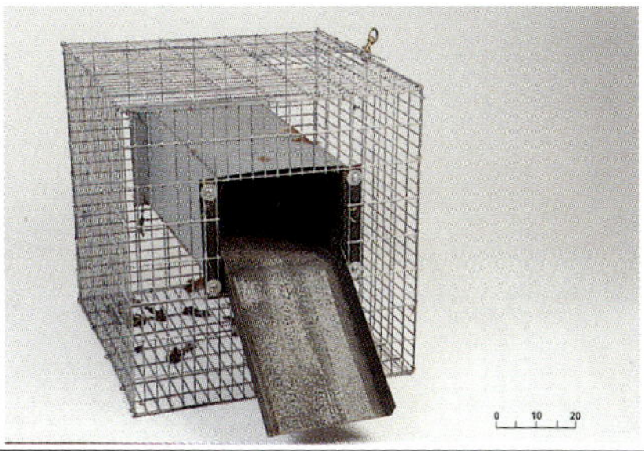

Snares

Snarem

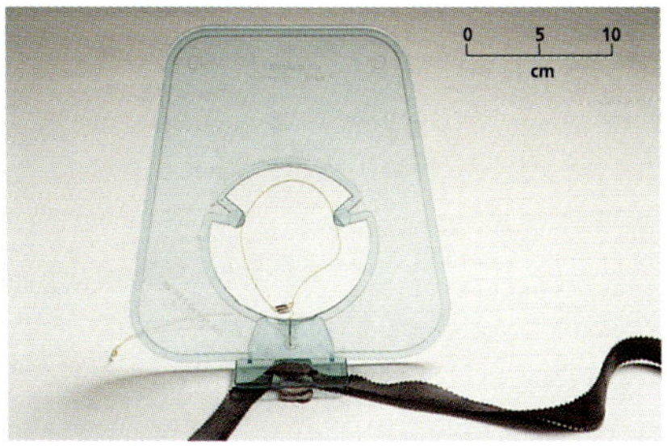

Marex

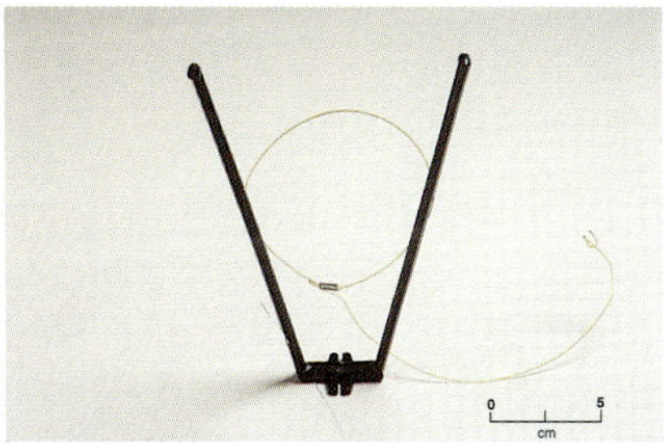

Poison trap

Stinger

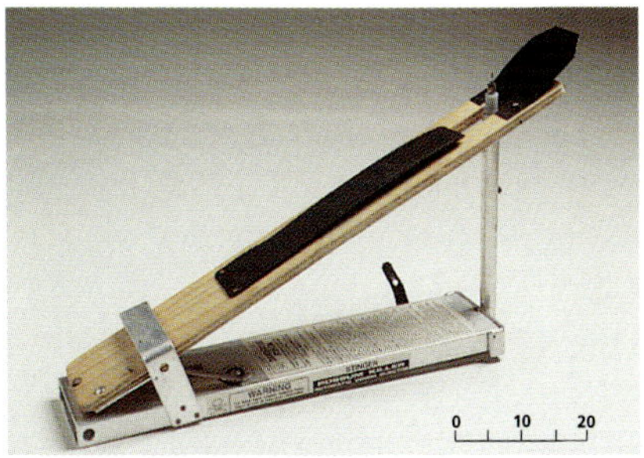

Fig. 15.1
Kill traps, leg-hold traps, live capture traps, snares, and other devices used to manage brushtail possums in New Zealand.

Multiple capture/killing devices

Several devices have been developed in New Zealand for achieving multiple captures or kills of possums. The Stinger, priced at about NZ$220 (Fig. 15.1) and made by Stinger Co., Warkworth, injects a poison into the possum as it leans out from a tree trunk to reach a bait, over weights the trigger and pushes its abdomen onto a needle that delivers 0.25 ml of an organophosphate poison. A similar design is being developed by HortResearch (McDonald *et al.* 1999).

The Electrostrike trap, priced at about NZ$600, kills possums by administering an electric shock that first immobilises the animal and then causes cardiac fibrillation and death. Once dead, the possum is ejected through a trapdoor and the trap automatically reset. Dix *et al.* (1994) showed that it was extremely difficult to kill possums using an electric shock, with some possums surviving 25 kV. To kill possums using electricity, it is essential that the heart is made to fibrillate otherwise some possums will recover from the electric shock. The Electrostrike trap has been further developed and renamed the Zap-trap. This model currently retails through the agricultural stockists Pyne Gould Guinness for NZ$720.

Several designs of multiple-live-capture traps have been developed, but few became commercially available or persisted on the market. The only one known to be currently available is the Jamieson trap (NZ$135), which tips possums into a wire-mesh cage and then automatically resets (Fig. 15.1).

Snares

Although snares are cheap and can be made from non-twist aeroplane cable, fishing nylon, or strong twine, they require skill to be used effectively. Snares have never been used extensively in New Zealand for capturing possums, although some of the early possum trappers who came to New Zealand from Tasmania used snares before adopting traps (L.T. Pracy pers. comm.). There have been two snare-frame devices developed in New Zealand to facilitate easier snaring, but even these did not manage to increase the popularity of snaring among potential users. The Marex snare and the Snarem snare frame (Fig. 15.1), are no longer commercially available. Diagrams and explanations of how to use snares for capturing possums are provided by Moresby (1984).

Lures for traps and snares

Although leg-hold traps can be set successfully on possum tracks, logs, and leaning branches without the need for lure, most trapping is done with a flour-based lure. The flour is usually mixed with icing sugar and a flavour such as cinnamon, aniseed, or cloves added. This lure is then placed behind the trap at the base of the tree, or on either side of the trap if the trap is set on a track (see Moresby 1984). Some trappers use jam, non-toxic cyanide paste, or peanut butter as a lure. Kill traps such as the Timms trap can be set with a piece of firm food such as apple or carrot, or a piece of cloth such as felt with an oil-based lure added.

Cowan (1987) assessed the effect of adding aniseed to apple when baiting cage traps and found that, although this increased the number of recaptures, the probability of initial capture was unaffected.

Shooting

Shooting possums at night using a spotlight to locate them, especially by their red eyes, is labour-intensive, and little evidence is available to suggest that shooting alone is a viable technique for the control of possums, except in small areas of orchards and stands of trees surrounded by pasture. Possums are shot after dark, typically using a shotgun, or a 0.22 or 0.22 magnum calibre rifle. Possum populations can become wary of spotlights when shot at over extended periods. The likelihood of developing a light-shy population can be minimised by not spotlighting too far ahead of the shooter, by twice sweeping the spotlight over an area or tree in quick succession to catch the eyes of possums that look away from the light with the first pass, and by shooting accurately. The importance of this along with other detailed advice on how to shoot possums is given in Swan (1996) and Nelson (1985).

One way to minimise the time needed to locate possums to shoot, and maximise the chance of shooting a high proportion of possums, is to shoot them as they feed at strategically placed feeding stations. This technique has been used by farmers who place bait feeders along accessible roads throughout their farms. After feeding possums for several nights, the farmer knows where the possums will be and thus improves his chances of successfully shooting them. However, it is unknown how effective such a technique is. The great appeal of shooting over other possum control techniques is that not only do some people find it a source of recreation, it is also very target-specific.

Repellents

The use of repellents for the control of possum-browse damage to young trees has been researched in New Zealand for some years now. Egg-and-paint formulations resulting from tests by Crozier & Ledgard (1988) with captive possums have been found to significantly reduce browse damage to pine seedlings. Morgan & Woolhouse (1997) have recently noted that predator odours can deter possums, and a repellent Pine Plus® associated with this research, which uses dog odours, has recently appeared on the New Zealand market. Three other repellents currently available that claim to repel possums and other herbivores are Treepel®, Liquid shotgun®, and Thiroprotect®. Treepel® and Liquid shotgun® are variants of the egg-and-paint formulation of Crozier & Ledgard (1988). Thiroprotect® is an organometallic formulation marketed by Kiwicare containing Thiram.

Application costs per seedling range from 7.5¢ to 27¢ (NZ) when each seedling is sprayed with 15–20 ml of repellent. The egg formulations are currently in widest use by forestry contractors because of their relatively cheap price. Repellents can be expected to protect seedlings for up to 3 months depending on rainfall and the rate at which new, unprotected foliage appears. None of the repellents currently on the market provides total protection. However, they can significantly reduce the amount of damage experienced relative to what would have occurred had the seedlings not been treated. All four repellents are foliar applications and none are considered systemic (i.e., taken up and circulated through the plant's vascular system). It is likely that the need for such repellents will decline in future as plant breeders develop crops less palatable to possums.

Tree guards, tree and pole bands

Tree guards are not used in New Zealand on any large scale for protection of seedlings from possum browsing. They are basically a physical barrier that excludes possums from contact with individual trees or seedlings (Fig. 15.2a and 15.2b). Tree guards typically consist of one or more stakes that support wire, or plastic mesh, or sheeting that surrounds each plant and excludes browsing mammals.

Although the merits of different tree guards have not been compared in New Zealand, research in Britain and Australia (Montague 1993; Pepper *et al.* 1985) suggests that planting seedlings inside rigid tubes 1.2 m tall (Fig. 15.2a) does provide some protection. However, the cost of such tubes is high at around NZ$5 each, excluding the cost of recovery and difficulty securing them in the high winds that frequently occur in New Zealand. There is also a question of aesthetics: 800–1000 tree guards/ha are visually intrusive.

Tree bands/sheaths, on the other hand, have been shown to be highly successful at preventing possum-browse damage to saplings and older trees (Jolly 1980; Thomas & Warburton 1985) (Fig. 15.3a and 15.3b). They come in a range of materials, shapes, and sizes, and most provide an impassable barrier to climbing possums. An inverted cone design shown in Nelson (1982) acts as a physical barrier, but is more complicated to manufacture and install than wrapping a sheet of plastic or metal around the tree so that it forms a cylinder about 600 mm high (Jolly 1980). One researcher, I. Payton (Landcare Research, Lincoln, pers. comm.), used 1.5-m-high metal tree bands to protect older tōtara trees on Banks Peninsula and rātā trees at Camp Creek (near Lake Brunner); and the Department of Conservation in Northland occasionally use bands to protect isolated pōhutukawa growing in coastal areas. Use of tree bands for conservation reasons can be problematic because trees need to be separate enough to exclude access via adjacent trees (Jones 1993). Nelson (1982) recommended using 450-mm-high bands. There seems to be some variation in the recommended height that bands should be installed up trees, but most recommend installing the bands higher than 1.0 m. We could not find any current comparative data on the availability, cost, and efficacy of tree guards to prevent possum damage.

Pole guards (Fig. 15.3c) have been used by the line maintenance company Connetics (Canterbury) for at least 20 years to prevent possums climbing power poles and interrupting electricity supplies (G. Wilcox, Connetics, pers. comm.). On the West Coast pole guards have been used for at least 30 years and are now routinely installed on any pole carrying over 11 kV as dictated by law (B. Wilson, Electronet, pers. comm.). Connectics use lichen green, powder-coated aluminium guards that form a cylinder 600 mm high. The aluminium sheeting is 0.7 mm thick and comes in sheets 600 mm by 1000 mm. Older pole guards were not coloured, but due to public requests, the more environmentally friendly coloured guards have now been adopted. Guards are nailed to poles 6–7 m above the ground and always

Fig. 15.2
Tree guards commonly used to protect against possum browse: (a) rigid tubes, (b) Vexar plastic mesh guards.

above any supporting stay attached to the pole. Pole guards cost approximately NZ$9 each. In rural areas guards are used on concrete poles but in urban areas they are not. A literature review on the use and design of possum pole guards can be found in Hess (1997) who noted that several different designs are in use, although there is little quantitative comparative data on their efficacy.

Fences – electric and others

Since 1980, the design of electric fences for the control of possums has improved. Jolly (1980) dismissed the use of electric fences for possum control and reported that possums seemed oblivious to electric shocks or were stimulated to jump higher, faster. Two years later, Nelson (1982) recommended using a three-strand electric fence to protect orchards and horticultural blocks close to scrub or within 500 m of patches of bush known to contain possums. In the early 1990s Cowan & Rhodes (1993) reported that a 650-mm-high, nine-strand fence energised at 5 kV reduced possum movements by 60–80% and the Department of Conservation published details of six fence designs they were using to exclude possums (Aviss & Roberts 1994). More recently Clapperton & Matthews (1996) noted success with a 600-mm-high wire netting fence fitted with three or four electrified outriggers (Fig. 15.4a is drawn from Clapperton & Matthews 1996). The outriggers were fitted half way up and at the top of the fence. They also found that while captive possums learned to avoid the electric fence, the aversion did not last long. Possums climbed over the fence within a day after the fence was turned off, unlike some domestic stock that avoid electric fences for up to a week (Clapperton & Matthews 1996). Although electric fences can be effective over short periods, maintaining them and preventing short outs requires continual vigilance, especially in the summer months when vegetation grows most vigorously and in areas where vegetation is likely to fall onto the fence.

A lower-maintenance alternative to electric fencing is a 1-m-high mesh fence (20–30 mm mesh) buried to 200 mm and fitted with a floppy top. This is a Tasmanian invention described in Coleman *et al.*

Fig. 15.3
(a) Tree bands, (b) sheaths and (c) pole guards used to exclude possums from older trees, saplings and power poles.

(1997) and Brown & Temby (1997). The floppy top consists of wire netting reinforced with Number 8 wire. It rolls over to the outside (Fig. 15.4b), and possums climbing it fall off when on the collapsing outrigger.

Sonic devices

While sonic devices have shown some potential for attracting possums towards traps (Carey *et al.* 1997), there are few data to suggest that sonic deterrents are consistently able to deter possums or other animals from inhabiting an area or damaging resources over the long term (Bomford & O' Brien 1990). In a trial of a sonic deterrent device called a Po-guard (manufactured by SHU-ROO, Australia) developed to repel possums, Coleman & Tyson (1994) concluded that possum foraging patterns were not influenced in any detectable way by Po-guard units. For these reasons sonic pest control devices should be viewed with skepticism by all seeking a non-lethal method for managing possums.

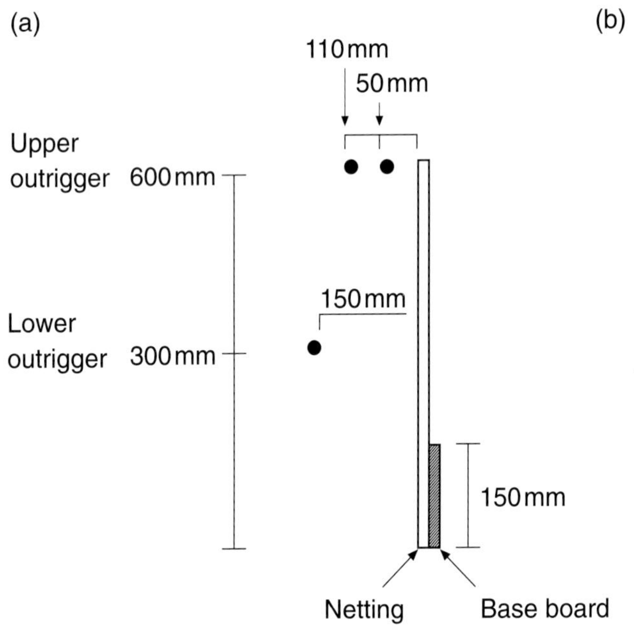

Fig. 15.4
Fences used to exclude possums from areas: (a) an electric fence, (b) a floppy-top fence.

Summary

- Non-toxic and non-lethal techniques have been developed to satisfy legislative, environmental, and practical constraints, as well as personal preferences for a non-toxic solution to the possum problem. Techniques developed include trapping, shooting, physical barriers, and chemical and sonic repellents.
- The Victor No.1 trap has now displaced the Lanes-Ace trap as the most popular leg-hold trap used to catch possums, with about 130 000 imported between 1995 and 1998.
- There have been few restrictions on using traps in New Zealand, with the only legislation (Animal Protection Act 1960) stipulating that all traps must be checked once every 24 hours. The new Animal Welfare Act 1999 will include more specific requirements.
- Kill traps have so far proved less capture-efficient than leg-hold traps.
- Use of cage and box traps is generally confined to regional councils and research organisations and multiple-capture/killing devices, although available, are comparatively expensive and rarely used. Snares are also used infrequently to capture possums.
- Shooting, while target-specific, is a time-consuming form of possum control. Few data have been collected on the efficiency of using this technique for reducing possum numbers.
- Repellents are effective at reducing possum browse to tree seedlings, but no repellent currently on the market provides total protection.
- Tree guards and tree and pole bands, electric and floppy-top fences can all be used to prevent access by possums.
- Sonic devices tested to date have been ineffective at repelling possums.

References

Aviss, M.; Roberts, A. 1994: Pest fences: notes and comments. *Threatened Species Occasional Publication No. 5.* Wellington, Department of Conservation. 39 p.

Bateman, J. A. 1971: Animal traps and trapping. Newton Abbot (UK), David & Charles. 286 p.

Bomford, M.; O'Brien, P. H. 1990: Sonic deterrents in animal damage control: a review of device tests and effectiveness. *Wildlife Society Bulletin 18*: 411–422.

Brown P.; Temby I. 1997: Living with possums. Melbourne, Australia, Natural Resources and Environment. 11 p.

Carey, P. W.; O'Connor, C. E.; McDonald, R. M.; Matthews, L. R. 1997: Comparison of the attractiveness of acoustic and visual stimuli for brushtail possums. *New Zealand Journal of Zoology 24*: 273–276.

Clapperton, B. K.; Matthews, L. R. 1996: Trials of electric fencing for restricting the movements of common brushtail possums, *Trichosurus vulpecula* Kerr. *Wildlife Research* 23: 571 –579.

Coleman, J. D.; Tyson, P. 1994: Re-evaluation of the Po-guard ultrasonic device as a deterrent for possums. Landcare Research Contract Report LC9394/98 (unpublished), 9 p.

Coleman, J.; Montague, T. L.; Eason, C. T.; Statham, H. L. 1997: The Management of problem browsing and grazing mammals in Tasmania. Browsing Animal Research Council, Tasmania. (unpublished), 73 p.

Cowan, P. E. 1987: The influence of lures and relative opportunity for capture on catches of brushtail possums, *Trichosurus vulpecula. New Zealand Journal of Zoology 14:* 149–161.

Cowan, P. E.; Rhodes, D. S. 1993: Electric fences and poison buffers as barriers to movements and dispersal of brushtail possums (*Trichosurus vulpecula*) on farmland. *Wildlife Research* 20: 671–686.

Crozier, E. R.; Ledgard, N. J. 1988: Animal repellents for tree seedlings. *What's New in Forest Research No. 162.* Rotorua, Forest Research Institute. 4 p.

Dix, G. I.; Jolly, S. E.; Bufton, L. S., Gardiner, A. I. 1994: The potential of electric shock for the humane trapping of brushtail possums, *Trichosurus vulpecula. Wildlife Research* 21: 49–52.

Hess, J. 1997: Possum Guard indoor study project. File Number 9007, Network South, Hobart, Tasmania, Australia. 47 p.

Jolly, J. N. 1980: Protecting erosion control plantings against brushtail possums. *What's New in Forest Research No. 90.* Rotorua, Forest Research Institute. 4 p.

Jones, C. 1993: Protection measures for mistletoes in Tongariro-Taupo Conservancy. *Ecological Management 1:* 1–3. Wellington, Department of Conservation.

Jotham, N.; Phillips, R. L. 1994: Developing international trap standards — a progress report. Proceedings: Sixteenth Vertebrate Pest Conference, Santa Clara, California, 28 February–3 March 1994. Davis, University of California. Pp. 308–310.

McDonald, R.; Short, J. A.; Cate, L.; Lyall, K.; Orchard, R. 1999: New technology for poison delivery. *New Zealand Journal of Ecology 23*: 289–292.

Miller, C. J. 1993: An evaluation of two possum trap types for catch-efficiency and humaneness. *Journal of the Royal Society of New Zealand 23*: 5–11.

Montague, T. L. 1993: An assessment of the ability of tree guards to prevent browsing damage using captive swamp wallabies (*Wallabia bicolor*). *Australian Forestry 56*: 145–147.

Moresby, D. J. 1984: Commercial opossum hunting. Te Kuiti, New Zealand, York-Pelorus Group Industries. 190 p.

Morgan D. R.; Woolhouse, A. D. 1997: Predator odours as repellents to brushtail possums and rabbits. Repellents in Wildlife Management a symposium. Denver, Colorado, 8–10 August 1995. Colorado, National Wildlife Research Center. Pp. 241–252.

Nelson, P. C. 1982: Possum control – Horticultural crops and orchards. Poisons, repellants [sic] and protective measures. *Aglink HPP 273.* Wellington, Ministry of Agriculture and Fisheries. 4 p.

Nelson, P. C. 1985: Possums: Control. Trapping, snaring, shooting. *Aglink HPP 274.* 1st revision. Wellington, Ministry of Agriculture and Fisheries. 6 p.

Pepper, H. W.; Rowe, J. J.; Tee, L. A. 1985: Individual tree protection. Aboricultural Leaflet No. 10. U.K., Forest Commission. 22 p.

Pracy, L. T.; Kean, R. I. 1969: The opossum in New Zealand (habits and trapping). New Zealand Forest Service Publicity Item No. 40. 52 p.

Swan, K. 1996: Goodbye Possum - How to deal with New Zealand's public enemy No 1. Auckland, Halcyon Press.

Thomas, M.; Warburton, B. 1985: Sleeves or slaying — protecting erosion plantings. *Soil & Water 21(3)*: 25–27.

Warburton, B. 1982: Evaluation of seven trap models as humane and catch-efficient possum traps. *New Zealand Journal of Zoology* 9: 409–418.

Warburton, B. 1992: Victor foot-hold traps for catching Australian brushtail possums in New Zealand: Capture efficiency and injuries. *Wildlife Society Bulletin* 20: 67–73.

Warburton, B. 1995: Setting standards for trapping wildlife. Proceedings: 10th Australian Vertebrate Pest Control Conference, Hobart, Tasmania, 29 May–2 June 1995. Tasmania, Department of Primary Industry and Fisheries. Pp. 283–287.

Warburton, B. 1998: Evaluation of escape rates by possums captured in Victor No. 1 Soft Catch traps. *New Zealand Journal of Zoology* 25: 99–103.

Warburton, B.; Orchard, I. 1996: Evaluation of five kill traps for effective capture and killing of Australian brushtail possums (*Trichosurus vulpecula*). *New Zealand Journal of Zoology* 23: 307–314.

Warburton, B.; Greg, N. G.; Morriss, G. 2000: The effect of jaw-shape in kill-traps on time and loss of palpabral reflexes in brushtail possum. *Journal of Wildlife Diseases*: in press.

CHAPTER SIXTEEN

Impacts of Possum Control on Non-target Species

Eric Spurr

Methods used for the control of brushtail possums (*Trichosurus vulpecula*) in New Zealand include trapping (see Chapter 15) and poisoning (see Chapter 13). Trapping is done using leg-hold traps, mainly Lanes-Ace (gin) and Victor No.1. Poisoning includes cyanide laid on the ground; apple-paste baits containing sodium monofluoroacetate (1080) laid on the ground; cereal-based baits containing 1080, brodifacoum, pindone, or cholecalciferol placed in bait stations; and cereal-based or diced-carrot baits containing 1080 spread from the air. Most possum control operations involve 1080-poisoning, three-quarters by ground application and one-quarter by aerial distribution (Parliamentary Commissioner for the Environment 1994).

All the above-mentioned methods of possum control cause some mortality of non-target species. This chapter reviews the effect of different possum control methods on non-target species, which may be other pests, domestic animals, or wildlife.

Impacts on non-target pests

Non-target pests caught in leg-hold traps used for possum control include hares (*Lepus europaeus*), rabbits (*Oryctolagus cuniculus*), goats (*Capra hircus*), pigs (*Sus scrofa*), hedgehogs (*Erinaceus europaeus*), rats (*Rattus* spp.), cats (*Felis catus*), ferrets (*Mustela furo*), and stoats (*Mustela erminea*) (Wodzicki 1950; Reid 1985, 1986; Warburton 1982). By far the most commonly caught non-target pests are rats. Non-target pests that have died as a result of 1080-poisoning for possum control include deer (*Cervus, Dama, Odocoileus* spp.), goats, pigs, rats, mice (*Mus musculus*), cats, ferrets, and stoats (Rammell & Fleming 1978). The ungulates and rodents died as a result of eating toxic baits, but the predators died from secondary poisoning after eating poisoned prey. The impacts of possum control on non-target pests have been monitored on only a few occasions. For example, red deer (*Cervus elaphus scoticus*) populations were reduced by up to 90% following four aerial 1080-poisoning operations for possum control (Fraser *et al.* 1995; P.J. Sweetapple, Landcare Research, Lincoln, pers. comm.). Ship rat (*Rattus rattus*) populations were reduced by 87–100% following six aerial 1080-poisoning operations for possum control, but recovered within 4–5 months (Innes *et al.* 1995). A stoat population was reduced by 100% following an aerial 1080-poisoning operation, but recovered within 9 months (Murphy *et al.* 1999). Deaths of non-target pests are considered a benefit by conservationists and some farmers, but a cost by hunters, who see animals such as deer and pigs as a resource.

Impacts on domestic animals

Domestic animals have seldom been caught in leg-hold traps, but those that have died as a result of 1080-poisoning for possum control include dogs (*Canis familiaris*), sheep (*Ovis aries*), cattle (*Bos taurus*), horses (*Equus caballus*), deer, goats, pigs, cats, and fowls (Rammell & Fleming 1978; Bruere *et al.* 1990; Orr & Bentley 1994). The domestic animals most commonly poisoned are dogs, which are extremely sensitive to 1080 and usually die from secondary poisoning after eating carcasses of poisoned possums or other animals. Possum carcasses may remain toxic to dogs for at least 75 days after 1080-poisoning operations (Meenken & Booth 1997). Deaths of domestic animals can be prevented by keeping animals away from poisoned areas and muzzling dogs to stop them from eating possum carcasses after 1080-poisoning operations.

Impacts on wildlife

Trapping

Birds of 23 native and five introduced species have been reported caught in leg-hold traps used for possum control (Table 16.1). Most of the birds caught have been introduced blackbirds and song

Table 16.1
Bird species found dead after possum control operations using traps, ground-laid cyanide, 1080 in bait stations, and aerial 1080-poisoning (+ species found dead; - species not found dead). Species listed in checklist order (Turbott 1990).

Common name	Scientific name	Control technique			
		Traps	Cyanide	Bait station 1080	Aerial 1080
Native species					
Brown kiwi	*Apteryx australis*	+	+	-	-
Little spotted kiwi	*Apteryx owenii*	-	-	-	-
Great spotted kiwi	*Apteryx haastii*	+	-	-	-
Blue penguin	*Eudyptula minor*	+	-	-	-
Paradise shelduck	*Tadorna variegata*	+	-	-	-
Blue duck	*Hymenolaimus malacorhynchos*	-	-	-	-
Grey duck	*Anas superciliosa*	-	-	-	-
Brown teal	*Anas aucklandica*	+	-	-	-
Harrier	*Circus approximans*	+	-	-	+
NZ Falcon	*Falco novaeseelandiae*	-	-	-	-
Weka	*Gallirallus australis*	+	+	+	+
Pūkeko	*Porphyrio porphyrio*	+	-	-	+
Takahē	*Porphyrio mantelli*	-	-	-	-
Black-backed gull	*Larus dominicanus*	+	-	-	+
Kererū	*Hemiphaga novaeseelandiae*	+	-	-	+
Kākāpō	*Strigops habroptilus*	+	-	-	-
Kākā	*Nestor meridionalis*	+	-	+	+
Kea	*Nestor notabilis*	+	-	+	+
Kākāriki	*Cyanoramphus* sp.	+	-	-	-
Shining cuckoo	*Chrysococcyx lucidus*	-	-	-	-
Long-tailed cuckoo	*Eudynamys taitensis*	+	-	-	-
Morepork	*Ninox novaeseelandiae*	+	-	-	+
NZ Kingfisher	*Halcyon sancta*	-	-	-	-
Rifleman	*Acanthisitta chloris*	-	-	-	+
Rock wren	*Xenicus gilviventris*	-	-	-	-
Welcome swallow	*Hirundo tahitica*	-	-	-	-
Pipit	*Anthus novaeseelandiae*	+	-	-	+
Fernbird	*Bowdleria punctata*	-	-	-	-
Whitehead	*Mohoua albicilla*	-	-	-	+
Yellowhead	*Mohoua ochrocephala*	-	-	-	-
Brown creeper	*Mohoua novaeseelandiae*	-	-	-	-
Grey warbler	*Gerygone igata*	-	-	-	+
Fantail	*Rhipidura fuliginosa*	+	-	-	+
Tomtit	*Petroica macrocephala*	+	+	-	+
Robin	*Petroica australis*	+	+	-	+
Silvereye	*Zosterops lateralis*	-	+	-	+
Stitchbird	*Notiomystis cincta*	-	-	-	-
Bellbird	*Anthornis melanura*	+	-	+	+
Tūī	*Prosthemadera novaeseelandiae*	+	+	-	+
Kōkako	*Callaeas cinerea*	+	-	-	+
Saddleback	*Philesturnus carunculatus*	-	-	-	-

Table 16.1 continued

Common name	Scientific name	Control technique Traps	Cyanide	Bait station 1080	Aerial 1080
Introduced species					
Mallard	*Anas platyrhynchos*	-	-	-	-
California quail	*Lophortyx californica*	+	-	-	+
Chukor	*Alectoris chukar*	-	-	-	+
Rock pigeon	*Columba livia*	-	-	-	-
Eastern rosella	*Platycercus eximius*	-	-	-	-
Little owl	*Athene noctua*	+	-	-	-
Skylark	*Alauda arvensis*	-	-	-	+
Hedge sparrow	*Prunella modularis*	-	-	-	+
Blackbird	*Turdus merula*	+	+	-	+
Song thrush	*Turdus philomelos*	+	-	-	+
Yellowhammer	*Emberiza citrinella*	-	-	-	+
Chaffinch	*Fringilla coelebs*	-	-	-	+
Greenfinch	*Carduelis chloris*	-	-	-	+
Goldfinch	*Carduelis carduelis*	-	-	-	+
Redpoll	*Carduelis flammea*	-	-	-	+
House sparrow	*Passer domesticus*	-	-	-	+
Starling	*Sturnus vulgaris*	-	+	-	-
Common myna	*Acridotheres tristis*	-	-	-	-
Magpie	*Gymnorhina tibicen*	+	+	-	+
Rook	*Corvus frugilegus*	-	-	-	-

Sources of information: Wodzicki 1950; Pracy unpublished NZFS reports 1956, 1958; Douglas 1967; Batcheler 1978; Harrison 1978; Spurr 1979, 1991, 1994a; Warburton 1982, 1992; Reid 1983, 1985, 1986; Warren 1984; Calder & Deuss 1985; Morgan & Warburton 1987; Morgan 1989; Cowan 1992; Sherley 1992; Spurr & Powlesland 1997; Powlesland *et al.* 1998, 1999.

thrushes, both of which are ground-feeding species (see Table 16.1 for scientific names of birds). Of the native birds caught, most have been kiwi and weka, also ground-feeding birds, and Australasian harriers and moreporks, which may scavenge on the ground. For example, in surveys of 53 trappers in 1936 and 59 trappers in 1946, more than 1200 birds were reported caught in gin traps, and 90% of those identified were introduced blackbirds and song thrushes (Wodzicki 1950). The most common native species caught were harriers, moreporks, and kiwi. In a 1984 survey of 66 trappers, 141 kiwi were reported caught in gin traps (Reid 1985, 1986). Unfortunately, the numbers of other native species caught were not reported.

Most birds have been caught in leg-hold traps set on the ground without surrounding barriers to reduce accidental capture of non-target species. During trapping to eradicate possums on Kapiti Island from 1980–87, all practicable precautions were taken to minimise bird captures (e.g., by placing traps on sloping boards about 0.7 m above ground; Sherley 1992). Only 181 birds were caught in 1.4 million trap-nights, equivalent to about one bird per 7730 trap-nights (Cowan 1992; Sherley 1992). Most were kererū (39%), moreporks (26%), weka (16%), and kākā (9%). No kiwi were caught despite being present.

Which bird species are caught in traps is affected not only by the species present and the trap site, but also by the method of setting. For example, some of the birds caught have been very small, such as fantails, which weigh only 8 g. Traps catching these birds must have been set very lightly. There is no evidence of any difference in the number of non-target species caught in different types of leg-hold traps currently in use (Warburton 1992).

To avoid captures of kiwi and weka the Department of Conservation (DOC) stipulates that, where these species are present, traps must be at least

0.7 m above ground and, if sloping boards or poles are used, they should be at least 38° to the ground. Recent research indicates that traps should be located even higher (at least 1 m above ground) and sloping boards should be steeper (55°) to avoid catching weka (Warburton *et al.* 1997). However, the placement of traps above ground level increases the risk of trapping birds such as kererū and kākā, as occurred on Kapiti Island (Cowan 1992; Sherley 1992). Little research has been carried out on the long-term impacts of possum trapping on populations of non-target species (Spurr 1991).

Ground-laid cyanide

Birds of six native and three introduced species have been reported killed by ground-laid cyanide (Table 16.1). This is many fewer species than killed by trapping, but cyanide is used less frequently than traps, and birds that die from cyanide poisoning are not always easy to find because they may die some distance from the baits. Native birds most commonly poisoned by cyanide have been the weka and kiwi. In 1947–48, extensive use of cyanide in Poverty Bay killed many thousands of possums, but only a small number of native birds, mainly weka (Pracy 1958). In a 1984 survey, 66 hunters reported 37 kiwi poisoned by cyanide, about a quarter of the number caught in traps (Reid 1985, 1986). As a consequence of this, DOC has stipulated that, where kiwi and weka are present, possum hunters must now lay cyanide baits off the ground. As with trapping, little research has been carried out on the long-term impacts of cyanide-poisoning on non-target species populations (Spurr 1991).

Apart from birds, the only other non-target wildlife species reported to have been killed by cyanide laid for possums is one short-tailed bat (*Mystacina tuberculata*) (Daniel & Williams 1984).

Ground-laid 1080-paste

Birds of only four native species have been reported dead after 1080-paste (a sweetened apple-based jam) was laid on the ground for possum control; namely, weka, silvereye, bellbird, and tūī. This is fewer than killed by ground-laid cyanide, but 1080-paste takes longer to kill birds than does cyanide, so a smaller proportion of the birds that are killed are likely to be found. Brown kiwi, kākā, kea, kākāriki , kererū, pūkeko , robins, and saddlebacks have all eaten non-toxic apple-paste in bait palatability trials (Morgan 1999; D. Morgan, Landcare Research, Lincoln, pers. comm.), and it is likely that most would have died had the baits contained 1080. However, none of 13 radio-tagged brown kiwi died after being exposed to apple-paste baits containing 1080 (Robertson *et al.* 1999).

Short-tailed bats, skinks, snails, and wētā have also eaten non-toxic apple-paste in bait palatability trials (Morgan 1999), but it is not known if they ate enough to receive a lethal dose had the baits contained 1080. Recently, a new gel formulation has been developed that is less attractive than apple-paste to birds, bats, lizards, and invertebrates (Morgan 1999).

Baits containing 1080 in bait stations

Birds of four native species have reportedly been killed by eating 1080-poisoned baits placed in bait stations for possum control (Table 16.1). The most common native bird reported to have died in this way is the weka. A kea died after eating cereal-based baits from Philproof bait stations (D. Butler, Department of Conservation, St. Arnaud, pers. comm.), and kākā have eaten apple-paste from modified Romark bait stations (Sherley 1992). Silvereyes and bellbirds have also died after eating 1080-paste from K.K. bait stations. Baits in bait stations are generally less accessible to non-target species than are baits on the ground. However, possums and rodents sometimes spill baits from bait stations onto the ground, where they become accessible to birds such as robins and tomtits. To reduce the accessibility of baits to weka and kiwi, where these species are present, bait stations should be placed at least 0.7 m above ground (G.A. Morriss & E.B. Spurr unpubl. data).

Other non-target wildlife that might eat 1080-poisoned baits from bait stations include lizards and invertebrates, but there has been no research on this topic.

Baits containing brodifacoum, pindone, or cholecalciferol in bait stations

There are no reports of birds found dead after use of baits containing brodifacoum, pindone, or cholecalciferol in bait stations for possum control, possibly because these alternatives to 1080 are not yet widely used, and because they do not cause death until several days after ingestion of the baits. However, there are reports of weka dying after eating cereal-based baits containing brodifacoum from bait stations used for rodent control, and of robins dying after eating cereal-based baits containing brodifacoum removed from bait stations by rats (Eason & Spurr 1995).

Aerial 1080-poisoning

Birds

Birds of 19 native and 13 introduced species have been found dead after aerial broadcasting of carrot or cereal-based baits containing 1080 (Table 16.1). Most of the birds found dead were found after just four aerial 1080-poisoning operations in 1976–77, and their deaths were attributed to the use of unscreened, poor quality carrot bait containing a lot of very small pieces ("chaff" or "fines") (Harrison 1978). Of 748 birds that were found dead after these four operations, most (68%) were introduced species, mainly blackbirds and chaffinches, but 240 were small native insectivorous birds, including tomtits, robins, whiteheads, grey warblers, riflemen, fantails, and silvereyes (Harrison 1978). Before 1976, few birds had been reported dead after aerial 1080-poisoning operations. For example, only one bird, a blackbird, was found dead after two operations using cereal-based baits in 1956–57 (Pracy 1958), and only one bird, also a blackbird, was found dead after 11 operations using unscreened carrot baits from 1958–1961, despite a thorough search after each trial (Daniel 1966). Few birds have been reported dead after aerial 1080-poisoning operations since 1977 (Spurr 1994a; Spurr & Powlesland 1997). For example, only 83 birds (including 34 native) were found dead after 70 aerial 1080-poisoning operations using screened carrots or cereal-based baits from 1978 to 1993 (Spurr 1994a). Again, the most common species found dead was the blackbird, and the most common native species was the tomtit. Significantly more birds were found dead after operations using carrot baits than after operations using cereal-based baits.

The impact of aerial 1080-poisoning for possum control on bird populations was initially assessed from 5-minute counts of birds heard or seen in poison areas before and after poisoning in relation to counts made at the same time in non-poison areas (Spurr 1981, 1991, 1994a; Pierce & Montgomery 1992; Fanning 1994). Overall, in 24 aerial 1080-poisoning operations using either carrot or cereal-based baits between 1978 and 1993, the numbers of common bird species counted in poison areas did not change 2–8 weeks after poisoning in relation to the numbers counted in non-poison areas (Spurr 1991, 1994a). Uncommon species of birds were present in too few of the operations to adequately assess the impacts of aerial 1080-poisoning on their populations.

Mapping bird territories and locating the occupants of those territories is an alternative technique to 5-minute counting that has been used to monitor impacts of aerial 1080-poisoning for possum control on populations of four uncommon bird species (Table 16.2). The only species in which

Table 16.2
Impact of aerial 1080-poisoning for possum control on bird populations as determined by territory mapping of unmarked birds.

Bird species	Territories occupied before	Territories vacant after	Bait type	References
NZ Falcon	>7	0	cereal	Calder & Deuss 1985; I. Flux pers. comm.
	3	0	carrot	T. Greene pers. comm.
Fernbird	12	0	cereal	Pierce & Montgomery 1992
Robin	10	0	carrot	Powlesland *et al.* 1998, 1999
Kōkako	322	4	cereal	J. Innes pers. comm.
	44	0	carrot	S. Marsh, H. Speed pers. comm.

Personal communications from I. Flux, Department of Conservation, Wellington; T. Greene, Department of Conservation, Auckland; J. Innes, Landcare Research, Hamilton; S. Marsh and H. Speed, Department of Conservation, Hamilton.

territories have become unoccupied after 1080-poisoning is the kōkako. One kōkako missing from its territory was found dead and contained residues of 1080 (Innes & Williams 1990). In all these studies, the territory occupants were unidentifiable as individuals. Where territories remained occupied, it is assumed that they were occupied throughout by the same birds, but it is possible that the original occupants could have been poisoned and their places filled quickly by new birds. Recent studies on kōkako have indicated that birds that die or leave a territory can be replaced within a few days without it being apparent that they are new birds (I. Flux, R. Powlesland, Department of Conservation, Wellington, pers. comm.; J. Innes, Landcare Research, Hamilton, pers. comm.).

The most accurate estimates of the impact of aerial 1080-poisoning operations on individual birds have been obtained when birds have been banded with coloured leg-bands or tagged with radio-transmitters. Colour-bands have been fitted to 125 birds of four species present during aerial 1080-poisoning operations (Table 16.3). Two species, robins and tomtits, suffered high mortality. Twelve of 22 colour-banded robins (55%) disappeared within 2 weeks of aerial application of 1080 in carrot baits in Pureora Forest in September 1996 (Powlesland *et al.* 1998, 1999), and three of 31 colour-banded robins (10%) disappeared within 2 weeks of aerial 1080-poisoning using carrot baits in Pureora Forest in August 1997 (Powlesland *et al.* 1999). None of the 24 colour-banded robins in the non-poison area disappeared in 1996 and only one of 42 (2%) disappeared in 1997. The high incidence of robin deaths in the poison area in 1996 was attributed to the use of poor quality carrot bait containing a high proportion of "chaff". Two out of two colour-banded tomtits in 1996 and 11 of 14 colour-banded tomtits in 1997 also disappeared in Pureora Forest (Powlesland *et al.* 1998). Both robin and tomtit populations seem able to recover from this level of mortality. The surviving robins in Pureora Forest bred more successfully in the poison areas than in the non-poison areas, as a result of reduced predation by rats and possums, so that there were more robins in the poison areas 1 year after poisoning than before poisoning (Powlesland *et al.* 1999). The tomtit populations were not monitored in Pureora Forest. However, in Cone State Forest, where no tomtits were seen or heard 2 weeks after aerial application of 1080 in unscreened carrot baits in September 1977, 5-minute counts of tomtits returned to pre-poison levels within 3 years (Spurr 1981).

Radio-transmitters have been attached to 130 birds of six species present during aerial 1080-poisoning operations (Table 16.4). Only one of 32 radio-tagged weka and one of 13 radio-tagged moreporks died. Both contained residues of 1080. None of the radio-tagged brown kiwi, great-spotted kiwi, blue ducks, or kākā died within 1 month. These data indicate that aerial 1080-poisoning operations for possum control pose little risk to these bird species.

Most birds that have died from 1080-poisoning have probably eaten baits. Even predominantly insectivorous birds eat fruit (Spurr 1979; Moeed & Fitzgerald 1982), and some (e.g., whiteheads and robins) have been seen eating baits (Spurr 1979; Spurr & Powlesland 1997). Also, most dead birds have been

Table 16.3
Impact of aerial 1080-poisoning for possum control on colour-banded birds.

Bird species	Birds banded	Birds died	Bait type	References
Fernbird	7	0	cereal	Pierce & Montgomery 1992; Walker 1997
Robin	2	0	cereal	Walker 1997
	53	15	carrot	Powlesland *et al.* 1998, 1999
Tomtit	16	13	carrot	Powlesland *et al.* 1998
Kōkako	47	0	cereal	I. Flux pers. comm.

Personal communication from I. Flux, Department of Conservation, Wellington

Table 16.4
Impact of aerial 1080-poisoning for possum control on radio-tagged birds.

Bird species	Radio-tagged	Birds died	Bait type	References
Brown kiwi	29	0	cereal	Pierce & Montgomery 1992; Robertson *et al.* 1999; C. Speedy pers. comm.
Great-spotted kiwi	16	0	cereal	Walker 1997; C. Miller pers. comm.
Weka	32	1	cereal	Walker 1997; C. Miller pers. comm.
Morepork	7	0	cereal	Walker 1997
	6	1	carrot	Powlesland *et al.* 1998
Blue duck	19	0	carrot	Greene 1995
Kākā	21	0	carrot	Greene 1995

Personal communications from C. Speedy, Department of Conservation, Taupo and C. Miller, Department of Conservation, Hokitika.

found after 1080-poisoning operations using carrot baits containing a high proportion of small pieces (or "chaff") of a size that is likely to be eaten by small birds (Harrison 1978). Carrot bait has been found in the gizzards of blackbirds and chaffinches, but the gizzards of poisoned insectivorous birds have usually been empty at autopsy (Spurr 1979; Spurr & Powlesland 1997; Powlesland *et al.* 1999). It has been suggested that some insectivorous birds may regurgitate 1080-poisoned bait (Powlesland *et al.* 1999).

Secondary poisoning of insectivorous birds is theoretically possible, depending upon the sensitivity of the birds to 1080 and the amount of 1080 found in sub-lethally poisoned invertebrates. The LD_{50} (dose of 1080 lethal to 50% of the population) is not known for any native insectivorous birds, but for Australian insectivorous birds it ranges from 3.38 to >18 mg/kg (McIlroy 1984). Concentrations of 1080 in invertebrates collected after aerial 1080-poisoning operations have ranged from zero to 130 mg/kg (Eason *et al.* 1993; Lloyd & McQueen 1998). Assuming a possible worst-case scenario, in which the LD_{50} of 1080 for a bird was 3 mg/kg and the concentration of 1080 in invertebrates was 130 mg/kg, a 10-g bird (e.g., a tomtit) would receive a lethal dose of 1080 from ingestion of 0.23 g of contaminated invertebrates (e.g., 115 ants weighing 2 mg each). This is theoretically possible. However, the same bird would receive a lethal dose from eating only 0.05 g of bait containing 0.15% 1080. If the LD_{50} of 1080 for birds was much greater (e.g., 10 mg/kg) and the concentration of 1080 in invertebrates was much lower (e.g., 10 mg/kg), then birds are unlikely to be able to consume a lethal dose of 1080. Although direct 1080-poisoning is most likely, the possibility of secondary 1080-poisoning of insectivorous birds has not been disproved. It has been suggested that insectivorous birds may have died as a result of eating 1080-poisoned ants in the US (Hegdal *et al.* 1986). Predatory and scavenging birds (e.g., harriers and moreporks) that have been found dead after 1080-poisoning operations have probably died from secondary poisoning after eating dead or sub-lethally poisoned possums, rodents, or small birds.

As a result of concerns about the risks to birds, baits used for possum control are dyed green to make them less attractive to birds (Caithness & Williams 1971). Since 1978, carrot baits have been screened to remove the small pieces or "chaff", and cereal-based baits have been used more frequently in DOC operations because they kill fewer birds (Harrison 1978; Spurr 1991, 1994a). Since 1983, cinnamon oil has been added to both carrot and cereal-based baits, partly to mask the smell and taste of 1080 from possums (Morgan 1990) and partly to repel birds (Udy & Pracy 1981). Application rates of carrot baits have declined from more than 30 kg/ha in the 1970s to about 12 kg/ha in the 1990s, and cereal-based baits from 10–20 kg/ha in the 1980s to 5–10 kg/ha in the

1990s (Morgan 1994; Spurr 1994a). However, despite these precautions, bird deaths still occur. Cinnamon oil has been shown to have only a limited repellency to birds (Spurr 1993). In trials to find an alternative bird repellent, a compound called cinnamamide reduced bait consumption by captive weka and kea by 83% and 89%, respectively, without affecting bait consumption by possums (Spurr & Porter 1998). This indicates that addition of bird repellents to baits used for possum control could make baits more target-specific and reduce bird deaths. Unfortunately, cinnamamide also reduced bait consumption by rats. Since rat mortality is often a desired side effect of possum control, further research is needed to find a bird repellent not repellent to possums or rats.

Mammals

New Zealand's only native terrestrial mammals are two species of bats, both of which live in areas where there has been aerial 1080-poisoning for possum control. The long-tailed bat (*Chalinolobus tuberculatus*) is considered entirely insectivorous and therefore at risk only from secondary 1080-poisoning, by eating invertebrates that have eaten baits, and this risk is considered minimal (Spurr & Powlesland 1997). However, long-tailed bat populations have not been monitored during any aerial 1080-poisoning operations. The short-tailed bat (*Mystacina tuberculata*) is primarily insectivorous, but sometimes eats fruit, nectar, and pollen. However, captive short-tailed bats at Wellington Zoo and wild short-tailed bats in Pureora Forest and on Codfish Island did not eat either carrot or cereal-based baits of the types used in aerial 1080-poisoning operations for possum control (Eckroyd 1993; Lloyd 1994). Also, no impacts were detected on a population of short-tailed bats after an aerial 1080-poisoning operation using cereal-based baits in Rangataua Forest in August 1997 (Lloyd & McQueen 1998). Although these results are reassuring, further research is needed to confirm their generality.

Lizards

New Zealand has 17 species of native geckos and 22 species of native skinks (Pickard & Towns 1988), but it is not known how many of these live in areas where aerial 1080-poisoning for possum control has been carried out. The impact of 1080-poisoning on lizard populations has never been monitored. New Zealand lizards are predominantly insectivorous and so could be at risk from secondary 1080-poisoning. However, they also eat soft fruit, honeydew, and nectar, and so may eat carrot and cereal-based baits that contain sugars. Captive McCann's skinks (*Oligosoma maccanni*) ate non-toxic cereal-based baits, especially when the baits were wet, but the level of consumption was probably insufficient for the animals to have received a lethal dose of 1080 had the baits been toxic (Freeman *et al.* 1996). Research in Australia has shown that lizards are more tolerant of 1080 than most other groups of animals, and would need to eat large quantities of poisoned bait to receive a lethal dose (McIlroy *et al.* 1985). Furthermore, even if lizards fed entirely on insects or other animals poisoned with 1080, they could never ingest enough 1080 to receive a lethal dose (McIlroy & Gifford 1992).

Frogs

The three species of native frogs have a very restricted distribution, and are rarely exposed to aerial 1080-poisoning operations. They are not known to eat carrot or cereal-based baits but, being insectivorous, could be at risk from secondary 1080-poisoning. However, there was no evidence of any impact of aerial 1080-poisoning using cereal-based baits for possum control on populations of Hochstetter's frog (*Leiopelma hochstetteri*) in the Hunua Ranges in 1994 (McNaughton & Greene 1994; Greene *et al.* 1995) or on populations of Hochstetter's frog and Archey's frog (*L. archeyi*) on the Coromandel Peninsula in 1995 (Bell 1996; Perfect 1996). These results indicate that aerial 1080-poisoning for possum control is unlikely to have any deleterious impacts on native frog populations. This is supported by Australian research that has shown that frogs, like lizards, are more tolerant of 1080 than most other groups of animals, and would need to eat large quantities of poisoned bait or poisoned insects to receive a lethal dose of 1080 (McIlroy *et al.* 1985; McIlroy & Gifford 1992).

Invertebrates

More than 100 species of ground-dwelling invertebrates have been seen feeding on carrot and cereal-based baits containing 1080 (Notman 1989; Lloyd 1997; Lloyd & McQueen 1998; Spurr & Drew 1999; Sherley *et al.* 1999). Most belong to seven taxa; namely, ants (*Formicidae*), beetles (*Coleoptera*), springtails (*Collembola*), amphipods (*Amphipoda*), harvestmen (*Opiliones*), mites (*Acari*), and wētā (*Orthoptera*). The species found on baits vary with the species present and the bait type used; for example, ants feed more on cereal-based baits and beetles more on carrot baits. The impact of aerial 1080-poisoning on invertebrate populations has been monitored in four studies, by pitfall trapping in poison and non-poison areas before and after application of the poison. One study purported to show a short-term reduction in the total number of ground-dwelling invertebrates, especially beetles and collembolans, caught in pitfall traps after an aerial 1080-poisoning operation using cereal-based baits in Whitecliffs Conservation Area in July 1991 (Meads 1994), but bait density within 1 m of the pitfall traps was 10 times higher than usual. In two other studies, no impacts were detected in the numbers of amphipods, ants, beetles, collembolans, millipedes, mites, slugs, snails, spiders, or wētā caught in pitfall traps up to 6 months after aerial 1080-poisoning operations using cereal-based baits in Puketi Forest Park in March 1992 and Titirangi Scenic Reserve in June 1992 (Spurr 1994b). A fourth study, monitoring the impacts of aerial application of 1080 in carrot bait in Waihaha Forest in August 1994, found no impacts on the numbers of ground-dwelling invertebrates caught in pitfall traps up to 1 year afterwards (Aspin *et al.* 1999).

The impact of aerial 1080-poisoning on invertebrate populations has also been assessed in two studies by counting the number of invertebrates feeding on non-toxic cereal-based baits in poison and non-poison areas before and after application of toxic cereal-based baits. In a study at Rangataua Forest in August 1997, there was no detectable reduction in the total number of invertebrates feeding on non-toxic baits after aerial 1080-poisoning (Lloyd 1997). In a second study in forest near Ohakune in 1997, the total number of invertebrates eating or in contact with non-toxic baits declined significantly within 20 cm of toxic baits containing 1080, but recovered within 3 days of removing the toxic baits (Sherley *et al.* 1999). However, invertebrate numbers did not decline at all on non-toxic baits more than 20 cm from toxic baits. Together, three of the four pitfall-trapping studies and the two direct-counting studies indicate that aerial 1080-poisoning for possum control is unlikely to have any direct deleterious long-term impacts on ground-dwelling invertebrate populations.

Conclusions

Most mortality of non-target wildlife species resulting from possum control has occurred after trapping and aerial 1080-poisoning operations, especially those using unguarded ground-set traps or poor-quality carrot bait. The native non-target species most at risk from both techniques are ground birds such as kiwi, weka, robins, and tomtits. Improvements in trapping (e.g., setting traps off the ground) and poisoning (e.g., better quality baits) have reduced non-target mortality in recent years. Further improvements could be made, for example, by adding a more effective bird repellent to baits (Spurr & Porter 1998). Although the Parliamentary Commissioner for the Environment (1994) concluded that the risks of using 1080 for possum control are "acceptable" in relation to the benefits of its use, non-target mortality is still one of the factors reducing public acceptability of 1080-poisoning as a method of possum control (Fitzgerald *et al.* 1996).

Despite the loss of some individuals, there is no evidence of deleterious long-term impacts of 1080-poisoning for possum control on populations of any non-target species that have been adequately monitored. Mortality from poisoning either is occurring in place of mortality from other causes such as predation and winter starvation, or is compensated for by increased breeding success of survivors (Powlesland *et al.* 1999). However, further research is required on the impacts of 1080-poisoning for possum control on some species of birds, bats, and lizards. Further research is also required on the effects of sub-lethal doses of 1080, which have been reported to affect the physiology and behaviour of non-target species (Spurr 1994b).

In assessing the impacts of possum control on non-target species, the losses of individuals must be balanced against the benefits to the population (Parliamentary Commissioner for the Environment 1994). It has generally been assumed that the benefits of possum control outweigh the risks to non-target species (Spurr 1991), but there is relatively little published evidence in support of this assumption (see Chapters 21 and 22). Often it has not been possible to differentiate whether the responses in non-target species populations are the result of a reduction in the numbers of possums or of other pest species (e.g., rats) killed incidentally, or of both. For example, the greater number of robins in Pureora Forest 1 year after an aerial 1080-poisoning operation than before was attributed to a reduction in the numbers of possums and rats, both of which prey on birds' eggs and chicks (Powlesland *et al.* 1999).

Summary

- All methods of possum control cause some mortality of non-target species (other pests, domestic animals, and wildlife), particularly birds.
- Trapping and aerial 1080-poisoning cause mortality of more birds than do other methods of possum control. The native bird species most at risk from trapping and/or aerial 1080-poisoning are ground birds such as kiwi, weka, robins, and tomtits.
- Improvements in trapping (e.g., raising traps off the ground) and poisoning techniques (e.g., better quality baits) have reduced non-target mortality in recent years.
- Despite the loss of some individuals, there is no evidence of deleterious long-term impacts of trapping or 1080-poisoning for possum control on populations of any non-target bird species that have been adequately monitored.
- Invertebrates feeding on baits used in aerial 1080-poisoning for possum control include ground-dwelling species of ants, beetles, collembolans, amphipods, harvestmen, mites, and wētā. As with birds, there is no evidence of long-term deleterious impacts of possum control on populations of any ground-dwelling invertebrate species that have been adequately monitored.
- Two studies on frogs and one on bats indicate that aerial 1080-poisoning for possum control has no deleterious impacts on their population numbers. There have been no studies of the impacts of possum control on lizards.

Acknowledgements

I thank the Foundation for Research, Science and Technology for funding; and J. D. Coleman and C. T. Eason for commenting on drafts of the manuscript.

References

Aspin, P.; Stringer, I.; Potter, M. 1999: Invertebrate abundance in pitfall traps before and after aerial sowing and bait station presentations of 1080. Oral paper presented at meeting organised by the New Zealand Ecological society on "Ecological consequences of poisons used for mammalian pest control", Christchurch, 9–10 July 1998 : Programme and abstracts. P. 17.

Batcheler, C. L. *Comp.* 1978: Report to Minister of Forests and Minister of Agriculture and Fisheries on compound 1080, its properties, effectiveness, dangers, and use. Wellington, New Zealand Forest Service (unpublished) 68 p.

Bell, B. D. 1996: Aspects of the ecological management of New Zealand frogs: conservation status, location, identification, examination and survey techniques. *Ecological Management 4:* 91–111.

Bruère, A. N.; Cooper, B. S.; Dillon, E. A. 1990: Fluoroacetate. *In:* Veterinary clinical toxicology. *Publication No. 127, Veterinary Continuing Education.* Palmerston North, New Zealand, Massey University. Pp. 96–104.

Caithness, T. A.; Williams, G. R. 1971: Protecting birds from poisoned baits. *New Zealand Journal of Agriculture 122(6):* 38–43.

Calder, B.; Deuss, F. 1985: The effect of 1080 poisoning on bird populations in Motere, Pureora Forest Park, winter 1984. New Zealand Forest Service internal report (unpublished). 39 p.

Cowan, P. E. 1992: The eradication of introduced Australian brushtail possums, *Trichosurus vulpecula,* from Kapiti Island, a New Zealand nature reserve. *Biological Conservation 61:* 217–226.

Daniel, M. J. 1966: Early trials with sodium monofluoroacetate (compound 1080) for the control of introduced deer in New Zealand. *Forest Research Institute, Technical Paper No. 51.* Wellington, New Zealand Forest Service. 27 p.

Daniel, M. J.; Williams, G. R. 1984: A survey of the distribution, seasonal activity and roost sites of New Zealand bats. *New Zealand Journal of Ecology 7:* 9–25.

Douglas, M. H. 1967: Control of thar (*Hemitragus jemlahicus*): evaluation of a poisoning technique. *New Zealand Journal of Science 10:* 511–526.

Eason, C. T.; Spurr, E. B. 1995: Review of the toxicity and impacts of brodifacoum on non-target wildlife in New Zealand. *New Zealand Journal of Zoology 22:* 371–379.

Eason, C. T.; Gooneratne, R.; Wright, G. R.; Pierce, R.; Frampton, C. M. 1993: The fate of sodium monofluoroacetate (1080) in water, mammals, and invertebrates. *Proceedings of the forth-sixth New Zealand Plant Protection Conference:* 297–301.

Eckroyd, C. E. 1993: Testing whether wild short-tailed bats will consume non-toxic carrot baits. *Conservation Advisory Science Notes No. 30.* Wellington, New Zealand, Department of Conservation. 1 p.

Fanning, J. 1994: Effects of an aerial 1080 operation on kokako (*Callaeas cinerea wilsoni*) in the Hunua Ranges 1994. *Auckland Regional Council Parks Technical Publication Series No. 8.* 19 p.

Fitzgerald, G.; Saunders, L.; Wilkinson, R. 1996: Public perceptions and issues in the present and future management of possums. *Ministry of Agriculture Policy Technical Paper 96/4.* 36 p.

Fraser, K. W.; Spurr, E. B.; Eason, C. T. 1995. Non target kills of deer and other animals from aerial 1080 operations. *Rod and Rifle 16(5):* 20–22.

Freeman, A. B.; Hickling, G. J.; Bannock, C. A. 1996: Response of the skink *Oligosoma maccanni* (Reptilia: Lacertilia) to two vertebrate pest-control baits. *Wildlife Research 23:* 511–516.

Greene, B.; McNaughton, A.; Singh, A. 1995: Hochstetter's frog (*Leiopelma hochstetteri*) survey in the Hunua Ranges 1995. *Auckland Regional Council Parks Technical Publication Series No. 10.* 16 p.

Greene, T. 1995: The effect of 1080 poison on kaka and blue duck populations – Pureora 1994. *Maniapoto Hunters' Newsletter No. 5:* 3–4.

Harrison, M. 1978: The use of poisons and their effect on birdlife. *In:* Seminar on the takahe and its habitat: proceedings, Te Anau, 5–6 May 1978. Invercargill, New Zealand, Fiordland National Park Board. Pp. 203–221.

Hegdal, P. L.; Fagerstone, K. A.; Gatz, T. A.; Glahn, J. F.; Matschke, G. H. 1986: Hazards to wildlife associated with 1080 baiting for California ground squirrels. *Wildlife Society Bulletin 14:* 11–21.

Innes, J.; Williams, D. 1990: Do large-scale possum control operations using 1080, gin traps, or cyanide kill North Island kokako? Forest Research Institute Contract Report FWE 90/26 (unpublished). 9 p.

Innes, J.; Warburton, B.; Williams, D.; Speed, H.; Bradfield, P. 1995: Large-scale poisoning of ship rats (*Rattus rattus*) in indigenous forests of the North Island, New Zealand. *New Zealand Journal of Ecology 19:* 5–17.

Lloyd, B. D. 1994: Evaluating the potential hazard of aerial 1080 poison operations to short-tailed bat populations. *Conservation Advisory Science Notes No. 108.* Wellington, New Zealand, Department of Conservation. 12 p.

Lloyd, B. 1997: Evaluating the impact of 1080 on invertebrate food sources for bats. *In:* Wright, D.E. *ed.* Report from the possum and bovine tuberculosis control National Science Strategy Committee, October 1997. Wellington, New Zealand. P. 35.

Lloyd, B.; McQueen, S. 1998: Evaluating the impacts of 1080 pest control operations on short-tailed bats. *In:* Lloyd, B. *comp.* Proceedings of the second New Zealand bat conference, Ohakune, New Zealand, 28–29 March 1998. *Science and Research Internal Report No. 162.* Wellington, New Zealand, Department of Conservation. P. 16.

McIlroy, J. C. 1984: The sensitivity of Australian animals to 1080 poison. VII. Native and introduced birds. *Australian Wildlife Research 11:* 373–385.

McIlroy, J. C.; King, D. R.; Oliver, A. J. 1985: The sensitivity of Australian animals to 1080 poison. VIII. Amphibians and reptiles. *Australian Wildlife Research 12:* 113–118.

McIlroy, J. C.; Gifford, E. J. 1992: Secondary poisoning hazards associated with 1080-treated carrot-baiting campaigns against rabbits, *Oryctolagus cuniculus. Wildlife Research 19:* 629–641.

McNaughton, A.; Greene, B. 1994: The effect of 1080 on the Hochstetter's frog (*Leiopelma hochstetteri*) population in the Hunua Ranges 1994. *Auckland Regional Council Parks Technical Publication Series No. 7.* 25 p.

Meads, M. 1994: Effect of sodium monofluoroacetate (1080) on non-target invertebrates of Whitecliffs Conservation Area, Taranaki, June 1994. Landcare Research Contract Report LC9394/126 (unpublished). 26 p.

Meenken, D.; Booth, L. H. 1997: The risk to dogs of poisoning from sodium monofluoroacetate (1080) residues in possum (*Trichosurus vulpecula*). *New Zealand Journal of Agricultural Research 40:* 573–576.

Moeed, A.; Fitzgerald, B. M. 1982: Foods of insectivorous birds in forest of the Orongorongo Valley, Wellington, New Zealand. *New Zealand Journal of Zoology 9:* 391–403.

Morgan, D. R. 1989: Comparison of the effectiveness of hunting and aerial 1080 poisoning for controlling possums; and an evaluation of a navigation guidance system for improving aerial sowing of possum baits. *Fur Facts 10(38):* 21–37.

Morgan, D. R. 1990: Behavioural response of brushtail possums, *Trichosurus vulpecula,* to baits used in pest control. *Australian Wildlife Research 17:* 601–613.

Morgan, D. R. 1994: Improved cost-effectiveness and safety of sodium monofluoroacetate (1080) possum control operations. *In:* Seawright, A. A.; Eason, C. T. *ed.* Proceedings of the science workshop on 1080. *The Royal Society of New Zealand Miscellaneous Series 28:* 144–150.

Morgan, D. R. 1999: Risks to non-target species from use of a gel bait for possum control. *New Zealand Journal of Ecology 23:* 281–287.

Morgan, D. R.; Warburton, B. 1987: Comparison of the effectiveness of hunting and aerial 1080 poisoning for reducing a possum population. *Fur Facts 8(32):* 25–49.

Murphy, E. C.; Robbins, L.; Young, J. B.; Dowding, J. E. 1999: Secondary poisoning of stoats after an aerial 1080 operation in Pureora Forest, New Zealand. *New Zealand Journal of Ecology 23:* 175–182.

Notman, P. 1989: A review of invertebrate poisoning by compound 1080. *New Zealand Entomologist 12:* 67–71.

Orr, M.; Bentley, G. 1994: Accidental 1080 poisonings in livestock and companion animals. *Surveillance 21 (1):* 27–28.

Parliamentary Commissioner for the Environment 1994: Possum management in New Zealand. Wellington, New Zealand, Office of the Parliamentary Commissioner for the Environment. 196 p.

Perfect, A. J. 1996: Aspects of the ecology of the native frogs *Leiopelma archeyi* and *L. hochstetteri,* and the impact of compound 1080. Unpublished MSc thesis, Victoria University of Wellington, Wellington, New Zealand. 167 p.

Pickard, C. R.; Towns, D. R. 1988: Atlas of the amphibians and reptiles of New Zealand. *Conservation Sciences Publication No. 1.* Wellington, New Zealand, Department of Conservation. 59 p.

Pierce, R. J.; Montgomery, P. J. 1992: The fate of birds and selected invertebrates during a 1080 operation. *Science and Research Internal Report No. 121.* Wellington, New Zealand, Department of Conservation. 17 p.

Powlesland, R.; Knegtmans, J.; Marshall, I. 1998: Evaluating the impacts of 1080 possum control operations on North Island robins, North Island tomtits and moreporks at Pureora – preliminary results. *Science for Conservation No. 74.* Wellington, New Zealand, Department of Conservation. 23 p.

Powlesland, R. G.; Knegtmans, J. J. W.; Marshall, I. S. J. 1999: Costs and benefits of aerial 1080 possum control operations using carrot baits to North Island robins (*Petroica australis longipes*), Pureora Forest Park. *New Zealand Journal of Ecology 23:* 149–159.

Pracy, L. T. 1958: Baiting trials in relation to birds. New Zealand Forest Service internal report (unpublished). 2 p.

Rammell, C. G.; Fleming, P. A. 1978: Compound 1080: Properties and use of sodium monofluoroacetate in New Zealand. Wellington, New Zealand, Ministry of Agriculture and Fisheries. 137 p.

Reid, B. 1983: Kiwis and opossums, traps and baits. *Fur Facts 4 (17):* 17–26.

Reid, B. 1985: The opossum trappers — our maligned conservators. *Fur Facts 6 (21/22):* 18–23.

Reid, B. 1986: Kiwis, opossums and vermin: a survey of opossum hunting; and of target and non-target tallies. *Fur Facts 7 (27):* 37–49.

Robertson, H. A.; Colbourne, R. M.; Graham, P.; Miller, P. J.; Pierce, R. J. 1999: Survival of brown kiwi exposed to 1080 poison used for control of brushtail possums in Northland, New Zealand. *Wildlife Research 26:* 209–214.

Sherley, G. H. 1992: Eradication of brushtail possums (*Trichosurus vulpecula*) on Kapiti Island, New Zealand: techniques and methods. *Science and Research Series No. 46.* Wellington, New Zealand, Department of Conservation. 31 p.

Sherley, G.; Wakelin, M.; McCartney, J. 1999: Forest invertebrates found on baits used in pest mammal control and the impact of sodium monofluoroacetate ("1080") on their numbers at Ohakune, North Island, New Zealand. *New Zealand Journal of Zoology 26:* 279–302.

Spurr, E. B. 1979: A theoretical assessment of the ability of bird species to recover from an imposed reduction in numbers, with particular reference to 1080 poisoning. *New Zealand Journal of Ecology 2:* 46–63.

Spurr, E. B. 1981: The effect of 1080-poisoning operations on non-target bird populations. *What's New in Forest Research No. 94.* Rotorua, Forest Research Institute. 4 p.

Spurr, E. B. 1991: Effects of brushtail possum control operations on non-target bird populations. *Proceedings of the XX International Ornithological Congress:* 2534–2545.

Spurr, E. B. 1993: Feeding by captive rare birds on baits used in poisoning operations for control of brushtail possums. *New Zealand Journal of Ecology 17:* 13–18.

Spurr, E. B. 1994a: Review of the impacts on non-target species of sodium monofluoroacetate (1080) in baits used for brushtail possum control in New Zealand. *In*: Seawright, A. A.; Eason, C. T. *ed.* Proceedings of the science workshop on 1080. *The Royal Society of New Zealand Miscellaneous Series 28:* 124–133.

Spurr, E. B. 1994b: Impacts on non-target invertebrate populations of aerial application of sodium monofluoroacetate (1080) for brushtail possum control. *In*: Seawright, A. A.; Eason, C. T. *ed.* Proceedings of the science workshop on 1080. *The Royal Society of New Zealand Miscellaneous Series 28:* 116–123.

Spurr, E. B.; Drew, K. W. 1999: Invertebrates feeding on baits used for vertebrate pest control in New Zealand. *New Zealand Journal of Ecology 23:* 167–173.

Spurr, E. B.; Porter, R. E. R. 1998: Cinnamamide as a bird repellent for baits used in mammalian pest control. *Australian Vertebrate Pest Conference 11:* 295–299.

Spurr, E. B.; Powlesland, R. G. 1997: Impacts of aerial application of 1080 on non-target native fauna: review and priorities for research. *Science for Conservation 62.* Wellington, New Zealand, Department of Conservation. 31 p.

Turbott, E. G. 1990: Checklist of the birds of New Zealand and the Ross Dependency, Antarctica. 3rd edition. Auckland, Random Century in assoc. with the Ornithological Society of New Zealand. 247 p.

Udy, P. B.; Pracy, L. T. 1981: Baits, birds and field operations. *Counterpest 6:* 13–15.

Walker, K. 1997: Effect of aerial distribution of 1080 for possum control on weka, great spotted kiwi, morepork and fernbird. *Ecological Management 5:* 29–37.

Warburton, B. 1982: Evaluation of seven trap models as humane and catch-efficient possum traps. *New Zealand Journal of Zoology 9:* 409–418.

Warburton, B. 1992: Victor foot-hold traps for catching Australian brushtail possums in New Zealand: Capture efficiency and injuries. *Wildlife Society Bulletin 20:* 67–73.

Warburton, B; Thomson, C.; Moran, L. 1997: Trapping possums – not weka or kiwi. Landcare Research, Lincoln, New Zealand. *He Kōrero Paihama – Possum Research News 7:* 1–3.

Warren, A. 1984: The effects of 1080 poisoning on bird populations in Tihoi, Pureora State Forest Park, winter 1983. New Zealand Forest Service internal report (unpublished). 34 p.

Wodzicki, K. 1950: Introduced mammals of New Zealand: an ecological and economic survey. *DSIR Bulletin No. 98.* Wellington, New Zealand, Department of Scientific and Industrial Research. 255 p.

CHAPTER SEVENTEEN

Public Perceptions and Issues in Possum Control

Gerard Fitzgerald, Roger Wilkinson, and Lindsay Saunders

To the vast majority of New Zealanders, possums present a problem. While the problem may not involve a direct threat to their personal well-being, most New Zealanders are aware of possums in the environment and the threat they pose to the nation's plants and animals. They also recognise that this problem requires active management. Public opinions of current and potential methods for controlling possum numbers, and accordingly the negative environmental and animal health impacts of possums, reflect an attempt to weigh the harms caused by possums against the actual and potential harms caused by attempting to control them, and the risks involved. People's opinions about possums and possum control are therefore strongly influenced by their perceptions of these risks and their views on what constitutes "good" and "harm".

However, perceptions of risk are not formed in isolation, nor are they fixed. Rather, they are socially constructed: people form their own perceptions of the risks within the context of their social and cultural environment. This environment includes prevailing beliefs about science and scientists, decision-making processes and experts, and even the nature and purpose of life. People's perceptions, therefore, need to be considered in that context (Douglas 1985).

Regardless of whether experts think the public's perceptions are accurate, such perceptions are real, and play a crucial role in determining whether a particular technology is able to be developed and adopted. Public and interest-group perceptions of potential biological control technologies for possums, their acceptability, their potential risks, and the management of these risks, therefore, need to be appreciated by possum researchers, control technologists, and decision makers alike. Yet compared with other aspects of possums in New Zealand, the area of public perceptions has received very little research.

To anticipate the issues of developing and introducing biological techniques for controlling possum numbers, MAF Policy and Landcare Research in 1994 commissioned the authors of this chapter to conduct a study of public attitudes to possums and their control in New Zealand. Our research represents the only major study undertaken on this topic in New Zealand. Previous related research includes Sheppard & Urquhart's (1991) survey of public attitudes to pest control methods, Couchman & Fink-Jensen's (1990) survey of perceptions of genetic engineering in New Zealand, Macer's (1994) international survey of bioethics, and a 1995 Roy Morgan Gallup Poll on rabbit control (which covered New Zealand and Australia). Of these, only Sheppard & Urquhart (1991) surveyed members of the public specifically about their attitude to possums, but possums represented only part of their study (which also included rabbits, wasps, fruit flies, flies, termites, grass grubs, and rats). The only other published report of public perceptions of possums is a brief mention of the nature of farmers' support for destroying possums in the Agricultural Pests Destruction Council's 1980 Opossum Survey Report. Because of the lack of directly relevant research, our review of public perceptions of possums and their control in this chapter draws heavily on our own research (Fitzgerald *et al.* 1996).

Our research involved 11 focus-group discussions and a nationwide telephone survey of a randomly selected representative sample of 1127 New Zealanders. Focus groups are moderated group discussions involving strategically selected participants, where interactions between participants are encouraged to stimulate discussion and thereby elicit beliefs and values in depth (Morgan 1988). Focus groups were held with primary producers (two groups), the rural public (two groups), the urban public (two groups), relevant government agencies, primary sector agencies, forestry producers, animal welfare interests, and members of environmental and conservation groups. The focus groups discussed and defined the issues relating to possum control and explored the attitudes to, and perceptions of, biological control among the various stakeholders. Issues raised by the groups assisted in the design of the nationwide telephone survey. This chapter summarises the results and the key issues that emerged from both the focus groups and the telephone survey. For more detailed information see Fitzgerald *et al.* (1996).

Perceptions of possums

All the focus groups thought possums were a major pest, a destroyer of native forest, a carrier of bovine tuberculosis (Tb), an economic threat, and an unwanted nuisance. At the same time, the groups saw possums as an aesthetically pleasing animal (using language such as "soft", "cuddly", "fluffy", "cute", and "nice brown eyes"). Some groups identified the utility value of possums as a source of fur, employment and income, or as a recreational opportunity. Others saw them as out of control.

The main impact of the possum was seen to be on the environment, although focus-group participants were generally unaware of the scale and nature of the problem. The more participants learned about the possum problem through group discussion and the more information we provided, the less favourable their perceptions of the animal became and the more their level of concern about its impact increased.

The respondents to the telephone survey generally perceived possums in a similar way to the focus-group participants. For example, possums were typically seen as a threat to livestock, a threat to trade, and a threat to native birds and forest (Table 17.1). The number of survey respondents concerned about possums was far greater than the number with direct experience of the animal (through having hunted or trapped possums, had a problem with possums, seen possum control operations, or been involved with possum control).

Interestingly, a substantial minority (12%) of the survey respondents thought the possum was native to New Zealand and a further 9% did not know. In addition, 17% did not know whether possums carried Tb. Even though many respondents may have had favourable or even erroneous views of possums (in that they saw possums as cute, basically harmless, or native to New Zealand), more than three-quarters of these respondents still considered possums to be a problem or a threat to New Zealand. Most of this group were not personally concerned about possums.

Goal of control

In the focus groups, the public participants believed the main objective for possum control should be stopping the destruction of New Zealand's natural environment. While primary-sector participants considered that controlling Tb was paramount (for which a Tb vaccine would be an effective solution), they also considered that it was important to control possum numbers to prevent Tb from spreading to livestock, and to reduce environmental problems.

Apart from the methods used to control possum populations, focus-group participants raised a number of other issues related to the broader social goals of possum management. These included employment creation; maintaining an opportunity for a future fur, fibre, and meat industry; and facilitating the public's role in decision making on possum management. In addition, the participants in the public focus groups recognised that increasing efforts to control possum numbers and Tb raised issues of how much it would cost, who would pay, and what would be the most appropriate organisational and managerial arrangements.

Acceptability of possum control technologies

The focus groups discussed the present and possible control technologies and issues associated with their use, including the future of the possum fur and meat industry. The perceptions of possum control technologies were remarkably consistent across the various focus groups. In the telephone survey, respondents were asked to rate the acceptability of manual (shooting and trapping), poisoning (aerial and ground baiting), and biological methods for "killing possums". For each of the specific manual, poison, and biological control technologies suggested, acceptability was not related to the respondent's previous direct experience of possums. Females were less accepting of all the technologies listed than males.

Manual methods

Shooting and trapping were rated acceptable by 82% and 67% of the survey respondents, respectively, and were the most acceptable of all the suggested methods for killing possums (Table 17.2). The level of acceptance of shooting and trapping possums obtained in our survey is higher than that recorded by Sheppard & Urquhart (1991): 69% for shooting and 57% for trapping. However, their survey asked about the "suitability" of the different methods, which also implies "efficacy", whereas we asked about "acceptability". The acceptability of the methods could therefore be considered greater than their perceived suitability.

Table 17.1
Perceptions of possums (n=1127).

Statement	% of respondents			
	agree	neither agree nor disagree	disagree	don't know
Possums are a threat to NZ's native bush	95	1	3	1
Possums are a problem in NZ	93	2	4	1
Possums are a threat to NZ bird life	80	3	7	9
Possums carry bovine tuberculosis	80	2	1	17
Possums are a threat to NZ's overseas trade	70	5	12	13
Possums are a concern to me	64	10	26	<1
Possums are cute, furry animals	35	10	55	<1
Possums are native to NZ	12	3	76	9
Possums are basically harmless	10	8	80	3

The focus groups recognised that the world fur market has been depressed, but felt that the economic and employment potential of possum harvesting should not be abandoned. The New Zealand deer industry was mentioned as an example of the transformation of "pest into product", where a productive industry was built from an animal that, as a pest, had once been subject to control campaigns (see Caughley 1983). The development and use of more effective technologies to control possum populations was thought likely to result in the loss of current and possibly unforeseen future economic opportunities. Participants expressed a desire to maximise the social and economic benefits of possum control by using ground hunters. It was believed that hunting, as opposed to large-scale poisoning or biological control, generated employment in rural or provincial areas, where unemployment is high.

Manual technologies were considered more friendly towards the environment than other technologies. Shooting was perceived to be humane, because it resulted in a "clean" and quick death, and highly specific, in that the risk to other species was considered to be low. Shooting was recognised as a labour-intensive method capable of generating employment. However, it was recognised that the intensity of effort required meant that shooting could not serve as the principal means of possum eradication or control.

Poisons

The poisoning methods included in the survey were ground-laying of 1080 (sodium monofluoroacetate) poisoned bait, aerial application of 1080 poisoned bait, other poisons (e.g., Talon® or cyanide), and a hypothetical poison that kills only possums. Of these, a poison that kills only possums was considered the most acceptable (68% of the respondents rated it acceptable), and was rated third most acceptable of all the technologies. Of the poisoning methods, aerial application of 1080 baits and the use of other poisons were the least acceptable, with aerial 1080 acceptable to 27% and unacceptable to 54%, and other poisons acceptable to 27% and unacceptable to 52%. Aerial application of 1080 baits and the use of other poisons were also considered the least acceptable of all the control methods listed, contrasting with the high level of acceptability of a poison that kills only possums (Table 17.2). The acceptability of such a poison might be explained by its exclusivity of action, implying that the public has concerns about the lack of specificity of 1080 and other currently used poisons. Our survey findings are consistent with those of Sheppard & Urquhart (1991), who reported that 44% of respondents rated 1080 as suitable and 44% rated cyanide as suitable.

The public focus groups confirmed an apparent growing concern over the use of poisons for pest control, especially the use of 1080. Over the course of the study the issue of the aerial application of 1080 baits had been given considerable media attention, and focus-group participants regularly referred to the talkback sessions on radio. Experts' reassurances about the safety of 1080 were simply not trusted, especially by female participants.

The members of the public focus groups appeared to be poorly informed about the nature and use of 1080 and other poisons. Compared with the urban public groups, members of the rural public groups had more experience of the use of 1080 and

Table 17.2
Acceptability of various methods for killing possums (n=1127).

Method	% of respondents			
	unacceptable	neutral	acceptable	don't know or don't understand
Manual methods				
shooting	8	10	82	<1
trapping	17	7	67	<1
Poisoning methods				
possum-specific poison	18	12	68	2
ground laying of 1080	43	18	36	3
aerial drops of 1080	54	17	27	2
other poisons	52	17	27	4
Biological methods				
genetically engineered organism (GMO)	30	14	45	11
imported possum-specific parasite	35	16	35	14
imported possum-specific bacterium	37	18	29	15
imported possum-specific virus	39	17	29	15

Respondents rated the methods on a 5-point scale, from "very unacceptable" (1) to "very acceptable" (5). "Unacceptable" ratings were 1 and 2; "acceptable" ratings were 4 and 5.

the problems with its use; to these groups 1080 was less "unknown", and therefore seen as less of a threat. As well as the fear of toxins and chemicals in the human food chain and water supplies, people also referred to the effect of 1080 on non-target native species and to some extent on farm animals. However, participants admitted they knew little about the extent and nature of such effects. In the environmental-interests focus-group discussion, the benefits of preserving New Zealand's remaining forest heritage from the ravages of the possum through the use of 1080 were considered greater than the problems arising from its use. Concerns over 1080 were also compared with the risks of introducing new control methods. Statements such as "it is better to use what we know than what we don't" were made in reply to concerns about the aerial application of 1080 compared with biological forms of possum control.

Biological controls

The survey respondents were also asked to rate the acceptability of four potential biological means for killing possums: an imported, naturally occurring possum-specific parasite; an imported, naturally occurring possum-specific bacterium; an imported, naturally occurring possum-specific virus; and a genetically modified or engineered organism (GMO) that would kill only possums. Generally the biological methods for killing possums attracted high proportions of neutral and "don't know" responses, indicating higher levels of ambivalence, uncertainty, or lack of knowledge on the part of respondents about such methods than about manual and poisoning methods. Use of a GMO was considered the most acceptable of the biological methods (rated acceptable by 45%). More respondents rated the use of a parasite, bacterium, or virus unacceptable than acceptable (Table 17.2). Overall, biological control, in a generic sense, was more acceptable to the public for killing possums than poisoning (currently New Zealand's main control technology).

Telephone survey respondents were also asked to rate the acceptability of four specific types of action of biocontrol agents for possums. The most acceptable of these was a biocontrol agent that "stops possums breeding" (which would include methods such as

Table 17.3
Acceptability of specific biological control methods (n=1127).

Method	% of respondents			
	unacceptable	neutral	acceptable	don't know or don't understand
stops possums breeding	8	8	83	1
immunises possums against Tb	23	14	60	3
kills young in the pouch	33	18	47	2
makes more susceptible to natural disease	37	21	39	3

Respondents rated the methods on a 5-point scale, from "very unacceptable" (1) to "very acceptable" (5). "Unacceptable" ratings were 1 and 2; "acceptable" ratings were 4 and 5.

immunocontraception and sterilisation). This was considered acceptable by 83% of the respondents (Table 17.3). The least acceptable biocontrol agent was one that "makes possums more susceptible to natural diseases". Both increasing the possum's disease susceptibility and killing its young in the pouch were significantly less acceptable to female than to male respondents. Our results show a lower level of acceptability of biological controls than that found 3 years earlier by Sheppard & Urquhart (1991). In their survey, 51% of respondents rated as "suitable" the "introduction of diseases, e.g. viruses, which will affect only possums" for "reducing the number of possums". Asked how they would feel "if a disease could be identified and introduced to kill possums in New Zealand", 57% of Sheppard & Urquhart's respondents felt it was "a good idea" or "ok", and 43% said they felt "opposed" to such introductions.

The uncertainties of using biological controls that involve the use of a new organism (either introduced or modified), were a major concern in all focus groups. Comments such as "if you cannot manage the pests now, how can you manage a virus that gets into the pests" and "a 99.9% guarantee is required before I would support this sort of thing" expressed a moderate to high degree of such concern. Use of new organisms as vectors for spreading biocontrol agents was considered unacceptable.

The clear message from the focus groups about biological control was that, where possible, killing possums should be avoided and, if this were not possible, then their death should be humane. Immunocontraception was therefore considered the most acceptable form of potential biological control for possums in New Zealand. However, as the focus groups discussed the perceived risks associated with different control technologies, participants became increasingly negative about the use of what were seen as new or exotic technologies such as biological controls.

Perceived risks of possum control technologies

The acceptability of possum control technologies is tied up with what the public and interest groups see as practical and ethical issues around the relative harms and benefits. Among the practical issues are questions of risks and their management, of which the specificity of the control agent is central. In both the focus groups and the telephone survey, the risk of not controlling possums was seen to be much greater than the use or introduction of any of the control agents suggested. The focus groups tended to see three domains of risk posed by possum control technologies: risk to the environment, the economy, and human health. The telephone survey, therefore, asked people to specifically rate the level of risk of each of the various control technologies to each of these domains. Overall, the respondents saw the risks to people from the aerial use of 1080 baits as being greater than the risks posed by biological controls (Table 17.4). A possum-specific GMO was considered to be the least risky of all options suggested, but even then, the majority (58%) of the survey respondents were not prepared to say it presented low or no risk to the environment, the economy, or people's health.

In each of the risk domains, the aerial use of 1080 baits was rated as presenting the greatest risk of the listed control methods, with males rating the risks lower than females. In each risk domain, an imported possum-specific virus was rated the second most risky technology, and the riskiest of all the listed

Table 17.4
Perceived risks of possum control technologies (n=1127).

Risk domain	Control method	% of respondents		
		no or low risk	moderate or high risk	don't know or don't understand
Environmental	no control	3	95	2
	imported possum-specific virus	17	56	27
	imported possum-specific bacteria	19	53	28
	imported possum-specific parasite	21	52	27
	aerial use of 1080 bait	21	64	14
	GMO specific to possums	40	39	21
Economic	no control	9	87	4
	imported possum-specific virus	26	45	29
	imported possum-specific bacteria	26	44	30
	imported possum-specific parasite	26	44	30
	aerial use of 1080 bait	33	52	15
	GMO specific to possums	41	36	23
Health	no control	28	65	7
	imported possum-specific virus	27	44	29
	imported possum-specific bacteria	26	44	30
	imported possum-specific parasite	28	42	30
	aerial use of 1080 bait	31	61	8
	GMO specific to possums	42	34	24

biocontrols, while a GMO was considered to present the lowest risk.

It seems, therefore, that the respondents considered a GMO might present lower risks than other forms of biocontrol agent put to them. The closeness of the risk ratings given to imported parasites, bacteria, and viruses suggests that the respondents did not distinguish well between these kinds of organisms, and that the perceived risks were more to do with the origin of the organism than its type. The perceived higher risks of these organisms to the environment than to the economy and human health also seems to point to a view that these organisms represent yet another unwanted or potentially dangerous introduction to New Zealand.

Even though GMOs are probably less well understood by the public than parasites, bacteria, and viruses, and are therefore more novel, they are perceived as presenting much less risk to New Zealand. This seems to indicate that the public believe that "engineering" an organism may make it more specific to its purpose, and therefore less risky. It should be noted that there was a relatively high level (approximately 30%) of self-professed ignorance about biocontrols and their potential risks, compared with 1080 poison (10% to 15%).

Males generally rated the risks of each of the biocontrols lower than females. The gender differences in risk ratings were most marked in the environment and health domains for a parasite, bacterium, and virus. Of the three risk domains, the environment was seen as at greatest risk from attempts to control possums, with the economy and human health seen as being at equal but lower risk.

In each of the public focus groups, participants raised public health issues associated with various control technologies. There were concerns about the use of biocontrols and the risks of introduced or mutated genetic material appearing in food chains. Human health fears were greatest for the use of chemicals, followed by introduced and modified organisms. A number of analogies, including DDT, thalidomide, and nicotine, were provided as examples of chemicals previously considered "safe" being found subsequently to have unforeseen disastrous consequences.

Safeguards in the development of biological control methods

The focus group discussions showed that the public wanted high levels of guarantee about the specificity of any organisms to be used to infect possums or employed as a vector to spread an infection. The key concerns were that an organism might spread directly or indirectly to the human population, or to native and domestic animals, including livestock (as in possums spreading Tb to cattle). The theme of potential mutation of an introduced or modified organism was raised consistently in the focus groups, with the public groups also raising the spectre of a "time bomb" or delayed effect.

Many focus-group participants felt that biological control methods should be researched and developed to the point where all doubt or risk was eliminated and, even then, they believed that science could not necessarily guarantee the ongoing specificity or immutability of a biocontrol organism. Even with such guarantees, formal probability-based risk assessment alone may not resolve many people's concerns. Part of the problem lies in commonly felt doubts about the credibility and trustworthiness of scientists and their funding agencies — doubts that cannot easily be resolved. The risk of the unknown, fear of catastrophe, and scepticism towards science are common themes in social research on new technologies (Couchman & Fink-Jensen 1990).

Issues in the acceptability of biological controls

Risks tend to be rated according to characteristics that can be grouped as "dread" and "unknown" factors (Slovic *et al.* 1980). These were evident in the New Zealand public's assessment of biological control and will need to be appreciated by possum managers, researchers, and policy makers. Lack of specificity of a biocontrol agent or vector, the possibility of mutation, including a potential delayed effect, were perceived to be the greatest threats.

Concerns about specificity were twofold. The first was a fear that introduced genetic material may directly or indirectly (through animal pathways) transfer to humans, albeit in the future, resulting in unforeseen and/or potentially catastrophic outcomes. The second concern about specificity was the possibility of a direct or indirect impact on non-target species, whether native or introduced domesticated species. Here the threat was seen to be to heritage and economic values, rather than to the human genetic base, and in particular to important iconic species (mainly native birds), and to animals on which New Zealand's agricultural production is based. However, participants were aware of the irony that to do nothing about the possum problem is to also threaten New Zealand's heritage, economic base, and "clean, green" international image. As indicated in the introduction, the question thus becomes one of balancing the risks or potential costs of the possum problem and technologies for its management against the potential benefits of controlling possums and eradicating Tb. However, as we have seen, the public is not well informed about the current impacts of the possum and Tb on the environment and the economy, and is not yet in a position to weigh up the costs and benefits of additional or novel control technologies, such as biological control, for vertebrates. In this regard, formal risk assessment of the control options may assist in the development of an understanding of the potential costs and benefits under various impact scenarios.

Along with the "dread risks" — such as potential risk to future generations, involuntariness of exposure, difficulty of mitigating unwanted effects (i.e., uncontrollability), widespread impact, and possible human fatality or health impacts — risks of the "unknowns" featured in people's perception of biological control. Relevant "unknown" risk factors include the invisibility of the risk (such as a micro-organism at work on possums), the possibility that those exposed to organisms would not know that they were being exposed, possible delayed effects, and that the risks may not be appreciated or known to science. Controllability of biocontrols is a major issue for the public and particular interest groups. The irony of the current problem was well recognised by those interviewed for this study: that, as they fear for biocontrols, both the possum and Tb are themselves introduced organisms and both are often out of control, or verging on it, in New Zealand. People also referred to high-profile examples such as giardia and AIDS to illustrate their concerns about the uncontrollability of micro-organisms.

A major concern with biocontrols is the ethics involved. One aspect is the right of the possum as a being to be dealt with humanely. People would prefer that, where possible, control should avoid causing the death of the possum, hence fertility control is very acceptable. If the action of the biological control, or any other control agent, must result in the death of the possum, then the issue becomes the quality and the visibility of the death. The possum is a sentient animal, which does not often directly bother people, attack them, or directly compete for food and, indeed, at one time was welcomed for the economic benefits it generated (Kirk 1920). Its current pest status merely reflects the conflict between its success at colonisation, and the values, and to a lesser extent the well-being, of the human population of New Zealand. These human values also include the ethic of stewardship of the environment and other species. The central concern is that pest status does not remove the rights of an animal — particularly to a "quality" death — if it is to die at our hands. A "quality" death is one that does not degrade the animal, is painless, and is quick. Also, because the public generally does not like being confronted with dead and dying animals, people would prefer that, regardless of the control method used, the (humane) death of possums took place away from public view.

Another important ethical concern of the public relates to the risks of a failure in the specificity of a biocontrol, either as agent or vector. The issue is twofold: do we, in seeking to do good for ourselves and other species that are already being negatively affected (in this case by possums), have the right to expose others (humans and non-humans, present and future) to potential but unforeseen significant harm? On the other hand, do we not also do harm by failing to attempt to do good (which appears to be unacceptable to the public)? The latter has been referred to as the sin of omission (Macer 1994). The dilemma of having to do harm to do good is one which people and policy makers regularly confront, but do not often easily resolve. However, as Macer (1994) states "we should not take our hands away from a situation and say this is 'out of my hands' ... Both options, to act or not to act, are ethical decisions". Possum management decision making, especially about the development and introduction of a new technology such as a biocontrol for possums, is therefore unavoidably about ethical matters as well as the technical and practical, and involves attempting to understand and weigh the risks, as perceived in a particular time, place, and social context.

At present, scientists and policy makers are unable to provide sufficient information about a specific biocontrol agent and its risks to aid in the resolution of this ethical dilemma and, ironically, may not be able to do so without first risking the development and testing of such a control. In the meantime, possum managers are being confronted with issues about the possum control technologies currently being used, in particular 1080 poison. As the survey shows, the public finds poisons the least acceptable of the control technologies. The issue here is that, despite the research findings and reassurances, the public (especially females) fear that 1080 is poisoning the environment, harming other species, and possibly endangering human health. Nevertheless, if a possum-specific poison could be developed, most of the public would probably find it acceptable.

In addition to questions of risk and ethics, the public are concerned about maximising the direct social and economic benefits of undertaking control. These include the extent to which a particular approach to control will help meet goals such as maintaining or providing needed economic and employment opportunities. Recent well-publicised public outcry over perceived negative impacts of the aerial application of 1080 also involved concerns that locally needed employment opportunities were lost through the choice of the particular control technology. Given that a biocontrol for possums might be 10 years away (T. Fletcher, ex-Landcare Research, pers. comm.), and that the social context for possum-control decision making is dynamic, it is difficult to predict what the future issues and situation for the introduction of some form of biological control might be. Ongoing monitoring of relevant trends, issues, and prevailing views will assist in the framing of relevant and appropriate pest management strategies, and in communicating these to the various publics.

Decision making, public participation, and communication

Because they often lack necessary information, the lay public may exaggerate or underestimate the risks of proposals. In the case of the risks of technologies for controlling possums, such information would include the impacts of possums on native forests if not controlled, or the impact of Tb on New Zealand's trading position and domestic economy. The public may overestimate risks, especially when the technologies have been the subject of previous public concern, such as with 1080. However, although the lay public's perceptions may be technically incomplete, their conceptualisation of risk is generally more complex than that of experts. It may reflect a number of legitimate public concerns that experts overlook or prefer to omit in risk assessments. Formal probability-based assessments alone are unlikely to meet the public's needs when it comes to making a decision on, for example, biological controls. As Chess *et al.* (1989) advise in the resolution of risk problems, "merely hammering away at the scientific information will rarely help." A carefully devised two-way communication process and participatory decision-making approach will therefore be necessary, when considering the introduction of organisms for controlling possums in New Zealand (Chess *et al.* 1989; Slovic 1986). Unfortunately, the new Hazardous Substances and New Organisms (HSNO) Act 1996 does not require such an approach, but provides instead for public notification of applications, written submissions and, if required, public hearings, rather than proactive consultation.

The development of a better understanding of the types of good that may arise from controlling possums (and also other pests) is a prerequisite for developing greater acceptance of the new biotechnologies (which, at present, may be seen by a significant proportion of the public as primarily doing harm). In the focus groups, when information on the scale and extent of the problem was introduced to the discussion, individual positions on the acceptability of control technologies were often recast. Opinions shifted towards unacceptability when the perceived risks of introducing the technologies were raised and discussed by the focus group members.

The key understanding that needs to be developed is that the harm done to the animal is more than offset by the good that will result from taking a particular course of action. Already the public believes that not controlling possums presents the greatest of risks to the environment and the economy. However, the public seems more able to articulate the issues and risks associated with various forms of control than those associated with insufficient control. The ethical trade-off, in particular, is one experienced by few members of the public, and at present they lack information on possum impacts with which to make the trade-off. Given a realistic degree of uncertainty about a biological control technology, the images of impacts and benefits must be portrayed correctly so the public can identify the good attached to reducing possum numbers through the use of such technology. If this is not done, constructive participation in the debate, let alone public acceptance of biological control, may not eventuate.

In the process of making the trade-off, debate can be expected on which technology or combination of technologies generates the greatest good. As in the call for the use of labour-intensive methods rather than aerial sowing of 1080, the greater good, and indeed what is considered "good", may be perceived differently by different groups in the community.

Public acceptance of a proposal such as biological control for pests, before its implementation, is likely to depend on the public seeing a fair process of decision making at work, and seeing that there is to be fairness in the outcome, both in terms of benefits and risks, and that these are assessed fairly. The HSNO legislation requires those proposing to introduce a new organism to state "all the possible adverse effects of the organism on the environment", and that assessment of proposals be based on minimum stated criteria, which cover potential negative impacts on the environment and human health. However, the public's perception of the justice of the decision to introduce such biocontrols will depend on the relative power positions of the participants, the degree to which information is provided to the public, the credibility and transparency of the decision making, and the level of consultation undertaken and stakeholder participation provided for.

Limited public consultation such as that provided for in the HSNO Act is unlikely to be able to identify genuine responses to the introduction of biological controls or associated agents, especially following the unofficial release of rabbit haemorrhagic disease (RHD) into New Zealand to control rabbits. The initial response, as identified by this survey, may only partially describe the manner in which the public will respond as part of a future debate, since the communication process is complex and by itself can change people's responses in unforeseen ways. In addition, as noted earlier, the debate may take place in a different social and economic context from that of this study, and people's perceptions of the issues and the risks involved in particular technology and management options may be different. Moreover, the circumstances of the introduction of RHD may have already altered the context, and people's perceptions of the issues.

At the national level, we suggest that a series of policy panels be established with members who are provided with a significant amount of educational material and the opportunity to interact with experts. This interaction could take place in a forum environment or as a delphi process (involving an iterative series of surveys of experts) that would allow the public to frame the questions to which the experts would respond. The conclusion of this process would be to evaluate the acceptance of possum control technologies, or more importantly, the nature of the concern about them.

A national possum management strategy should include the development of community and regional processes that can assist in the design of appropriate strategies. Such a process would involve the public to a greater degree than that provided in the HSNO Act, although the lack of a legal requirement should not be used to constrain substantive community involvement. At present there are multiple national objectives for possum control, including environmental protection and Tb control. There is a risk that technology for the management of Tb, such as a vaccine, may become available before more effective controls for possums become available. The extent that this would limit the ability to introduce, for example, a biocontrol agent to reduce the impact of possums on native plants and animals needs to be carefully considered.

Images, information, and processes used in public education will require careful consideration. Visual transmission of information on pests and risks should portray the impacts and the benefits of removing them, and not the animal. In the focus groups it was the use of photographs of possums (currently used for public education displays) that induced the contradictory perceptions of the possum as both a pest and a cute furry animal. Other materials designed to warn about Tb were regularly misinterpreted by focus group participants. However, pest managers and decision makers need to be aware that the provision of information may not always have the desired impact (Slovic 1986). There is evidence that science-based data is not readily accepted in New Zealand (see Couchman & Fink-Jensen 1990), and the recent public debate on 1080 shows that despite scientific evidence that 1080 is a safe poison, public acceptance of its use decreased during the debate. The development of a communication programme on the biological control of possums should ideally be designed with the participation of the public and key interest groups. Risk communication research has shown that the communication process itself is risky. As Morgan *et al.* (1992) warn, "one should no more release an untested communication than an untested product."

We consider that the process by which the technologies are introduced may ultimately prove more effective in gaining acceptance of a biocontrol agent than the information provided to the process by experts, the reasoning being that the process should enable an agreement on the degrees of good and harm to be assessed at the local and national levels so risk trade-offs can be undertaken from an informed and agreed position. The ethical and practical issues facing the control of possums cover the animal as a living being, the animal as a pest, the impacts of the animal and the flow-on effects of these impacts, the control technologies, and the effects and uncertainties of these technologies. The future challenge for policy makers is to approach each of these issues in its own terms, rather than attempt to cover all in a single and potentially inappropriate, even if legally satisfactory, strategy.

Summary

- There has been little research on public perceptions of possums in New Zealand. The most extensive study, conducted by the authors, involved group discussions with 11 groups of New Zealanders, and a national survey of 1127 members of the public. This chapter summarises the findings of that study.
- People see possums in several ways at once: they are a major pest, a destroyer of native forest, a nuisance, a carrier of bovine tuberculosis, a potential source of employment and recreation, "cute and furry", and sentient beings deserving of being treated humanely.
- Manual methods of possum control (shooting and trapping) are the most acceptable. They are seen as humane, safe, and employment-creating.
- Poisoning is the least favoured control method, with aerial application of 1080 poison particularly unacceptable. People fear poisons getting into the food chain and water supplies, and are concerned about the effects on native species and farm animals. However, they recognise that 1080 is the most effective form of possum control currently available in New Zealand.
- The public acceptability of biological control for possums lies somewhere between that of manual methods and poisons. However, the public has limited understanding of biological control, and people find it difficult to distinguish between different methods. The main concerns about biological control methods are that they are new and may have unknown long-term effects, and that they might not be specific to possums.
- Among biocontrols, a genetically modified organism has greater public acceptance than a bacterium, parasite, or virus. A biocontrol that stops possums breeding is the most publicly acceptable.
- Men and women differ in their attitudes to possum control. Men are generally more accepting of possum control methods than women. Women are particularly unaccepting of poisons, especially 1080.
- To form opinions about current and potential methods of controlling possum numbers, members of the public weigh the harms caused by possums against the harms caused by controlling them, and assess the risks involved. However, for the public to form a considered opinion about the development and use of biological control of possums, a carefully devised two-way communication process and participatory decision-making approach will be necessary.

References

Agricultural Pests Destruction Council 1980: Opossum survey report. 39 p.

Caughley, G. 1983: The deer wars: the story of deer in New Zealand. Auckland, Heinemann.

Chess, C.; Hance, B.; Sandman, P. 1989: Improving dialogue with communities: a short guide for government risk communication. Environmental Research Program, New Jersey Agricultural Experiment Station. New Brunswick, Rutgers University.

Couchman, P. K; Fink-Jensen, K. 1990: Public attitudes to genetic engineering in New Zealand. Christchurch, New Zealand, *DSIR Crop Research Report 138.*

Douglas, M. 1985: Risk acceptability according to the social sciences. New York, Russell Sage Foundation.

Fitzgerald, G.; Saunders, L.; Wilkinson, R. 1996: Public perceptions and issues in the present and future management of possums. *MAF Policy Technical Paper 96/4.* Wellington, New Zealand, Ministry of Agriculture. 36 p.

Kirk, H. B. 1920: Opossums in New Zealand: Report on Australian opossums in New Zealand. *Appendix to the Journals of the House of Representatives of New Zealand H28*: 1–12.

Macer, D. R. J. 1994: Bioethics for the people by the people. Christchurch, New Zealand, Eubios Ethics Institute.

Morgan, D. L. 1988: Focus groups as qualitative research. *Sage University Paper Series on Qualitative Research 16.* Beverly Hills, Sage Publications.

Morgan, M. G.; Fischhoff, B.; Bostrom, A.; Lave, L.; Atman, C. 1992: Risk communication. *Environmental Science and Technology 26*: 2048–2056.

Roy Morgan Research Centre 1995: Rabbit problem survey. Melbourne, Roy Morgan Research Centre.

Slovic, P. 1986: Informing and educating the public about risk. *Risk Analysis 6*: 403–415.

Slovic, P.; Fischhoff, B.; Lichtenstein, S. 1980: Facts and fears: Understanding perceived risk. *In:* Schwing, R. C.; Albers, W. A. *ed.* Societal risk assessment: How safe is safe enough? New York, Plenum Press.

Sheppard, R.; Urquhart, L. 1991: Attitudes to pests and pest control methods. *Agribusiness and Economics Research Unit, Research Report 210.* Lincoln, New Zealand, Lincoln University.

CHAPTER EIGHTEEN

Economic Analysis of Possum Management

Ross Cullen and Kathryn Bicknell

Efforts to control possum numbers or possum densities are costly and in 1993/94 approximately NZ$58 million was spent on possum research and management activities (Parliamentary Commissioner for the Environment 1994). Current expenditure levels will allow effective possum management on only 17% of conservation lands over the next decade (Parkes *et al.* 1997). The benefits obtained from these possum management activities are not easily quantified, but they provide the economic justification for the expenditures. This chapter illustrates ways in which economists can assist in making better decisions about possum management by (a) providing a framework for considering the damage caused by possums; (b) analysing the case for management of possum numbers; (c) providing methods for measuring and comparing the costs and benefits associated with possum management; (d) examining the incentives and disincentives for possum management; and (e) appraising the merits of various forms of delivery of possum management.

Incentives to control possums

It is useful to compare the economic incentives to control possums with the incentives to capture a marine fish such as snapper (*Chrysophrys auratus*). Both are a wild animal species, and the individual animals of both species are unowned. Harvest or capture of individuals of both species involves effort and cost. Snapper have a high commercial value, and are widely sought after for their direct value to individuals, or for their commercial value. As a consequence of this value there is sustained fishing pressure for snapper, and stocks of snapper in many regions are greatly reduced below their natural levels. Possums, on the other hand, have either a very low value for their pelts, or they have a negative value because they cause damage or are a threat to some economic activity. Only in situations where possums are a serious pest are there likely to be concerted efforts to control their numbers. The outcomes are an interesting contrast. Unless fishing effort is controlled snapper stocks will become depleted and the survival of snapper populations threatened. Because possums have low or negative economic value, possum numbers are likely to increase to carrying capacity levels, except where they cause such severe damage that they are hunted in attempts to restrict the damage.

The incentives to individuals to hunt possums whether for skins, for Tb control, for agricultural productivity, or for conservation reasons, are determined by the pay-offs individuals will receive from those efforts. In many situations control of possum numbers provides pay-offs to a diffuse group of individuals, including commercial hunters, farmers, and conservationists. Because those who control possums directly can not collect payments from all members of this wide-ranging group of beneficiaries, they will under-provide possum control services.

Because possums are unowned, because possum management is typically a defensive activity, and because it is often conducted by Government, it can not be analysed in the way that many activities are by looking at the market demand and supply of possums. Alternative perspectives are required to understand the economics of possum management.

Varying levels of possum damage

Possums now inhabit almost all areas of New Zealand (Parliamentary Commissioner for the Environment 1994). The density of possums per hectare varies across the country from near zero up to 24 per hectare (Cowan 1990) and the damage they cause varies with both vulnerability of species present and possum density. We can first consider how the damage varies with changes in possum density per hectare. Within similar habitats or for the same resource species, the lower the possum density, the less damage they will cause. However, as possum density increases, they will cause increasing amounts of damage, whether through browse pressure on urban plants, pasture, crops, or indigenous flora and fauna, or through an increased probability of spreading Tb to cattle and deer. If this damage is measured in dollars, we can calculate how total damage varies with possum density. The change in total damage as possum density changes is defined as *marginal damage*.

The marginal damage caused to high- and low-value resources (Fig. 18.1) illustrates the additional costs that occur as possum densities increase, assuming proportional increase in damage with density. Below density level D' possums are economically unimportant. Above that density level they become a pest — they cause economic losses — and the marginal damage increases up to a maximum. The precise shape and position of the damage function will vary with the habitat of the possums. Possums will cause little damage in the arid Mackenzie Basin where pasture species have low vulnerability to possum browse, pasture productivity is low, and there are few cattle or farmed deer. By contrast possums found on a South Otago beef or dairy farm will compete for pasture, and pose a threat of Tb transmission. Similarly possums living in Mapara Forest are known predators of eggs and nestling native birds including kōkako, and may compete with some species for food and occupy potential nest sites (Innes *et al.* 1999).

Two possum damage functions are illustrated (Fig. 18.1), one for possums living in a "low value" area, and the second for possums in a "high value" area. The higher the value of the resource or ecosystem that possums damage, the closer the damage function will be to the vertical axis. In some situations the damage function may be discontinuous if marginal damage changes sharply with small changes in possum density.

Costs of managing possums

Possums can cause damage, but do they cause enough damage to provoke individuals, firms, or a nation to attempt to reduce possum densities? Managing possum densities requires effort and the use of various inputs, including planning, labour, transport, poison, equipment, and monitoring. It is therefore costly. Resources that are used on one possum management activity are not available for use elsewhere in the economy, i.e., they have an opportunity cost. Recognition of the opportunity costs of possum management actions is crucial as the resources available are limited compared to the potential scale of the task.

There are several techniques available to manage possums, including trapping, shooting, ground-based poisoning, aerial application of poisons, bait stations, and metal bands on trees. Chapters 13 and 15 provide good explanations of these techniques. The precise objectives of possum control activities will vary by site. Many possum management actions are designed to reduce possum densities to a 5% residual trap catch, where no more than five possums are caught per 100 traps per night. The cost of knockdowns of high-density possum populations, via aerial control operations, ranges between approximately NZ$20 and NZ$30 per hectare (Warburton & Cullen 1993). A representative breakdown of the costs for an aerial control operation are bait 50%, aircraft 14.5%, monitoring 16.8%, planning and overheads 10% (Warburton *et al.* 1992).

Several studies have attempted to determine if there are significant differences in the costs of various possum management techniques (Warburton & Cullen 1993; Cowan & Pugsley 1995; Montague 1997). These studies typically compare the average cost per hectare of achieving one-off (knockdown) control of possum numbers, through aerial application of poisoned baits versus ground-hunting techniques such as trapping or poisoning. Implicit in these cost comparisons is an assumption of similarity of outcomes in terms of reductions in possum numbers. In reality average costs of possum management per hectare are influenced by several factors, including total area to be managed, terrain, location of the area to be managed, initial possum density, and density reduction target (Cullen 1992). In many regions control programmes are aimed at maintaining possum densities at low levels. The cost of maintenance programmes are usually lower than for knockdown programmes, as labour inputs and numbers of animals killed per hectare are much lower. Some representative data on average cost per hectare for three types of possum management programmes are provided in Table 18.1.

Recent data suggest that ground-based possum management can be accomplished for similar average costs per hectare as aerial application of poisoned baits (Montague 1997). There are, however, few examples of ground hunters achieving possum knockdown from high densities (>7/ ha) on large areas. The limited data available suggest that for large areas (>5000 ha) ground hunters take much longer to complete possum knockdown from high densities than do aerial operators (Montague 1997). In ground-based operations, poisons or traps must be carried and spread over an area manually. For large sites, travel

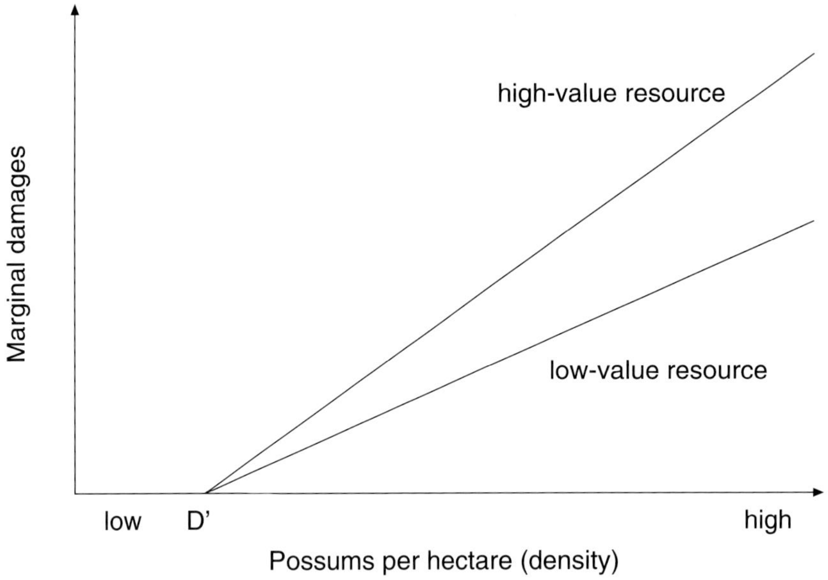

Fig. 18.1
Hypothetical change in total damage (marginal damage) to resources resulting from increased possum density. Density levels below D' are economically unimportant. Damage is assumed to increase proportionally with density.

Table 18.1
Costs for a range of possum management operations.

Type of operation	area (ha)	% reduction	approximate cost 1998 NZ$/ha
Ground control, maintenance operations with self-feeding bait stations, 16 sites, 1991–93	3659	78	5.30
Knockdown, aerial 1080 cereal baits, 10 sites, 1990–93, overhead costs excluded	6473	76	22.60
Knockdown, ground hunting, 2 sites, 1997, direct costs.	2059	83	22.40
Inclusive of monitoring costs			31.30
Knockdown, aerial 1080, 2 sites, 1997, direct costs.	5073	78	23.00
Inclusive of monitoring costs			31.00
Eradication, 1 site, 1980–87	1965	100	c. 462.00
Eradication, 1 site 1990–92	2300	100	c. 61.00

From: Montague 1997; Parliamentary Commissioner for the Environment 1994; Parkes 1991.

times on the ground, difficulty of carrying poisons and traps, and the logistics of organising labour to cover a large area in a finite time, limit the area that can be managed in that way. A recent study of two ground-hunting operations shows labour costs comprised 61–70%, and materials 27–31%, of total costs (Montague 1997). Monitoring these operations adds another NZ$6–10 per ha to costs.

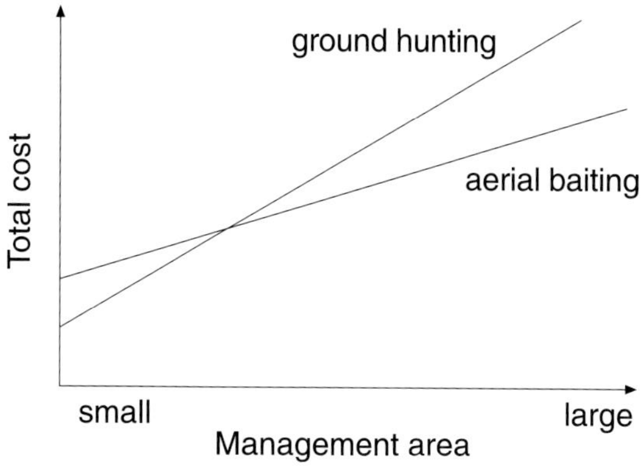

Fig. 18.2
Comparative costs of two possum management methods with increase in area managed.

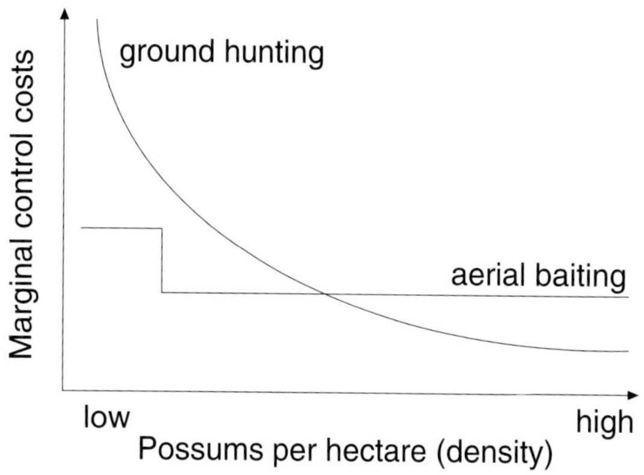

Fig. 18.3
Increasing costs per hectare (marginal control costs) of alternative possum management methods as possum densities are lowered.

Aerial application of poisoned baits is much less limited by location, total area to be managed, or steepness of terrain. Figure 18.2 provides a representation of how total costs for ground hunting and aerial poisoning vary with increase in area.

For small areas and for maintenance operations at low possum densities, ground hunting or use of bait stations will provide possum management at lower cost per hectare than will aerial application of poisoned baits. Aerial operations incur significant fixed costs (e.g., planning, public notification of operations, having medical personnel available) to be spread over the area managed, so the larger the area, the lower the average fixed cost.

Achieving very low possum densities is possible, but the lower the density sought, the higher will be the costs per hectare. Increasing costs per hectare as possum densities are lowered can be defined as *marginal control costs* (MCC) of possum management. The rate at which MCC increase as possum densities are lowered depends on a number of factors: the type of terrain, the location of the managed area, and the control technique. Particularly for ground-hunting techniques, MCC increase sharply as possum density is reduced to low levels (Fig. 18.3). Possum management via aerial application of poisoned baits is likely to result in MCC increasing in steps, as successive operations are required to achieve lower possum densities (Fig. 18.3). Empirical evidence of sharply increasing MCC of pest management is provided by Parkes (1991), Warburton & Frampton (1991), and Cowan (1992). On Kapiti Island where possums have been eradicated, costs of control of the last 1% (80 possums) were approximately NZ$235,000 in 1998 dollars (Cowan 1992).

The important point about both control techniques is that MCC increases as lower possum densities are achieved. This is obviously a barrier to achieving and maintaining low possum densities, particularly when the resources available are limited. An ambitious goal, such as eradication of possums at mainland sites, is likely to be extremely costly. In cases where possums have been eradicated, the costs per hectare have been significantly greater than in conventional possum management operations. Costs, in 1998 dollars, of eradication of possums on Kapiti Island during the 1980s, without inclusion of overhead costs, were NZ$462/ha (Cowan 1992), and on Rangitoto Island during the early 1990s, also without inclusion of overhead costs, were NZ$60.95/ha (Parliamentary Commissioner for the Environment 1994). Recent advances in control technology have clearly reduced the cost of possum eradication, and control operations can now simultaneously control possums, rats, and stoats.

Eradication of possums can be justified if the present value of eradication costs is lower than that of a continuing series of possum management operations (Warburton *et al.* 1992). Present values are calculated by selecting a discount rate and weighting costs from differing years to make them comparable to a base year, such as the year a possum management programme commenced (Hanley & Spash 1993). Eradication also brings greater certainty of improved forest condition than does a continuing series of possum operations.

How far should possum densities be reduced?

We can combine the concepts marginal damage (MD) and marginal control costs (MCC) to illuminate some key points regarding economically optimum levels of possum management. Figure 18.4 shows that, for a single site, and for one year, where there is no uncertainty about outcomes, the economically optimum level of possum density is at D* where MD = MCC. To attempt to reduce possum density below that level will mean MCC > MD and the funds would be better used at an alternative site. If the funds available for possum management are limited, it may not be economically possible to reduce possum density as low as D* and managers may have to accept higher possum densities.

If we recall the points made regarding different MD functions, the higher the damage level for any given possum density, and for a given MCC function, the lower the optimum level of possum density. When the MCC vary, for a given MD function, the higher the MCC, the higher the optimum possum density. If landowners or managers have information about MD and MCC for each level of possum density, they can compare their magnitudes and determine the economically justifiable level of possum density.

The MD and MCC functions we have graphed (Fig. 18.1–18.4) show the marginal damage and marginal control costs that might be incurred by a single landowner. In reality possums are unowned, mobile, and have the potential to cause harm on more than one property. The functions shown are therefore "private" MD and MCC functions. If possums impose damage on more than one property, we can represent this (Fig. 18.4) by an "external damages" function. To capture both the "private" damage plus the "external" damage caused to neighbouring properties, we must combine the private and external marginal damages to obtain a "social" MD function. The impact of including this external damage is to increase the marginal damage possums cause for a given possum density. As social marginal damages are higher than private marginal damages, the optimum possum density will be lower for a given MCC (Fig. 18.4 shows D** as the economically optimum level). Because individuals are not able to capture all of the benefits from possum management operations, they will under-invest in possum management, and reduce possum densities only as far as D* (Fig. 18.4).

Are optimum possum densities pursued and achieved?

The likelihood of landowners aiming for and achieving optimum possum densities depends on their objectives, and on the financial constraints they face. In areas where Tb is a problem, for example, landowners may focus on the elimination of Tb from their herds. Control of possum numbers is one means to that objective. Beef, dairy, and deer farmers, who focus on the profitability of their businesses, may aim for possum density level D* (Fig. 18.4). A group of landowners who recognise there are both private and social marginal damages, can act collectively and aim to reduce possum densities to

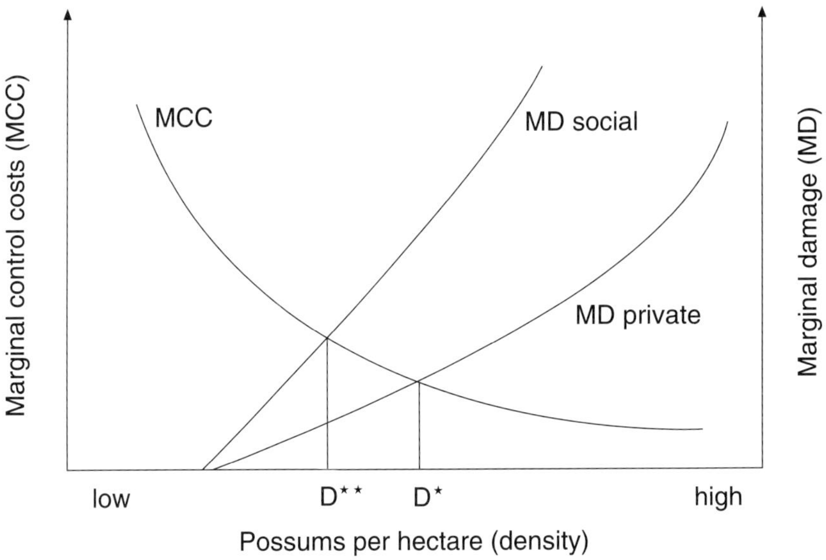

Fig. 18.4
Marginal damages (MD) and marginal control costs (MCC) of possum management.

D**. Their ability to determine those optimum possum densities is dependent on how well informed they are about MD and MCC.

Our analysis assumes that possum management involves annual comparisons of MD and MCC. In reality possum management is an ongoing task, and attempting to determine the best programme of possum management over a long time period, such as 30 or 50 years, requires discounting of damage and costs, and calculation of present values. Once the present values of damage and costs from differing years have been established, optimisation techniques can be used to determine which sequence of actions is likely to provide the lowest present value of possum management costs for a region (Bicknell *et al.* 1997).

Many possum control operations occur on conservation land, and involve control programmes once every 4–6 years or longer. The objective pursued in those programmes is to reduce possum densities by some target amount such as 80%, or to achieve a designated residual-trap-catch figure such as 5%. Cost-effectiveness analysis is a useful technique in those cases. If there is a range of possum management techniques available each with differing costs per hectare, cost-effectiveness analysis can be used to determine how the greatest output can be achieved for the funding available. Warburton *et al.* (1992) suggest that to maintain possum densities below a chosen level (such as 6/ha) on a site with low rates of possum migration, 6-yearly aerial control and low bait application per hectare may provide the lowest present value of possum control costs over a 30-year period.

Accurate data on costs and effectiveness of the various possum management techniques are essential for cost-effective possum management. Achieving optimum possum densities also requires such information. Methods to estimate the damage caused by possums and quantify the output from possum management programmes are vital.

Measuring outputs and benefits from possum management

Possums cause considerable damage and can contribute to species extinction. Possum management is not an end in itself; it occurs because farmers, conservation area managers, the Animal Health Board, and regional councils want to reduce the harm that possums cause, and diminish the threat that bovine Tb poses to human health and to exports.

Possum management is a long-term activity, so to usefully assess the benefits from control operations we need measures that report both short-term and long-term effects. Also, the benefits obtained from possum management activities diminish over time as possum densities in an area rebound following a one-off possum management activity. These facts mean that sophisticated measures of outcomes from possum management are needed.

In cases where damage by possums is confined to farms, the present values of short- term and long-term damage can be estimated. Damage includes less pasture available for livestock grazing, higher incidence of bovine Tb, and so lower beef, dairy, or venison production, elevated risk of lost access to export markets, higher herd-testing costs, and a greater need for replacement stock. Successful one-off possum management operations reduce this damage, but the magnitude of the damage prevented will fall in each year following the possum management action, as possum numbers rebound. Bioeconomic research can estimate the dollar values of marginal damage prevented by possum management activities on farms (Bicknell *et al.* 1997; Lambie & Bicknell 1998). A recent study reveals that, for a representative beef farm, the marginal benefit per hectare from control of possums is worth NZ$1.64 (Lambie & Bicknell 1998). This marginal benefit can be compared to the marginal cost of achieving lower possum density on the farm, to judge if further possum density reduction is warranted.

Episodic possum management provides short-term protection of conservation areas. Here the damages prevented are not easily measured and valued, as there are few markets for the output from conservation lands. Determining if cost-effective possum management is occurring requires a new tool to measure output. One way to measure the output from possum management actions is to use a unit called *conservation protection years* (COPY) Cullen *et al.* (1998). A *conservation asset security* (CAS) score can be estimated for the conservation status of a region, with a range from 6 = completely free from damage by possums, to 0 = extreme damage by possums. The CAS scores can be estimated annually for a conservation area and, compared to the CAS score assessed prior to the commencement of possum management, scores can be expected to decrease each year if the possum density rebounds following a one-off possum management action. Table 18.2 provides hypothetical CAS scores for a site where possum damage and therefore conservation status is in assumed equilibrium. In cases where the conservation status would have fallen without possum management, the CAS scores would also fall.

Table 18.2
Hypothetical conservation asset security (CAS) scores at a site (with a stable conservation status).

Year	0	1	2	3	4	5	6	7
Without possum management	2	2	2	2	2	2	2	3
With sustained possum management	2	2	3	3	3	4	4	4
Gain due to possum operation	0	0	1	1	1	2	2	1

The output from possum management is the average of the sum of the CAS scores for each year, less the initial CAS score in year zero prior to the commencement of the action. If this output is denoted conservation protection years or COPY, it can be described for the second year following a possum management action, as

$$COPY = \Sigma_i\ [(CAS_i + CAS_{i-1})/2 - CAS_0] \qquad (1)$$

Where:
CAS_i = Conservation Asset Security score in year i
CAS_{i-1} = Conservation Asset Security score in year i – 1

All species and ecosystems are not of equal importance or merit. The current system used in New Zealand to rank areas for possum management estimates primary scores for each region by calculating *rarity* × *vulnerability* scores (Department of Conservation 1994). These rarity scores, or some other merit score, could be used together with COPY estimates to calculate a cardinal measure of value of output from conservation protection actions:

$$COPYR_j = R_j \times \Sigma_i\ [(CAS_i + CAS_{i-1})/2 - CAS_0] \qquad (2)$$

Where: R_j = merit score for the conservation area.

Conservation protection years, as a measure of outcome from possum management actions, seem well suited for cost-effectiveness analysis of possum management, and other evaluations of conservation programmes (Fairburn 1998). A possum management action on a conservation area can be evaluated by dividing the number of COPY it delivers, by the cost per hectare of the protection action. This calculated "cost per hectare COPY" can be compared to another, calculated for an alternative conservation protection action on that area. After weighting by merit scores where necessary, the decision on which possum management approach to select can be determined by comparing costs per hectare COPY for, say, annual ground hunting versus 5-yearly aerial poisoning. Least cost per hectare COPY possum management activities provide the most cost-effective actions.

In contrast to cost-effectiveness analysis, cost-benefit analysis of possum management requires measurement of benefits in dollars to allow comparison with costs measured in dollars. However, evaluating the dollar benefits obtained from possum management on conservation lands is difficult. Non-market-valuation techniques such as the *contingent valuation method* (CVM) have been developed to provide estimates of "willingness to pay" for items that are not sold in markets. Use of CVM to measure benefits from possum management has been attempted, and there is evidence that people are willing to pay for possum management to obtain use and existence values (Kerr & Cullen 1995). For CVM to be used successfully, respondents must be well informed about the nature and duration of the outcomes from possum management. The COPY analysis seems well suited to measure outcome from possum management on conservation lands, and would therefore be useful in CVM studies.

Allocating a possum management budget

Once optimal possum density has been estimated for a site by comparing MD and MCC, and a choice made between alternative possum management techniques based on cost-effectiveness analysis, a major issue remaining is how to allocate a possum management budget across many sites. The total budget available for possum management on conservation lands is insufficient to cover all of the at-risk areas. Choices are necessary about which areas will receive possum management, and how much expenditure is economically justified on each area.

The marginal damages caused by possums and the marginal control costs of possum management can be estimated for each of the potential areas for possum management. Budgetary constraints on possum management will likely mean that possum density levels where MD = MCC cannot be achieved. The allocation rule to aim for is to decrease possum densities until the budget is exhausted, and MD – MCC is equal across all sites, as this will ensure that net benefits are maximised and could not be increased by adjusting the site of expenditures. Where there is a tight budgetary constraint, not all potential sites will qualify for possum management. Sites where possums cause little damage, or sites that are very costly to manage, should not receive possum management because costs will be high relative to any benefits obtained.

In practice possum management decisions are increasingly focusing on long-term programmes rather than one-off actions. A second budget allocation issue is how to schedule possum management over many years. *Dynamic optimisation techniques* can be used to determine the most beneficial programme of pest management over long periods, such as 30 years (Bicknell & Lambie 1996). The rule to aim for here is to ensure that the marginal net benefits from pest management are equal across the total period.

Incentives and delivery of possum management

Allocating a possum management budget can also involve spreading costs over several stakeholders. Possums are unowned, although landowners are responsible for the animals on their land. Because possums are mobile, they have the potential to inflict damage on more than one property. Possum management on one property can therefore also provide benefits to neighbouring properties. But landowners are likely to consider only the benefits *they* will derive from possum management, and the external benefits of those actions will not play a part in their calculations of how much expenditure on possum management is justified. If external benefits occur, there is likely to be under-provision of possum management when decisions are left entirely to individuals or to firms. Referring back to Figure 18.4 we can see that individuals may choose density level D*, when D** is the appropriate density from a social or collective standpoint. Where these external benefits are significant even at the lower possum density levels being achieved, a prima facie case for collective action exists.

If farmers receive all of the benefits from possum management, they should handle and bear all of the costs. But bovine Tb poses a threat to the export of meat and meat products, and the employment of many people apart from farmers will be harmed, at least temporarily, if bovine Tb halts those exports. Where possums play a role in hosting and transporting bovine Tb, farmers may not be the only people who receive benefits from possum management. The cost of managing possums in those regions should be shared between farmers and the general public, as the latter also receive benefits from the possum management.

But an important economic balance to achieve is to ensure that landowners are confronted with the full marginal damages associated with each bovine Tb reactor. If these damages are masked from landowners, there will be a reduced incentive for them to pursue Tb vector (e.g., possums) management (Animal Health Board 1995a). Masking of these marginal damages will occur if the costs of each Tb reactor and its consequences are met collectively from industry levies or from government subsidies.

The Animal Health Board and regional councils play major roles in the management of possums, and their activities are funded through a combination of industry levies and grants from Government. Indicative funding contributions for Tb vector control are: Crown as exacerbator NZ$10.6m (34%), Crown/industry as national beneficiaries NZ$13.1m (42%), landowner/local industry NZ$7.5m (24%) (Animal Health Board 1995b).

Management of possums should only be a collective action if the external damages associated with possums are significant at the level of possum management that individuals choose to provide. An important conclusion is that collective and general taxpayer funding of possum management on farms may be justifiable in many instances, but the funding arrangements should not blunt the price signal to landowners to take action to reduce possum densities.

It has long been recognised that possum harvesting, and possum management, are rarely profitable activities, and financial incentives have been provided to ensure more effort is applied to possum management. Bounties are often proposed to increase the number of possums harvested, but bounties provide crude financial signals and are inappropriate tools to increase possum management (see Chapter 23). As we noted, the marginal damage caused by possums depends on their location. Possums living on grass and shrubs by a country roadside may cause little economic damage compared to the damage caused by a possum on a deer farm, or a possum eating kōkako eggs in Mapara Forest.

A simple bounty system will reward equally the killing of a possum by the roadside and the killing of a possum on a deer farm or in a key conservation area. If the costs of possum control are low for roadside possums because of easy access, but high in conservation areas where access is difficult, bounties will result in the wrong possums being targeted. This is why the Department of Conservation specifies the precise areas where it wishes to manage possums and hires contractors to complete that work.

Supply of possum management actions on conservation lands has been subjected to considerable study, particularly to determine if least-cost possum management techniques can be identified (Warburton & Cullen 1993; Cowan & Pugsley 1995; Montague 1997). Possum management on conservation lands is typically provided by contractors, who have been selected via a tendering system. Concern has been expressed over the incentives facing contractors, and the suggestion made that they have an incentive not to harvest all possums for fear of destroying their source of income (Eriksen 1991). This fear does not seem justified if full payment is strictly contingent upon achieving prespecified abundance of possums, determined by trap-catch monitoring at the end of the control operation.

Possum management techniques vary in cost, effect, risk, and public acceptability. Choice of technique to manage possums can be constrained by requirements for employee and public health and safety, effects on non-target species, social acceptability, and the need to control other species. Once those requirements have been met, tenders can be called for delivery of possum management on conservation lands, for example, and lowest-cost tenders selected. For least-cost possum management, the choice of technique to manage possums should not be constrained by criteria such as local employment effects.

Summary

- Management of possums can be usefully analysed using an economic framework that compares marginal damage caused by possums with the marginal control costs of managing possum densities. This approach suggests that optimum levels of possum management will occur where marginal damage equals marginal control costs.
- Indicative costs of possum management range from NZ$5 to $10/ha for annual culling operations at low densities, NZ$20 to $30/ha for knockdown of high possum densities, to over NZ$61/ha for eradication of possums from islands.
- In many instances possum management is aimed at achieving specified reductions in possum densities. In these cases cost-effectiveness analysis is a valuable tool to determine if best value for money is being achieved from possum management.
- Units of output must be established and measured to apply cost-effectiveness analysis. Conservation protection years (COPY) provide a way to measure outcomes for possum management on conservation lands.
- A major issue in possum management is how to allocate a budget when the potential management area is much greater than is achievable. Economics, with its focus on the allocation of scarce resources, provides rules to ensure the best use of possum management resources.
- Possums impose both private and collective costs. Funding of possum management requires a judicious mix of private and collective contributions to ensure signals for private action are maintained, and the costs are shared equally among beneficiaries from possum management.

References

Animal Health Board 1995a: Time to decide: Tb beyond 2000. Wellington, New Zealand, Animal Health Board.

Animal Health Board 1995b: National Tb strategy: Proposed national pest management strategy for bovine tuberculosis. Wellington, New Zealand, Animal Health Board. 116 p.

Bicknell, K. B.; Lambie, N. R. 1996: An optimal control model for pest management under bait shyness. *Department of Economics and Marketing, Lincoln University, Discussion Paper No. 27,* Lincoln University, Canterbury, New Zealand.

Bicknell, K. B.; Wilen, J. E.; Howitt, R. E. 1997: Public policy and private incentives for livestock disease control. *Department of Economics and Marketing, Lincoln University, Discussion Paper No. 40.* Lincoln University, Canterbury, New Zealand.

Cowan, P. 1990: Brushtail possum. *In:* King, C. M. *ed.* The handbook of New Zealand Mammals. Auckland, Oxford University Press. P. 89.

Cowan, P. E. 1992: The eradication of introduced Australian brushtail possums, *Trichosurus vulpecula*, from Kapiti Island, a New Zealand nature reserve. *Biological Conservation 61*: 217–226.

Cowan, P.; Pugsley, C. 1995: Monitoring the cost effectiveness of aerial 1080 and ground hunting for possum control. *Science for Conservation 2.* Wellington, New Zealand, Department of Conservation. 35 p.

Cullen, R. 1992: Vegetation protection: Some economic analysis. New Zealand Association of Economists Conference paper (unpublished).

Cullen, R.; Fairburn, G.; Hughey, K. 1998: COPY: A new technique for evaluation of biodiversity protection programmes. NZARES Conferencc proceedings (unpublished).

Department of Conservation 1994: Department of Conservation National possum control plan 1993–2002: a strategy for the sustained protection of native plant and animal communities. Wellington, New Zealand. 86 p.

Eriksen, K. 1991. A new approach to possum control. New Zealand Association of Economists Conference paper (unpublished).

Fairburn, G. 1998: A recognition and analysis of the economic tradeoffs between biodiversity conservation projects. Unpublished MCM thesis, Lincoln University, Canterbury, New Zealand.

Hanley, N.; Spash,C.L. 1993: Cost-benefit analysis and the environment. Aldershot, Edward Elgar.

Innes, J.; Hay, R.; Flux, I.; Bradfield, P.; Speed, H.; Jansen, P. 1999: Successful recovery of North Island kokako *Callaeas cinerea wilsoni* populations, by adaptive management. *Biological Conservation 87*: 201–214.

Kerr, G. N.; Cullen, R. 1995: Public preferences and efficient allocation of a possum control budget. *Journal of Environmental Management 43*: 1–15.

Lambie, N. R.; Bicknell, K. B. 1998: A dynamic microeconomic analysis of bovine Tb movement control policy. New Zealand Agricultural and Resource Economics Society conference paper (unpublished).

Montague, T. L. 1997: The relative cost and effectiveness of hunting and aerial 1080 poisoning for reducing possum populations in native forest. Landcare Research Contract Report LC9697/82 (unpublished) 25 p.

Parkes, J. P. 1991: Pest management for conservation in New Zealand. Forest Research Institute Contract Report FWE 91/10 (unpublished) 28 p.

Parkes, J.; Baker, A. N.; Ericksen, K. 1997: Possum control by the Department of Conservation: background, issues, and results from 1993 to 1995. Wellington, New Zealand, Department of Conservation. 40 p.

Parliamentary Commissioner for the Environment 1994: Possum management in New Zealand. Wellington, New Zealand, Office of the Parliamentary Commissioner for the Environment. 196 p.

Warburton, B.; Cullen, R. 1993: Cost-effectiveness of different possum control methods. Landcare Research Contract Report LC9293/101 (unpublished) 21 p.

Warburton, B.; Frampton, C. 1991: Bennett's wallaby control in South Canterbury. Forest Research Institute Contract Report FWE 91/59 (unpublished) 23 p.

Warburton, B.; Cullen, R.; McKenzie, D. 1992: Review of Department of Conservation possum control operations in West Coast Conservancy. Forest Research Institute Contract Report FWE 91/62 (unpublished) 37 p.

CHAPTER NINETEEN

Models for Possum Management

Nigel Barlow

Models predicting the dynamics of wildlife populations are central to the discipline of wildlife management. Models can be used to consider many aspects of a species' demography, the most common use being to predict the effects of management on population density. Different models can be selected to answer different questions. This chapter begins with an introduction to ecological models in general and a brief historical review of possum models, with an emphasis on Tb management but with reference to the effects of possums on vegetation. Epidemiological models are introduced, then the different approaches used for modelling bovine Tb in possums are considered in more detail. I then focus on a particular "pragmatic approach", involving a suite of four different models addressing four distinct questions, and draw conclusions about future needs and the limited use of models in management.

Why model?

If it is to have other than decorative value, any model must represent something found in nature. In other words, its structure and outputs must satisfy a criterion of realism that is consistent with the understanding or management questions demanded of it. It must also be comprehensible, it should be integrated with research to obtain the best out of both model and research, and it must genuinely advance understanding rather than present an obvious or already well-known answer in a different form. Advancing understanding is probably the most important use of models, but they also aid communication and research by conveying information and biological principles, enforcing rigour in debate, and identifying data needs and promising directions for future research. Lastly, and most obviously, they help in the evaluation of management policies by answering "what if?" questions without the need for extensive, costly and sometimes impossible real-world experimentation.

The first models for possum management addressed the question of whether possums could be harvested for sustainable fur yield in habitats, such as pine plantations, which were not ecologically sensitive (Clout & Barlow 1982). Maximum sustainable yields were estimated for a range of habitat types, based on their carrying capacities and possum intrinsic rates of increase of around 0.3/year. Likely annual returns for pelts from exotic forests were in the order of \$2/ha. Observed changes in possum age-structure, based on the size of pelts harvested, were used in a simple age-class model to deduce that survival of young and/or reproductive rate had increased as a result of lowered densities through harvesting. This work was expanded to include a more detailed consideration of possum population behaviour and ecology (Barlow & Clout 1983). In particular, the authors used detailed submodels for two different population regulation mechanisms, namely, food limitation and competition for den sites, to show that both led to asymmetric, rightward-peaked logistic models for the whole population. These are referred to as theta-logistics where θ is the parameter governing the asymmetry of the dome-shaped curve relating the rate of increase (possums/ha/year) to current density. These asymmetries are important because, for example, they may imply a smaller reduction in density given an additional mortality from disease or from culling, compared to the symmetrical curve of the ordinary logistic growth model. A range of 3-parameter logistic-type population models were compared (Barlow & Clout 1983) and the possum parameters derived from them allowed development of the first possum–Tb models (Barlow 1991a,b, 1994a).

Indeed, the possum modelling itself led naturally to that for Tb. How this happened provides a useful and cautionary tale in this era of detailed research prioritising and planning. It involved no planning, it was serendipitous and "bottom-up", and it arose out of a social conversation between two scientists with knowledge of the area and a perspective on the likely future importance of Tb. This allowed the modelling work to be initiated at least 2 years before external funding agencies or research managers of the relevant organisations at the time became interested. Such a lead time was extremely valuable given how serious the problem became and how difficult the system is to understand.

Following these early possum–Tb models came subsequent variants along similar lines (Roberts 1992, 1995, 1996), together with spatial versions that considered Tb spread (Barlow 1993; Barlow & Kean 1995; Louie *et al.* 1993), and one that was similar in structure but made radically different assumptions about disease development and transmission (Kalmakoff *et al.* 1995). These were complemented by two, more detailed and individual-based approaches (Pfeiffer 1994; Pfeiffer *et al.* 1995; Efford 1996). The first included a major field study, which provided the first real data on possum–Tb dynamics, including durations of infection, prevalence, incidence, seasonality, spatial distribution of disease, birth rates, death rates, and behaviour of possums (Pfeiffer 1994). More recent developments include a continued quest to understand how the disease operates, based on a steady improvement in models as new data become available (e.g., Barlow 2000), and the development of models addressing the particular question of immunocontraception as a control for the future (Barlow 1994b, 1997).

Taken together, the variety of models and modelling approaches to possum–Tb represent one of the largest contributions to modelling of a wildlife disease system worldwide. At the same time they have helped in our understanding and management of the disease. Their demonstration that Tb could be eliminated by sustained control that kept possums below a threshold of about 50% of the carrying capacity contributed to "maintenance control" regimes, which have been largely successful in reducing Tb in areas where they have been applied (G. Hickling, Lincoln University, Lincoln, pers. comm.). Although the criteria for control have subsequently evolved through practice, as pest control agencies became able to achieve higher rates of suppression, there remains considerable potential for further model input to possum–Tb management. This relates particularly to the evaluation of spatial containment using "buffer zone" strategies, and to the use of alternative strategies like possum harvesting, vaccination, and immunocontraception.

Compared with modelling possums as Tb vectors, few models are designed to help understand and manage possum impacts on conservation lands. Clearly, the possum dynamics embodied in the Tb models, and the results from the models on the impacts of population controls like culling and sterilisation on possums themselves, are directly relevant to mitigation of their conservation impacts. In addition, the spatial GEOPOSS model of Efford (1996), which does not include Tb, provides a possible basis for tactical, spatial, and habitat-explicit control policies targeted at possum "hot spots" (areas of locally high possum density) or sensitive environments. However, these models lack a critical element, namely, the ecological impact of possums on vegetation and, where appropriate, feedback from vegetation to the possums; in other words, an interactive possum/vegetation model linked with models for forest succession. A starting point was provided by Barlow (1995a), who attempted to develop the simplest possible structure for such a model, embodying the key features of production and consumption of specified vegetation types, relative palatability of the different types, and feedback to possum rate of increase. However, there is an urgent need to develop this possum model further, link it with succession models, and possibly build in to the same general framework the effects of deer as well.

Epidemiological models

The approaches generally used in epidemiological modelling are either "population-based", or "individual-based". Population-based models separate the population into categories according to their disease status (Fig. 19.1a), the models are deterministic (i.e., contain no random elements), and they take the form of mathematical equations, which may be solved analytically (e.g., using a calculator or pen and paper) or which are run in a series of time steps on a computer (simulation). As a simple example, if we know that the infectious part of a population initially doubles each year and that the initial density of infectious animals is N_0, then after 10 years this density will be

$$N_{10} = N_0 \times 2^{10}$$

This equation represents an analytical model that can be solved with a pocket calculator. Alternatively, a computer can be instructed in a program to multiply N_0 by 2, repeat the operation a total of ten times representing 10 years, then present the answer. This is a simulation model with a time step of 1 year. "Individual-based" or "Monte Carlo" models store the status of individual animals on a computer and simulate their changes over time using probabilities and pseudo-random numbers from the computer to decide if events occur or not; they are therefore stochastic (i.e., contain a random element), and the model has to be run several times to gain an average answer.

Both approaches have problems. Population-based analytical models can generate unrealistically low numbers of infectious animals (e.g., 10^{-18} foxes in past rabies models!), which can give rise to spurious new epidemics, because numbers are expressed as densities, and fractions of infectious animals can persist in the models. In practice it is possible to overcome this by introducing arbitrary lower cut-off values for densities. Such models are also prone to oversimplification when attempting to massage complex biology into manageable mathematics. On the other hand, computer simulation allows the luxury of indefinite complexity so simulation models can become over-complex. When they do, they also become less general and more specific to the problem at issue, and their ability to reveal general insights becomes limited. Because they have many more parameters, they also become extremely data-hungry. Often these data are not available, nor is there any realistic hope of obtaining them, so the detailed model begins to embody more and more doubtful assumptions. These problems are shared by individual-based models, since they too use computer simulation to predict changes over time. Published wildlife/disease models of all these types were recently reviewed with particular emphasis on their realism and biological relevance (Barlow 1994c).

Modelling approaches

The possum and Tb modelling work during the 1990s has involved four rather distinct approaches. Two of these are individual-based stochastic simulations, one is mathematical and deterministic, and one is best described as pragmatic, meaning that the approach involves different models for different questions.

One of the individual-based approaches deals only with management of possums, not Tb. This model is still under development, but is designed to assist in determining the "combination of localised control measures that will most cost-effectively reduce possum numbers in sensitive areas" (Efford 1996). Possums in the model are subject to birth, death, and dispersal rules, and their spatial locations are stored together with habitat maps as raster GIS files.

The second individual-based approach (Pfeiffer 1994; Pfeiffer *et al.* 1995) involves an even more detailed consideration of possum behaviour at the individual level, together with the dynamics of Tb. It was initially designed to explain the possum–Tb system in a particular study site of 21 ha subject to immigration and emigration of possums, but can potentially be upgraded to larger spatial scales (D. Pfeiffer, Massey University, Palmerston North, pers. comm.). As an example of the model's detail and its behaviour in relation to simpler models, possum denning behaviour and its translation to density-dependent population regulation provide some interesting insights. The model (Pfeiffer 1994) simulates searching for dens by each possum each night, where every animal has a memory of previously visited den sites and qualities, and there are rules for occupancy each night, which depend partly on the presence of existing residents. The model assumes that if no den is found, or if the den is of low quality, the animal loses body weight and this eventually affects its survival probability. There are two interesting outcomes of these complex effects in the model. Firstly, not all dens are occupied, even if there is a large excess of possums over dens. Secondly, the resulting relationship between population rate of increase and density is not what one might expect. Den sites do not act like a bucket that allows a population to increase exponentially until the bucket is full, at which point the rate of increase suddenly becomes zero. Rather, as found by Barlow & Clout (1983), there is a smoothly curved relationship between the rate of increase and density. Depending on the density of den sites and the effective searching ability of the possums, this relationship ranges from a symmetric logistic (given very low searching abilities) to the asymmetric theta-logistic assumed by, and used as input into, the simple population models of Barlow (1991a,b, 1993, 1994a, 1995b) and Roberts (1992, 1995, 1996). The equilibrium density of possums may be considerably less than the density of den sites if searching ability is low (Barlow & Clout 1983). Thus, detailed mechanistic models can help "explain" simpler, more empirical ones. Experimental confirmation of both is required, but the value of the models is that they show clearly what data are required at both levels of study.

The mathematical approach of Roberts (1992, 1995, 1996) and Louie *et al.* (1993) represents the opposite extreme to the highly detailed, individual-based simulations, and encapsulates the whole possum–Tb system in a set of up to three differential equations. Solutions are obtained analytically in some cases (i.e., by calculation from a formula) and numerically in more complex ones (a process more or less equivalent to simulation, using a computer). Although not presented in the same way, these mathematical models, and that of Kalmakoff *et al.* (1995), are similar to the ones discussed in the next section, and will be considered alongside them. The exception is the spatial model of Louie *et al.* (1993), which represents the only attempt so far to apply mathematical equations for a diffusion process to mimic the spread of Tb in space. Although the authors were primarily concerned with determining the stability criteria for the spatial model (i.e., whether the predicted equilibrium situations were stable or unstable and whether they gave rise to oscillations over time), the model also yielded the important result that disease transmission in space or through migration of infected animals could not generate the sustained patchiness of Tb that has been observed in the field.

The pragmatic approach

A pragmatic modelling approach to issues like possum–Tb has been proposed by Barlow (1991a,b, 1993, 1994c, 1995b). This has four key features: it uses whatever kind of model is most appropriate for the particular question asked of it; the models are realistic, as defined above; the models are as simple as possible; and results are presented in a simple way that makes them accessible to interested users. To satisfy the criterion of realism for possum–Tb, the models aim to recreate (a) low overall prevalence of Tb (but local "hot-spots"), (b) little effect of Tb at low prevalence on possum density, and (c) recovery of disease in 10–12 years after a single control operation. The last point is less certain since we have no direct measure of possum Tb recovery after control, but it is based on patterns of recovery in cattle reactor rates (Barlow 1991a). Furthermore, unless the disease can recover this rapidly, it proves in models to be very much easier to eliminate than is the case in practice. The criterion of simplicity dictates that the models are those of "intermediate complexity" (Godfray & Waage 1991; Barlow & Wratten 1996). That is, they seek to avoid the potential problems and inaccuracies associated with the two extremes of very simple analytical (mathematical) models and extremely detailed computer simulation models. The potential problems arise from incorrect structure and incorrect parameter values respectively, the first dictated by the needs of the mathematics, the second by the large numbers of parameters relative to available data. The pragmatic approach has involved four different possum models, which are discussed in turn.

Model 1: Modified Anderson–May model for Tb epidemiology and control in endemic areas

The simplest starting point for possum–Tb modelling was the Anderson–May population-based models (Fig. 19.1a), named after the authors who first wrote profusely about them (e.g., Anderson & May 1979). After discovering that these models could not represent reality as defined above, and that this appeared to be because they could not handle disease patchiness, a modification was introduced to allow for this (Fig. 19.1b; Barlow 1991a). The model contained only four possum parameters: carrying capacity, birth rate, minimum death rate, and a parameter governing the non-linear relationship between realised death rate and density (Figs. 19.2a and b). There were four disease parameters: additional death rate from disease, latent period, transmission coefficient (proportion of susceptibles infected per infectious possum per unit time), and a measure of observed (Hickling 1995) disease patchiness. Initially the contact rate (number of potentially infectious contacts per infectious animal per unit time, where a potentially infectious contact is one that results in infection if the contact is with a susceptible) was assumed to change linearly with density, but the effect of non-linear relationships was investigated more recently (Fig. 19.2b; Barlow 1994c, 1995b).

The model was parameterised from a thorough review of the ecological literature (Barlow & Clout 1983; Barlow 1991a), and lack of knowledge about Tb compared with that on possum ecology was partly overcome by examining a range of possible disease scenarios (Barlow 1991a). Finally the model was rewritten in user-friendly interactive form (see, for example, Fig. 19.3), then used to investigate the epidemiology of Tb in possums (Barlow 1991a) and the likely impact of control measures including culling, harvesting, sterilisation, and vaccination (Barlow 1991b).

Although it treats Tb patchiness phenomenologically and does not attempt to explain its origin, the model is consistent with all observations so far on possum–Tb dynamics. For example, it predicts that reducing a possum population to 20% of its pre-control density (assumed to be the carrying capacity) will eliminate Tb within 10 years (defined as a 99% probability that no tuberculous possums are present

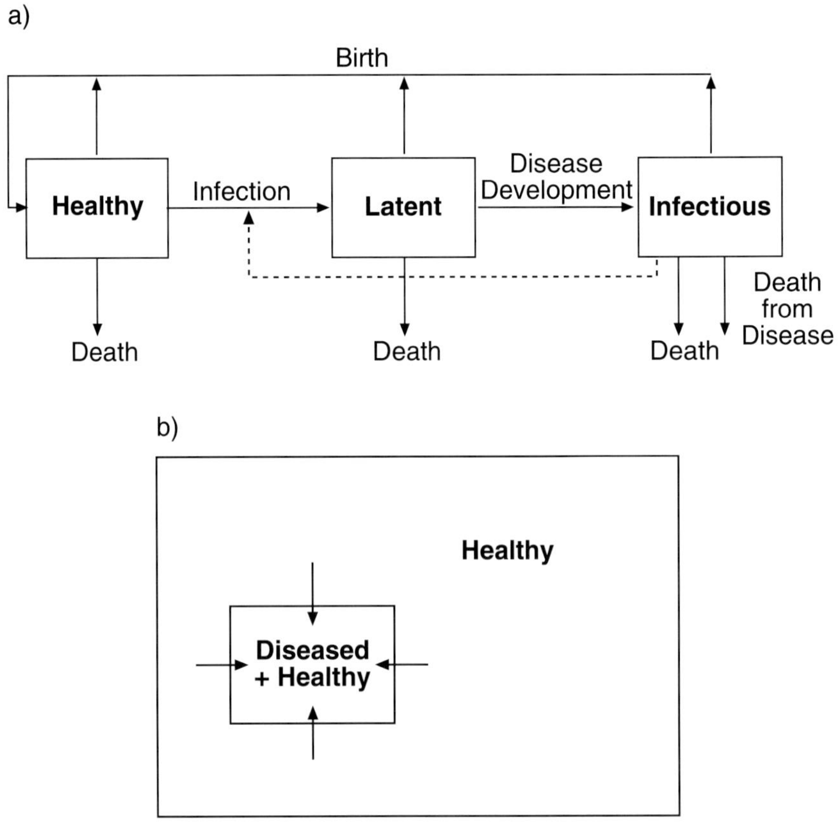

Fig 19.1
(a) The Anderson–May possum–Tb model (Model 1), showing the parameters used.
(b) The modification allowing for patchiness of disease: within the patch the model in Fig. 19.1a applies, with density-dependent immigration of healthy possums, but prevalence is averaged over the whole area.

in a given 1 km^2 of the original habitat), in the absence of immigration of diseased animals. As Figure 19.3 shows, this appears to agree with what has happened when such a control regime was implemented at Hohotaka in the central North Island of New Zealand (Caley *et al.* 1995; P. Caley, Landcare Research, Palmerston North, pers. comm.). More specifically, no Tb-infected possums have been recorded since 1994 in the 1200-ha study site (P. Caley, pers. comm.). This compares with the model's predictions that the number of tuberculous possums was only about 1.4% of pre-control levels in 1996 (year 7 in Fig. 19.3) and 0.3% of pre-control levels in 1998 (year 9 in Fig. 19.3). These correspond to probabilities of 70% and 94%, respectively, that there were no possums with gross Tb lesions in the study site at the two times. Note that the predictions assume no immigration of Tb-infected possums (cf. Caley *et al.* 1995). The model also accommodates the situation described by Coleman *et al.* (1994) for Flagstaff Flat in the western South Island of New Zealand, where Tb was widespread throughout the possum population at a high prevalence of around 60%. This would be represented in the model by a patchiness parameter (θ in Barlow 1991a) close to 1, which would lead to a much higher predicted prevalence than the normal 3–4% with gross lesions.

The model provided formulae for the basic reproductive rate of Tb in possums (R_0, or number of new infections produced in the lifespan of an infectious animal > 2); the possum threshold density needed for disease elimination (about half the carrying capacity, less if the contact rate declines non-linearly with density — see below); and for the sustained rates of control needed to eliminate disease (0.18/year for culling and sterilisation, 0.20 for vaccination). Most of these results were analytical (i.e., formulae could be developed predicting rates of control and threshold densities as functions of the possum and disease parameters), but computer simulation allowed investigation and comparison of more realistic control strategies over time. For example Figure 19.4a shows the relative effects of culling, sterilisation, and vaccination on numbers of Tb-infected possums, assuming the same rates of control in each case (30% per year). Sterilisation is least effective because its impact on susceptible possum numbers is indirect rather than direct. Figure 19.4b shows the same comparison but with a contact rate declining less than linearly with density; under

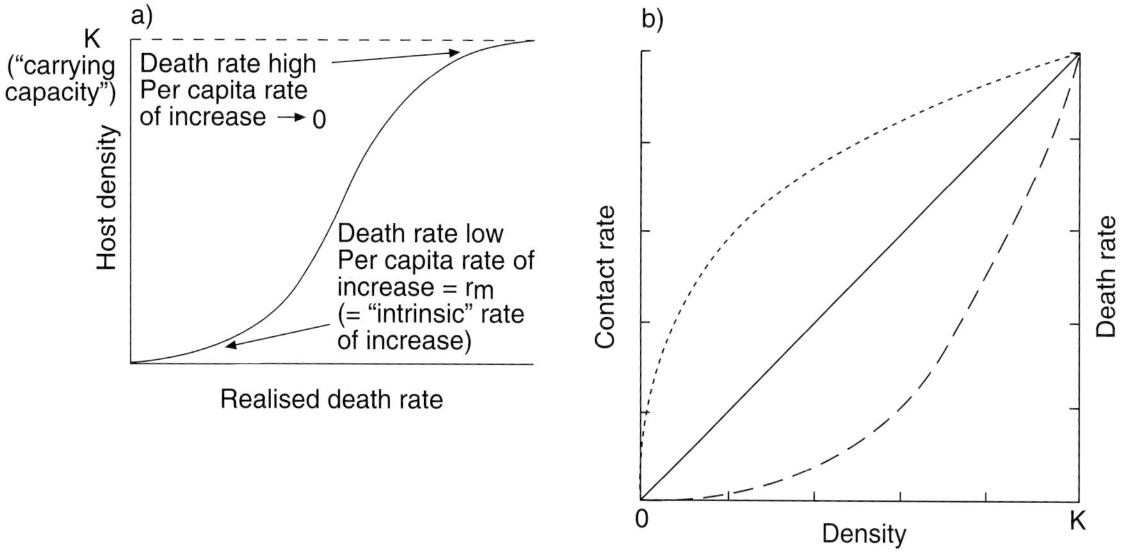

Fig. 19.2

(a) The standard, logistic growth assumption for the host population and host density-dependence, including the two key parameters r_m (= intrinsic rate of increase = maximum birth rate − minimum death rate) and K (carrying capacity or equilibrium population density). Non-linear rate – density relationships would make the curve flatten off less gradually as it approaches K.

(b) The non-linear host death rate/density relationship assumed in the model (dashed line) and the linear contact rate/density relationship (solid line). An alternative, non-linear contact rate/density relationship was also tested (dotted line) and leads to lower possum density thresholds for Tb elimination. Contact rate is the number of potentially infectious contacts made per infectious possum per unit time.

these circumstances culling and sterilisation become less effective, but vaccination remains largely unaffected and therefore becomes a practically more attractive alternative to culling.

Still more realistic controls involve episodic operations with different timing and intensity and with different degrees of immigration of diseased animals. Such control scenarios were evaluated and helped develop a control policy for bounded endemic areas: an initial operation giving a 75% kill followed by repeated ground control to maintain possums below 40% of their pre-control density. The model suggests that this should eliminate Tb if there is no immigration of disease. Although these original targets were apparently successful in causing cattle reactor rates to decline, current controls adopt a much more stringent target of around an 80% reduction (as in Fig. 19.3) for three reasons. Firstly, high rates of kill have proved feasible in practice and have become a matter of professional pride among pest control operators; secondly, if the original model is correct, the lower threshold will cause Tb to be eliminated more quickly; and thirdly, it allows for the possibility of a non-linear contact rate/density relationship, which would make the model's threshold possum density for Tb elimination lower than that originally predicted. As implemented, the current controls actually use a fixed possum density index as the target for eliminating Tb, which assumes that the threshold for doing so is independent of the habitat's carrying capacity. However, it is by no means certain that this is the case (Barlow 1993) and the use of fixed density indices as control targets therefore is being tested experimentally.

The mathematical models of Roberts (1992, 1996) also identified threshold-susceptible possum densities for Tb elimination and compared culling, vaccination, and sterilisation strategies. Initial results were similar to those of Barlow (1991b), except those for sterilisation where the lower control rates predicted as necessary for disease elimination reflect a questionable assumption in Roberts' (1992) equations for sterilisation. In addition, there is the important difference that Roberts' models do not include disease patchiness. Consequently, they predict that the threshold possum density for Tb elimination is close to the natural density or carrying capacity, that Tb at low prevalence suppresses

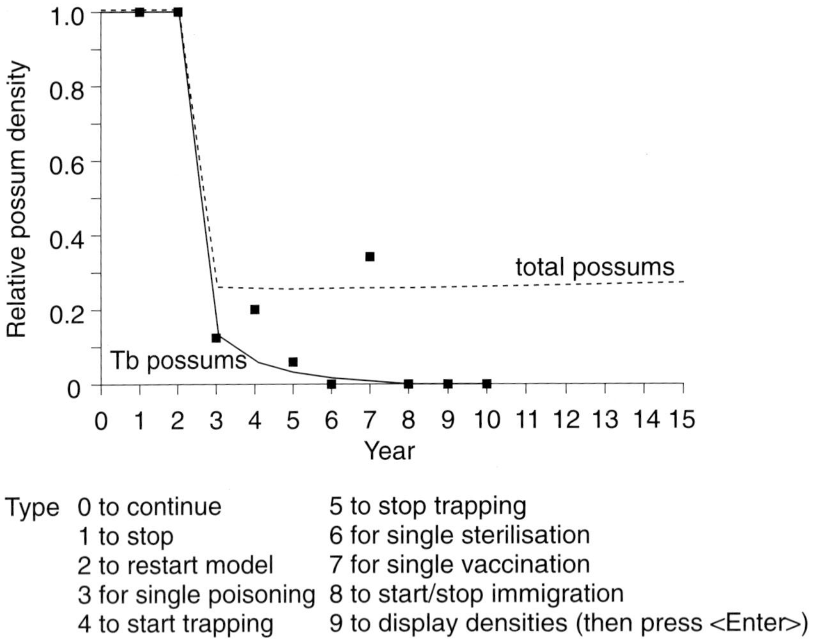

Fig. 19.3
Example output of the basic possum–Tb model (Barlow 1991b), showing a comparison between predicted results of sustained possum control (a 22% reduction in density) and those observed when such a strategy was implemented at Hohotaka, New Zealand (data from Caley *et al.* 1995 and P. Caley, Landcare Research, Palmerston North, pers. comm.). The dotted line is total possum density, as observed (averaged over the years following control) and as implemented in the model by repeated 17–18% kills each year following an initial 80% reduction. The resulting model-generated density of Tb-infected possums relative to that before control is shown by the solid line, compared with the observed relative densities shown by the squares.

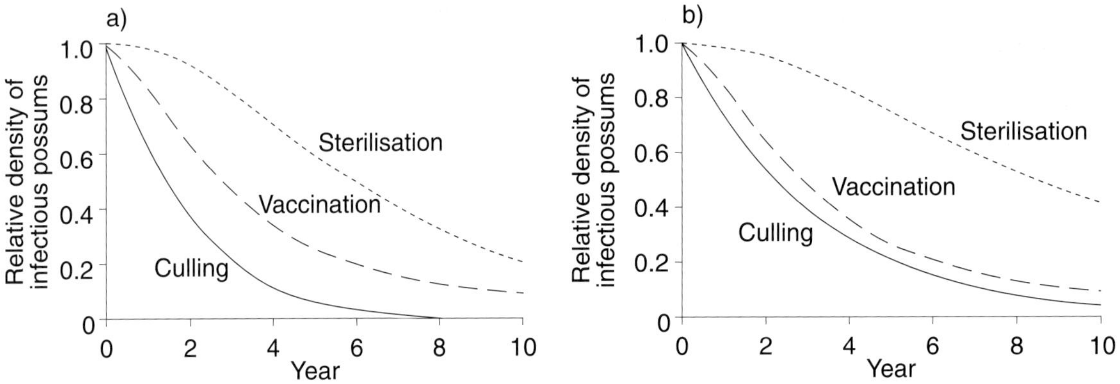

Fig. 19.4
(a) Predicted reduction in Tb-infected possums over time, given 30% culling, vaccination, or sterilisation per year.

(b) As for (a) but with a non-linear contact rate–density relationship (dotted line in Fig. 19.2b).

possum densities by up to 40%, and that Tb does not recover significantly after a single possum control so can readily be eliminated. None of these features appear to match realised observations. Similarly, vaccination of possums is predicted to cause almost a doubling in possum densities in Tb-infected areas, whereas the models with disease patchiness suggest little effect of vaccination on possum density. The differences between the models' predictions are further confounded by an anomalous value for the intrinsic rate of increase for possums assumed in Roberts (1996). This is half the accepted value (e.g., Hickling & Pekelharing 1989) and half that in the earlier models (Barlow 1991a,b; Roberts 1992), so the quantitative predictions for rates of control in Roberts (1996) must be viewed with caution. These and other issues are covered in the most recent model of the Anderson–May type (Barlow 2000).

The other possum–Tb model of the Anderson–May type is that of Kalmakoff *et al.* (1995), which adopted the interesting assumption that Tb in possums operated similarly to that in humans. In other words, the incubation period was very long (2 years) and there was an immune class. In addition, pseudo-vertical transmission was assumed between latent mothers and their susceptible young. Because a relatively high proportion of the population was in the latent state, this provided a sufficiently large number of contacts to sustain the disease. In the other mathematical models pseudo-vertical transmission, which is assumed to be from infectious not latent mothers, plays little part in disease dynamics since, by definition, it accounts for a maximum of around half a susceptible possum contacted per infectious possum per year. This compares with the total of 7 to 10 contacts per infectious possum per year that are needed to sustain observed Tb prevalences in the models. The problems with the Kalmakoff *et al.* model are, firstly, that there is no obvious mechanism for transmission from a latent mother to a susceptible offspring, as assumed in the model, and secondly, that, contrary to the statement in Kalmakoff *et al.* (1995), the model again predicts a significant reduction in possum density due to low prevalences of Tb, which contravenes one of the criteria of realism.

Recent work with simple Anderson–May models like Model 1 has taken two main directions. Firstly, the models investigate more closely the relative merits of culling and vaccination. Such analyses (e.g., Barlow 1996) have identified the critical ecological factors that might make vaccination more effective than culling, given the caveat that simple Anderson–May models take no account of spatial effects (territoriality, "vacuums" etc). The criteria are: low R_0; low host death rate (hence little loss of vaccinated animals); high intrinsic rate of increase (hence rapid recovery after culling); recruitment rather than mortality density-dependent; birth/death rate — density relationship non-linear; and contact rate — density relationship non-linear. Future, more detailed, comparisons of culling and vaccination for disease elimination will require spatially explicit disease/host models that can take account of host territoriality and density-dependent, age-specific movement. Territoriality inhibits movement and contacts between individuals, while density-dependent movement causes the opposite effect, in which locally depleted areas rapidly become recolonised from the surrounding population (the so-called "vacuum effect").

The second major direction considers the possible use of immunocontraception for controlling possums and eliminating Tb. Biological control using disseminating or vectored immunocontraception is considered in more detail below, but a necessary first stage was to consider non-vectored immunocontraception using baits, and the general issue of sterilisation as a population control methodology. How effective is it? To answer this, logistic-type models, without disease, were used to assess the factors (such as birth rate, death rate, and the degree and form of density-dependence) determining the effectiveness of sterilisation, measured in terms of the impact of a given proportion of females sterile on population density, and the level of sterilising effort (e.g., baiting frequency) necessary to achieve a given proportion sterile. It was concluded that the impact of sterilisation could be predicted for any population, given knowledge of the maximum instantaneous birth rate and intrinsic rate of increase, the carrying capacity or equilibrium population density, whether density-dependence acts on birth rate or death rate, the age at first reproduction, and the mating system. Except where both sexes are sterilised and a population has a monogamous or "harem" breeding system, culling gives a more rapid reduction in density than sterilisation, but the final level of suppression is the same for the same level of control effort (proportion sterilised or killed per unit time).

Work is still in progress, but models suggest that sterilisation is most effective relative to other control options when the population's intrinsic rate of increase (r) is high, and that a given proportion sterile gives greatest suppression when (a) the birth rate is high relative to r, (b) density-dependence acts on mortality rather than recruitment, (c) density-dependence is linear rather than non-linear, and (d) animals reproduce at an early age (the pre-reproductive period is short) (Barlow 1997; Barlow *et al.* 1997). On this basis, possums are not among the best candidates for sterilisation.

Model 2: Macro-spatial population model for Tb spread

The two main objectives of Model 2 were (a) to predict immigration rates of diseased and healthy animals, as percentages of the surrounding densities each year, into areas of different sizes; and (b) to compare the effectiveness of different buffer-zone widths around areas such as the Hauhungaroa Range in the central North Island. To achieve these objectives a simulation model was developed based on Model 1 but with three alterations. Firstly, it represents possum dynamics more realistically over time by having a birth pulse (compared with the assumption in Model 1 of continuous reproduction throughout the year), and by having sex- and age-structures, which were absent in Model 1 but which are necessary to simulate differential dispersal of males and females, juveniles and adults. Secondly, it is replicated over a grid of 1-km squares with the addition of disease transmission between squares, short-range density-dependent movement, and long-range density-independent juvenile dispersal, all quantified from the existing ecological literature (Barlow 1993). Thirdly, it is partially stochastic, using whole numbers of animals and probabilities rather than rates when numbers are small. This was necessary to allow for rare but significant events such as the founding of new disease foci by infected juveniles.

The first result was an immigration rate/area relationship (% immigration = $27.5 \times \text{Area}^{-0.4}$ (km^2); Barlow 1993), which was used to improve Model 1 by including the effect of immigration on control policies applied in limited areas, given the size of the area to be controlled.

The second result was a comparison of controlled buffer zone policies for Hauhungaroa Forest, as shown in Table 19.1. The results in the Table suggest that (a) a 3-km controlled buffer zone would give a considerable reduction in movement of Tb possums onto farmland, and (b) repeated control of the whole forest may be as cost-effective or more so than controlling a 3-km fringe, on the basis of the average total area controlled per year and the reduction in movement of Tb possums onto farmland.

Vaccinated classes of possums were then added to the model in order to evaluate vaccinated as well as culled buffers. This followed a suggestion at a National Science Strategy workshop on vaccination that such buffers may be more effective than culled ones because they obviate any "vacuum effect". This consists of a higher settlement probability or higher survival and establishment probability for animals dispersing into an empty, as opposed to an occupied, area. Rather surprisingly, the model suggested that vaccinated buffers were no more effective than culled ones (Barlow & Kean 1995). However, as these authors emphasised, more data are urgently needed on possum dispersal, particularly relating to the establishment success of those dispersing.

Model 3: Micro-spatial individual-based model for Tb spread

Model 3 was designed to underpin and explain in mechanistic terms the phenomenological treatment of disease patchiness used in Model 1. It is an individual-based, stochastic model again comprising a grid of squares, each of which is either empty or holds one individual susceptible or infected/infectious possum (Barlow & Kean 1995). Possums and disease obey exactly the same rules as in Model 1 except that this model's rates are now interpreted as probabilities, and disease transmission has to occur between adjacent neighbours rather than over

Table 19.1

Comparison of Hauhungaroa buffer control culling strategies, on the basis of the reduction in average number of Tb-infected possums dispersing onto farmland per kilometre of the bush/pasture fringe per year (an index of benefit), and the average number of kilometre squares controlled per year (an index of cost).

Control strategy	Tb possums/ km/year	km^2/year controlled
No control	4.16	0
3-km buffer 2-yearly	0.38	83
3-km buffer 4-yearly	0.59	41
Whole forest 6-yearly	0.17	154
Whole forest 10-yearly	0.19	77

the whole universe as assumed in the Anderson–May models. So far, this model can account for spatial patchiness as the outcome of stochastic (random) processes, but can only account for low overall prevalence combined with rapid disease increase after culling if (a) disease patches are environmentally determined (contact rates or contact radiuses are higher in some places than in others), or (b) disease does not actually increase after control (other than by immigration).

Model 4: Immunocontraception model

The aim of this model (Barlow 1994a,b, 1997) was initially to predict whether successful biological control of possums using immunocontraception and a suitable vector (a sexually transmitted herpes virus) was either (a) possible or (b) extremely unlikely. The model suggested that it is possible in ecological or epidemiological terms, but that the necessary criteria for success are stringent. The genetically engineered vector must persist, and attain a high prevalence that results in a high proportion of animals being sterilised, and, finally, this must lead to significant suppression of the population. Persistence of a disease vector is enhanced by the (local) absence of competing vector strains or a higher basic reproductive rate for the new strain if a wild-type already exists. High prevalence is associated with a high basic disease reproductive rate, but is prejudiced by the spatial patchiness that appears to characterise many diseases. A sexually transmitted vector is likely to be superior to a non-sexually transmitted one, both because the contact rate is higher at low densities and because multiple matings following sterilisation may increase the contact rate, the basic reproductive rate, and hence the vector's competitive ability. Sexual transmission also adds the additional safeguard of increased host specificity.

Criteria identified by the models for successful immunocontraception of possums are qualitatively similar to those applying to any population, and can be expressed relatively simply. To reduce possum populations to very low levels, at which density-dependent mortality is insignificant, the birth rate (0.305/year) must be reduced to match the minimum or density-independent death rate (0.105/year). This represents a 66% reduction in the birth rate, achieved if 66% of females are sterile at mating. If a microbial vector such as a herpes virus has a prevalence of, say, 90%, then 100 x 66/90 = 73% of the females carrying the vector must be successfully sterilised by it. This increases by a further 10% if there is a year's lag, from one mating to the next, between infection and sterilisation. If density-dependence is non-linear and these targets are not met, the level of population suppression achieved drops dramatically.

The impact of immunocontraception is significantly enhanced if the vector causes mortality of the host, but is compromised to varying extents by either the existence of an immune class of hosts or by spatial aggregation of the vector. Even slow-acting or partially effective sterilisation methods can contribute to integrated control by eroding a pest's compensatory abilities. Currently, the models are being adapted to test the effects of non-random, non-polygamous mating due to changes in possum density or the direct effect of immunocontraception, and the possibility that sustained immunocontraception in an individual is dependent on repeated exposure to the active vector. Models are also being developed to assess the rate of development of genetic resistance to an immunocontraceptive (D. Cooper, Macquarie University, Sydney; P. Bayliss, Environment ACT, Canberra, pers. comm.).

The conclusion, therefore, is that success is theoretically possible in ecological terms, but that the prerequisites for success are demanding and, even if these are met, the effect of immunocontraception is likely to be so slow that it would be best used in an integrated control strategy combined with initial culling.

Conclusions

The possum–Tb models have clearly contributed substantially to our ecological understanding of possums, Tb, and the management of both. This much is evident from the above account. There have been new insights into possum rates of immigration into cleared areas of different sizes, possum movement through buffer zones, Tb rate of increase, and the nature and sustainability of Tb-infected patches. The models have predicted results of a variety of realistic punctuated or continuous control options on possum densities and Tb, and the likely problems and advantages of novel strategies like possum vaccination and immunocontraception. In the process, the models have led to more generally relevant understanding of the ecological and epidemiological factors governing the effectiveness of sterilisation and vaccination, whatever the target species.

These findings and understandings are accessible and reasonably readable in the ecological literature. They have had some influence on research, stimulating certain lines of experimental work or analysis (e.g., Hickling 1995), often to fill gaps in the data required by the models. However, although the "adaptive management" experiment at Hohotaka, of a sustained reduction in possum density to 20% of carrying capacity (Caley *et al.* 1995), allows a valuable comparison of model predictions and observed rate of elimination of Tb, the single most critical research need for the models has not been addressed. This is the effect of a single possum control operation on Tb recovery in the possums over an area large enough to limit immigration. Such a result would show which of the various models is appropriate (those with or without spatial patchiness, and with different contact rate/density relationships) and would enable much more reliable estimates of the disease transmission coefficient.

In spite of the models' contribution to research and understanding, they have had relatively little impact on management. Even a result such as the likely cost-effectiveness of controlling possums in the whole of the Hauhungaroa Forest (Barlow & Kean 1995) was arrived at apparently independently and 2 years later by the relevant pest management agencies. The only obvious case where a model result contributed significantly to a management practice was the initial threshold concept for Tb persistence from the model of Barlow (1991b), and the suggestion that keeping possums below 40% (theoretically 50%) of their natural density would eliminate the disease in isolated endemic areas free from Tb invasion. The main distinguishing feature of this model was that its results were consciously "sold" to potential users. This required two things. Firstly, the model itself had a graphical display that could readily be understood, largely because it was so simple and because managers could relate to it. Secondly, an effort was made to demonstrate it to the relevant users. This leads to the conclusion that, however scientifically worthwhile they may be, for models to have an impact on management they must, firstly, address a real management question and, secondly, be proactively "sold" to managers. They must also be intelligible enough for this to be possible.

Summary

- Modelling enhances understanding, communication, and management, the latter through its ability to answer "what if?" questions that are difficult, costly, or impossible to do in the real world. The answers should not be final but should guide the field work necessary to confirm them.
- Models for possums have addressed sustained-yield harvesting of possums for fur and their control as pests and Tb reservoirs, through culling, poisoning, vaccination, sterilisation (chemical or immunocontraception), and vectored immunocontraception. More general analyses, applicable to all wildlife, have shown the ecological factors determining the relative merits and impacts of culling and vaccination as disease-control options, and culling and sterilisation for population control.
- For possum Tb several different modelling approaches have been employed, including population-based and individual-based, stochastic and deterministic, analytical and simulation. Together they represent a significant body of ecological modelling work on a wildlife disease system, probably only exceeded by that on rabies.
- A pragmatic approach is advocated here, which uses models that are (a) tailored to the question asked, (b) give biologically realistic results, (c) are of intermediate complexity, and (d) produce outputs accessible and comprehensible to interested users.
- In spite of their relevance, few of the model results have been used for Tb management in New Zealand. Experience suggests that not only must the model results be comprehensible and accessible, but they must be proactively "sold" to managers if they are to influence policy and practice.

References

Anderson, R. M.; May, R. M. 1979: Population biology of infectious diseases: Part I. *Nature 280*: 361–367.

Barlow, N. D. 1991a: A spatially aggregated disease/host model for bovine Tb in New Zealand possum populations. *Journal of Applied Ecology 28*: 777–793.

Barlow, N. D. 1991b: Control of endemic bovine Tb in New Zealand possum populations: results from a simple model. *Journal of Applied Ecology 28*: 794–809.

Barlow, N. D. 1993: A model for the spread of bovine Tb in New Zealand possum populations. *Journal of Applied Ecology 30*: 156–164.

Barlow, N. D. 1994a: Bovine tuberculosis in New Zealand: epidemiology and models. *Trends in Microbiology 2*: 119–124.

Barlow, N. D. 1994b: Predicting the effect of a novel vertebrate biocontrol agent: a model for viral-vectored immunocontraception of New Zealand possums. *Journal of Applied Ecology 31*: 454–462.

Barlow, N. D. 1994c: Critical evaluation of wildlife disease models. *In*: Grenfell, B. T.; Dobson, A. P. *ed.* Ecology of infectious diseases in natural populations. Cambridge University Press. Pp. 230–259.

Barlow, N. D. 1995a: Towards a possum/vegetation model. *In:* O'Donnell, C. F. J. *comp.* Possums as conservation pests. Proceedings of an NSSC Workshop . . . 29–30 November 1994. Wellington, Department of Conservation. Pp. 41–45.

Barlow, N. 1995b: Simple models for TB and possum control. *In*: Griffin, F.; de Lisle, G. *ed.* Tuberculosis in wildlife and domestic animals. *Otago Conference Series No. 3.* Dunedin, University of Otago Press. Pp. 161–164.

Barlow, N. D. 1996: The ecology of wildlife disease control: simple models revisited. *Journal of Applied Ecology 33*: 303–314.

Barlow, N. D. 1997: Modelling immunocontraception in disseminating systems. *Reproduction, Fertility and Development 9*: 51–60.

Barlow, N. D. 2000: Non-linear transmission and simple models for bovine tuberculosis. *Journal of Animal Ecology.* in press.

Barlow, N. D.; Clout, M. N. 1983: A comparison of 3-parameter, single-species population models, in relation to the management of brushtail possums in New Zealand. *Oecologia (Berlin) 60*: 250–258.

Barlow, N.; Kean, J. M. 1995: Spatial models for tuberculosis in possums. *In*: Griffin, F.; de Lisle, G. *ed.* Tuberculosis in wildlife and domestic animals. *Otago Conference Series No. 3.* Dunedin, University of Otago Press. Pp. 214–218.

Barlow, N. D.; Wratten, S. D. 1996: Ecology of predator-prey and parasitoid-host systems: progress since Nicholson. *In*: Floyd, R. B.; Sheppard, A. W.; De Barro, P. J. *ed.* Frontiers of population ecology. Collingwood, CSIRO Publishing. Pp. 217–243.

Barlow, N. D.; Kean, J. M.; Briggs, C. J. 1997: Modelling the relative efficacy of culling and sterilisation for controlling populations. *Wildlife Research 24*: 129–141.

Clout, M. N.; Barlow, N. D. 1982: Exploitation of brushtail possum populations in theory and practice. *New Zealand Journal of Ecology* 5: 29–35.

Caley, P.; Hickling, G. J.; Cowan, P. E. 1995: Sustained control of possums to reduce bovine tuberculosis in cattle and possum populations in New Zealand. Proceedings: 10th Australian Vertebrate Pest Control Conference, Hobart, Tasmania, 29 May–2 June 1995. Tasmania, Department of Primary Industry and Fisheries. Pp. 276–281.

Coleman, J. D.; Jackson, R.; Cooke, M. M.; Grueber, L. 1994: Prevalence and spatial distribution of bovine tuberculosis in brushtail possums on a forest-scrub margin. *New Zealand Veterinary Journal 42*: 128–132.

Efford, M. G. 1996: Simulating spatially distributed populations. *In*: Floyd, R. B.; Sheppard, A. W.; De Barro, P. J. *ed.* Frontiers of population ecology. Collingwood, CSIRO Publishing. Pp. 409–418.

Godfray, H. C. J.; Waage, J. K. 1991: Predictive modelling in biological control: the mango mealy bug (*Rastrococcus invadens*) and its parasitoids. *Journal of Applied Ecology 28*: 434–453.

Hickling, G. 1995: Clustering of tuberculosis infection in brushtail possum populations: implications for epidemiological simulation models. *In*: Griffin, F.; de Lisle, G. *ed.* Tuberculosis in wildlife and domestic animals. *Otago Conference Series No. 3* . Dunedin, University of Otago Press. Pp. 174–177.

Hickling, G. J.; Pekelharing, C. J. 1989: Intrinsic rate of increase for a brushtail possum population in rata/kamahi forest, Westland. *New Zealand Journal of Ecology 12:* 117–120.

Kalmakoff, J.; Mackintosh, C.; Griffin, F. 1995: A pseudovertical transmission model for the epidemiology of tuberculosis in New Zealand possums. *In*: Griffin, F.; de Lisle, G. *ed.* Tuberculosis in wildlife and domestic animals. *Otago Conference Series No. 3* . Dunedin, University of Otago Press. Pp. 225–227.

Louie K.; Roberts, M. G.; Wake, G. C. 1993: Thresholds and stability analysis of models for the spatial spread of a fatal disease. *IMA Journal of Mathematics Applied in Medicine and Biology 10*: 207–226.

Pfeiffer D. U. 1994: The role of a wildlife reservoir in the epidemiology of bovine tuberculosis. Unpublished PhD thesis, Massey University, Palmerston North, New Zealand. 439 p.

Pfeiffer, D.; Cochrane, T.; Stern, M.; Morris, R. 1995: The use of a geographical model in setting directions for future control methods. *In*: Griffin, F.; de Lisle, G. *ed.* Tuberculosis in wildlife and domestic animals. *Otago Conference Series No. 3* . Dunedin, University of Otago Press. Pp. 219–221.

Roberts, M. G. 1992: The dynamics and control of bovine tuberculosis in possums. *IMA Journal of Mathematics Applied in Medicine and Biology 9*: 19–28.

Roberts, M. 1995: Tuberculosis control in possums: culling or vaccination? *In*: Griffin, F.; de Lisle, G. *ed.* Tuberculosis in wildlife and domestic animals. *Otago Conference Series No. 3* . Dunedin, University of Otago Press. Pp. 222–224.

Roberts, M. G. 1996: The dynamics of bovine tuberculosis in possum populations, and its eradication or control by culling or vaccination. *Journal of Animal Ecology 65*: 451–464.

CHAPTER TWENTY

Fewer Possums: Less Bovine Tb

Jim Coleman and Paul Livingstone

Bovine tuberculosis (Tb) in possums puts at risk New Zealand's international trade in dairy, beef, and deer products. This chapter reviews the nature of that threat, through a description of the level of infection in livestock, and the distribution of infected possums and other wildlife. It describes the impact of Tb on the health of livestock and humans, strategies for its local and national control, and the likely costs and benefits of seeking its eradication.

Significance of bovine Tb

Bovine tuberculosis is a disease caused by *Mycobacterium bovis*. It affects a wide range of domestic and wild animals in New Zealand (Morris & Pfeiffer 1995). It is the most important disease of cattle (*Bos taurus*) and deer (*Cervus* spp.), and occurs in their populations at relatively high incidences by international standards (Alspach 1996). In contrast, the vast majority of New Zealand's trading partners and competitors (North America, Australia, most Western European and South-east Asian countries) are classified as being officially free from Tb (Animal Health Board 1995), although a few also have wildlife infected with Tb.

In New Zealand, attempts to eradicate Tb from infected herds is based on a test and slaughter programme (see below). While this is successful in many areas, herds in some defined areas become reinfected after contact with tuberculous wild animals. Wild animals that spread Tb to other wild and domestic animals are known as Tb vectors. The major Tb vector in New Zealand is the possum (*Trichosurus vulpecula)*, but ferrets (*Mustela furo*), wild deer, and possibly feral pigs (*Sus scrofa*) and cats (*Felis catus*) may also act as vectors (Morris & Pfeiffer 1995). Because of the vector-related Tb problem, an intense, ongoing effort was directed at controlling vectors, costing NZ$28.4 million in 1998/99.

Risk to trade

The reason for such vector control is that New Zealand's Tb problem has the potential to adversely affect future trade in dairy, beef, and venison products to profitable overseas markets, with losses estimated in a worst-case scenario to be worth NZ$5 billion over 10 years (Parliamentary Commissioner for the Environment 1994, p. 12). The Ministry of Agriculture and Forestry (MAF) believes that the presence of the existing Tb vector control programme, together with herd testing, product inspection, and pasteurisation of milk, will ensure New Zealand's primary products continue to be safe for human consumption. This should allay any reasonable concerns from trading partners over the Tb status of New Zealand's livestock products (O'Neil & Pharo 1995). Further, under the WTO (World Trade Organization) Sanitary and Phytosanitary (SPS) Agreement, countries have to prove that importation of a product puts their human or livestock populations at risk before they can ban its importation.

However, with the escalation in the number of cases of human Tb (*M. tuberculosis*) being diagnosed in the western world and growing concerns over food quality, consumers may show a market preference for primary produce from countries that are free of Tb, and this preference may be exploited by importing companies. This would be a market ploy and, as such, outside WTO control.

Comparison of New Zealand's Tb problem with its trading partners

In New Zealand, the Tb status of a herd is determined from the disease surveillance programme (see below). International standards of disease freedom are set by the Office International des Epizooties (OIE). According to these standards a country, or region of a country, is considered free of Tb when 99.8% of all herds present have been officially tested free of the disease each year for a minimum of 3 years. In contrast, approximately 97% of New Zealand's herds have been classified as officially free of Tb during the last 3 years (1997–99). The Tb standards set by the OIE means that Tb is a significant problem of livestock in a small number of countries, including New Zealand, that trade in livestock products (Table 20.1).

Significance of cattle and deer industries to New Zealand

Agriculture produces 55% of New Zealand's total exports (M. Doak, Ministry of Agriculture and Forestry, Wellington, pers. comm.). The dairy industry with 4.2 million head of stock (statistics from the New Zealand Official Year Book, Anon. 1998) is easily the most valuable of all of the livestock industries with exports in 1996/97 totalling NZ$4.2 billion or about 20% of all foreign trade. Some of these exports may be at risk; in 1990, the European Union, then New Zealand's most valuable market for butter (Allison 1992), made moves to introduce regulations requiring that milk and milk products be free of all pathogens (including *M. bovis*) prior to pasteurisation. These regulations were never promulgated, but they indicate the direction that some importing countries are heading and their potential to affect New Zealand's trade. Hence the urgency for New Zealand to clear Tb infection from its remaining infected dairy herds.

The beef industry is based on 4.9 million head of cattle and had export returns of NZ$1 billion in 1996/97 (Anon. 1998), largely from markets in North America (US and Canada) and northern Asia (Japan and the Republic of Korea). At present, because of New Zealand's Tb control programme and inspection procedures, these markets are satisfied with the quality of the product purchased. However, as all these countries surpass the international levels for freedom from Tb (see below), importers may place stricter quality requirements on beef products and the herds from which they come. Already, some markets are trying to link herd Tb status with the produce they will accept, which is in excess of OIE requirements.

The deer industry, with a national herd of 1.5 million animals, provided export returns for venison and velvet of NZ$177 million in 1997/98. This rapidly growing industry sends most of its venison to Germany, France, and the US. These markets are also reliant on New Zealand having a high-quality meat inspection service and a national Tb control programme.

Due to the Tb problem in New Zealand, only 5 400 cattle and 12 deer with a combined value of NZ$14.7 million were exported live from New Zealand in the 1997/98 financial year, and none were able to be exported to Australia or North America. Improving the Tb status of New Zealand's cattle and deer herds would facilitate an increase in this trade.

Tb classification of New Zealand

For Tb management purposes, New Zealand is divided into Tb Vector Risk Areas (VRAs) and Tb Vector-Free Areas (VFAs). In VRAs, some wild animals have either been found, or are suspected, to act as Tb vectors for cattle and farmed deer. In contrast, no Tb-vector-related infection is found in cattle or deer in VFAs. The VFAs contain a Fringe Testing Zone, which

Table 20.1
Herd Tb incidence and Tb reactors per 10 000 cattle for a sample of countries, including New Zealand, that are categorised as having Tb under "full control" (from 1996 FAO-OIE-WHO statistics).

Country	No. of cattle	No. of herds	Average herd size	Herd incidence (%)	Reactors per 10 000
New Zealand	9 272 325	48 541	191	1.78	3.45
Australia	25 736 007	90 077	286	0.01	0.00
Belgium	3 158 682	45 798	69	0.09	1.48
France	19 150 000	381 000	50	0.15	8.30
Greece	536 672	39 010	14	0.43	86.36
Republic of Ireland	7 422 800	153 877	48	5.76	40.95
Taipei China	151 922	10 543	14	0.50	35.94
UK	11 913 000	115 000	103	0.64	1.25

surrounds the VRA. Cattle and deer testing in the fringe testing zone together with disease surveys of wild animals are used to monitor the spread of tuberculous wild animals from the VRA.

In June 1998, there were 28 discrete VRAs scattered throughout New Zealand, which collectively covered 6.25 million ha or 23.6% of New Zealand's land area (Animal Health Board 1998; Fig. 20.1). Tb-infected possums have been identified in 23, ferrets in 17, and feral deer in 8 of these areas. The current major VRAs include the central North Island, south-eastern North Island, the West Coast of the South Island, North Canterbury/Marlborough, south-western Canterbury, coastal and central Otago, and south-east Otago (Fig. 20.1).

Most (76.4%) of New Zealand is classified as vector-free. This includes Northland, Auckland, Taranaki, Bay of Plenty, Gisborne, and Hawke's Bay which, together, contain 36.6% of New Zealand's cattle and deer herds. In June 1998 these regions possessed 0.07% of infected cattle and deer herds (P. G. Livingstone, unpubl. data).

Present-day Tb situation, herd and cattle numbers

In June 1998, 924 cattle herds and 118 deer herds (1.4% and 2.1% of the national herds respectively) were classified as infected with Tb (Animal Health Board 1998). In 1997/98, gross lesions of Tb were

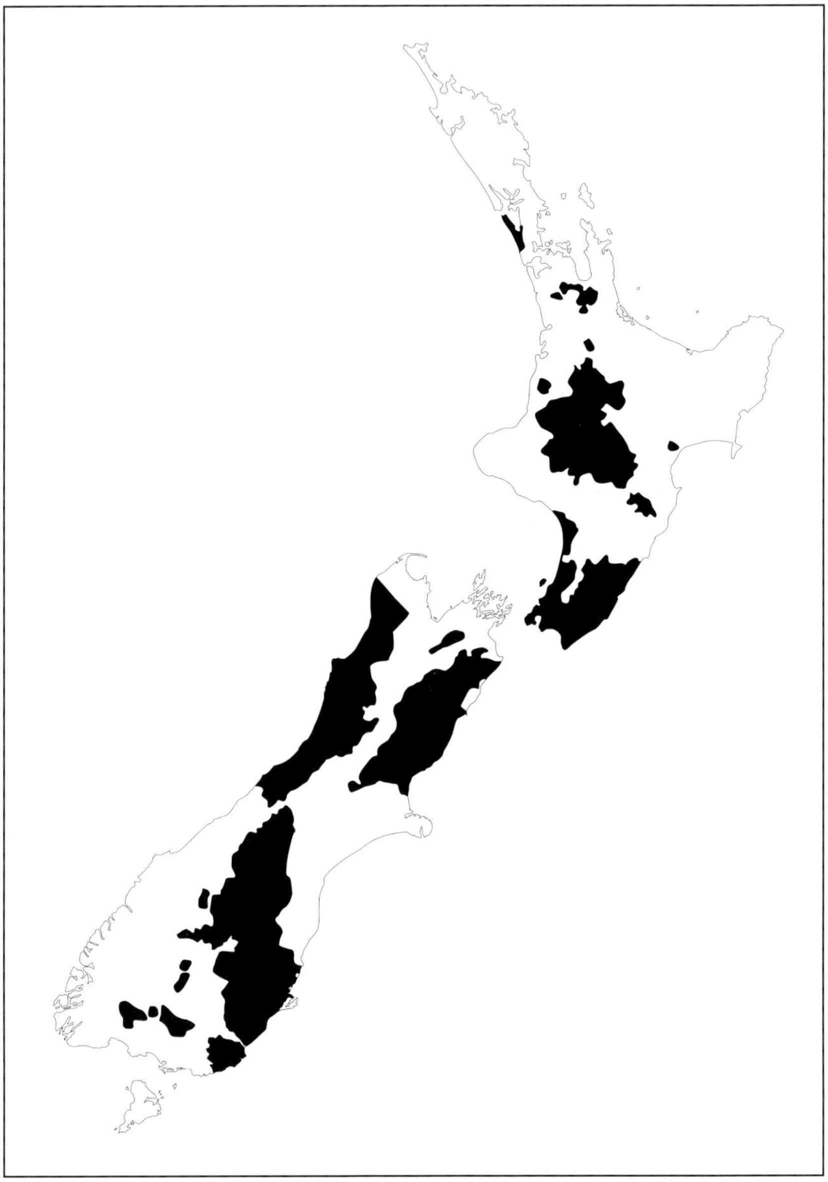

Fig. 20.1
Tb Vector Risk (shaded areas) and Vector-Free Areas in New Zealand, as at November 1998 (T. Ryan, MAF, pers comm.).

found in 2546 cattle and 352 deer. These totals included both lesioned Tb reactors and animals found with Tb lesions during routine slaughter. Most herds containing these cattle (85%) and deer (74% in the North Island, 81% in the South Island) came from VRAs, and they accounted for over 80% of all livestock showing gross lesions at slaughter in 1997/98. Further, over half of the infected herds in VFAs arise, we believe, from the movement of infected stock (through sale or from grazing) that have been in contact with Tb vectors in VRAs or on the fringe of VFAS.

Role of Tb vectors

Species involved, distribution, and abundance

The elimination of Tb from New Zealand livestock is made more difficult by the involvement of possums, deer, and possibly ferrets (Livingstone 1996) acting as vectors, and of possums acting as self-sustaining reservoirs of the disease (Pfeiffer 1994). Possums, in particular, appear to play a role analogous to that played by badgers (*Meles meles*) in the maintenance of Tb in livestock in Ireland and England (Krebs 1997). Tuberculosis has also been identified in other wild animals in New Zealand, including pigs, stoats (*Mustela erminea*), feral goats (*Capra hircus*), feral cats (Morris & Pfeiffer 1995), hares (*Lepus europaeus*; Cooke *et al.* 1993), and hedgehogs (*Erinaceus europaeus*; Lugton *et al.* 1995). However, present information indicates that, apart from ferrets (Caley 1998), these species are not involved in the transmission of Tb to livestock (Morris & Pfeiffer 1995).

Possums now occur throughout most farming areas in New Zealand. Their numbers have been estimated at 60–70 million (Batcheler & Cowan 1988), and their highest densities often occur on farm/forest margins (up to 25/ha, Coleman *et al.* 1980) where they have the opportunity to interact with livestock. Numbers of possums are generally lower on farmland away from forest cover; i.e., 1/ha on scrubby farmland to 5–10/ha about streamside willows and swamps (Cowan 1990).

Approximately 250 000 wild deer are distributed throughout most forested areas in New Zealand (Nugent & Fraser 1993), except Taranaki and Northland, and even there, new populations are being established through illegal liberations (Fraser *et al.* 1996). Deer living close to the forest edge regularly graze farm/forest margins and may interact with livestock. Their role in transmitting the disease is, however, unclear, although scavengers such as ferrets, feeding on infected deer carcasses, may subsequently be in contact with and transmit the disease to livestock. Recent information (Nugent *et al.* 1998) suggests that wild deer are principally being infected by possums and that, in the wild, deer-to-deer transmission occurs infrequently.

Ferrets prey primarily on rabbits and are found in relatively high densities in rabbit-prone country. In the South Island they are largely confined to the drier eastern grasslands, where they appear to reach their highest numbers (Ragg & Walker 1996). Ferrets are more widely distributed in the North Island, but they are rare in deep forest or in the wetter areas of New Zealand (Lavers & Clapperton 1990). Their total numbers are unknown. Ferrets with Tb have been incriminated as a source of infection for cattle and deer in some areas (Ragg *et al.*1995), and ferret-to-ferret transmission of Tb has been shown to occur under pen conditions (Ragg 1997).

Tuberculosis occurs at high prevalences in feral pigs in some VRAs of New Zealand, but feral pigs are considered to be end hosts of Tb (McInerney *et al.* 1995). However, the release of feral pigs by hunters has been postulated as the source of the infection in other wild animals in two VRAs (Wakelin & Churchman 1991).

Reasons for possum control

A compulsory national Tb control programme was introduced for factory-supply dairy cattle in 1961 and for beef cattle in 1971 (Anon. 1986). This programme successfully cleared infection from herds throughout most of New Zealand. In a few discrete areas (central North Island, Wairarapa, and the West Coast of the South Island), however, herds became reinfected after Tb had been eradicated from them. This happened in spite of an increased herd-testing frequency (see below) and, in some instances, the removal of other livestock vectors. In 1971, epidemiological investigations found an association between herds that became reinfected and the local presence of Tb-infected possums (Stockdale 1976). Reducing the possum population on and around these properties enabled herds to become, and to remain, free of Tb for varying lengths of time, depending on the extent of the initial reduction in possum density (therefore infected possum density), and whether this was kept low through ongoing maintenance control. Unless infected possum populations were severely reduced (>75% reduction), unacceptably high levels of infection were recorded in adjacent cattle herds, and the population of infected possums expanded. The realisation that Tb was capable of being maintained independently in possum populations in at least three discrete farming areas came as a major blow to the predicted progress of New Zealand's Tb control programme (Allison 1992). Since 1972, possums with Tb have been identified as a source of infection for cattle and farmed deer in a further 20 discrete locations.

Veterinary epidemiological information indicates that the most difficult diseases to control in livestock are those where wildlife are involved as vectors or sources of infection (O'Reilly & Daborn 1995). Because tuberculous possums are unique to New Zealand, there has been no opportunity to learn from other countries' experiences with this problem. Given the concern over the threat posed by Tb in cattle to trade, New Zealand attacked their Tb problem on four fronts. These included a programme of test and slaughter of cattle, and since 1990, of farmed deer populations; the control of off-farm movements of stock from infected herds and herds in Declared Movement Controlled Areas within VRAs; vector control (possum and latterly ferret control); and research. Vector control, though, is pivotal to New Zealand's ability to control Tb.

Research has provided information that has assisted in both the management of the disease in infected cattle and farmed deer herds, and of infected possum populations. Initial modelling indicated that, in the absence of immigration, Tb could be eradicated from a possum population on forest margins if the population was reduced and held at <40% of the pre-control carrying capacity of the occupied habitat (Barlow 1991). This finding was supported by historical possum control data from the West Coast, where ongoing maintenance control of possum populations, generally undertaken at annual or biannual intervals, has provided long-term freedom from infection in herds within the controlled area (Anon. 1986). Consequently, since 1987 all possum control operations have included an ongoing maintenance programme to ensure that possum density is maintained at low levels.

Research has also identified that the risk of infection to cattle from tuberculous possums is a result of a number of aspects of possum ecology and behaviour. These include relatively high possum density on forest/pasture margins, their strong preference for feeding on pasture (Coleman *et al.* 1985), the high prevalence of Tb in possums denning within 500 m of forest margins (Coleman 1988; Hickling 1991), and the inquisitive behaviour of dominant livestock towards terminally ill (tuberculous) possums and ferrets (Paterson & Morris 1995; Sauter & Morris 1995). Possums and ferrets in the terminal stages of Tb appear to be oblivious of their surroundings, and their movements attract cattle and deer, which sniff, lick, and mouth them. This is considered the most likely means whereby cattle and deer become infected from Tb vectors.

Possums are now recognised as having been the most important vector of Tb for most infected herds in New Zealand over the last 25 years. In VRAs, sustained local control of the disease in livestock only occurs following substantial reductions in possum numbers. Local eradication of Tb from wild and domestic animals is possible, and the expense of eradication is likely to be less than continued maintenance control in perpetuity.

Nature of the Tb problem

Effect on animal production

Bovine tuberculosis causes a chronic wasting disease in cattle, but the disease may be more acute in deer as illustrated by its pathology, namely, a tendency towards the formation of abscesses (Morris *et al.* 1994). In both species, Tb can cause production losses including failure to gain weight or loss of weight, infertility, and death, while in cattle, milk production may decline. These factors may also result in production losses arising from premature culling. However, since the introduction of the national compulsory Tb control programme (see below), Tb-related production losses have been insignificant as infected cattle are usually culled out before becoming seriously affected by the disease.

The mandatory slaughter of livestock reacting to the Tb test (reactors), with the consequent devaluation of carcasses with localised infections or condemnation of carcasses with generalised (widespread) Tb, affects the financial return received by farmers. Cattle farmers, through a compulsory national insurance scheme administered by the AHB, received 65% of Fair Market Value[1] for cattle taken as Tb reactors in 1997/98. Deer farmers do not have a national insurance scheme and thus only receive carcass proceeds for Tb deer reactors. In 1997/98, 49.5% of reactor cattle and 17.2% of reactor deer had visible

1 Fair Market Value is the estimated price that a non-reactor animal of similar type and condition to the reactor would fetch at a local public auction at the time the reactor was valued.

Tb lesions at slaughter. In addition, Tb was identified in a further 722 (0.03%) cattle and 156 (0.04%) deer during the routine slaughter of 2.6 million cattle and 380 000 deer, respectively. Farmers incurred additional costs in mustering and holding animals for testing, as well as through the disruption to farming practices caused by vector control.

Impact on human health

Tuberculosis caused by the human tubercle bacillus is one of the most widespread infections known to humans, and an estimated 2.7 million die from the disease annually (Pooley 1996). Despite new methods of treatment, Tb remains a serious affliction in humans. Bovine Tb is indistinguishable from human Tb clinically, radiologically, and in histopathology (O'Reilly & Daborn 1995), and is transmissible from livestock to humans (Zuckerman 1980). Bovine Tb was a serious public health problem to New Zealanders (and elsewhere where dairy herds were commonplace) in the first half of the 20th century, when children particularly were infected through drinking unpasteurised milk contaminated with *M. bovis* (Carter 1997). Infections in humans also arose from direct contact with infected animals or carcasses. However, *M. bovis* is now a minor cause of Tb in humans in New Zealand, largely as a result of pasteurisation of milk, aided by the compulsory Tb control programme for cattle. Two to 10 cases are diagnosed each year from the 50–67% of cases that are tuberculosis typed (Table 20.2). This represents between 1% and 5% of all cases of Tb diagnosed in New Zealand. However, Tb is diagnosed mostly in humans in middle to old age, suggesting that many cases have arisen from reactivation of old infections. It is likely that the risk of developing new infections is less than the current incidence. In Great Britain, about one in 60 cases of Tb is caused by *M. bovis* (Zuckerman 1980); the remainder arise from infections with *M. tuberculosis*. The current ratio of bovine Tb to human Tb in New Zealanders appears higher than that recorded in Great Britain (i.e., 1:37 to 1:50). This is probably related to the higher incidence of Tb in livestock and in wild animal populations in New Zealand, which together with the higher proportion of its population working with these animals, or processing their carcasses, presents humans with a greater risk of exposure to *M. bovis*. Even so, the low incidence of Tb in those most exposed to it suggests the risk of clinical disease in newly infected adults is also low.

The current risk to humans of being exposed to *M. bovis* by eating meat grown and processed through a licensed slaughter plant in New Zealand is believed to be zero. No case exists in the scientific literature of infection in humans derived from eating meat.

Strategies for the control of bovine Tb

The national bovine Tb strategy

The New Zealand Tb control scheme was promulgated as a National Pest Management Strategy (NPMS) by an Order in Council under the Biosecurity Act 1993, and came into force on 1 July 1998. Under this order, the Animal Health Board (AHB) is identified as the agency responsible for administering the Tb control programme in New Zealand. The mission of the AHB is "eradication of Tb from New Zealand." The key objectives of the NPMS (July 1996 – June 2001) are to:

- reduce the number of infected herds in Tb VFAs from 0.7% to 0.2 % of the total herds in those areas;
- prevent the establishment of new Tb VRAs and/or the expansion of existing Tb VRAs into farmland free of Tb vectors;
- decrease the number of infected herds in Tb VRAs from 17% to 11% of the total number of herds in those areas.

The strategy seeks to achieve these objectives through the joint action of controlling Tb in livestock and controlling Tb-vector populations.

Table 20.2
The cases of untyped tuberculosis and those caused by *M. bovis* in New Zealanders, 1985–97 (E. Galloway, Institute of Environmental Health and Forensic Sciences, pers. comm.).

Year	85	86	87	88	89	90	91	92	93	94	95	96	97
Tuberculosis	359	320	296	295	303	348	335	327	333	357	374	358	331
Bovine Tb	2	3	4	3	3	4	2	4	9	4	4	10	6

Members of the Animal Health Board (Inc.) include the New Zealand Dairy Board, Meat New Zealand, New Zealand Game Industry Board, New Zealand Deer Farmers Association, Dairy Farmers of New Zealand, Federated Farmers Meat & Wool section, and local government. Members collectively appoint directors to AHB. The AHB makes decisions on the expenditure of funds obtained from both beneficiaries and exacerbators[2] of the Tb control scheme, the size of levy payments, and operational policy matters. Cattle farmers (beneficiaries) fund the scheme via a levy on all cattle over 3 months of age that are slaughtered, or exported live. In addition, the Dairy Board and Game Industry Board contribute funds as beneficiaries, according to an agreed formula based on the cost of controlling Tb in their respective industries. All territorial authorities (regional councils and some district councils) with Tb-vector-related problems, except Otago, collect rates on behalf of the exacerbators in their region. This rate is used to fund vector control in their region. In Otago, the AHB levies all landholders owning more than 4 ha of land. The Government contributes funds both as an exacerbator, due to the fact that Tb vectors also reside on the Crown Estate, and as a beneficiary.

The aims of the national Tb strategy are underpinned by a research strategy developed and monitored by the Possum/bovine Tuberculosis Control National Science Strategy Committee (NSSC). This body seeks "to enhance co-ordination of all research programmes on possum/bovine Tb control", both for the control of Tb in livestock, and to ensure the sustainability of New Zealand's native plants, animals, and ecosystems (Elliott 1997). Both aims depend on the reduction of possum populations. The aims of the NSSC's research portfolio and the determinant of its allocation of research funds for the improved control of Tb are to:

- eliminate Tb through cattle and deer management;
- understand the relationship between the possum and the plant resources at risk;
- control possums through improving conventional control techniques;
- control possums using new and novel control techniques; and
- manage control strategies through understanding public attitudes to possum control methods.

Techniques for the control of bovine Tb in livestock

In most parts of the world, eradication of Tb from cattle herds has been achieved in a relatively straightforward manner through implementation of the "test and slaughter" model. In its simplest form, all cattle in each herd are injected intradermally with tuberculin (a protein derivative of *M. bovis*) and the change in skin thickness due to their immune response at the injection site is determined 72 hours later. Depending upon their reaction to tuberculin, cattle are classified as "test-positive" or "test-negative". Test-positive animals are either retested or slaughtered as reactors, depending on the Tb history of their parent herd.

By comparison, Tb control in cattle herds in a few countries, including New Zealand, Great Britain, and Ireland, has not been straightforward, due to reservoirs of infection in wild animal populations. As a consequence, New Zealand has implemented a range of procedures to assist farmers in identifying and eliminating Tb from their cattle and deer. These include:

- surveillance testing: compulsory routine tuberculin testing to identify infection in herds. The age of animals to be tested and the testing frequency depend upon the Tb status of the herd and the risk of it becoming infected via contact with Tb vectors or infected livestock;
- slaughter: the destruction and detailed post-mortem inspection of cattle or deer categorised as reactors to the herd test;
- area movement control: compulsory testing of all cattle and deer moving from herds within Declared Movement Controlled Areas, unless going directly for slaughter;
- herd movement control: restriction of movement of stock from infected or at-risk herds unless going direct for slaughter;
- abattoir surveillance: detection of Tb animals during routine post-mortem inspection of slaughtered livestock; and
- vector control.

Vector control: Strategies and tactics

As the spread of Tb in mammals is clearly density-dependent (Zuckerman 1980), it follows that Tb will

2 Landholders, who because of the land they manage, provide habitat suitable for the establishment of Tb vectors.

only be eradicated from possum populations if contact amongst possums or between possums and other vectors of Tb is reduced to levels (threshold densities) that break the transmission cycle. Fewer infected possums mean a lower likelihood of possum-to-possum spread of Tb, and a lower probability of livestock encountering infected possums.

Prevalence of infection in possums is determined from direct evidence (i.e., necropsy), or indirectly from the patterns of infection in livestock. Tuberculous possums occur most commonly on farmland and adjacent margins of forest and scrub (Caley & Coleman 1998). Annual testing of livestock and surveys of wild animals in the fringe areas surrounding VRAs assist in identifying the spread of Tb in wildlife and the early targeting of control to contain it.

Current management of Tb in possum populations involves first reducing possum populations and then maintaining their density on and adjacent to farmland in VRAs below the perceived threshold for transmission of Tb. In addition, control zones are established about VRAs to minimise the spread of infection. Where wildlife is thought to be a potential source of newly identified herd infection in VRAs, wild animal populations are controlled to contain the disease. In practical terms, this means reducing and maintaining possum densities to levels at or below 5% as indexed by trap-catch rate (between 5% and 25% of carrying capacity of most forest habitats). Initial reductions of possum populations generally involve aerial dispersal of 1080-poisoned baits over large (>1000 ha) or inaccessible areas of forest and scrub, complemented by ground control of possums on adjacent pasture and forest/pasture margins. Initial control is followed by regular (annual to biennial) ground control (maintenance) to hold the population at this imposed low density. Such control operations normally achieve immediate reductions in the incidence of Tb in herds contiguous with infected possum populations. Without ongoing maintenance control of possums, Tb levels in herds usually return to pre-control levels in 5–8 years (P. G. Livingstone, unpubl. data) as possum numbers increase. Thus far, Tb has been eradicated from possum populations in six small areas.

The effectiveness of annual maintenance control of possums to reduce the incidence of Tb in dairy cattle is well illustrated by results from an ongoing 15-year study at Hohotaka in the King Country (Fig. 20.2; Caley 1997; P. Caley, Landcare Research, Palmerston North, pers. comm.). There, the mean annual incidence of Tb in six herds has been reduced by 88% over 10 years, and most if not all of this reduction is attributed to a reduction in the number of possums through annual control. At the same time the post-control abundance of grossly lesioned possums (the primary source of infection in the cattle) fell by 92%. No Tb-infected possums have been found since 1993.

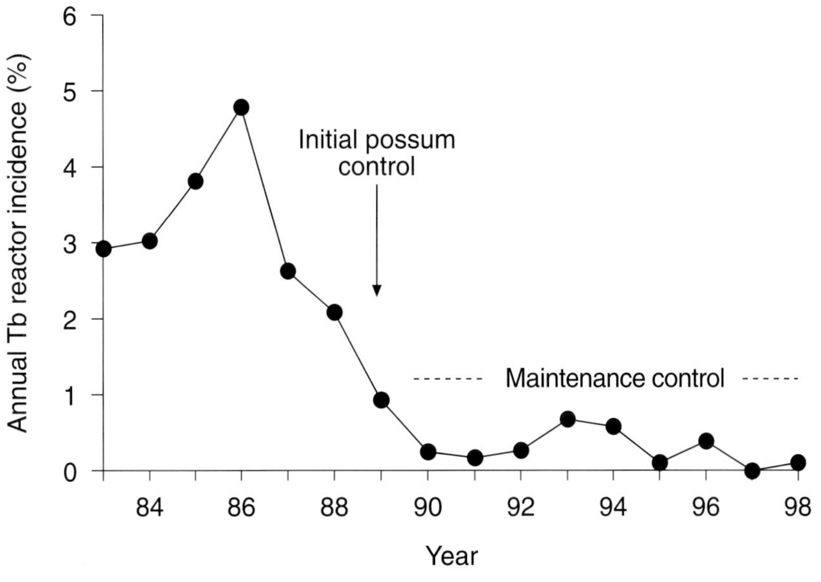

Fig. 20.2
The reduction in incidence of Tb in six dairy herds near Hohotaka following control of infected possums (after Caley 1997).

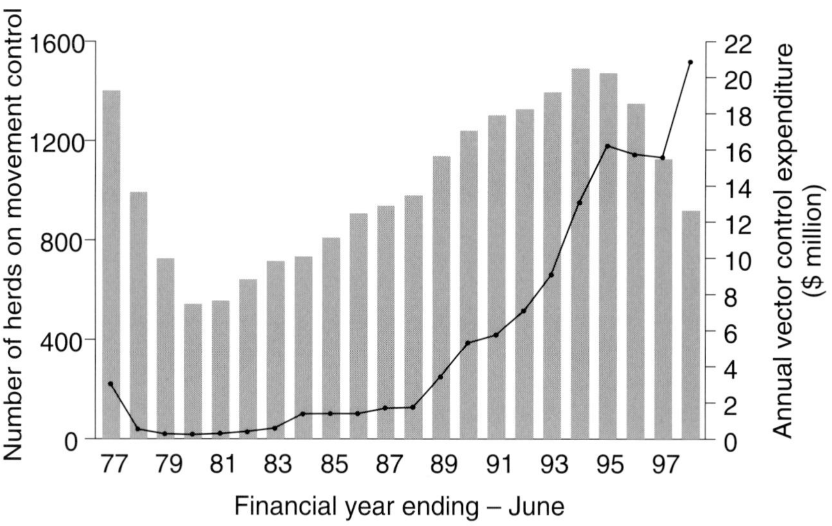

Fig. 20.3
The number of infected cattle and deer herds (bars) and the expenditure (solid line) on vector control standardised to 1990 dollars from 1977 to 1998.

In 1997/98, 41 initial control operations or extensions to existing operations aimed primarily at possums were undertaken over 283 000 ha, and 260 maintenance control operations undertaken over 2.15 million ha (Animal Health Board 1998). Approximately 10% of New Zealand's land area is thus under some form of sustained Tb vector control.

Trends in infected herds

The number of infected cattle and deer herds and the amount of money (standardised to 1990 dollars) spent on vector control between 1976/77 and 1997/98 has varied greatly (Fig. 20.3). Initially, the number of infected cattle herds declined rapidly over the period 1977–1980. This resulted largely from the expenditure of c. NZ$3 million/year on possum control from 1974 to 1978, together with the implementation of movement control restrictions on infected herds in 1977 (Tweedle & Livingstone 1994). Possum control funds were largely spent in three VRAs : central North Island, south-eastern North Island, and the West Coast of the South Island. A rapid decrease in the number of infected herds followed, and the possum control operations were considered very successful. The amount of funding for ongoing possum control was then reduced in 1977/78 to between NZ$200,000 and NZ$500,000 per year. It was not until 1989 that vector control funding once again rose to NZ$3 million. During the intervening period, tuberculous wild animals became established in a further seven areas, relative to the eight that existed in 1978. Of more importance during this period, herd and within-herd infection returned to pre-control levels and numbers of Tb-infected wild animals increased within existing VRAs. As a consequence of the reduced possum control from 1978 to 1988, the number of infected cattle herds continued to increase through to October 1994 when they peaked at 1527. Infected deer herds peaked in 1987 at 361, most of which were due to non-vector-associated infection.

Since 1989, the funding for vector control increased steadily until 1993/94, when the Government provided additional funding for controlling possums along a 3-km-wide strip on conservation land adjacent to farmland with a Tb vector problem. Since the start of the AHB's Pest Management Strategy in June 1996, funding has again increased dramatically. In 1995, the ongoing vector maintenance programme that began in 1987 started to affect the number of infected cattle herds. Since then, infected cattle and deer herds have declined in number by more than 40%, and in June 1998 there were 924 and 118 infected cattle and deer herds, respectively.

Costs and benefits of Tb management

The establishment of a National Pest Management Strategy (NPMS) carries with it a requirement of the Biosecurity Act 1993: that of economic evaluation of its likely costs and benefits. For the NPMS on Tb, economic evaluations have been undertaken for the AHB by Nimmo–Bell and Company (1995) using a cost–benefit framework based on the expected economic benefits arising from each of two options:

(a) A "do nothing" option, in which all expenditure on the management of Tb would cease. Farmers would no longer take precautions against the disease occurring in their livestock, and New Zealand's trading partners would quickly realise New Zealand had stopped its Tb management programme. The evaluation of this option indicated it had a very high risk of being economically unacceptable (trade restrictions would be likely), socially unacceptable (job losses would follow), and politically unacceptable.

(b) An option that compared marginal and trade protection benefits likely under the NPMS with those obtained under the previous (1992/93) Tb strategy. This evaluation indicated that the clear benefits and the potential loss of trade (see sections 1.2 & 1.3) fully justify the current strategy, as its costs discounted over 15 years amount to only 0.94% of any potential trade loss.

Costs

The income required to manage Tb and the associated vector control programme in New Zealand is derived from a number of sources, such as levies and grants derived from the farming community, from regional and district councils, and from Government. In addition, there are indirect costs associated with the programme, such as on-farm production losses including additional stock management and decreased land values; loss of overseas investment; cost impacts of vector control on farm management; and secondary poisoning and social concerns, such as reduced recreational opportunities for hunting groups (Nimmo–Bell and Co. 1995).

In 1997/98, the AHB spent over NZ$43.7 million on controlling the disease (Animal Health Board 1998), and this level of expenditure is likely to be needed for the foreseeable future. New Zealand has been trying to resolve the problem in cattle for about 50 years and in wildlife for about 30 years, but the disease has not yet been eradicated from livestock in VRAs due to the continuing infection in wildlife. Eradication of Tb from both wild and domestic animals with current funding and techniques is by necessity a slow and relatively expensive process. Local containment of Tb accompanied by a progressive "rolling back" of the peripheries of VRAs seem the most optimistic outcomes of the current strategy. However, hopes are high that research will provide new tools, such as Tb vaccines for wildlife and methods for reducing the fertility of possums, that will assist vector management with a consequential benefit to cattle and deer Tb statistics. In addition, research on developing a vaccine for use in cattle and farmed deer that will not sensitise them to skin testing is advancing and may provide a means of preventing disease in areas where no or limited vector control is being undertaken.

Disease control of livestock cost NZ$12.7 million in 1997/98, with about 70% of this being spent on testing cattle (deer farmers pay for their own testing, estimated at NZ$1.4 million in 1997/98). The balance of costs were associated with scheme and database management, laboratory testing, and compliance. Direct losses to beef and deer farmers in 1997/98, due to the downgrading of tuberculous cattle and deer carcasses (with Tb reactor cattle receiving 65% of Fair Market Value), is estimated at NZ$1.15 million.

Approximately NZ$24.7 million was spent on vector management in 1997/98, of which $13.3 million came from the Crown, $5.8 million from regional and district councils, and $5.6 million from industry (Animal Health Board 1998). This represents a threefold increase (from $6.9 million) since 1990/91 (Coleman 1993). Research costs for possum/Tb control research have also risen sharply in recent years, from about NZ$3.2 million in 1990/91 (Allison 1992) to NZ$15.5 million in 1998/99 (NSSC Annual Report, 1998), with $7.3 million of this coming from the Government via the Public Good Science Fund (PGSF). Other major providers of funds for research included MAF (NZ$3.2 million), AHB (NZ$2.8 million), and the Department of Conservation (NZ$1.5 million).

Benefits

The major benefits (i.e., items which incrementally improve human welfare) of the Tb control programme are as follows: minimising the risk to trade, a reduction in future costs (such as reduced carcass value), increased conservation of native flora and fauna, minor improvements in human health, reduced pasture losses to pest species, and gains for regional economies (Nimmo–Bell and Co. 1995). The major beneficiaries are the farming interests involved.

Summary

- Tb in wild animals in New Zealand has the potential to affect trade in dairy, beef, and deer products.
- Wild animal vectors occur in 28 discrete areas of New Zealand, encompassing 24.5% of its land area.
- Possums are probably the only self-sustaining reservoir of Tb, but ferrets and wild deer are vectors of Tb.
- The production losses sustained by farmers with infected herds is minimal under the current Tb control programme.
- *Mycobacterium bovis* causes between 1% and 5% of all cases of human Tb in New Zealand.
- A National Pest Management Strategy for the control of Tb came into force in July 1998.
- Control of tuberculous possum populations is based on their initial reduction, generally using aerially delivered 1080 baits on large areas of forests, followed by a variety of ground-control techniques on farm land at annual or biennial intervals to maintain them at low densities.
- National cattle and deer herd testing results have varied over the last three decades, depending on the intensity of tuberculous possum control. Latest test data for both cattle and deer show encouraging trends and represent the results of high investment in vector control since 1994.
- Cost–benefit studies indicate that the present National Pest Management Strategy on Tb has clear benefits for New Zealand.

References

Allison, A. J. 1992: Towards a national strategy for science: the problem of possums, and bovine and cervine tuberculosis. Ministry of Research, Science and Technology, Report No.7. 88 p.

Alspach, R. 1996: Towards a national pest management strategy for bovine tuberculosis. *In:* Pest Summit 95. *The Royal Society of New Zealand Miscellaneous Series 31*: 49–52.

Animal Health Board 1995: National Tb strategy: Proposed national pest management strategy for bovine tuberculosis. Wellington, New Zealand, Animal Health Board. 116 p.

Animal Health Board 1998: Annual report for the year ending 30 June 1998. Wellington, New Zealand, Animal Health Board. 43 p.

Anonymous 1986: History of Tb control scheme. *Surveillance 13(3)*: 4–8.

Anonymous 1998: New Zealand official year book 1998. Wellington, GP publications. 607 p.

Barlow, N. D. 1991: The role of modelling in policy and control decisions. *In:* Jackson, R. *convenor.* Symposium on tuberculosis. *Publication No. 132, Veterinary Continuing Education.* Palmerston North, New Zealand, Massey University. Pp. 251–265.

Batcheler, C. L.; Cowan, P. E. 1988: Review of the status of the possum (*Trichosurus vulpecula*) in New Zealand. Unpublished FRI contract report to the Technical Advisory Committee (Animal Pests) for the Agriculture Pests Destruction Council, Department of Conservation, and Ministry of Agriculture and Fisheries. 129 p.

Caley, P. 1997: Effects of eight years of possum maintenance control on levels of bovine tuberculosis in cattle and possums — the Hohotaka study. Landcare Research Contract Report LC9697/141 (unpublished). 18 p.

Caley, P. 1998: Broad-scale possum and ferret correlates of macroscopic *Mycobacterium bovis* infection in feral ferret populations. *New Zealand Veterinary Journal 46*: 157–162.

Caley, P.; Coleman, J. 1998: Possum Tb infection in the Hohonu Range — left alone it won't go away. *He Kōrero Paihama – Possum Research News 9*: 7–8.

Carter, C. E. 1997: Bovine tuberculosis — a New Zealand perspective. *In:* Proceedings of Epidemiological Programme, 10th Federation of Asian Veterinary Association (FAVA) Congress, Cairns, Australia. Pp. 21–29.

Coleman, J. D. 1988: Distribution, prevalence, and epidemiology of bovine tuberculosis in brushtail possums, *Trichosurus vulpecula*, in the Hohonu Range, New Zealand. *Australian Wildlife Research 15*: 651–653.

Coleman, J. D. 1993: The integration of management of vertebrate pests in New Zealand. *New Zealand Journal of Zoology 20*: 341–345.

Coleman, J. D.; Gillman, A.; Green, W. Q. 1980: Forest patterns and possum densities within podocarp/mixed hardwood forests on Mt Bryan O'Lynn, Westland. *New Zealand Journal of Ecology 3*: 69–84.

Coleman, J. D.; Green, W. Q.; Polson, J. G. 1985: Diet of brushtail possums over a pasture-alpine gradient in Westland, New Zealand. *New Zealand Journal of Ecology 8*: 21–35.

Cooke, M. M.; Jackson, R.; Coleman, J. D. 1993: Tuberculosis in a free-living brown hare (*Lepus europaeus occidentalis*). *New Zealand Veterinary Journal 41*: 144–146.

Cowan, P. E. 1990: Brushtail possum. *In*: King, C.M. *ed.* The handbook of New Zealand mammals. Auckland, Oxford University Press. Pp. 68–98.

Elliott, R. W. E. 1997: Assessment of the Possum/bovine Tb Control National Science Strategy Committee. Wellington, Ministry of Research, Science and Technology, Report No. 63 (unpublished). 43 p.

Fraser, K. W.; Cone, J. M.; Whitford, E. J. 1996: The established distribution and new populations of large introduced mammals in New Zealand. Landcare Research Contract Report LC9697/22 (unpublished). 64 p.

Hickling, G. 1991: Ecological aspects of endemic bovine tuberculosis infection in brushtail possum populations in New Zealand. *In:* Proceedings: 9th Australian Vertebrate Pest Control Conference, Adelaide. Pp. 332–335.

Krebs, J. R. 1997: Bovine tuberculosis in cattle and badgers: report to the Rt Hon. Dr Jack Cummingham, MP. MAFF Report No PB3423, London, Ministry of Agriculture, Fisheries and Food.

Lavers, R. B.; Clapperton, B. K. 1990: Ferret. *In:* King, C.M. *ed.* The handbook of New Zealand mammals. Auckland, Oxford University Press. Pp. 320–330.

Livingstone, P. G. 1996: Overview of the ferret problem. *In:* Ferrets as vectors of tuberculosis and threats to conservation. *The Royal Society of New Zealand Miscellaneous Series 36*: 2–6.

Lugton, I. W.; Johnstone, A. C.; Morris, R. S. 1995: *Mycobacterium bovis* infection in New Zealand hedgehogs (*Erinaceus europaeus*). *New Zealand Veterinary Journal 43*: 342–345.

McInerney, J.; Small, K. J.; Caley, P. 1995: Prevalence of *Mycobacterium bovis* in feral pigs in the Northern Territory. *Australian Veterinary Journal 72*: 448–451.

Morris, R. S.; Pfeiffer, D. U. 1995: Directions and issues in bovine tuberculosis epidemiology and control in New Zealand. *New Zealand Veterinary Journal 43*: 256–265.

Morris, R. S.; Pfieffer, D. U.; Jackson, R. 1994: The epidemiology of *Mycobacterium bovis* infections. *Veterinary Microbiology 40*: 153–177.

National Science Strategy Committee (NSSC) for Possum and Bovine Tuberculosis Control. 1998: Annual report. Wellington, New Zealand, NSSC. 79 p.

Nimmo–Bell and Co. 1995: Economic evaluation and effectiveness of the National Pest Management Strategy for bovine tuberculosis. Unpublished report to AHB. 25 p.

Nugent, G.; Fraser, K.W. 1993: Pests or valued resources? Conflicts in management of deer. *New Zealand Journal of Zoology 20*: 361–366.

Nugent, G.; Whitford, J.; Coleman, J. D.; Fraser, K. W. 1998: Effect of possum (*Trichosurus vulpecula*) control on the prevalence of bovine tuberculosis in wild deer. *In:* Proceedings of the OIE International Congress with WHO co-sponsorship on Anthrax, brucellosis, CBPP, clostridial and mycobacterial diseases. Pretoria, Sigma Press. Pp. 462–466.

O'Neil, B. D.; Pharo, H. J. 1995: The control of bovine tuberculosis in New Zealand. *New Zealand Veterinary Journal 43*: 249–255.

O'Reilly, L. M.; Daborn, C. J. 1995: The epidemiology of *Mycobacterium bovis* infections in animals and man: a review. *Tubercle and Lung Disease 76 (Supplement 1)*: 1–46.

Parliamentary Commissioner for the Environment 1994: Possum management in New Zealand. Wellington, New Zealand, Office of the Parliamentary Commissioner for the Environment. 196 p.

Paterson, B. M.; Morris, R. S. 1995: Interactions between beef cattle and simulated tuberculous possums on pasture. *New Zealand Veterinary Journal 43*: 289–293.

Pfeiffer, D. U. 1994: The role of a wildlife reservoir in the epidemiology of bovine tuberculosis. Unpublished PhD thesis, Massey University, Palmerston North, New Zealand. 439 p.

Pooley, R. B. 1996: *Mycobacterium bovis* : an old disease in a new era? A review of the epidemiology and public health importance of human *Mycobacterium bovis* infection in New Zealand. Unpublished PhD thesis, University of Otago, Dunedin, New Zealand. 73 p.

Ragg, J. R. 1997: Tuberculosis (*Mycobacterium bovis*) epidemiology and the ecology of ferrets (*Mustela furo*) on New Zealand farmland. Unpublished PhD thesis, University of Otago, Dunedin, New Zealand.

Ragg, J.; Walker, R. 1996: The association of tuberculous (*Mycobacterium bovis*) ferrets (*Mustela furo*) with Tb infected cattle and deer herds in the South Island; a review of four predator surveys and the case study of a Tb problem area in East Otago. *In:* Ferrets as vectors of tuberculosis and threats to conservation. *The Royal Society of New Zealand Miscellaneous Series 36*: 7–14.

Ragg, J. R.; Moller, H.; Waldrup, K. A. 1995: The prevalence of bovine tuberculosis (*Mycobacterium bovis*) infections in feral populations of cats (*Felis catus*), ferrets (*Mustela furo*) and stoats (*Mustela erminea*) in Otago and Southland, New Zealand. *New Zealand Veterinary Journal 43*: 333–337.

Sauter, C. M.; Morris, R. S. 1995: Behavioural studies on the potential for direct transmission of tuberculosis from feral ferrets (*Mustela furo*) and possums (*Trichosurus vulpecula*) to farmed livestock. *New Zealand Veterinary Journal 43*: 294–300.

Stockdale, H. G. 1976: Possums as a source of tuberculosis infection for cattle. Animal Health Division technical report AH26.1175. 7 p.

Tweedle, N. E.; Livingstone, P. G. 1994: Bovine tuberculosis control and eradication programmes in Australia and New Zealand. *Veterinary Microbiology 40*: 23–29.

Wakelin, C. A.; Churchman, O. T. 1991: Prevalence of bovine tuberculosis in feral pigs in Central Otago. *Surveillance 18*: 19–20.

Zuckerman, Sir S. 1980: Badgers, cattle and tuberculosis. Report to the Right Honourable Peter Walker, MBE, MP. London, Her Majesty's Stationery Office. 106 p.

CHAPTER TWENTY-ONE

Benefits of Possum Control for Native Vegetation

David Norton

The role of possums in changing the composition, structure, and dynamics of New Zealand's forests has been extensively documented. Possums are opportunistic feeders, not only browsing foliage, flowers, and fruit, but also fungi, and preying on birds and invertebrates (Green 1984; Cowan 1990; Brown *et al.* 1993). The changes in forest composition and structure that result from prolonged possum browse have also been well documented (Brockie 1992; Rose *et al.* 1992; Bellingham *et al.* 1996; Rogers & Leathwick 1997; Allen *et al.* 1997) and have lead to extensive possum control operations in order to reduce possum numbers and enhance forest condition (Parkes *et al.* 1997).

While a considerable area of native forest is now subject to possum control (c. 13 000 km^2; Parkes *et al.* 1997), surprisingly little has been published on vegetation response to control. The reasons for this reflect the relative recency of consistent widespread possum control, methodological difficulties associated with assessing vegetation response, the undertaking of these operations by conservation managers who are perhaps less inclined to publish their results in the scientific literature, and the lag between recent control and measurable vegetation response.

The possum–vegetation interaction is not a simple one, with possums targeting certain species, or even certain individuals, within a forest. Furthermore, the factors that influence food selection by possums are poorly understood, as also is the response in diet choice of the possums left after a control operation. In this chapter I review four case studies that provide information on the response of vegetation to a reduction in possum numbers and consider some of the factors that modify the vegetation response. Finally, I outline some considerations that are necessary for better understanding how vegetation responds to possum control. The examples used come primarily from forested ecosystems as this is where most of the research has been undertaken. However, similar points can probably be made about other vegetation types (e.g., shrublands and grasslands).

Case studies

Waipoua Forest, Northland (Payton et al. *1997a)*

The warm temperate forests of Waipoua are dominated by several angiosperm and conifer trees including northern rātā (*Metrosideros robusta*), kohekohe (*Dysoxylum spectabile*), tawa (*Beilschmiedia tawa*), taraire (*Beilschmiedia tarairi*), makamaka (*Ackama rosifolia*), kauri (*Agathis australis*), and rimu (*Dacrydium cupressinum*), and are amongst the most species-rich in New Zealand. Possums began colonising this area in the 1960s with obvious canopy damage visible by the late 1980s. In July 1990 an aerial poison operation was undertaken, which resulted in c. 87% possum mortality, and leg-hold traps have been used subsequently to maintain possum numbers at low levels (7–9% trap catch index, cf. 12–32% before control).

Eight plant indicator species (northern rātā, tōtara, and Hall's tōtara (*Podocarpus totara/ P. hallii*), kohekohe, māhoe (*Melicytus ramiflorus*), tōwai (*Weinmannia silvicola*), five-finger (*Pseudopanax arboreus*), patē (*Schefflera digitata*), and raukawa (*Raukaua edgerleyi*) were monitored at the time of the 1990 poison operation and annually for 4 years after control at three sites in both the treatment (possum control) and non-treatment areas. All indicator species showed evidence of possum browse at the time of possum control, although the overall levels of defoliation were generally low (Fig. 21.1). Although possum browsing declined significantly in the treatment area after control compared to the non-treatment area, where there was little change, there was little evidence of increasing foliage cover, with only one species, kohekohe, showing a significant increase after poisoning. However, all eight indicator species showed significant increases in defoliation in the non-treatment area (Fig. 21.1). This study showed that while some severely browsed trees did improve in health after possum control, there was no general improvement in canopy condition over the 4-year study period. However, possum control did halt the continuing decline in canopy condition evident in the non-treatment area.

Kapiti Island (Atkinson 1992)

The vegetation of Kapiti Island ranges from mixed shrub-grassland on the steep western slopes to mixed angiosperm forests on the gentle eastern slopes dominated by kohekohe and tawa, with five-finger and kānuka (*Kunzea ericoides*) also common. Possums were introduced in 1893 and were controlled intermittently between 1910 and 1968 when control ceased. A possum eradication programme started in 1980 with the last animal killed in December 1986 (Cowan 1993).

Increased defoliation and in some cases mortality, presumably due to possum browse, were observed in several plant species during the period of no possum control. In particular, tawa suffered 60% mortality in one area while fuchsia (*Fuchsia excorticata*) was reduced to a few isolated clumps. After possum control was resumed in 1980 and especially after eradication in 1986, most species adversely affected by possums showed rapid improvements in condition. For example, within 2 years of renewed trapping, no browsing attributable to possums was evident in monitored northern rātā trees, while flowering and fruiting in kohekohe became apparent again having been virtually absent before this.

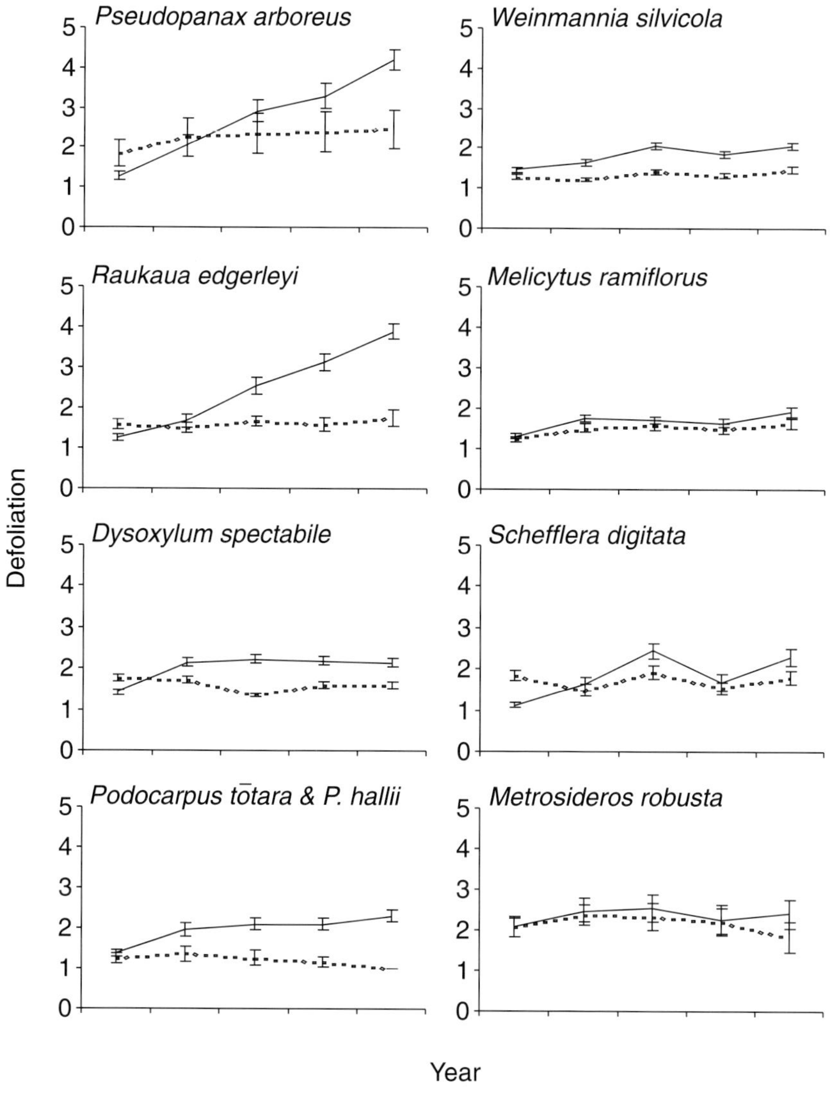

Fig. 21.1
Changes (mean ± SE) in defoliation level for eight indicator plants in treatment (- - - - -) and non-treatment (———) areas between 1990 and 1994 for Waipoua Forest, Northland (Payton *et al.* 1997). Defoliation classes: 1, < 1/3 foliage lost; 2, 1/3 – 2/3 foliage lost; 3, > 2/3 foliage lost; 4, few leaves remaining; 5, completely defoliated.

Otira-Deception Valleys, Westland (Smale et al. *1993, Tubbs 1997)*

The conifer-broadleaved forests of central Westland are dominated by the angiosperm trees southern rātā (*Metrosideros umbellata*) and kāmahi (*Weinmannia racemosa*) with smaller numbers of Hall's tōtara, and are relatively species-poor compared with the northern forests discussed in the two previous examples. Possums spread into these forests in the 1950s with extensive dieback being observed by the late 1960s and early 1970s. Possum control was undertaken in the early 1970s but with relatively low kills (55–56%). Further dieback in the 1980s resulted in the implementation of intensive possum control in these two valleys from 1988 with an initial c. 70% reduction in possum numbers. After further possum control operations in 1995 and 1996, possum numbers had been reduced to residual trap-catch-index levels of 1.8% and 3.5% for the Otira and Deception valleys, respectively.

Analysis of vegetation response to possum control occurred in two phases. Permanent vegetation plots were established in 1988 and remeasured in 1993. This included assessment of canopy condition and mortality (Smale *et al.* 1993). Over this 5-year period there was no detectable improvement in canopy condition for the four major possum-preferred species (southern rātā, kāmahi, Hall's tōtara, fuchsia) and, despite possum control, Hall's tōtara and southern rātā showed a significant decline in canopy condition in both catchments. Four permanent foliar-browse-index lines were established in 1994 in the Otira and Deception valleys for māhoe, fuchsia, wineberry (*Aristotelia serrata*), and patē, and remeasured in 1997 (Tubbs 1997). Overall canopy condition appeared to improve over the study period, but not significantly. There was also an indication of reduced stem use by possums for fuchsia and patē, and an increase in use for wineberry, but the difference was again not significant.

Loranthaceous mistletoes (de Lange & Norton 1997)

There has been considerable debate about the interaction between possums and mistletoes in New Zealand, although it is clear that, at least in some situations, possums are important browsers of mistletoe foliage (see papers in de Lange & Norton 1997). While still limited, there is a growing body of information that quantifies the response of mistletoes to possum removal. In several parts of New Zealand, host trees with mistletoes have been banded to prevent possum access to mistletoe plants. In most cases, banding has resulted in a rapid regrowth of the mistletoe, including resprouting in *Peraxilla tetrapetala* and *Tupeia antarctica* from mistletoe tissue within the host tree (Ogle 1997).

Possum poison operations have also resulted in a marked response in mistletoes in some areas. For example, in beech (*Nothofagus* spp.) forests in the South Branch of the Hurunui River, north Canterbury, 87% of marked mistletoes (*Alepis flavida* and *P. tetrapetala*) showed an improvement or no change in health in a possum control (treatment) area compared with 29% in a non-treatment area 1 year after possum control began (A. Grant, Department of Conservation, Christchurch, unpubl. data). Wilson (1984) observed a similar response to possum control in beech forests at Mt Misery, 45 km north-east of the Hurunui site. In Waihaha Forest, central North Island, rapid resprouting of *T. antarctica* from marble leaf (*Carpodetus serratus*) host trees was observed following aerial poisoning of possums in an area where mistletoes had not been previously detected (P. Sweetapple & G. Nugent, Landcare Research, Lincoln, unpubl. data). A similar response in *T. antarctica* was observed on Kapiti Island following possum eradication (Ogle 1997).

Factors influencing vegetation response

The case studies reviewed above suggest that the response of vegetation to possum removal is not necessarily straightforward and that a variety of factors can modify the response. These modifiers relate both to the dynamics of the possum–forest system and to the methods used in assessing vegetation response. The following are some of the likely modifiers of vegetation response to possum control.

Possum population history

Populations of large herbivores introduced into previously unoccupied habitat experience irruptive fluctuations where numbers increase to a peak and then decline rapidly to a lower level, which may or may not be followed by further irruptions depending on resource availability (Caughley 1970). Such fluctuations are thought to be a consequence of the time lag between the increase in population size and food availability, with extensive mortality occurring when herbivore numbers exceed the carrying capacity of the system. Although possums are much smaller than herbivores known to experience irruptive fluctuations, (2.2–3.5 kg cf. > 30 kg), it appears that they also experience a similar initial fluctuation in their abundance (Thomas *et al.* 1993), although numbers may stabilise with time (Brockie 1992).

The timing of possum control in relation to an irruptive fluctuation appears to have important implications for subsequent vegetation recovery. In rātā–kāmahi forests on the west coast of the South Island, Pekelharing & Batcheler (1990) assessed vegetation condition in two areas for 11–15 years after possum poisoning. In one area, possums were poisoned at close to peak density and the basal area of possum-palatable species declined by 21% over the following 15 years. In the second area possum control was delayed for 4 years, and the basal area of possum-palatable plant species declined by 47% over a comparable period. Pekelharing & Batcheler (1990) suggested that possum control at or before possums reach peak densities is critically important for limiting canopy mortality. Clearly the timing will also be critical in influencing the ability of the vegetation to respond to possum control. Areas subject to possum control before possums reach peak numbers and cause widespread dieback of established trees would seem far more likely to return to near the original state than areas in which most of the original canopy has gone. In this latter situation succession is likely to lead to vegetation of a composition different from that existing before possum colonisation (Rose *et al.* 1992; Rogers & Leathwick 1997).

Number of possums left after control

As well as vegetation response being influenced by the density of possums before control, it will also be strongly affected by the density of possums after control (i.e., the success of the control operation). Strong evidence for a density-dependent effect of possums on populations of fuchsia has been shown (Fig. 21.2; Pekelharing *et al.* 1998) and similar relationships may occur for other species. However, it has been suggested (Payton *et al.* 1997a) that even when possum numbers are reduced to low levels, this may still be insufficient for vegetation recovery (both regrowth of existing plants and recruitment of new individuals) as the remaining possums continue to affect the remaining individuals of the most-preferred food species.

There is unfortunately little information on the maximum possum densities at which forest regeneration will occur, especially for different forest types or even different plant species (i.e., target densities for control); the observed slow response to possum removal in some situations may be because residual possum numbers are still too high (Smale *et al.* 1993; Payton *et al.* 1997a). However, a slow response to possum control may also reflect a lag in plant response (e.g., the time taken for new individuals to become established). The ability of even low possum numbers, potentially, to continue to limit vegetation recovery is an important issue as, unlike islands such as Kapiti where eradication is possible, eradication is not at present a realistic

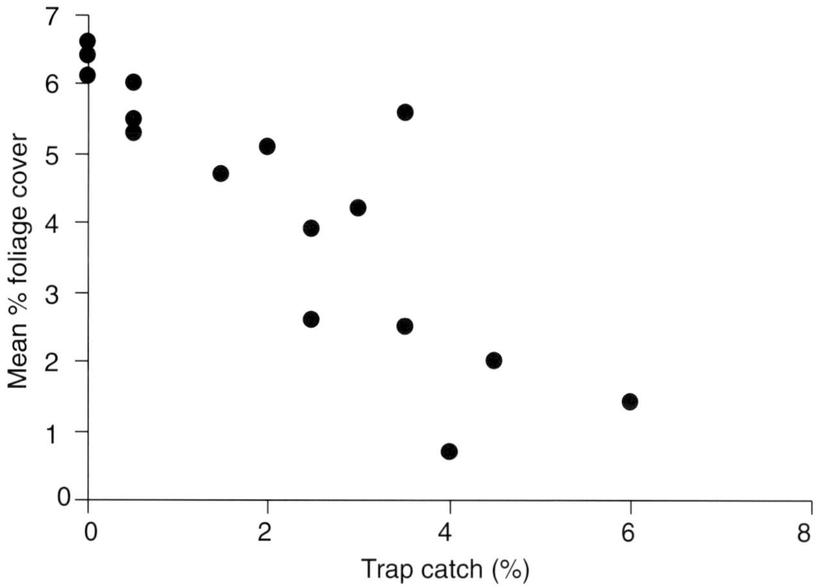

Fig. 21.2
Relationship between *Fuchsia excorticata* foliage cover and possum trap-catch rates for five study sites over 4 or 5 years (Pekelharing *et al.* 1998).

option at most mainland forest sites and possum control therefore aims to hold numbers at low levels. Determining these levels is essential for conservation management as it influences the amount and type of control required (cf. Hickling 1995).

Frequency of possum control

Vegetation response is also likely to differ between situations where possum control is undertaken only once every several years and situations where sustained annual control is used. In west coast mixed angiosperm forest it took 10 years after a one-off control operation for possum numbers to recover to within 20% of pre-poison densities (Pekelharing & Batcheler 1990). In that study, canopy mortality appeared to be ongoing despite previous possum control, suggesting that infrequent control is not sufficient to ensure vegetation recovery. Bellingham *et al.* (1996) also suggest that irregular possum control has little apparent effect in arresting the decline of some palatable species. In contrast, regular sustained control appears to result in sustained vegetation recovery, although there are few data to quantify this. Hickling (1995) has suggested that the relationship between control frequency and vegetation response is dependent on the threshold densities for instigating control and the target densities for control; he suggests that when possum control is infrequent, threshold densities for instigating control can be higher so long as control target densities are lower than for annual control operations. However, this relationship is likely to be dependent on what components of vegetation recovery are considered and what the condition of the vegetation was before control.

Vegetation composition and structure before possum control

While the history of possum impacts, the number of remaining possums, and the frequency of control, all affect vegetation response, a number of aspects of the vegetation itself are likely to have as great or greater influence on vegetation response to possum control. These include the composition and structure of the vegetation before control, presence of propagule sources for recolonisation, and the autecology of individual plant species affected by possums.

Possums do not affect all forests equally and hence the response of the vegetation to possum removal is likely to differ between different forest types. Dieback also appears to be less at sites with higher rainfalls and cooler temperatures (Rogers 1997). Vegetation response is also likely to depend on the stage of stand development. For example, in west coast rātā–kāmahi forests, young even-aged stands, especially those lacking seral species, appear more resilient to dieback than stands containing abundant seral angiosperm species and a high proportion of old canopy trees (Payton 1988; Stewart & Rose 1988).

Reducing possum density in an area is not necessarily sufficient in itself for vegetation recovery especially if some previously abundant species are now scarce or absent. In Orongorongo Valley, Wellington, species such as fuchsia and tītoki (*Alectryon excelsus*) that were important in possum diet in 1946/47 were unimportant or absent in possum diet 21 years later as they had been virtually eliminated from the area (Fitzgerald 1976). Re-establishment of these species will require dispersal from more distant seed sources, which may take some time. Similar problems appear to face mistletoes, especially in many North Island and northern South Island beech forests where they appear to have largely disappeared as a result of possum browse (Ogle 1997). Re-establishment after possum removal will again require dispersal from distant or sparse seed sources.

Different plant species also show different responses to possum browse, and hence different abilities to respond when possums are controlled. For example, dieback of New Zealand cedar (*Libocedrus bidwillii*) appears to be rapid in the face of possum browse as terminal buds are eliminated with little potential for regrowth after possum control (Rogers 1997). In contrast, kāmahi appears able to resprout after repeated possum browse and is more likely to reoccupy sites once possums have been controlled (Bellingham *et al.* 1996). Fuchsia, although often killed by possums at high densities, does recover after possum control even when plants have been almost completely defoliated (Pekelharing *et al.* 1998). The mistletoes *Tupeia antarctica*, *Peraxilla colensoi*, and *P. tetrapetala* have tissue within the host tree that appears to remain physiologically active even in the presence of repeated possum defoliation (Fig. 21.3; D. Norton & B. Fineran, University of Canterbury, unpubl. data). This tissue is able to resprout after possum removal and these species are more likely to respond to possum control than the mistletoes *Alepis flavida* and *Ileostylus micranthus*, which lack such tissue. The size and persistence of species in seed banks and

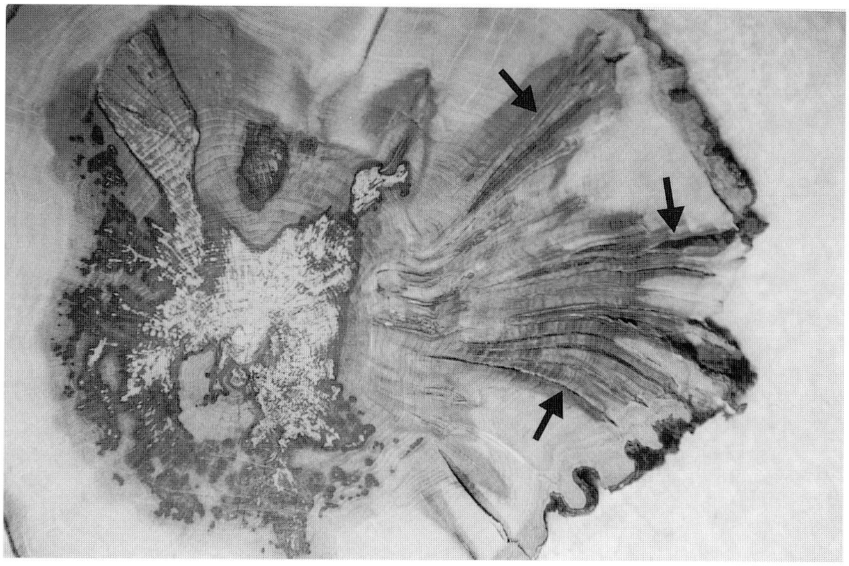

Fig. 21.3
Peraxilla tetrapetala tissue (marked) within the host *Nothofagus solandri* stem.

the different dispersal mechanisms of seeds will also influence their ability to re-establish after possum control, especially for those species that have been most affected by possums.

The regeneration ecology of individual species is also likely to influence vegetation response to possum removal. Regeneration strategies of forest plants range from catastrophic, through gap-phase, to continuous (Veblen 1992) but, typically, forest trees, especially canopy dominants, typically do not normally regenerate under their own canopies. Canopy species such as southern rātā (Stewart & Veblen 1982) and many conifers (Ogden & Stewart 1995) typically regenerate in even-aged cohorts after disturbance to the previous forest canopy. Rogers (1997) suggests that the scale of stand replacement processes for New Zealand cedar is correlated with vulnerability to dieback, with greater dieback in stands that have a larger-scale regeneration regime. The absence of regeneration in species like New Zealand cedar and southern rātā, even after possum control, may therefore simply reflect the stage of stand dynamics rather than an absence of response to possum control.

Presence of other pest species

An additional factor complicating the response of vegetation to possum removal is the impact of other pest species on the forest ecosystem. Even if possums are removed from a forest or reduced to very low levels, ongoing browse by deer on the forest floor has the ability to prevent the re-establishment of many species, especially canopy and seral species (Rogers & Leathwick 1997). Many forest species affected by possum browse are reliant on birds for dispersal of their fruit (e.g., mistletoes). Reduced bird densities as a result of predation by mustelids and rodents have the potential to reduce the dispersal of these species (Ladley & Kelly 1996) even when possums have been reduced to very low numbers.

Widespread canopy collapse in forests that have also been affected by ungulate browse (Rogers & Leathwick 1997) suggests that, even in the absence of both possums and ungulates, vegetation recovery may be very slow or that vegetation succession may proceed towards plant communities that are new to the site. In this latter situation, the vegetation has been forced across a threshold from one metastable state to another, and rapid return to the original state is unlikely without massive management input (e.g., through planting; Hobbs & Norton 1996).

Monitoring methods and time

The limited response of vegetation to a reduction in possum density in some of the case studies reviewed above may simply reflect the limited time that has elapsed since possum control was instigated. This is further compounded by a lack of statistical power in some monitoring programmes (see below), which makes it difficult to detect significant vegetation responses over short time periods. With the widespread adoption of the foliar browse index (Payton *et al.* 1997b) as the main method for monitoring the success of possum control operations in improving canopy condition, the selection of monitored species becomes critical. Possum diet is not uniform in space or time (Fitzgerald 1976; Green 1984; Cowan 1990) and species choice needs to be based on good information on possum diet at the site in question; it should not necessarily be extrapolated from other sites. Poor selection of indicator species may limit the ability to detect vegetation response to possum control.

How can we better study vegetation response to possum control?

To better understand the response of vegetation to possum control it is essential that monitoring programmes are properly designed (Norton 1996). Key problems with much monitoring are a lack of sufficient statistical power to detect changes in the vegetation and a lack of sufficient information on conditions before possum control and at sites without control with which to compare changes at the possum control site. Statistical power is the probability that the null hypothesis will be rejected when it is false (Fairweather 1991) and can result in no statistically significant effect being found in monitoring despite an effect actually occurring. The power of a monitoring programme is strongly related to the sample size and power analysis can be used a priori to determine appropriate sample sizes for monitoring (Fairweather 1991). Use of a BACI (before-after, control-impact) design (Underwood 1993) makes it possible to separate out effects that might be common to both the treatment (impact) and non-treatment (control) sites (e.g., due to climate change) from those that relate to possum control.

There is also an urgent need to understand better the relationship between possum density after control and vegetation response. The Department of Conservation often uses five catches per 100 trap nights as an arbitrary target for possum control operations (C. Robertson, Department of Conservation, Hokitika, pers. comm.), but this number has little ecological basis. Properly replicated studies are urgently required to provide conservation managers with information on the maximum tolerable possum densities for vegetation recovery. These densities are likely to vary with forest type and with vegetation condition before control. Without this information it is very difficult for conservation managers to make informed decisions about the levels of control that are required.

Finally, it is important that choice of indicator species used for monitoring the health and status of a forest is based on a good understanding of possum diet at the site in question rather than through extrapolation from other sites. This is especially important when using foliar browse index as the main method for quantifying vegetation response to possum control as it focuses on only a small number of species. However, foliage is only one aspect of the vegetation that is affected by possums. It is important that monitoring programmes also consider other aspects of the vegetation, such as flowering and fruiting, and fungal biomass, to increase our understanding of vegetation response to possum removal.

Conclusions

If we are to protect or restore forest ecosystems in mainland New Zealand it is essential that we obtain good information on the densities of possums that are compatible with vegetation recovery. To achieve this the Department of Conservation should (a) undertake a comprehensive audit of some of its current possum control operations to ensure that monitoring methods are adequate to provide statistically meaningful information on vegetation response, and (b) establish a series of properly designed and replicated studies in a range of different forest types to determine maximum tolerable possum densities for vegetation recovery.

Adaptive management (Holling 1978; often called "research by management" in New Zealand), whereby management and research are undertaken in a heuristic manner, is essential if we are to increase our understanding of possum–vegetation interaction after possum control. Key questions that require further research include: how much initial control is needed? how few possums can be tolerated? how should control be distributed spatially? and how can we optimise the benefits without increasing the costs too far?

To obtain the full benefits of possum control, it is essential that it not be undertaken in isolation from other animal control operations. For example, control of deer is important to ensure that species that have declined because of possum browse are able to successfully regenerate and re-establish in to the forest canopy. Control of predators, especially stoats and rats, is also important to ensure that dispersal of species with bird-dispersed fruit occurs successfully. Integrated pest control, especially within the framework of the mainland island concept, is essential if we are to achieve comprehensive ecosystem restoration in New Zealand.

Summary

- Vegetation response to possum control is not straightforward and a variety of factors can modify the response including:
  - the density of possums prior to and subsequent to control;
  - the frequency of control and resultant fluctuation in possum numbers;
  - the composition and structure of the vegetation before control, the presence of propagule sources for recolonisation, and the autecology of individual plant species affected by possums; and
  - influence of other herbivores on vegetation response.
- The limited response of vegetation to a reduction in possum density in some situations may simply reflect the limited time that has elapsed since possum control was instigated.
- To understand better the response of vegetation to possum control it is essential that monitoring programmes are properly designed and replicated, and have sufficient statistical power to detect change.
- There is an urgent need to better understand the maximum tolerable possum densities for vegetation recovery.
- It is important that choice of indicator species used for monitoring the health and status of a forest is based on a good understanding of possum diet at the site in question rather than through extrapolation from other sites.

Acknowledgments

This chapter is dedicated to John Holloway who did so much to ensure that conservation management was based on good science. I am very appreciative of the helpful discussions with and or comments from Peter de Lange, Ian James, Craig Miller, Tom Montague, Graham Nugent, John Parkes, Campbell Robertson, and Mark Smale in the preparation of this chapter, and to Andrew Grant and Peter Sweetapple for use of their unpublished data.

References

Allen, R. B.; Fitzgerald, A. E.; Efford, M. G. 1997: Long-term changes and seasonal patterns in possum (*Trichosurus vulpecula*) leaf diet, Orongorongo Valley, Wellington, New Zealand. *New Zealand Journal of Ecology 21*: 181–186.

Atkinson, I. A. E. 1992: Effects of possums on the vegetation of Kapiti Island and changes following possum eradication. DSIR Land Resources Contract Report 92/52 (unpublished) 68 p.

Bellingham, P. J.; Wiser, S. K.; Hall, G. M. J.; Alley, J. C.; Allen, R. B.; Suisted, P. A. 1996: Impacts of possum browsing on the long-term maintenance of forest biodiversity. Landcare Research Contract Report LC9697/47 (unpublished) 56 p.

Brockie, R. E. 1992: A living New Zealand forest. Auckland, David Bateman.

Brown, K.; Innes, J.; Shorten, R. 1993: Evidence that possums prey on and scavenge bird's eggs, birds and mammals. *Notornis 40*: 169–177.

Caughley, G. 1970: Eruption of ungulate populations, with emphasis on Himalayan thar in New Zealand. *Ecology 51*: 53–72.

Cowan, P. E. 1990: Brushtail possum. *In*: King, C. M. *ed.* The handbook of New Zealand mammals. Auckland, Oxford University Press. Pp. 68–98.

Cowan, P. E. 1993: Effects of intensive trapping on breeding and age structure of brushtail possums, *Trichosurus vulpecula*, on Kapiti Island, New Zealand. *New Zealand Journal of Zoology 20*: 1–11.

de Lange, P. J.; Norton, D. A. *ed.* 1997: New Zealand's loranthaceous mistletoes. Proceedings of a workshop hosted by Threatened Species Unit, Department of Conservation, Cass, 17–20 July 1995. Wellington, Department of Conservation. 220 p.

Fairweather, P. G. 1991: Statistical power and design requirements for environmental monitoring. *Australian Journal of Marine and Freshwater Research 42*: 555–568.

Fitzgerald, A. E. 1976: Diet of the opossum *Trichosurus vulpecula* (Kerr) in the Orongorongo Valley, Wellington, New Zealand, in relation to food-plant availability. *New Zealand Journal of Zoology 3*: 399–419.

Green, W. Q. 1984: A review of ecological studies relevant to management of the common brushtail possum. *In:* Smith, A. P.; Hume, I. D. *ed.* Possums and gliders. Chipping Norton, NSW, Surrey Beatty in assoc. with the Australian Mammal Society. Pp. 483–499.

Hickling, G. J. 1995: Action thresholds and target densities for possum pest management. *In:* O'Donnell, C. F. J. *comp.* Possums as conservation pests. Proceedings of an NSSC Workshop . . . 29–30 November 1994. Wellington, Department of Conservation. Pp. 47–52.

Hobbs, R. J.; Norton, D. A. 1996: Towards a conceptual framework for restoration ecology. *Restoration Ecology 4*: 93–110.

Holling, C. S. *ed.* 1978: Adaptive environmental assessment and management. *International series on applied systems analysis No. 3.* Chichester, John Wiley. 377 p.

Ladley, J. J.; Kelly, D. 1996: Dispersal, germination and survival of New Zealand mistletoes (Loranthaceae): dependence on birds. *New Zealand Journal of Ecology 20*: 69–79.

Norton, D. A. 1996: Monitoring biodiversity in New Zealand's terrestrial ecosystems. *In:* McFadgen, B.; Simpson, P. *comp.* Biodiversity. Wellington, Department of Conservation. Pp. 19–41.

Ogden, J.; Stewart, G. H. 1995: Community dynamics of the New Zealand conifers. *In:* Enright, N. J.; Hill, R. S. *ed.* Ecology of southern conifers. Carlton, Victoria, Melbourne University Press. Pp 81–119.

Ogle, C. C. 1997: Evidence for the impacts of possums on mistletoes. *In:* de Lange, P. J.; Norton, D. A. *ed.* New Zealand's loranthaceous mistletoes. Proceedings of a workshop hosted by Threatened Species Unit, Department of Conservation, Cass, 17–20 July 1995. Wellington, Department of Conservation. Pp. 141–147.

Parkes, J.; Baker, A. E.; Erickson, K. 1997: Possum control by the Department of Conservation: Background, issues, and results from 1993 to 1995. Wellington, Department of Conservation. 40 p.

Payton, I. J. 1988: Canopy closure, a factor in rata (*Metrosideros*) – kamahi (*Weinmannia*) forest dieback in Westland, New Zealand. *New Zealand Journal of Ecology 11*: 39–50.

Payton, I. J.; Forester, L.; Frampton, C. M.; Thomas, M. D. 1997a: Response of selected tree species to culling of introduced Australian brushtail possums *Trichosurus vulpecula* at Waipoua Forest, Northland, New Zealand. *Biological Conservation 81*: 247–255.

Payton, I. L.; Pekelharing, C. J.; Frampton, C. M. 1997b: Foliar browse index: a method for monitoring possum damage to forests and rare or endangered plants. Landcare Research Contract Report LC9697/60 (unpublished) 64 p.

Pekelharing, C. J.; Batcheler, C. L. 1990: The effects of control of brushtail possums (*Trichosurus vulpecula*) on condition of a southern rata/kamahi (*Metrosideros umbellata*/*Weinmannia racemosa*) forest canopy in Westland, New Zealand. *New Zealand Journal of Ecology 13*: 73–82.

Pekelharing, C. J.; Parkes, J. P.; Barker, R. J. 1998: Possum (*Trichosurus vulpecula*) densities and impacts on fuchsia (*Fuchsia excorticata*) in south Westland, New Zealand. *New Zealand Journal of Ecology 22*: 197–203.

Rogers, G. M. 1997: Trends in health of pahautea and Hall's totara in relation to possum control in central North Island. *Science for Conservation 52*. Wellington, Department of Conservation. 49 p.

Rogers, G. M.; Leathwick, J. R. 1997: Factors predisposing forests to canopy collapse in the southern Ruahine Range, New Zealand. *Biological Conservation 80*: 325–338.

Rose, A. B.; Pekelharing, C. J.; Platt, K. H. 1992: Magnitude of canopy dieback and implications for conservation of southern rata-kamahi (*Metrosideros umbellata* - *Weinmannia racemosa*) forests, central Westland, New Zealand. *New Zealand Journal of Ecology 16*: 23–32.

Smale, M. C.; Rose, A. B.; Frampton, C. M.; Owen, H. J. 1993: The efficacy of possum control in reducing forest dieback in the Otira and Deception catchments, central Westland. Landcare Research Contract Report LC9293/110 (unpublished) 13 p.

Stewart, G. H.; Rose, A. B. 1988: Factors predisposing rata-kamahi (*Metrosideros umbellata*- *Weinmannia racemosa*) forests to canopy dieback, Westland, New Zealand. *Geojournal 17*: 217–223.

Stewart, G. H.; Veblen, T. T. 1982: Regeneration patterns in southern rata (*Metrosideros umbellata*) - kamahi (*Weinmannia racemosa*) forest in central Westland, New Zealand. *New Zealand Journal of Botany 20*: 55–72.

Thomas, M. D.; Hickling, G. J.; Coleman, J. D.; Pracy, L. T. 1993: Long-term trends in possum numbers at Pararaki: Evidence of an irruptive fluctuation. *New Zealand Journal of Ecology 17*: 29–34.

Tubbs, M. 1997: Report on possum damage to selected plant indicators in the Otira and Deception Valleys, West Coast 1997. Unpublished West Coast Conservancy Report. Hokitika, Department of Conservation.

Underwood, A. J. 1993: The mechanics of spatially replicated sampling programmes to detect environmental impacts in a variable world. *Australian Journal of Ecology 18*: 99–116.

Veblen, T. T. 1992: Regeneration dynamics. *In:* Glenn-Lewin, D. C.; Peet, R. K.; Veblen, T. T. *ed.* Plant succession, theory and prediction. London, Chapman & Hall. Pp. 152–187.

Wilson, P. R. 1984: The effects of possums on mistletoe on Mt Misery, Nelson Lakes National Park. *In:* Dingwell, P. R. *ed.* Protection and parks. Essays in the preservation of natural values in protected areas. Proceedings of Section A4e, 15th Pacific Science Congress, February 1983. *Information Series No. 12*. Wellington, Department of Lands and Survey. Pp. 53–60.

CHAPTER TWENTY-TWO

Do Native Wildlife Benefit From Possum Control?

Clare Veltman

Possums consume snails and insects, eat birds' eggs, prey on chicks (see Chapter 11) and change the forests that support native animals (Chapter 10) so possum control is carried out on conservation lands to reduce these direct and indirect threats to native wildlife. Once possum control is underway to reduce this consumption, the demographic statistics of animal populations affected by possums can be expected to change. For example, survival or reproductive success might rise. A beneficial change in either the death rate or birth rate could lead to higher local densities or lowered risks of extinction in the treated patch. As a result of such changes, other patches might be colonised by native animals, once they have been cleared of possums, by immigration from a rapidly expanding source population. Any of these responses may be scaled to the magnitude of possum control achieved by conservation managers, making it difficult to detect benefits at some residual possum densities, but fewer possums should mean more native animals in the long run. After all, possums accounted for about 5% of the total estimated animal biomass in the Orongorongo Valley and made up 99% of the biomass of mammals living there (Brockie & Moeed 1986). If a figure of 25 kg/ha is representative of the possum biomass in other broadleaf-podocarp forests in New Zealand, then reduction in consumption by this mass of omnivorous marsupials liberates tonnes of forage for other herbivores, and removes a predator from the community.

Even though data of the sort I described above help in advocating a continued investment in reducing possum numbers (Hickling 1995), there are few case studies to draw on. Several strategies for obtaining evidence that populations of native animals could or do benefit from possum control can be identified. One might measure the abundance of a species of interest at many sites, and look for a negative correlation with the density of possums at those sites (a "spatial correlation" strategy). If some of the sites have never had possums — an increasingly unlikely scenario — so much the better (Innes 1995). Alternatively, if time-series data were available, we might correlate variations in abundance of some native animals at one site with variations in possum abundance at the same site, (a "temporal correlation" strategy). If we observed reductions in numbers or reproduction by native wildlife in years of high possum densities, we could form directional predictions for the outcomes of possum control. A third approach is to quantify the abundance of native wildlife before a possum control operation and continue afterwards to give a before-and-after comparison (a "response-to-management" strategy). Without a non-treatment area, however, inference is limited. For strong inference, we require surveillance at several sites, followed by possum control at half the sites chosen randomly, and continued measurements of animal abundance at all sites (an "experimental" strategy).

In this chapter, I review cases in which native animals were studied to provide evidence that possum control benefits their populations. It turns out that most examples are from response-to-management research, and an experimental study of the question seems not to have been published. The question of whether native fauna benefit from possum control may become academic, because ongoing control of possums to low numbers at some sites is achieved using tools that also reduce the numbers of rodents. At such sites, an increase in numbers of native animals cannot be attributed to control of any single pest.

Evidence from spatial correlations

Possum densities vary between sites for all sorts of reasons. Differences in climate, forest type, time since invasion, previous control efforts, and chance, will all contribute to instantaneous density estimates. A strict reading of this chapter's role in the book requires us to concentrate on studies in which possum densities varied because of management actions. Not only does this directly address the question of what happens after possum control, but it also lessens the risk of attributing increases in wildlife performance to decreases in possum numbers when an underlying difference between forest communities caused the apparent correlation.

Desirable though it may be, a correlation based on sites where possum densities are the outcome of management has not yet been reported. However, researchers have tracked a possum invasion wavefront on the west coast of the South Island, and investigated bird populations in colonised and uncolonised areas. It was assumed that the uncolonised areas were not systematically different in any way, but simply free of possums for historical reasons.

Kākā in South Westland

To assess the impact of possums as they invaded forests in South Westland, the New Zealand Wildlife Service counted kākā (*Nestor meridionalis*) at 15 study sites between Cook River and Big Bay during the summers of 1983 to 1986 (Rose *et al.* 1990). Observers walked along defined transects, counting all the kākā seen and heard during 5-minute listening periods from listening points at 500-m intervals. A total of 1692 counts were gathered. Afterwards, the sites were classified by their history of possum occupation into four groups: possums absent; possums present less than 10 years and at low density; possums present 10–30 years and at moderate density; and possums present more than 30 years and at high densities. Kākā abundance was greatest at sites not yet colonised by possums and those with the shortest history of possum occupation, and declined with increasing possum densities (Fig. 22.1). Canopy damage was also evident in the sites with post-peak possum densities (Rose *et al.* 1993), consistent with a pattern of forest change that could imperil the parrots.

The possibility that other threats to kākā may have covaried with possum abundance cannot be ignored. Research in beech (*Nothofagus*) forest near Nelson has uncovered the seriousness of predation by stoats (*Mustela erminea*) at kākā nests (Wilson *et al.* 1998). If possum and stoat numbers varied in similar ways between the sites, then either of them might have been responsible for the observed pattern of kākā abundance. Whether kākā in South Westland were directly affected by competition and predation by possums, or indirectly through possum-mediated stoat predation, cannot be determined in retrospect. Inference could be drawn from counts of kākā after possum control in forests lacking a stoat population. This was the case on Kapiti Island, and I return to this point later in the chapter.

Evidence from temporal correlations

Only two long-running studies of possum abundance exist that might allow use of the temporal correlation strategy to investigate how native animals responded to cyclical changes in possum abundance. Unfortunately, no surveillance of native wildlife was done in the Pararaki catchment of Haurangi State Forest Park where possums have been monitored every year since 1965 (Thomas *et al.* 1993). The

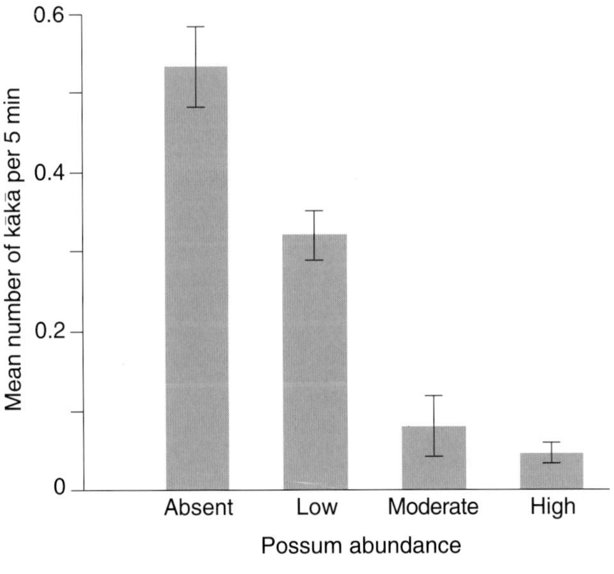

Fig. 22.1
Mean number of kākā counted during 5-minute intervals at observation points in South Westland forests ranging from sites where possums were absent to sites with high possum densities. Standard errors shown. Data provided by C. O'Donnell, Department of Conservation, Christchurch.

second time-series of possum densities was measured in the Orongorongo Valley in Rimutaka State Forest Park (Brockie 1992). The prediction is that in years when possum numbers were relatively high, affected species of native animals were less abundant. The relationship might be lagged, such that a response to rising numbers of possums was delayed by months or years. Although forest birds and insects were surveyed in Orongorongo Valley by various researchers at different times, no dataset was available for examining any temporal correlation with possum abundance.

Evidence from response-to-management studies

Studies of animal abundance before and after possum control are surprisingly rare considering the number of operations now undertaken. In 1996–1997, 139 possum control operations were carried out on conservation lands, and 172 were planned for 1997–1998 (Department of Conservation 1997). There are at least four reasons (aside from cost and the limited usefulness of results gained by this research strategy) to explain the dearth of examples.

First, some operations to control possums were organised so quickly that there was insufficient time to measure numbers of native animals beforehand. This happened at Rangitoto Island, where forest birds were counted on the day before 1080 poison was dropped and counted again at 6-week intervals for 1 year (Miller & Anderson 1992). Because bird numbers in New Zealand forests fluctuate from month to month, and also from year to year (e.g., Spurr *et al.* 1992), it was not possible to infer from Miller & Anderson's data that possum control caused the observed increases in tūī (*Prosthemadera novaeseelandiae*) and silvereye (*Zosterops lateralis*) numbers on the island. A reduction in rat numbers may have had additional effects.

Second, in many forest restoration projects pest control was directed at several species that share similar ecological effects. For example, on Chatham Island, parea (*Hemiphaga novaeseelandiae chathamensis*) numbers rose steadily between July 1990 and December 1994 (Grant *et al.* 1997) after control of cats, rats, and possums began in 1989. Although possums are nest predators of pigeons on the mainland (James & Clout 1996), the improvement in parea numbers could not be related to reduction in possums because control procedures killed all predators. At Wenderholm Regional Park, north of Auckland, possums have been controlled since 1986 using trapping and cyanide poisoning. Rat control using the anticoagulant brodifacoum in wax baits placed in tunnels began in 1992. Wētā droppings were caught from 1996 in large canvas funnels placed under trees at Wenderholm and also at nearby Loch Amber, where there was no pest control. The effects of possum control could not be separated from those of rat control (A. Dijkgraaf, Department of Conservation, Wanganui, pers comm.). The total number of wētā droppings caught below trees was higher at Wenderholm, suggesting that reduction in rat and possum numbers was beneficial for these insects (Fig. 22.2).

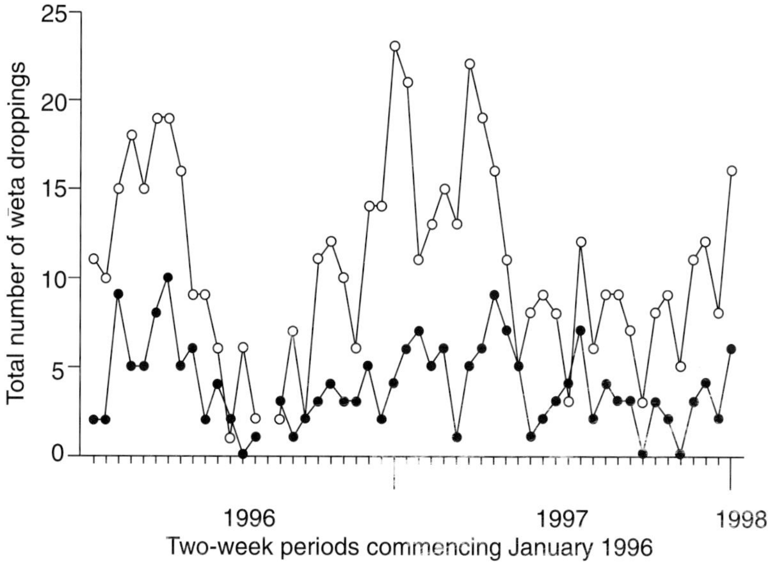

Fig. 22.2
Total numbers of wētā droppings collected in litterfall traps placed under trees in remnant forest reserves near Auckland (o — Wenderholm; • — Loch Amber). Data span the period 22.1.96 to 6.1.98 and were provided by A. Dijkgraaf, Department of Conservation, Wanganui.

Third, the bait+toxin combinations commonly employed to reduce possum numbers also kill stoats, rats, and deer. When 1080 and brodifacoum in cereal baits were used to kill possums and ship rats (*Rattus rattus*) in three forest patches (of which Mapara is perhaps the best known) in the North Island, kōkako (*Callaeas cinerea wilsoni*) raised more chicks. Both possums and rats are predators of kōkako nests (Brown *et al.* 1993). Chick production fell again when poisoning was halted as part of the experiment (Innes *et al.* 1999). As was the case with parea, clear benefits accrued from multi-pest control but the particular benefit (if any) from controlling possums cannot be isolated from the overall benefit.

A final reason for the lack of evidence from response-to-management studies may be the possibility that the procedure used to kill possums itself directly reduced numbers of native forest animals (Spurr 1993; Innes 1995; Chapter 16). Counts have been made of birds, insects, bats, and frogs before and after use of toxins to quantify the danger of non-target poisoning, and much effort is now directed at detecting and mitigating non-target deaths (e.g., Powlesland *et al.* 1999). However, mortality over a relatively short time after presentation of the toxin would have to be extremely large to have a deleterious effect on native animal populations in the long term. Just the opposite can be expected from density-dependent responses amongst survivors.

Because modern methods for pest control do not allow outcomes to be partitioned between possum and rat reduction, most opportunities to use the response-to-management strategy to measure changes in native fauna come from operations conducted prior to adoption of broad spectrum anticoagulant poisons. Forest birds were counted in the Callery catchment of Westland National Park for 3 years before and 3 years after an aerial 1080 poison operation in June 1983 that killed 82% of the possums. Any reduction in rat numbers after distribution of 1080 in carrot baits would have been short-lived relative to the timescale of the bird counts. Similar counts were also completed 2 years before and after use of 1080 in the Copland catchment in June 1986, in an operation that killed 62% of possums. In neither catchment was any change in bird abundance able to be attributed to possum control (Spurr 1989), though re-analysis with multivariate approaches would permit closer scrutiny of patterns in the data.

What follows are four accounts of research in which native animals were counted before or at the time possum control was applied, and again afterwards. In each case, the pest control was targeted at possums or probably affected possums most of all. In the first example, forest bird numbers were measured from the time intensive possum control began on Kapiti Island until well after possums were eradicated. The second example involves before-and-after counts of native snails at Charming Creek and other South Island sites that received possum control, and may indicate release from predation after a reduction in possum numbers. The third example raises the intriguing possibility that herbivorous insects may have multiplied in a block of Whirinaki Forest after possums were knocked down to 7% of their initial abundance. In the fourth example, Australian researchers observed higher egg and nestling survival rates at glossy black-cockatoo (*Calyptorhynchus lathami*) nests protected from possums on Kangaroo Island.

Kapiti Island

The history of possum control and the story of how they were eradicated from Kapiti Island was told in Cowan (1992). Possum trappers removed 15 284 possums from the island between 1960 and 1968, at which point trapping ceased. From 1975 researchers estimated possum abundance and measured damage to plants and effects on animals, and it was decided in 1980 that possum control should begin again. A decision was made to use leg-hold traps set above the ground on wooden ramps, because other trap configurations and poisons risked killing little spotted kiwi (*Apteryx owenii*). This meant that the control procedure killed only a few rats, and any changes on the island after a reduction in possum abundance would not be confounded with the effects of removing other pests.

As many possums were killed between February 1980 and October 1982 (15 631) as had been removed between 1960 and 1968. From the disappearance rates of marked possums, Cowan estimated that the island's population was reduced by 70–75% during this period. In the ensuing 3 years, possums were eradicated.

Tim Lovegrove arrived on Kapiti Island in 1981 to follow the fates of North Island saddlebacks (*Philesturnus carunculatus rufusater*), which had been released on the island to start a new population. In February 1982 he set up three transects along Trig Track, each one 500 m long, on which to measure saddleback abundance. The first transect was in tall seral and coastal forest from 25 to 75 m a.s.l., with high species diversity. Transect 2 followed Trig Track through tall seral forest including kānuka, rātā, tawa, and māhoe from 210 to 395 m a.s.l.. Transect 3 was near the apex of the island in forest comprising kāmahi, rātā, māhoe, raukawa, and five-finger at 410–500 m asl. Unfortunately saddlebacks suffered predation by Norway rats (*Rattus norvegicus*) and their abundance remained very low. However, Lovegrove counted all birds he saw and heard within 10 m on either side of the transects as he walked up and then down the track on 5 different days per field trip (T. Lovegrove, Auckland Regional Council, pers. comm.). From the 10 counts obtained like this for each transect, it was possible to estimate bird abundance in 1 ha of forest. This research took place in February and March every year from 1982 to 1988.

Bird numbers on all three transects increased during the years of intense possum control (Fig. 22.3a), but were always highest on Transect 1 near the coast. There were marked increases after the 1983 count, and again after the 1984 count, and transects exhibited similar abundance patterns. By 1988 there were approximately 30 birds per hectare on Kapiti Island, about twice the average density that was estimated in 1982. The most abundant of the 16 bird species counted during Tim Lovegrove's study were whiteheads (*Mohoua albicilla*), followed by robins (*Petroica australis*), tūī, and bellbirds (*Anthornis melanura*).

Pigeon (*Hemiphaga novaeseelandiae*) numbers increased sixfold on Transect 2 in the mid-section of the island after possum control (Fig. 22.3b). The absence of ship rats from Kapiti Island may have permitted this recovery, since elsewhere ship rats rob pigeon nests (Clout *et al.* 1995). There were too few kākā away from the coastal forest to test the hypothesis that possums affected their abundance, but along Transect 1 their numbers increased (Fig. 22.3c). Given the absence of stoats on Kapiti Island, this observation adds some support to the inference that low kākā numbers at sites with high possum densities in South Westland were directly due to possum impacts.

Lovegrove's research began after 74% (11 535 of 15 631 from Cowan 1992, table 1) of the 1980–82 possum tally had already been taken. Thus, the initial bird densities he measured may already have benefited from reductions in possum densities on the island.

Charming Creek

Possums consume snails in what may be a culturally transmitted foraging pattern (K. Walker, Department of Conservation, Nelson, pers. comm.). At Charming Creek in Mokihinui Forest in North Westland, *Powelliphanta* land snails almost disappeared from an area of beech forest in which the density of live

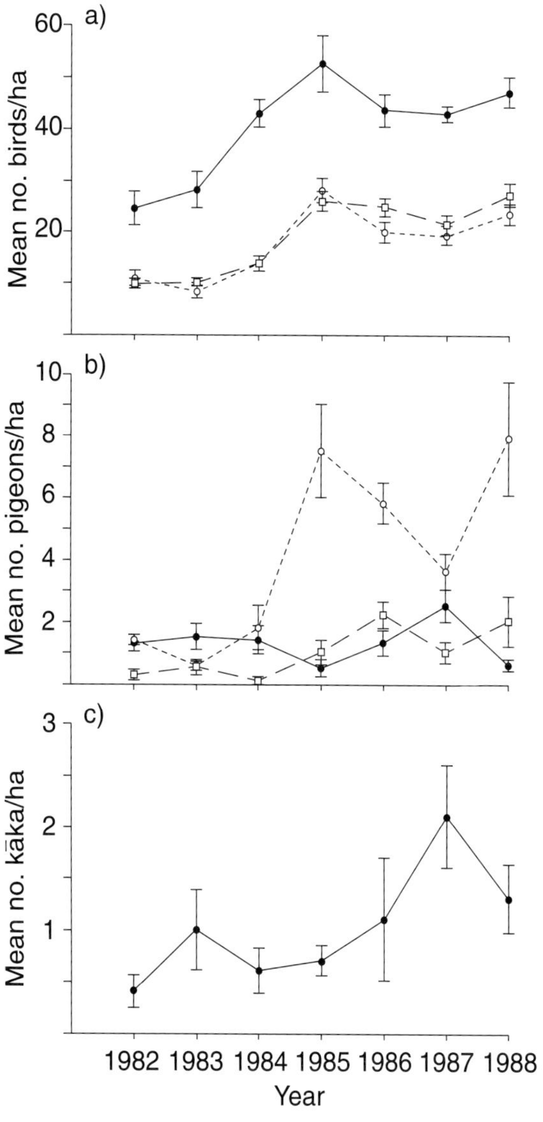

Fig. 22.3
(a) Mean total bird numbers per hectare on Kapiti Island from 1982 to 1988. Possum eradication was completed in 1986. (● — Transect 1; ○ — Transect 2; □ — Transect 3). (b) Mean number of New Zealand pigeons per hectare on Kapiti Island from 1982 to 1988. (● — Transect 1; ○ — Transect 2; □ — Transect 3). (c) Mean number of kākā per hectare along Transect 1 in coastal forest on Kapiti Island from 1982 to 1988. Data provided by T. Lovegrove, Auckland Regional Council.

snails was monitored from 1984 (Fig. 22.4). Leaf litter on the forest floor was carefully searched, and live snails were tagged, mapped, and replaced. The plot size was 225 m^2 in 1984, 300 m^2 in 1985, and 500 m^2 every year after that. The decline in snail abundance coincided with a decline in commercial interest in possums in the area (i.e., a probable increase in possum density). From January 1993 possums were killed using traps, cyanide, and hand-laid 1080 pollard baits over a 1-km^2 area around the plot. Indices of possum abundance before and after control work in the first 2 years confirmed that these control measures were lowering possum abundance at the plot (K. Walker, Department of Conservation, Nelson, pers. comm.). Snail numbers on another 375-m^2 plot at Charming Creek were measured from 1993 to provide a non-treatment comparison. No snail counts were made on either plot in 1994.

By the autumn of 1997 snail numbers appeared to be recovering in the treated area (Fig. 22.5). However, any comparison with the untreated area is overly affected by two values: the short-lived rise seen in the treated area in 1995, and the relatively high initial number of snails in the non-treatment area. A conclusion that possum control was beneficial for Charming Creek land snails from these data alone would be premature, but tempting.

Snails were also counted at other South Island localities undergoing possum control. Following possum control, snail densities multiplied at three sites in high-altitude silver beech (*Nothofagus menziesii*) forest near Flora Stream on Mt Arthur,

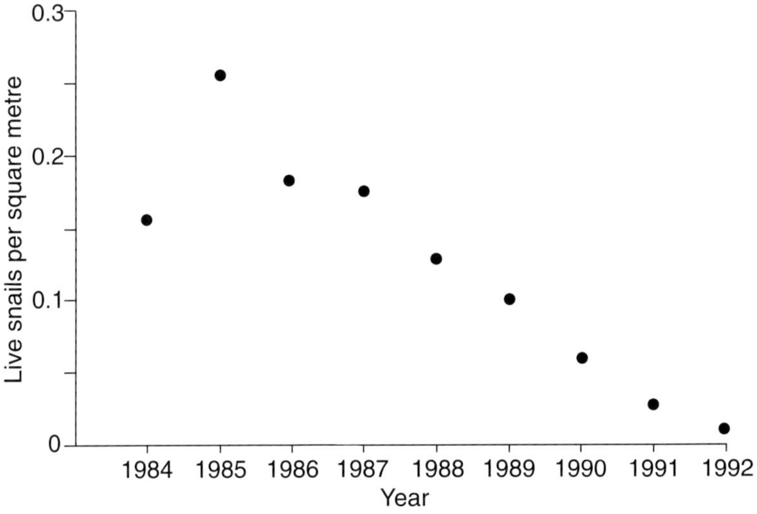

Fig. 22.4
Estimated density of snails on one plot at Charming Creek. Data provided by K. Walker, Department of Conservation, Nelson.

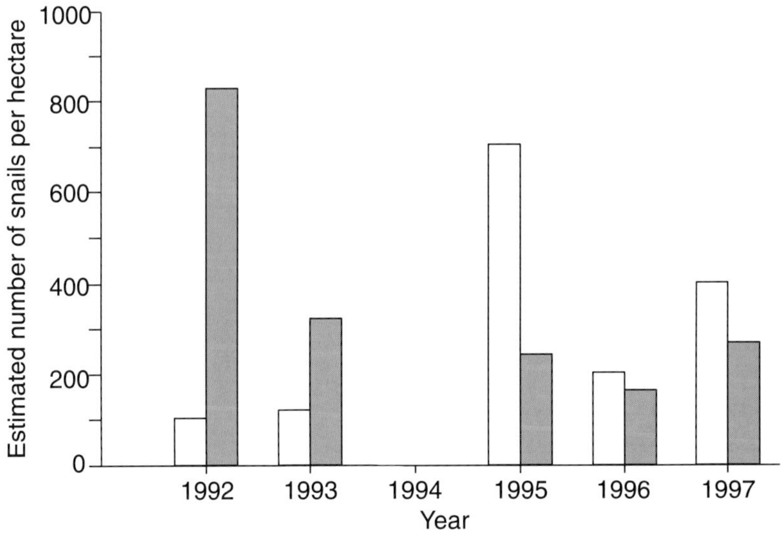

Fig. 22.5
Estimated number of snails per hectare at two sites near Charming Creek. Possums were trapped and poisoned on the treatment plot (open bars) from 1993, but were left unmanaged on the control plot (solid bars). Data provided by K. Walker, Department of Conservation, Nelson.

Kahurangi National Park (Table 22.1). Annual possum control was carried out using trapping and cyanide in the first two seasons, and cholecalciferol, brodifacoum, or encapsulated cyanide, in bait stations after that (K. Walker, pers. comm.).

Whirinaki Forest Park

Early in 1994 litterfall traps were placed under nine northern rātā trees (and one tawa tree) in an area of about 5 ha near Kakiraoa Stream in Whirinaki Forest Park. All the material collected at monthly intervals in three periods (March 1994 to February 1995, August 1995 to July 1996, October 1996 to September 1997) was inspected for possum and insect damage (Fig. 22.6) (M. Numata, Otago University, Dunedin, pers. comm.). Some commercial hunting of possums happened in the winters of 1994 and 1995, but an all-out effort to knock down possums in September 1996 using 1080 in bait stations reduced the population by 90%. Leaf fall was greatest in summer, and consistent between years (Fig. 22.7). The proportion of undamaged leaves declined, as did the number of possum-damaged leaves, after possum control. Almost all the rātā leaves recovered in the traps after the possum control had been damaged by herbivorous insects (Fig. 22.7). Indirect though they are, these results offer circumstantial evidence of benefits to the insect leaf-browser guild in Whirinaki Forest Park from control of possums, and warrant a follow-up study.

Kangaroo Island

In Australia, glossy black-cockatoos lay a one-egg clutch in large hollows in sugar gum trees (*Eucalyptus cladocalyx*). Only about 200 birds remain in South Australia, in a single population on Kangaroo Island. Breeding by these birds was investigated from 1995 to 1997, and possums were discovered to be taking eggs and nestlings (Garnett *et al.* in press). From March 1996, most nests were protected from possums by fitting a circlet of corrugated iron around the tree trunk and then pruning nearby branches to prevent possums from gaining access from other trees.

The benefits to glossy black-cockatoos were clear-cut. Of 19 eggs lost from 29 nests in unprotected trees, 12 (63%) were taken by possums. In contrast, 30 eggs were lost from 75 protected nests and none of those eggs were consumed by possums. Cockatoos produced 10 nestlings in unprotected trees and lost

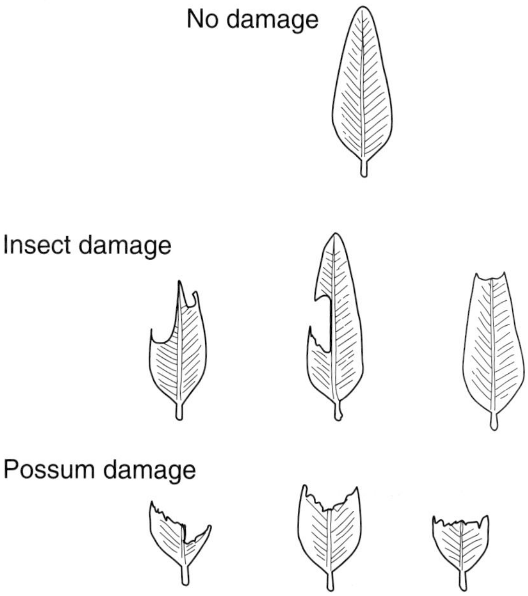

Fig. 22.6
Patterns of damage on rātā leaves, by possums and insects. Figure drawn by M. Numata, University of Otago, Dunedin.

Table 22.1
Densities of live snails per square metre in permanent plots at three sites in Kahurangi National Park, before and after possum control. Data provided by K. Walker, Department of Conservation, Nelson.

	Snails per square metre	
	Before possum trapping (1993)	After possum trapping (1997)
Horseshoe Creek	2	13
Clouston's Mine	2	20
Ryan's Camp	1	9

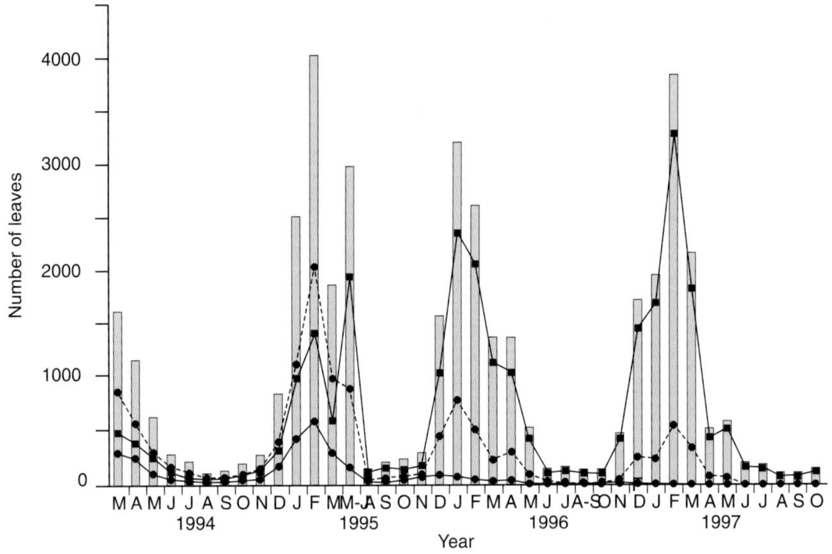

Fig. 22.7
Seasonal patterns of leaf fall and leaf damage in rātā leaves collected in Whirinaki Forest (Bars – total number of leaves; ● – possum damage; ■ – insect damage; Points joined by dotted line show leaves with no damage). Data and figure provided by M. Numata, University of Otago, Dunedin.

3 of them, 2 to possums. On the other hand, none of the 9 nestlings lost from 45 that hatched in protected trees was killed by possums (Garnett *et al.* in press).

Evidence from experiments

A designed experiment with measurements of native wildlife from treated and untreated sites before and after possum control seems never to have been reported. An all-fields search of 1271 references to possum research in the bibliography [online] at URL http://possum.massey.ac.nz retrieved 107 references that used the term "experiment", most of which were physiological experiments or papers published in the Australian *Journal of Experimental Biology and Medical Science.* The terms "bird" and "insect" also did not locate experimental studies of changes in animal abundance caused by manipulations of possum densities.

Do native fauna benefit from possum control?

The evidence that possum control helps native forest animals is weak, perhaps because workers were constrained by logistical and financial reasons to gather data only from treated areas. It appears there was an emphasis on gathering information during management actions, and on learning by doing. The result may be smarter pest control, but how forest food webs change due to possum control has yet to be measured.

It is unlikely that the question can be answered during future pest control projects, since managers increasingly aim to control several pest species at the same sites using bait stations that control both possums and rats. Consumption of poisoned possums and rats, in turn, may lower densities of other predators by secondary poisoning (Alterio 1996). Unless investigators explicitly test for the effects of possums on populations of native wildlife, it appears we will have to find the answer from as yet unpublished historical sources.

This is a pity because the question has not lost its intrinsic interest. Our best test of possums' roles in food webs come from their removal. Over what time period is the fauna of a forest ecosystem perturbed by the "echo" of possums past and by their control? Is a smaller perturbation predicted for beech forests, in which possum populations reach much lower densities (Owen & Norton 1995), or will possum control in beech forests actually lead to larger increases in invertebrate abundance than elsewhere as a result of their relatively large importance as prey for possums? How much vegetation recovery is needed to cause detectable increases in native herbivore populations?

Finally, managers are always under pressure to justify expenditure on possum control and a demonstration of benefits to native fauna would help their case. An insistence that expenditure be targeted at killing possums and not at measuring the results over ecologically meaningful time frames has not helped to amass the evidence they need.

Summary

- Four strategies for obtaining evidence that possum control is beneficial for native wildlife are described.
- A spatial correlation found kākā numbers were highest in forests not colonised by possums, and kākā became more numerous on Kapiti Island after possum eradication.
- No temporal correlations have been carried out to see whether native animals become more abundant in years when possum numbers are lower than average.
- Response-to-management studies may not have sufficient surveillance beforehand, may not have singled out the effects of possums, may have employed control methods that also killed rodents, and must run long enough to overcome any initial by-kill of native animals.
- Evidence from bird counts on Kapiti Island, snail counts at Charming Creek and elsewhere, leaves collected in Whirinaki Forest, and from a study of glossy black-cockatoos in Australia, is consistent with the hypothesis that native animals benefit from possum control.
- A strong test of the hypothesis using an experimental approach has not been accomplished (or, at least, published).

Acknowledgements

Data were generously provided by Colin O'Donnell, Astrid Dijkgraaf, Mihoko Numata, Stephen Garnett, Tim Lovegrove, and Kath Walker. Discussions with Bob Brockie, Mike Fitzgerald, Richard Sadleir, Chris Richmond, Ben Bell, Joe Hansen, and Lindsay Canham helped my understanding of particular cases. I'm grateful to Dave Paine and Richard Sadleir for bringing useful examples to my attention and for Jim Henry's help about Rangitoto Island. Comments by Craig Gillies, Tim Lovegrove, Richard Sadleir, and Kath Walker on a draft of the chapter helped to improve it.

References

Alterio, N. 1996: Secondary poisoning of stoats (*Mustela erminea*), feral ferrets (*Mustela furo*) and feral house cats (*Felis catus*) by the anticoagulant poison, brodifacoum. *New Zealand Journal of Zoology 23*: 331–338.

Brockie, R. E. 1992: A living New Zealand forest. Auckland, David Bateman. 172 p.

Brockie, R. E.; Moeed, A. 1986: Animal biomass in a New Zealand forest compared with other parts of the world. *Oecologia 70*: 24–34.

Brown, K.; Innes, J.; Shorten, R. 1993: Evidence that possums prey on and scavenge birds' eggs, birds and mammals. *Notornis 40*: 169–177.

Clout, M. N.; Denyer, K.; James, R. E.; McFadden, I. G. 1995: Breeding success of New Zealand pigeons (*Hemiphaga novaeseelandiae*) in relation to control of introduced mammals. *New Zealand Journal of Ecology 19*: 209–212.

Cowan, P. E. 1992: The eradication of introduced Australian brushtail possums, *Trichosurus vulpecula*, from Kapiti Island, a New Zealand nature reserve. *Biological Conservation 61*: 217–226.

Department of Conservation 1997: Conservation action. Wellington, Department of Conservation. 68 p.

Garnett, S. T.; Pedler, L. P.; Crowley, G. M. in press: The breeding biology of the glossy black-cockatoo *Calyptorhynchus lathami* on Kangaroo Island, South Australia. *Emu.*

Grant, A. D.; Powlesland, R. G.; Dilks, P. J.; Flux, I. A.; Tisdall, C. J. 1997: Mortality, distribution, numbers and conservation of the Chatham Island pigeon (*Hemiphaga novaeseelandiae chathamensis*). *Notornis 44*: 65–77.

Hickling, G. J. 1995: Action thresholds and target densities for possum pest management. *In:* O'Donnell, C. F. J. *comp.* Possums as conservation pests. Proceedings of an NSSC Workshop . . . 29–30 November 1994. Wellington, Department of Conservation. Pp. 47–52.

Innes, J. 1995: The impacts of possums on native fauna. *In:* O'Donnell, C. F. J. *comp.* Possums as conservation pests. Proceedings of an NSSC Workshop . . . 29–30 November 1994. Wellington, Department of Conservation. Pp. 11–15.

Innes, J.; Hay, R.; Flux, I.; Bradfield, P.; Speed, H.; Jansen, P. 1999: Successful recovery of North Island kokako *Callaeas cinerea wilsoni* populations, by adaptive management. *Biological Conservation 87*: 201–214.

James, R. E.; Clout, M. N. 1996: Nesting success of New Zealand pigeons (*Hemiphaga novaeseelandiae*) in response to a rat (*Rattus rattus*) poisoning programme at Wenderholm Regional Park. *New Zealand Journal of Ecology 20*: 45–51.

Miller, C. J; Anderson, S. 1992: Impacts of aerial 1080 poisoning on the birds of Rangitoto Island, Hauraki Gulf, New Zealand. *New Zealand Journal of Ecology 16*: 103–107.

Owen, H. J.; Norton, D. A. 1995: The diet of introduced brushtail possums *Trichosurus vulpecula* in a low diversity New Zealand *Nothofagus* forest and possible implications for conservation management. *Biological Conservation 71*: 339–345.

Powlesland, R. G.; Knegtmans, J. W.; Marshall, I. S. J. 1999: Costs and benefits of aerial 1080 possum control operations using carrot baits to North Island robins (*Petroica australis longipes*), Pureora Forest Park. Proceedings of a conference on the ecological consequences of poison use for mammalian pest control. *New Zealand Journal of Ecology 23*: 149–159.

Rose, A. B.; Pekelharing, C. J.; Platt, K. H.; O'Donnell, C. F. J.; Hall, G. M. J. 1990: Impact of brush-tailed possums on forest ecosystems, South Westland. Forest Research Institute Contract Report FWE90/52 (unpublished) 35 p.

Rose, A. B; Pekelharing, C. J; Platt, K. H; Woolmore, C. B. 1993: Impact of invading brushtail possum populations on mixed beech-broadleaved forests, South Westland, New Zealand. *New Zealand Journal of Ecology 17*: 19–28.

Spurr, E. B. 1989: Bird populations before and after 1080-poisoning of possums in Westland National Park. Forest Research Institute Contract Report to Department of Conservation (unpublished).

Spurr, E. B. 1993: Feeding by captive rare birds on baits used in poisoning operations for control of brushtail possums. *New Zealand Journal of Ecology 17*: 13–18.

Spurr, E. B.; Warburton, B.; Drew, K. W. 1992: Bird abundance in different-aged stands of rimu (*Dacrydium cupressinum*) — implications for coupe-logging. *New Zealand Journal of Ecology 16*: 109–118.

Thomas, M. D.; Hickling, G. J.; Coleman, J. D.; Pracy, L. T. 1993: Long-term trends in possum numbers at Pararaki: Evidence of an irruptive fluctuation. *New Zealand Journal of Ecology 17*: 29–34.

Wilson, P. R.; Karl, B. J.; Toft, R. J.; Beggs, J. R.; Taylor, R. H. 1998: The role of introduced predators and competitors in the decline of kaka (*Nestor meridionalis*) populations in New Zealand. *Biological Conservation 83*: 175–185.

CHAPTER TWENTY-THREE

Possums as a Resource

Bruce Warburton, Gordon Tocher, and Neil Allan

The brushtail possum was introduced into New Zealand to establish a fur trade, and this chapter provides an overview of the history of the trade, the characteristics of possum fur and how these position this fur in the international market, the contribution the industry makes to possum control, and finally a discussion on bounties.

History of the possum fur industry in New Zealand

Hunting of possums for skins in New Zealand dates back to the 1890s, although they could not be taken legally until 1912 when their status and protection as "imported game" was removed (Thomson 1922). Although official records of fur exports did not start until 1921 (Stewart 1991), the Otago Acclimatisation Society reported that not less than 60 000 skins were taken from the Catlins district in Southland in 1912 (Thomson 1922). Although this number seems unlikely given only 12 possums were reportedly liberated in the Catlins district in 1895, it is evidence of an early fur trade in New Zealand.

The fur trade has always been export based and earned New Zealand NZ$830 million (1997 dollars) in foreign exchange from the official export of 58 million skins. Like the rest of New Zealand's primary produce exports, demand patterns and prices for possum fur show a marked cyclic pattern with a wide range of factors affecting the product's value. Key markets were initially the UK and US, with Eastern Europe coming to the fore prior to World War II. In the 1950s and 60s, the UK, Australia, and US again dominated the buying, but were eclipsed by the Republic of Korea (South Korea) in the 1980s. Currently, China is emerging as a major processing and consuming nation and will almost certainly replace South Korea as the largest buyer of possum skins.

Possum skins are also exported from Tasmania (Callister 1991) where, because possums are classified as partly protected wildlife, they can only be harvested (by shooting only) during winter. The number of skins exported annually from Tasmania is considerably less than that from New Zealand reaching a maximum of 294 046 skins in 1978 (Callister 1991). Some of these skins went through the New Zealand fur auction system.

A price-driven industry

Hunting possums for skins has always been subject to the classic price/supply relationship, with the number of skins sold increasing as the value of the skins rise and vice versa (Fig. 23.1). The export statistics for possum skins show marked fluctuation from year to year with a peak in 1981 when 3.2 million skins were exported (Fig. 23.2). Since 1983 significant numbers (up to 230 000) of dressed possum skins have also been exported (Parkes *et al.* 1996), (Table 23.1).

The number of possum hunters also rises and falls with changes in the value of skins (C. Taylor, Taimex Trading, Dunedin, pers. comm.). In good years many thousands of full-time and part-time hunters work in the industry, many earning a substantial income. Nugent (1992) estimated 10 000 hunters were involved in the industry in 1988, but only 15% of these sold more than 1000 skins. During the peak periods of the possum fur industry access to land for hunting became competitive, possum populations in many areas (especially around farm edges) were reduced substantially, and hunters could make an acceptable income while hunting more remote areas, taking a larger proportion of possums from an area, and hunting lower-density areas. When possum skin prices decrease, hunters divert their efforts to more lucrative employment or become unemployed.

Before 1989/90, hunters sold their skins either to a merchant or at auction. The merchants often travelled to the hunters' bases, collected skins, and paid the hunter immediately. The auction system was similar to that for wool, giving hunters the market value for their skins. The latter option entailed a delay

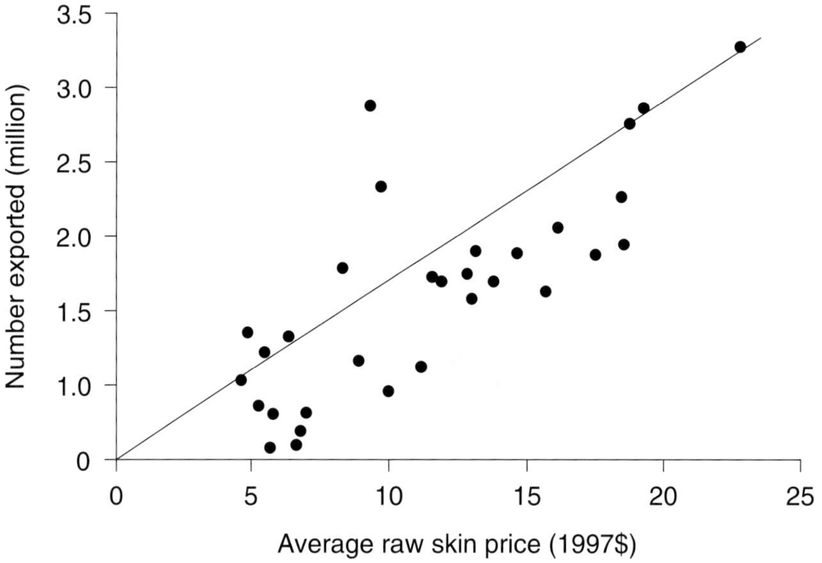

Fig. 23.1
Relationship between raw possum skin prices and numbers exported between 1967 and 1997.

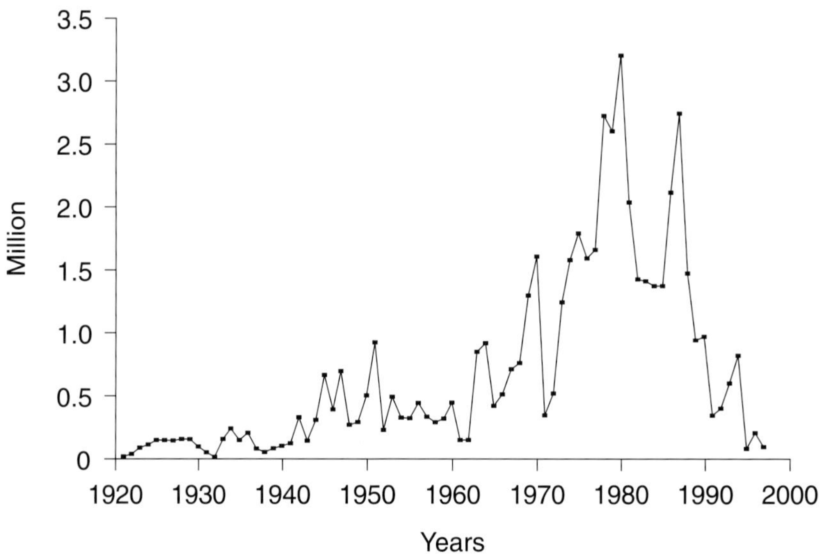

Fig. 23.2
Number of possum skins exported from 1921 to 1997.

in payment and the risk that skin prices might fall between consignment and sale. New Zealand's major possum auction house (Wrightson Dalgetys) left the industry in 1989, and the one other auction house closed in 1990 because of low skin volumes and prices.

There have been a large number of merchant skin buyers and exporters over the history of the possum trade, of whom only a relatively small number remained in the trade longer than 5 years because of its cyclic nature and unacceptable changes in profitability and liquidity. These merchants compete within New Zealand first for skins and then for overseas sales. Merchants may buy to order or for stock, the latter sometimes resulting in skins having to be stored for several years before resale.

Recent years have seen two periods of high skin prices, one in the early and one in the mid-to-late 1980s. Both periods were followed by a depression in skin prices. The last decline was the most severe, causing a large exodus of professional and part-time hunters. This decline, which caused the demise of the auctions, also resulted in some large companies, such as Wilson Neill, abandoning the industry.

Possum farming

Several possum farming ventures, such as the Kiwi Bear Company, were established during the 1980s to provide quality skins, and some meat for human consumption (Laurenson 1987; Keber 1985a). These ventures varied in size from small hobby farms to large publicly listed companies, but despite research

Table 23.1
The number and value (1997NZ$ Free On Board) of raw and dressed possums skins exported between January to December from 1967 to 1997 (Statistics New Zealand).

Year	Raw skins[1] Number	Value (NZ$)	Price/skin (NZ$)	Dressed skins Number	Value (NZ$)	Price/skin (NZ$)
1967	710 086	668,268	11.13			
1968	587 202	452,521	8.80			
1969	1 292 475	1,862,660	15.66			
1970	1 604 896	2,052,331	13.10			
1971	345 523	262,223	6.99			
1972	517 043	595,362	9.96			
1973	1 240 112	2,020,912	12.95			
1974	1 579 782	3,869,470	17.46			
1975	1 788 299	4,637,670	16.10			
1976	1 589 929	4,384,058	14.61			
1977	1 659 102	6,632,451	18.50			
1978	2 724 138	12,557,662	19.20			
1979	2 599 401	13,454,100	18.72			
1980	3 201 981	23,421,550	22.74			
1981	2 741 228	19,840,131	19.49			
1982	2 032 242	13,906,647	12.78			
1983	1 426 626	7,897,915	12.13	124 955	2,218,895	17.76
1984	1 409 677	7,406,023	10.76	145 962	2,736,863	18.75
1985	1 371 051	9,231,257	11.85	231 857	3,959,820	17.08
1986	1 910 830	12,633,461	9.70	136 873	3,143,021	22.96
1987	2 478 142	18,811,382	9.30	206 311	4,380,384	21.23
1988	1 475 219	9,538,323	8.30	121 913	2,066,873	21.76
1989	941 738	5,004,000	6.37	100 058	1,502,472	17.99
1990	972 386	4,128,217	4.84	75 833	1,182,284	17.79
1991	341 914	1,773,708	5.79	76 879	1,407,564	20.45
1992	401 269	1,911,607	5.27	130 772	1,920,581	16.24
1993	599 557	2,545,841	4.63	106 662	1,514,844	15.48
1994	820 744	4,197,663	5.48	86 678	1,487,702	18.38
1995	80 262	440,716	5.68	19 910	386,509	20.08
1996	205 046	1,375,999	6.78	55 973	763,281	13.78
1997	96 474	639,856	6.63	70 836	975,373	13.77

1 Before 1983 export statistics of raw and dressed skins were grouped.

undertaken to determine the best husbandry practices in order to improve fur quality and maximise profits (Fisk 1985; Pearson 1987), all ventures failed. Failure was primarily a result of the length of time required to produce a saleable product and the high mortality of possums due to captivity stress.

Processing of possum fur

Processing of possum skins in New Zealand began in the 1920s when specialist fur skin dressing plants were established (Stewart 1991). Initially, possum fur was only a minor part of their work, but soon expanded to comprise the majority. Today, New Zealand's dressing capacity remains limited and most possum skins are still exported in their raw state.

Producing finished goods from fur skins (rabbit, seal, and imported furs) was a large industry in New Zealand from the 1920s until the 1980s, but possum fur was not used in large quantities until the 1960s. Historically, the trade was furrier-based and produced almost entirely ladies' outerwear for the domestic market. However, changes in fashion during the 1980s saw a marked decline in this trade. Today, a relatively small number of craft businesses manufacture garments and souvenirs using possum fur.

Legislation governing possum hunting

Government regulation of fur hunting has been relatively limited. Prior to 1911 possums were not protected but, because of increasing numbers being killed for fur, an Order in Council was issued declaring possums as "Imported game within the meaning of the Animals Protection Act 1908", thus making it illegal to catch or destroy them (Thomson 1922). However, some individuals still wanted to trap them for fur and representations to Parliament resulted in a Gazette notice in 1912 removing all protection from possums so they could be taken without restriction. This produced an over-reaction of protectionists and their pressure resulted in another Order in Council in 1913 declaring possums totally protected. Nevertheless, illegal trapping and export continued over the following years until, in 1921, legislation was passed to prohibit the harbouring and liberation of possums, and to allow for possums to be killed for their skins, which were to be stamped, a levy taken, and sold to licensed dealers. This situation remained until 1946 when the licensing of brokers, stamping of skins, and the levy were discarded (Pracy 1962). In 1947 further amendments to the legislation removed all restrictions on the taking of possums. In 1951 a bounty scheme was initiated, but proved ineffective at reducing possum numbers and was halted in 1961 (Pracy 1981).

Currently, little restricts the activity of possum hunters as long as they obtain permission to hunt from the landowner, have a poison licence if using cyanide, and make a daily check of their traps. An individual's income from possum skins was subject to 25% withholding tax from April 1980 to January 1992, but now income from skins is treated in the same manner as other personal income and assessed solely on the basis of the hunter's annual tax return.

The international fur market and possum fur

The world fur market is a large, geographically widespread, semi-commodity market. Retail sales in a number of countries exceed US$1 billion/year. The dominant fur type is mink, which is farmed in North America, China, UK, and Scandinavia. Major markets are Italy, Spain, Germany, North America, and Japan. Since the early 1980s South Korea, and more recently, China, and Russia have emerged as important buyers.

Fur skins are mostly used in quality outerwear such as the traditional fur coat or jacket, as fur trim (collars, cuffs, and hoods), as interior lining in textile or leather garments, or in combination with other products in reversible garments. In many countries both men and women wear fur garments.

The last major boom in the fur market occurred in the 1980s. World mink production rose from 19 million pelts in 1980 to 42 million at the end of the decade, and prices per pelt increased from US$18 to a peak of US$75. Demand then began to fall in 1987 and this continued until late 1992. Being primarily a farmed (ranched) product, there was a substantial lag before production began to fall in response to the reduced demand. This lag resulted in a period of significant oversupply and a sharp fall in mink values. By 1993 only 19.5 million pelts were offered for sale. This amount was now less than the market wanted, and prices have been rising in response to this increased demand, causing the whole cycle to begin anew. Mink prices are now well over double what they were at the lowest point of the downturn.

The 1987 to 1992 downturn was more severe than usual for the fur market, because many negative factors affected demand simultaneously. In addition to overproduction, particularly of mink, there were:

- the 1987 international share market "crash" and subsequent reduction in spending on luxury items and drop in consumer and business confidence;
- several consecutive warm winters in both North America and Europe, resulting in lower demand for new and replacement garments;
- changes in fashion after a long period of popularity;
- varying effects of animal rights – anti-fur lobbies on different markets. Because these groups gained a high public profile, outside observers tended to place more importance on this factor than it deserved.

Possum fur is a very small part of the world fur market, less than 1% by value. In the hierarchy of fur types, possum is near the middle, unable to command the premium of sable, mink, or fox, but considered superior to furs such as rabbit and muskrat. Possum fits into the market as an economically priced fur of variable quality.

Possum skins are a well-sized, versatile product capable of being shorn, sueded, napped, bleached, dyed, or used in its natural colour and hair length. Like most wild furs it is best suited to and traditionally used in the trimming business, where the fur trade supplies pre-made collars, cuffs, and strips to textile or leather garment manufacturers. This end-use requires better-quality (first, second, and some third) grades, fully furred skins of large size. These skins represent less than half of the normal possum skin harvest. Less than 5% of possum skins are used in the manufacture of traditional, ladies' fur jackets or coats. Only first- and some second-grade skins are used for this and these grades constitute only a small part (10–20%) of the overall possum harvest. During the 1980s possum fur became popular for lining purposes, being affordable and having the correct fur characteristics. Typical product use was in fur-lined raincoats and leather bomber jackets, often for male customers. Lower-quality skins (third and fourth grades) are typically used for linings.

Recent initiatives for the improved commercialisation of possum products

Up to the early 1990s possums were seen solely as a source of fur, primarily as a dried pelt export, but also to a small extent for use in the production of coats, warmers, souvenirs, or other products. Since 1993, there has been an increased interest in diversifying the range of possum-based products with both individual and collective initiatives being undertaken.

In 1996 the Possum Products Marketing Council was established with representatives from fur exporters, employment and training, international trade, possum hunters, and research. This council brought together the disparate sectors of the industry to initiate the setting of standards for fur grading, coordinating new initiatives, carrying out market research and developing a "brand" that could be used to place possum products in the international market place. It aimed to generate a stable and sustained demand for possum products by developing an industry based on a broad range of possum-based products to counteract the volatility of the fashion markets. This aim would, in turn, lead to sustained control pressure on some possum populations, and provide a significant employment opportunity, especially in rural areas of New Zealand.

Possum products

There are five major product groups derived from possums: pelts, leather, fibre, added-value products, and meat.

Pelts

The traditional pelt market is undergoing a significant change in direction, with the focus shifting to newer emerging markets, such as China and the former Soviet Union. It is recognised by the industry that some measure of technical assistance may have to be given in order to ease the entry of possum into these marketplaces. As well as that, there is a recognition that quality standards must be established in order to get possum fur re-established in the international fur market.

Leather

Research on possum leather carried out by the New Zealand Leather and Shoe Research Association has shown that, for its weight, it is the second-strongest leather in the world, and is particularly suited to the manufacture of gloves, shoe uppers, and innersoles. One disadvantage of possum leather is that the skins are scratched during mating and fights, which necessitates stamping or overprinting to attain a reasonable finish. Possum leather is also being used on a small scale for book binding and appears to be superior to goatskin, which is commonly used for this purpose.

Fibre

The fur fibre of possums has also been extensively researched and is now a commercial reality, with five companies (e.g., Snowy Peak Ltd) involved in processing it. The possums are either skinned and subsequently slipped (a chemical process to remove fur from skin), or plucked in the field. The fibre is then carded and mixed with other fibres and spun into yarn from which products such as pullovers, tam o' shanters, gloves, muffs, and cardigans are made for local sale and export. Further research is underway to produce material from possum fibre that is of dress suit quality. Another company, Wild Peter Products, has been developing fabrics using possum fur (branded kapua) and, in conjunction with Snowy Peak Knitwear Company, has created a new fabric called merino mink (Hay 1998).

Accessories

There has long been a small industry based on added-value utilisation of possum skins. By and large these businesses are small cottage-type enterprises, with a couple of exceptions. One such company, Possum Pam, based in Maruia, Westland, uses in excess of 6000 skins for producing high-quality accessories. This company produces a large range of products, including hats, mitts, coats, shoe liners, sunglass pouches, hot-water bottle covers, money pouches, rugs, soft-toy souvenirs such as kiwi, and other novelty items. The main outlets for these products are through Department of Conservation (DOC) offices and tourism-related shops.

Although the commercial use of possums is never going to control their numbers in New Zealand by itself, hunting for profit is without doubt an additional mechanism that pest control agencies can use to put pressure on possum populations, especially in areas that don't warrant official control measures.

Meat

The production and export of possum meat for human consumption has had a slow start with the first exports to Asia in the early 1980s. The initial possum meat trade met with problems because the 1981 Meat Act did not allow for processing of possum carcasses. However, in 1987, the Meat Act was amended to include farmed possums within the definition of "animal" under the Act. This amendment excluded possums killed in the wild because it required all possums to be held in captivity for at least 28 days prior to slaughter. The then Ministry of Agriculture and Fisheries (MAF) developed a protocol for preparation and processing of possum meat for export, mainly to address overseas concerns about the threat of bovine Tb and the possibility of possums having high levels of long-acting toxins in their systems prior to slaughter (Woodd 1987). At the time of writing, further amendments to the protocols for exporting possum meat have been made with wild-caught possums now being defined as animals rather than game, but with the restriction that possums must be harvested from only Tb-free areas (MAF Regulatory Authority 1998). There is now no withholding period imposed by New Zealand meat processing standards, and the one country that maintained the 28-day withholding period, Taiwan, is in the process of agreeing to waive this requirement. Consequently, the opportunity to export possum meat is therefore in the hands of entrepreneurs with the main markets being Taiwan, South Korea, and mainland China. A possum-meat trade is also being developed in Tasmania (Manigian 1996) where they do not have the bovine Tb or potential toxic residue problems that New Zealand exporters have to deal with.

Commercial value as a pest management tool

Historical perspective

Historically, commercial possum hunters have received little or no support for their efforts from official control agencies. Most Government pest control staff believed ground-based control methods would not contribute significantly to the control of possums. Nevertheless, a decline in the number of possum kills reported per DOC hunting permit over a 5-year period in the 1980s (Clout & Barlow 1982), suggests that commercial hunting did reduce possum densities in some areas. Further evidence of the impact of commercial hunting was reported by Brockie (1982), who found possum numbers in a hunted area were being held at about half of that in an adjacent unhunted area. Anecdotal evidence from hunters who were finding increasing difficulty in obtaining good numbers of possums during the 1980s also supports the notion that positive market forces were sufficient to reduce possum numbers.

Although hunting solely for commercial gain may not achieve the very low densities desired for areas with high conservation values or where Tb eradication is needed, it could provide some conservation or disease control benefits in areas where other control efforts are absent. Brockie (1982), found that the forest vegetation in a hunted block in the Orongorongo Valley had a higher abundance of northern rātā and kāmahi than an unhunted block, and suggested this may have been a result of the lower density maintained by the hunting.

Employment schemes and contract hunters

People on government employment schemes, particularly in the late 1980s and early 1990s, have been used to carry out possum control, but most of these operations have provided few skins for sale. Lack of training and often poor motivation have resulted in reports of low kills, inefficient worker performance and therefore high costs, loss of equipment, poisoning of stock and dogs, and a return to unemployment at the end of the scheme.

Part of the reason for these failures is the misconception that any able-bodied person can make a good hunter. Successful hunters are highly skilled, motivated, self-employed professionals who do not fit the personal profile of many of the unemployed. To achieve the full potential of using hunters for possum control, prospective hunters must receive high-quality training aimed at creating self-sufficiency, professionalism, and operational effectiveness. To this end several training courses have been established, such as that run by the Tairawhiti Polytechnic.

With the rising awareness of the possum problem in recent years, the increasing funds being made available for possum control, and the downturn in possum skin prices since 1989, many professional hunters made the transition from being commercial hunters to being possum control contractors. They developed new skills, such as contract pricing, tendering, population monitoring, and the ability to reduce possum numbers to the low levels required by their customers. Contract hunters are increasingly being used by DOC, private landowners, and most regional councils, especially as more operations shift from initial knockdown to maintenance control, and as an increasing number of hunters are being recognised for their professionalism and effectiveness.

In some areas, it has been suggested the low possum densities required to meet pest control objectives could be achieved by using performance-based possum control contracts (cf. Batcheler & Cowan 1988). These would allow control authorities to use the commercial value of possums to reduce control costs. For example, in an area where possums are at a density of 6/ha, and an 80% reduction is desired, contract hunters would have to remove 4–5 possums/ha. The financial return obtained for these skins, which is dependent on skin prices, could either fully or partially cover the control costs. Current official control costs are in the order of NZ$20/ha. When possum skin values increase, control costs where hunters are used would diminish.

Economics and population dynamics

Technical and financial limitations of current possum control methods mean eradication of possums from any mainland area is not possible at present. Thus, the only ecologically valid management practice is sustained control: maintaining the pest population at, or below, a level at which it does unacceptable damage.

Sustained control is mathematically indistinguishable from sustained harvesting. The distinction between the two actions is more in their objective than in the method used. That is, sustained control of possums endeavours to keep possum numbers at low levels where they do not cause unacceptable damage. Sustained harvesting is usually aimed at increasing the harvest of a valuable resource to a maximum. Both actions are sustainable harvests but they are unlikely to result in the same population density.

If hunter efficiency has not improved significantly since the 1980s, then to get the fur exports exceeding 3 million skins again would require average skin prices of NZ$20–$25 (Fig. 23.1). Taking account of possums that are not skinned because of damage, not found on a poison line (about 10% — Morgan & Warburton 1987; but currently nearer 20%), and those used within New Zealand, the total number killed, if 3 million were exported, would be about 4.8 million. Additionally, the NZ$30–$40 million spent on possum control on conservation land and in areas of endemic Tb would be removing an estimated additional 5 million a year. With many possums also killed by private landowners carrying out their own possum control, a total of about 10–11 million possums could be removed annually provided export returns of about NZ$20 per harvested skin were obtained.

The possum population in New Zealand has been estimated to be in the order of 70 million (Batcheler & Cowan 1988), with some local populations still increasing and some declining. If the 70 million estimate is used as the carrying capacity value (K) for New Zealand, maximum sustainable harvests (MSY) or any equilibrium density resulting from any sustained harvest can be calculated. Estimates of MSY for possums range from 5.3 to 10.7 million (Batcheler & Cowan 1988) depending on the r_m value (intrinsic rate of population growth) and the population growth model used. The higher MSY value of 10.7 million (Keber 1985b) is based on an r_m value of 0.59, which appears unrealistically high (it assumes most females are double breeders) in comparison to empirical measures of r_m of 0.22 obtained for a possum population recovering after an aerial poison operation (Hickling & Pekelharing 1989). Using an r_m value of 0.3 and a theta-logistic growth model, Barlow & Clout (1983) estimated a MSY of between 8 and 9.8 million possums/year. If 8 to 9 million possums could be removed by commercial hunting, a population of 70 million possums would decline on average by about 35%. If the kill could be increased above the MSY, the population would decline further as would the number of possums needed to be killed to maintain it at that level (Fig.

23.3). However, as the population declined, possums would become more difficult to hunt and therefore the price per skin or hunter efficiency would have to increase to maintain hunters income. The full potential of commercial hunting will depend on the upper limit of the prices paid for skins and the relationship between prices and hunter efficiency. Hunter efficiency, however, depends on accessibility, and hunters would have a significantly greater impact on possum numbers on easily accessible farmland than in remote areas of forest. Consequently, if the majority of harvested possums were taken from accessible areas of New Zealand, then it is likely that the localised reduction of possum numbers would be considerably higher than that suggested from the above analysis. That is, a harvest of 8–10 million possums might only reduce the national possum population by 35%, but could reduce local, accessible populations by considerably more.

Bounties

The incentive that appears to best motivate most people is money, and when a value is placed on a dead specimen of a particular species, people are encouraged to kill this species. Such an incentive system is called a bounty or token system. The effectiveness of such a system depends on the relationship between the price (token value) and the cost of getting the token. In principle, it is no different than commercial hunting for skins except that the price is paid by the tax payer. When the bounty or token price is high compared to the cost per possum kill, then many people will hunt the species, and if the differential between cost and price is held as the population declines (i.e., as cost increases), then population numbers of the pest species will be continuously reduced.

Unfortunately, most bounty systems have had token values in relation to the costs of catching pests that do not significantly reduce population numbers. For example, a bounty of two shillings and sixpence (NZ$5.0, 1997 value) was placed on possums in 1951, and over the following 10 years about 8.2 million possums were killed (Table 23.2). Nevertheless, there was no evidence that the continuing spread of possums was halted, and analysis of the token returns indicated that a large proportion of kills came from non-critical areas (Pracy 1981). Consequently the bounty was discontinued. The failure of this bounty was not so much in the bounty system per se, but in the value of the bounty being insufficient relative to the cost required to trap possums.

Because skin price largely determines the number of possums killed (Fig. 23.1) we can use the economic incentive (NZ$22.74/skin, 1997 dollars) that resulted in the peak exports of 3.2 million skins in 1980 (Table 23.1) to assess the likely effectiveness of a bounty. Killing possums for tokens requires far less time than removing, carrying, drying, and preparing the skins for selling. Moore (1997) estimated that for about 14 hours spent laying and checking a cyanide poison line, an additional 10 hours would be required for skinning and drying. This suggests the same hunting

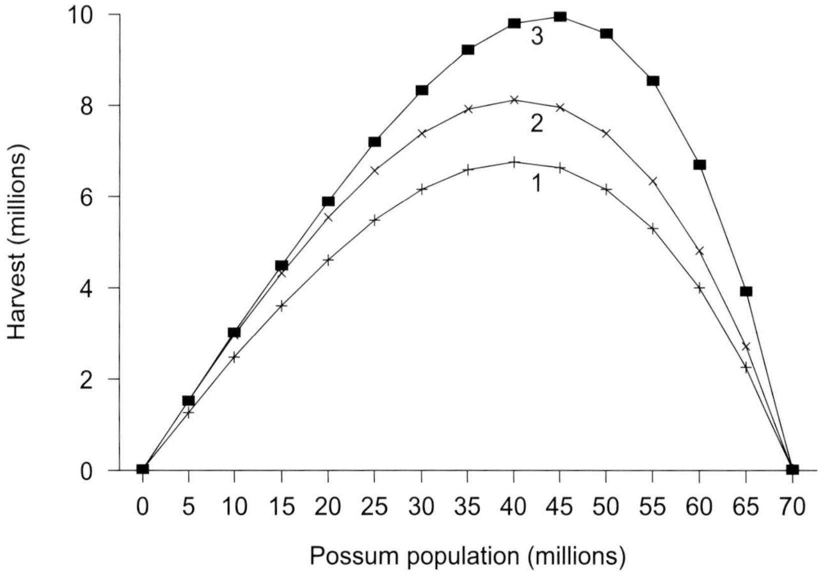

Fig. 23.3
Estimated sustained harvests from possum populations ranging in size from zero to 70 million. Curve 1 is generated from an r_m value of 0.25 and a theta value of 2; curve 2 has an r_m of 0.3 and a theta value of 2; curve 3 has an r_m of 0.3 and a theta value of 3. The inflexion point of the curves denotes the maximum sustained yields.

Table 23.2
Number of tokens paid for from 1951 to 1961 when a bounty was placed on possums.

Year	Number	Year	Number
1951	25 000	1957	1 080 000
1952	160 000	1958	900 000
1953	490 000	1959	960 000
1954	560 000	1960	1 100 000
1955	570 000	1961	1 390 000
1956	930 000		

effort could have been generated by a token price of NZ$13.26 or NZ$42 million for the 3.2 million possums. The total payment would, however, be higher because tokens would be taken from the possums that would normally be rejected because of substandard fur. Despite a high kill in the early 1980s, possum numbers were still higher than those required to eradicate Tb and protect the more vulnerable conservation values. It is unlikely, therefore, that bounties of NZ$2, a figure commonly mooted, will have any significant effect on possum numbers.

The other major flaw with bounties is that they cannot be easily targeted at critical areas. If the area where possums have significant impacts is small compared to their total range, then any bounty system would clearly be very expensive because most possums would come from areas not requiring control. However, if the size of the area needing control is close to the total range of the possums, and possum numbers need to be reduced to equally low levels everywhere, then the need to target control to specific areas is reduced. It is very unlikely that the resources will ever be available to control possums to very low levels everywhere, so control will always need to be carefully targeted, making bounties ineffective as an alternative option to official control. Some suggestions have been made for better targeting of bounty-based control, such as using "smart" bounties that rely on hunters recovering tagged or diseased animals for which they get paid. In the future possums from target areas may be able to be genetically identified and this will overcome the problem of the generalised nature of the bounty.

Some proponents of bounties suggest they should be used to support the poorer grades of fur that at present have little market value (New Zealand Opossum Fur Producers Association 1994). However, such a system, which will no doubt benefit possum hunters and create employment, will not effectively reduce possums to the low levels required in the target problem areas.

Summary

- Hunting of possums for skins in New Zealand dates back to the 1890s although official records of exports didn't start until 1921.
- The export statistics of possum skins show marked fluctuation from year to year with a peak in 1980 when 3.2 million skins were exported worth a value of about NZ$23.4 million.
- Several possum farming ventures, such as the Kiwi Bear Company, were established during the 1980s to provide quality skins, and some meat for human consumption.
- Possum fur is best suited to and traditionally used in the trimming business, where the fur trade supplies pre-made collars, cuffs, and strips to textile or leather garment manufacturers.
- Since 1993, there has been an increased interest in diversifying the range of possum-based products with both individual and collective initiatives being undertaken.
- In 1996 the Possum Products Marketing Council was established with representatives from fur exporters, employment and training, international trade, possum hunters, and research to develop a coordinated approach to possum marketing.
- There are five major product groups derived from possums: pelts, meat, leather, fibre, and added-value products (accessories).
- Historically, the official position taken by government agencies was that commercial hunters were ineffective at reducing possum populations to the low levels required to protect flora and fauna or prevent the transmission of bovine Tb.
- It appears that although hunting solely for commercial gain may not achieve the very low densities desired for areas with high conservation values or where Tb eradication is needed, commercial hunting of possums when prices are high could provide measurable ecological benefits in the rest of New Zealand where taxpayer funds are unavailable.
- A bounty system, while benefiting possum hunters and creating employment, will not effectively reduce possums to the low levels required in the target problem areas unless the fundamental problem of ineffective targeting can be overcome.

References

Barlow, N. D.; Clout, M. N. 1983: A comparison of 3-parameter, single-species population models, in relation to the management of brushtail possums in New Zealand. *Oecologia 60*: 250–258.

Batcheler, C. L.; Cowan, P. E. 1988: Review of the status of the possum (*Trichosurus vulpecula*) in New Zealand. Unpublished FRI contract report to the Technical Advisory Committee (Animal Pests) for the Agriculture Pest Destruction Council, Department of Conservation, and MAFQual Ministry of Agriculture and Fisheries. 129 p.

Brockie, R. E. 1982: Effect of commercial hunters on the number of possums, *Trichosurus vulpecula*, in Orongorongo Valley, Wellington. *New Zealand Journal of Ecology 5*: 21–28.

Callister, D. J. 1991: A review of the Tasmanian brushtail possum industry. *Traffic Bulletin 12*: 49–58.

Clout, M. N.; Barlow, N. D. 1982: Exploitation of brushtail possum populations in theory and practice. *New Zealand Journal of Ecology 5*: 29–35.

Fisk, T. 1985: Farming opossums — can it be done at a profit? *The New Zealand Farmer 106(21)*: 31–32.

Hay, D. 1998: Pest dressed. *North and South, November*: 20.

Hickling, G. J.; Pekelharing, C. J. 1989: Intrinsic rate of increase for a brushtail possum population in rata/kamahi forest, Westland. *New Zealand Journal of Ecology 12*: 117–120.

Keber, A. W. 1985a: New Zealand Opossum Fur Producers Association meat technical report No. 4. *Fur Facts 6 (24)*: 31–33.

Keber, A. W. 1985b: The role of harvesting in the control of opossum populations. *Fur Facts 6 (24)*: 47–51.

Laurenson, W. 1987: Playing possum on world markets — trying to turn a pest to profit. *Management 34(2)*: 19–25.

MAF Regulatory Authority 1998: Information for the Agriculture and Seafood Industries. *Food Focus issue 10*, November. 4 p.

Manigian, S. 1996: Tasmanian possums meat trade. *Habitat Australia 24(4)*: 35.

Moore, M. 1997: Can fur hunting pay in 1997? *Fur Facts 16(45)*: 21–22.

Morgan, D. R.; Warburton, B. 1987: Comparison of the effectiveness of hunting and aerial 1080 poisoning for reducing a possum population. *Fur Facts 8(32)*: 25–49.

New Zealand Opossum Fur Producers Association 1994: Submission to the Parliamentary Commissioner for the Environment. *Fur Facts 13(41)*: 26.

Nugent G. 1992: Big-game, small-game, and gamebird hunting in New Zealand: hunting effort, harvest, and expenditure in 1988. *New Zealand Journal of Zoology 19*: 75–90.

Parkes, J. P.; Nugent, G.; Warburton, B. 1996: Commercial exploitation as a pest control tool for introduced mammals in New Zealand. *Wildlife Biology 2*: 171–177.

Pearson, A. J. 1987: Development in the commercial exploitation of the opossum in New Zealand. *Fur Facts 8(30)*: 32–37.

Pracy, L. T. 1962: Introduction and liberation of the opossum (*Trichosurus vulpecula*) into New Zealand. *New Zealand Forest Service Information Series 45*: 1st ed. 28 p.

Pracy, L. T. 1981: Opossum survey. *Counterpest 5*: 5–16.

Statistics New Zealand: Overseas trade : year ended June. Wellington, New Zealand.

Stewart, D. W. 1991: From fur to fashion: the background story to the establishment of the New Zealand fur industry. D.W. Stewart, P.O. Box 54, Dunedin. 180 p.

Thomson, G. M. 1922: The naturalisation of animals and plants in New Zealand. London, Cambridge University Press. 607 p.

Woodd, R. 1987: Opossums: ripe for export bonanza? *Fur Facts 8(29)*: 29–31.

CHAPTER TWENTY-FOUR

Biological Control of Possums: Prospects for the Future

Phil Cowan

What is biological control?

Biological control is usually taken to mean the use of a natural enemy of a pest species (often a parasite, disease, or predator) that is imported into an area in order to control that pest. This is often referred to as *classical biological control* (Waage & Mills 1992). The introduction of stoats to New Zealand to control rabbits is an example, albeit ill-conceived and unsuccessful, of such an approach. More recently, with rapid advances in molecular biology and genetic engineering, the concept of biological control has widened to include the use of genetically modified organisms and biologically based substances, such as immunotoxins, that interfere with biological processes within the pest. Jolly (1994) suggested the term *biotechnological control* for such approaches. Viral-vectored immunocontraception where rabbits are rendered infertile after infection with a mildly pathogenic strain of the myxoma virus expressing rabbit contraceptive antigens (Jackson *et al.* 1998), is an example.

Why is biological control of possums necessary?

Despite all attempts to control them, introduced common brushtail possums (*Trichosurus vulpecula*) now occupy more than 90% of New Zealand, and in many habitats they occur at densities up to 20 times those in their native Australia (Cowan 1990). The threats they pose to both conservation values and agricultural production have been detailed in Chapters 8–11. Although conventional control of possums is effective locally (Chapter 13), integrated management of possums, using a combination of conventional control and biological control, offers the only long-term cost-effective solution to the national possum problem.

The development of biological control for possums is being driven by a wide range of factors. Public attitudes to killing as a means of controlling wildlife pests are changing, not just in New Zealand but also internationally and, where these attitudes influence trade in New Zealand's products, they become particularly important (Williams 1994). Coupled with this is increasing public concern, again both national and international, about the application to the environment of non-selective poisons for pest control (Chapter 1; Parliamentary Commissioner for the Environment 1994; Stewart *et al.* 1994). New Zealand and Australia are the only countries that make extensive use of 1080 poison (sodium monofluoroacetate) for vertebrate pest management, and New Zealand's annual use of 3000–4000 kg is an order of magnitude higher than that of Australia (Parliamentary Commissioner for the Environment 1994). Biological control has the potential to reduce toxin use for possum control and hence concerns about environmental contamination, particularly of water supplies.

Biological control also has the potential to reduce threats to non-target species. Although monitoring of many 1080 operations has demonstrated no lasting effects on non-target species (Chapter 16), there is still repeated public concern about the immediate small number of non-target deaths that accompany every operation. Attempts are being made to develop biological control that will attack systems specific to marsupials, and preferably only possums (Cowan 1996).

Finding a solution to New Zealand's possum problem is urgent. An annual expenditure of c. $45 million on conventional possum control using poisons and trapping (Morgan & Eason 1996) is barely sufficient to reduce the current expanse of bovine-tuberculosis-infected possum populations or alleviate damage in all but the highest priority conservation areas (National Science Strategy Committee 1997). The level of bovine Tb in New Zealand's cattle herds, although reducing, is still an order of magnitude higher than that of its major competitors, and there is a real threat of non-tariff trade barriers being put in place if New Zealand does not rapidly reduce its livestock Tb problems (Animal Health Board 1995). Canopy damage continues to occur in many areas of native forest, some individual plants are threatened with local extinction, and some native animals suffer unacceptable losses from possum predation (Cowan 1990; Innes *et al.* 1999). With its current level of funding the Department of Conservation will be able to protect only about 17% of national conservation lands from possum damage over the next decade (Parkes *et al.* 1997).

Significant progress is hampered by the scale of the problem. Possums occupy most of New Zealand with a population size estimated at 60 to 70 million (Cowan 1990). The methods used currently to manage possums are still not cheap, despite recent incremental cost efficiencies, costing from $5–$40/ha (Chapter 13). The cost of poison is negligible relative to the cost of baits and delivery. The present annual investment in possum control of c. NZ$45 million is probably approaching the maximum that government, regional councils, and farmers will bear, unless a major export market was seriously threatened. However, the cost of eradicating possums from the highest priority areas for conservation protection and tuberculosis management has been estimated at $NZ1 billion (Parliamentary Commissioner for the Environment 1994). Biological control systems that would spread naturally from possum to possum would offer major cost-efficiencies and allow the current expenditure on conventional control to be redeployed to maximise benefits for conservation and Tb management.

The expansion in conventional control over the last 5–10 years has produced a number of technical problems. The repeated use of 1080 and cyanide poisons has resulted in possums that are bait and poison shy (Chapter 14). Much effort is currently going into developing ways of overcoming shyness in field operations by changing operational procedures. New Zealand is also heavily dependent on a single poison, 1080, for most of its possum control. A major problem could arise if this poison becomes unacceptable to New Zealand's major trading partners. Biological control could alleviate both these problems.

Biological control by itself will not eradicate possums from New Zealand — the hope is either that it will reduce the possum problem to a scale manageable by sustained conventional control, or that in combination with conventional control it will provide opportunities for local or national eradication (Jolly 1993; Cowan 1996).

Strategy for development of biological control

Recognising the seriousness of the possum problem, in 1992 the New Zealand Government established a National Science Strategy Committee (NSSC) for the Control of Possums and Bovine Tuberculosis. The aim of the national science strategy is to reduce possum impacts to levels that are socially and politically acceptable using techniques that are target specific, humane, cost-effective, ecologically safe, and sustainable (Atkinson & Wright 1993). Since 1993 biocontrol of possums has been recognised as the key long-term goal of the science strategy (NSSC 1997). Two approaches are being taken. First, parasites and diseases are being sought that might reduce possum numbers (Heath *et al.* 1998; Obendorf *et al.* 1998; O'Keefe & Wickstrom 1998; Meers *et al.* 1998), as possums are known to have lost some of their specific parasites and diseases since their introduction from Australia (Chapter 7). Second, methods are being developed that attack key physiological processes and that can be incorporated genetically into possum-specific disseminating vectors to produce the desired control effects (Jolly 1994). At the same time, the potential for parasites and diseases to act as disseminating vectors is being assessed (Gruenberg & Bisset 1998).

One problem for the strategy is that although thresholds at which possum impacts become unacceptable and target densities of possums in relation to their impacts have been clearly identified for tuberculosis management (Barlow 1991, 1994), these largely remain to be identified for conservation of biodiversity, partly because different conservation resources will have different thresholds at which damage is considered unacceptable (Hickling 1995).

Approaches to biological control

Parasites and diseases

Possums in New Zealand have a very limited parasite fauna (Chapter 7). Intestinal parasites are patchily distributed in the North Island, rare in the South Island, and absent from Stewart Island. If such parasites have pathogenic effects, then their introduction into naive populations may affect possum numbers or productivity. Research is currently underway to assess the impact of parasites on possum populations (Heath *et al.* 1998). Although a survey of Australian possums identified a much larger parasite fauna (Obendorf *et al.* 1998), few of the additional species were possum specific, with the notable exception of the nematode *Adelonema trichosuri*, which may have potential both as a pathogen and as a vector.

Herpesviruses, adenoviruses , coronaviruses, and coronavirus-like particles have been identified in possum intestinal contents, although as yet, apart from one adenovirus, none have been isolated and cultured (Meers *et al.* 1998; D. Thomson, Massey University, Palmerston North, pers. comm.), and their specificity for possums is unknown. The adenovirus isolated from possums appears to be a previously undescribed species (D. Thomson, pers. comm.). Molecular techniques have also shown the likely presence of retroviruses (Baillie & Wilkins 1998).

Investigations of outbreaks of infectious disease in possums housed for experimental purposes identified a new Borna-like virus, wobbly possum virus, as the causative agent in one outbreak, and wild possums infected with this virus have since been identified (O'Keefe & Wickstrom 1998; Meers *et al.* 1998). The virus responsible for an infectious enteritis outbreak in another possum colony has not yet been isolated (O'Keefe & Wickstrom 1998). Both these diseases caused high mortality, but much more work is needed to define their potential as biological control agents.

The option remains to look further afield, such as among the phalangers of New Guinea and nearby islands, or even among American marsupials, for the possum equivalent of diseases like myxomatosis or rabbit haemorrhagic disease (RHD). However, despite its presence in Europe for more than 15 years and a lack of data suggesting a threat to any species other than rabbits, the enormous public and scientific debate in New Zealand about the introduction of RHD (O'Hara 1997) suggests that any proposal to introduce a new virus of unknown epidemiology to control possums would be subject to even more rigorous scrutiny and requirements for non-target species testing.

Physiological controls

In parallel with the search for classical biological control agents, recent advances in molecular biology are being applied to the development of genetically engineered methods for biological control of possums. These programmes are almost entirely focused on interfering with physiological processes involved in the production of young, i.e., fertility control. The potential of fertility control as a practical tool for the management of wild species has gained wide acceptance in recent years (see Kreeger 1997). This support comes largely from acceptance of fertility control as a more humane approach to wildlife management than many of the currently used methods (Loague 1993; Oogjes 1997; Singer 1997).

For possums, the main thrusts of research into biological control involve interference with:

- *Fertilisation*: immunisation against proteins from sperm and eggs such that fertilisation will be blocked and no young born (Duckworth 1998). Immunisation of female possums with either whole sperm or porcine zona pellucida (egg coat) resulted in a 75% reduction in fertility (Duckworth 1998; Duckworth *et al.* 1998). The possum zona pellucida-3 gene has recently been cloned and sequenced, and immunogenicity and fertility trials to assess the contraceptive potential of possum ZP3 are underway (Marsupial CRC 1998; J. Duckworth, Landcare Research, Lincoln, unpubl. data). A number of possum sperm proteins have been identified, cloned, and sequenced and are being assessed as candidate contraceptive antigens (Harris & Rodger 1998, M.S. Harris, Landcare Research, Lincoln, unpubl. data).
- *Embryonic development*: immunisation against extracellular matrix (ECM) proteins from the developing embryo, coat proteins produced by the reproductive tract, and a factor (LIF) important for implantation that will result in death of the embryo (Selwood *et al.* 1998), or against a pregnancy-specific protein (riboflavin carrier protein) that is essential for foetal viability (Eckery *et al.* 1998). Current research is directed at identifying, cloning, and sequencing genes for these proteins or the possum homologues so that recombinant proteins can be produced for vaccination trials. Interference with egg coat or ECM proteins has been shown to result in embryonic death *in vitro* (*Selwood et al.* 1998).
- *Post-natal development*: immunisation against proteins essential for sex differentiation (Deakin *et al.* 1998) or the normal development of the ovary or testis such that possums will be born sterile (Eckery *et al.* 1998). Various genes important in early development of the reproductive systems have been identified, sequenced, and cloned (androgen receptor, c-kit, and stem cell factor) and vaccination trials are underway.
- *Central endocrine control of reproduction*: destruction of pituitary cells controlling the

secretion of sex hormones or immunisation against sex hormones, thus inhibiting breeding (Eckery *et al.* 1998). The efficacy of gonadotrophin-releasing hormone-toxin conjugates for sterilising possums is currently being evaluated.

- *Passive immunity*: immunisation against a gut immunoglobulin receptor to prevent development of passive immunity and thus reduce survival of young (Eckery *et al.* 1998). The gene for the possum homologue of a receptor that mediates absorption of maternal immunoglobulin across the gut epithelium (FcRn) is currently being sought (F. Adamski, AgResearch, Ruakura, pers.comm.).
- *Lactation*: immunisation against milk proteins or regulating factors such that pouch young survival will be reduced, or the transfer of antibodies that will disrupt normal development, is enhanced (Demmer *et al.* 1998). Current research is directed at identifying, cloning, and sequencing genes for these proteins (e.g., ELP, LLP, prolactin, Stat5) so that recombinant proteins can be produced for vaccination trials, and determining the identity and temporal patterns of milk immunoglobulins and Ig-receptors, with the intention of targeting the very early stages of lactation.

In each approach, potential species-specificity for possums is likely to rest on the identification of particular epitopes in the key proteins in the same way, for example, that such epitopes have been identified in ZP3 protein of mice (Lou *et al.* 1996). Alternatively, for hormone-based approaches, analogues are available that show differing efficacies between species (Becker & Katz 1997; K. McNatty, AgResearch, Wallaceville, pers. comm.).

By themselves, these biological control methods require delivery to each possum and offer only the potential advantages of increased species-specificity and public acceptability. Poisoning would often be as cheap and would offer the advantages of instant reduction in numbers, greater reduction in numbers, and longer periods below target thresholds (Barlow 1994). The keys to achieving full benefits of biological control lie in the associated delivery systems.

Delivery systems

Initially, biological controls for possums are likely to be delivered by bait. Achieving bait acceptance by a sufficient proportion of the population does not appear to be a problem. With present aerial-baiting technology cereal baits can be delivered routinely to >90% of possums (Chapter 13). The bait will need to contain within it the immunising protein or protein/toxin complex either in recombinant form or in the form of a replicating but non-infectious microbial vector that will infect the host, produce the recombinant protein, and stimulate a sufficient mucosal immune response.

Vaccine delivery is a highly active research area, and many of the developments are likely to be applicable to delivery of possum biological control agents. A range of viral vectors (particularly poxviruses, adenoviruses, and herpesviruses) (Smith 1996) and bacterial vectors (particularly attenuated *Salmonella typhimurium*) (Lintermans & De Greve 1995) have been evaluated as delivery systems for heterologous antigens. An excellent example is the successful use of recombinant vaccinia (cowpox) virus expressing a part of the rabies virus genome to immunise populations of wild foxes (Brochier *et al.* 1990). Aerosol vaccination, as is being trialled to immunise possums with BCG for Tb management (L. Corner, Massey University, Palmerston North, pers. comm.), is another possible method of delivery of biocontrol vaccines. Synthetic, oral, particulate delivery systems are also available based on phospholipid constructs (microspheres, or liposome emulsions) (Miller 1997), although the costs of production are currently high.

Recent developments in delivery systems include bacterial ghosts, which are killed and evacuated bacterial membranes with associated recombinant antigen (Szostak *et al.* 1996), the use of transferrin and immunoglobulin fusion proteins to deliver cytotoxic peptides specifically into target cells that express the appropriate receptor or antigen (Melton & Sherwood 1996), and the use of plants for vaccine production (Mason & Arntzen 1995). Transgenic plants have been developed that express mammalian contraceptive antigens capable of inducing an immune response (Smith *et al.* 1997). Research is currently underway to develop carrots expressing possum-ZP3, a potential contraceptive antigen from possum eggs (Marsupial CRC 1998). A second approach to plant-derived vaccines uses modified plant viruses to carry and express the gene of interest in infected plants (Smith *et al.* 1997). During virus replication and spread, large amounts of target protein are produced, as well as viral protein. Infected plant material could be fed directly to possums or the target protein could be extracted from the leaves, stems, or fruits of infected plants, and repackaged in alternative oral delivery forms. Modified plant viruses can also be used to generate virus-like particles (VLPs). These are usually formed from structural proteins of animal viruses, such as rotavirus, that have bound to them the epitopes of interest. When expressed, the structural proteins self-assemble into particles similar in morphology to native virus and are capable of generating effective immune responses that provide protection against challenge (Jiang *et al.* 1999).

However, the key to the solution of New Zealand's possum problem lies in disseminating biological control, similar to viral-vectored immunocontraception being developed by the Vertebrate Biocontrol CRC for rabbits and mice (Tyndale-Biscoe 1994; Shellam 1994; Jackson *et al.* 1998). Natural transmission from animal to animal spreads the biocontrol agent through the population, so that the effects act at a much larger scale than can be achieved with conventional control and which, with integrated conventional control, offer real possibilities for local or regional eradication (Barlow 1994). A disseminating system requires a suitable possum-specific vector, which is capable of infecting a large proportion of the target population, can be genetically modified to include the genes for the possum antigens, and can express them to a sufficient extent in the host. It may also be necessary to incorporate in the vector genes for cell signals (such as Interleukin-6) that will enhance the immune response to infection with the vector (Alexander & Bialy 1994; Bradley 1994; Ramsey & Ramshaw 1997). A sexually transmitted vector would have the advantage of effective transmission even at low densities.

Suitable vectors are being evaluated from among the parasites and pathogens of possums identified in New Zealand and Australia (Day *et al.* 1997; Gruenberg & Bisset 1998; Meers *et al.* 1998). Use of a vector already present in New Zealand will require consideration of competition between the genetically modified vector and its field strains. This is an issue of concern in the rabbit immunocontraception research programme (Robinson *et al.* 1997), but New Zealand has the option of introducing possum-specific vectors from Australia that are absent from New Zealand, such as the nematode *Adelonema trichosuri*. Modelling of the invasion of existing parasite communities by new species is being used to assess the nature and magnitude of potential problems (Roberts & Dobson 1995). Selection for resistance to infection by the vector is also an issue that will need to be addressed once potential vectors have been identified.

One intriguing concept is the idea of inserting genes for contraceptive antigens and for protection against infection with bovine Tb into the same vector. Infected possums would be both infertile and resistant to infection with Tb.

Will biological control work?

The feasibility of biological control of possums is being demonstrated in three ways. First, computer modelling has identified that the levels of population suppression necessary for Tb management can be achieved by biological control under certain assumptions (Barlow 1991, 1994). Although thresholds for management and target densities for reduction of conservation impacts are not known (Hickling 1995), modelling has also demonstrated that the potential level of population reduction using a biocontrol agent may be as great as that currently achieved by conventional control (Barlow 1994; Parliamentary Commissioner for the Environment 1994).

At this stage modelling has concentrated on what is potentially the "ideal" biocontrol system — a sexually transmitted virus that induces permanent sterility without suppressing reproductive cycles (Barlow 1994, 1997). The key findings of the model are:

- the advantage of targeting females rather than males;
- the very high level of sterilisation (>70%) likely to be required to achieve population suppression;
- reduced efficacy associated with spatial patchiness of infection;
- the advantages of using a sexually transmitted vector;
- enhanced efficacy if there is vector-induced mortality and sterility;
- the benefits of combining biocontrol with conventional control, even if sterilisation is slow-acting or only partially effective.

As new field data become available, the models are being enhanced to include factors such as social structure (Taylor *et al.* 1998), lag periods between infection and sterility, non-response and the potential development of genetic resistance to vectors and/or antigens (N. Barlow, AgResearch, Lincoln, D. Cooper, ex Macquarie University, Sydney, pers. comm.).

Second, the proportion of females that need to be sterilised to reduce wild populations to the desired levels is being assessed in a large-scale field experiment (Ramsey 1998). In local populations, 0%, 50%, or 80% of adult female possums have been surgically sterilised by ligation of the oviducts, and effects on possum numbers, survival, and recruitment measured. The experiment will also provide information on effects on social organisation and behaviour, compensatory breeding, and changes in survival rates, which could counteract the effects of fertility regulation.

Third, studies in captivity have demonstrated the efficacy of vaccines from sperm and eggs in blocking reproduction in female possums (Duckworth *et al.* 1998; J. Duckworth, Landcare Research, Lincoln, unpubl. data) and the lack of effect on behavioural dominance of various methods of reproductive inhibition, such as ovariectomy or vaccination against gonadotrophin-releasing hormone (Jolly *et al.* 1998).

Social and political issues in biological control of possums

To succeed, biological control of possums has to be acceptable ethically, socially, and politically (Oogjes 1997; Singer 1997; Williams 1997). Public opinion was one of the major factors in the Government decision not to allow the importation of myxomatosis into New Zealand. Surveys of attitudes to possum management in New Zealand (Chapter 17; Fraser 1995) identified safety, humaneness, transmissibility, and specificity as key issues. The implementation of biological control of possums, whatever the method, will require careful risk assessment and risk management and, equally important, effective risk communication (Gough 1994) because attitudes and opinions to wildlife management vary widely within and between communities and countries (Gill & Miller 1997; Curtis *et al.* 1997).

It is important, however, to separate the risks associated with the development of biological control methods from those associated with delivery systems. For bait delivery, species specificity and biodegradability are of critical importance, and the main concerns are at the national level. If a transmissible recombinant vector is used, the specificity of the transmissible vector and the specificity of the mode of vector transmission also become major issues, and the concerns become international. To satisfy these concerns, the aim of possum biocontrol is to have twin safeguards of species (or at least marsupial) specificity through the choice of antigens, epitopes, and hormone analogues, and vector specificity through the use of possum-specific organisms.

International concerns will be mostly directed at the risk to the target species or related species in countries where they are indigenous (Tyndale-Biscoe 1997; Williams 1997). If a transmissible vector is used, biological control of possums in New Zealand must not pose an unacceptable risk for possums or other marsupials in Australia, Papua New Guinea and nearby islands, North and South America. These concerns can be addressed by risk assessment of the probability of escape from New Zealand, methods to prevent it, the likelihood of establishment in another country, and the development of contingency plans to eradicate any outbreak.

Tyndale-Biscoe (1997) discussed these issues in relation to the development of immunocontraception for rabbits in Australia, and much of his discussion is relevant to the biological control of possums. Current legislation and its implementation appear sufficient to prevent export of micro-organisms from New Zealand other than by illegal transfer. Legislative controls have been sufficient to prevent the myxoma virus, released in Australia in the 1950s and now widespread in rabbits there, from establishing in New Zealand. They were insufficient, however, to prevent the illegal introduction by farmers of RHD.

The risk of establishment of a possum biocontrol agent in another country will depend critically on competition with existing strains of the same or related vectors. Evidence from studies of the myxoma virus in Australia suggest that competition with existing strains may require careful management of the introduction of new strains if they are to establish and spread successfully (Tyndale-Biscoe 1997; L. Hinds, CSIRO Wildlife & Ecology, Canberra, pers. comm.). Limited events, such as accidental or illegal releases, are therefore likely to have low probability of establishment, although this will need to be established experimentally once candidate vectors for biological control of possums have been identified.

New Zealand, like Australia, has detailed contingency plans for containing outbreaks of major diseases of domestic livestock not present currently in the country, such as foot-and-mouth disease (Sanson 1994). Similar plans could be developed for possum biocontrol agents that would result in rapid containment and eradication of accidental or illegal introductions. Vaccines could also be developed to prevent or limit infection, as has been done to protect domestic rabbits against RHD.

Future challenges

The successful development of biological control for possums is likely to be a long and complex process, both scientifically and in ensuring acceptability of the technology in New Zealand and overseas. Although proof of concept has been demonstrated for sperm and egg antigens as blocks to fertilisation, the potential to compromise many of the other systems being targeted is still being explored. The issue of species specificity is of prime importance. Whether it is possible to provide double safeguards by building specificity into the system at both the antigen level and at the vector/delivery system level remains to be established.

Effective biological control will also depend critically upon the design of the delivery systems. For possums one of the key challenges will be to ensure that a biocontrol vaccine induces a long-lasting immune response in a sufficiently large proportion of the population. Much more needs to be known about possum immunology. But for delivery of baits at least, research investment over the last 5 years has resulted in effective aerial- and ground-baiting systems that achieve high uptake by possums (Morgan & Eason 1996).

Finally, we will need to combat the natural responses of the possums themselves. Pests persist by compensating in some way for attempts to manage them, increasing birth rates or reducing death rates in response to reduced density. Field trials currently underway will allow us to assess the importance of these effects (Ramsey 1998). With time we might also expect possums to show genetic selection for resistance to infection with the biocontrol vector or for non-response to the immunising antigens (P. Bayliss, Environment ACT, Canberra, D. Cooper, Macquarie University, Sydney, pers. comm.), so that strategies will need to be developed to avoid or overcome these problems.

Summary

- A new approach to possum management is needed urgently to complement ongoing incremental improvements in conventional control — biological control may offer such a solution.
- A range of approaches to biological control are currently being explored, most of them based on immunological blocks to production of young.
- We need first to establish whether such approaches can reduce possum numbers to the required levels to satisfy management goals and then explore the range of options for the use of the technique.
- The different risk profiles relating to the development of biological control and the development of delivery systems mean that to some extent these need to be treated separately in any discussions of the ethical, social, or political acceptability of biological control options.
- Like immunocontraception for rabbits and foxes, biological control for possums is a new approach with an uncertain outcome and considerable risk. If the risks, particularly those relating to species specificity and delivery systems, can be demonstrated to be acceptable at national and international levels, then the potential benefits to the New Zealand economy and to enhancing the condition of New Zealand's conservation lands are enormous.

Acknowledgements

Janine Duckworth and Terry Fletcher provided helpful comment on the manuscript. Research on biological control of possums is supported by the New Zealand Foundation for Research, Science and Technology Contract Nos C09602and C09801; Ministry of Agriculture, Policy Division; the Board of Landcare Research; Fletcher Forests; the New Zealand Lottery Science Grants Board; and the Cooperative Research Centre for Conservation and Management of Marsupials.

References

Alexander, N. J.; Bialy, G. 1994: Contraceptive vaccine development. *Reproduction, Fertility and Development 6*: 1–8.

Animal Health Board 1995: National Tb strategy. Proposed national pest management strategy for bovine tuberculosis. Wellington, Animal Health Board.

Atkinson, P. H.; Wright, D. E. 1993: The formulation of a national strategy for biological control of possums and bovine Tb. *New Zealand Journal of Zoology 20*: 325–328.

Baillie, G. J.; Wilkins, R. J. 1998: Retroviruses in the common brushtailed possum (*Trichosurus vulpecula*). *In:* Biological control of possums. *The Royal Society of New Zealand Miscellaneous Series 45*: 20–22.

Barlow, N. D. 1991: A spatially aggregated disease/host model for bovine Tb in New Zealand possum populations. *Journal of Applied Ecology 28*: 777–793.

Barlow, N. D. 1994: Predicting the effect of a novel vertebrate biocontrol agent: a model for viral-vectored immunocontraception of New Zealand possums. *Journal of Applied Ecology 31*: 454–462.

Barlow, N. D. 1997: Modelling immunocontraception in disseminating systems. *Reproduction, Fertility and Development 9*: 51–60.

Becker, S. E.; Katz, L. S. 1997: Gonadotrophin-releasing hormone (GnRH) analogs or active immunization against GnRH to control fertility in wildlife. *In:* Kreeger, T. J. *ed.* Contraception in wildlife management. *Technical Bulletin No. 1853, Animal and Plant Health Inspection Service.* Washington DC, USDA. Pp. 11-20.

Bradley, M. P. 1994: Experimental strategies for the development of an immunocontraceptive vaccine for the European red fox, *Vulpes vulpes. Reproduction, Fertility and Development 6*: 307–317.

Brochier, B.; Thomas, I.; Baudin, B.; Leveau, T.; Pastoret, P. P.; Languet, B.; Chappuis, G.; Desmettre, P.; Blancou, J.; Artois, M. 1990: Use of vaccinia-rabies recombinant virus for the oral vaccination of foxes against rabies. *Vaccine 8*: 101–104.

Cowan, P. E. 1990: Brushtail possum. *In:* King, C. M. *ed.* The handbook of New Zealand mammals. Auckland, Oxford University Press. Pp. 68–98.

Cowan, P. E. 1996: Possum biocontrol: Prospects for fertility regulation. *Reproduction, Fertility and Development 8*: 655–660.

Curtis, P. D.; Decker, D. J.; Stout, R. J.; Richmond, M. E.; Loker, C. A. 1997: Human dimensions of contraception in wildlife management. *In:* Kreeger, T. J. *ed.* Contraception in wildlife management. *Technical Bulletin No. 1853, Animal and Plant Health Inspection Service.* Washington DC, USDA. Pp. 247–255.

Day, T. D.; Waas, J. R.; O'Connor, C. E.; Carey, P. W.; Matthews, L. R.; Pearson, A .J. 1997: Leptospirosis in brushtail possums: is *Leptospira interrogans* serovar *balcanica* environmentally transmitted? *Journal of Wildlife Diseases 33*: 254–260.

Deakin, J. E.; Harrison, G. A.; Cooper, D. W. 1998: Androgen receptor as a potential target for immunosterilisation in the brushtail possum. *In:* Biological control of possums. *The Royal Society of New Zealand Miscellaneous Series 45*: 44–45.

Demmer, J.; Ginger, M. R.; Ross, I. K.; Piotte, C. P.; Grigor, M. R. 1998: Targets in lactation for biocontrol of the common brushtail possum (*Trichosurus vulpecula*). *In:* Biological control of possums. *The Royal Society of New Zealand Miscellaneous Series 45*: 53–58.

Duckworth, J. A. 1998: Reproductive immunology of the brushtail possum (*Trichosurus vulpecula*). *In:* Biological control of possums. *The Royal Society of New Zealand Miscellaneous Series 45*: 49–51.

Duckworth, J. A.; Buddle, B. M.; Scobie, S. 1998: Fertility of brushtail possums (*Trichosurus vulpecula*) immunised against sperm. *Journal of Reproductive Immunology 37*: 125–138.

Eckery, D.; Lawrence, S.; Greenwood, P.; Stent, V.; Ng Chie, W.; Heath, D.; Lun, S.; Vanmontfort, D.; Fidler, A.; Tisdall, D.; Moore, L.; McNatty, K. P. 1998: The isolation of genes, novel proteins, and hormones and the regulation of gonadal development and pituitary function in possums. *In:* Biological control of possums. *The Royal Society of New Zealand Miscellaneous Series 45*: 100–110.

Fraser, W. 1995: Public attitudes to introduced wildlife and wildlife management in New Zealand. Proceedings: 10th Australian Vertebrate Pest Control Conference, Hobart, Tasmania, 29 May–2 June 1995. Tasmania, Department of Primary Industry and Fisheries. Pp. 101–106.

Gill, R. B.; Miller M. W. 1997: Thunder in the distance: the emerging policy debate over wildlife contraception. *In:* Kreeger, T. J. *ed.* Contraception in wildlife management. *Technical Bulletin No. 1853, Animal and Plant Health Inspection Service.* Washington DC, USDA. Pp. 257–267.

Gough, J. 1994: Risk communication and risk management. *In:* Seawright, A. A.; Eason, C. T. *ed.* Proceedings of the science workshop on 1080. *The Royal Society of New Zealand Miscellaneous Series 28*: 39–44.

Gruenberg, A.; Bisset, S. 1998: *Parastrongyloides trichosuri* as a potential vector for genes expressing proteins which could interfere with reproductive success, growth, or longevity of brushtailed possums. *In:* Biological control of possums. *The Royal Society of New Zealand Miscellaneous Series 45*: 10–12.

Harris, M. S.; Rodger, J. C. 1998: Characterisation of fibrous sheath and midpiece fibre network polypeptides of marsupial spermatozoa with a monoclonal antibody. *Molecular Reproduction and Development 50*: 461–473.

Heath, D.; Cowan, P.; Stankiewicz, M.; Clark, J.; Horner, G.; Tempero, J.; Jowett, G.; Flanagan, J.; Shubber, A.; Street, L.; McElrea, G.; Chilvers, L.; Newton-Howse, J.; Jowett, J.; Morrison, L. 1998: Possum biological control — parasites and bacteria. *In:* Biological control of possums. *The Royal Society of New Zealand Miscellaneous Series 45*: 13–19.

Hickling, G. J. 1995: Action thresholds and target densities for possum pest management. *In:* O'Donnell, C. F. J. *comp.* Possums as conservation pests. Proceedings of an NSSC Workshop . . . 29–30 November 1994. Wellington, Department of Conservation. Pp. 47–52.

Innes, J.; Hay, R.; Flux, I.; Bradfield, P.; Speed, H.; Jansen, P. 1999: Successful recovery of North Island kokako *Callaeas cinerea wilsoni* populations, by adaptive management. *Biological Conservation 87*: 201–214.

Jackson, R. J.; Maguire, D. J.; Hinds, L. A.; Ramshaw, I. A. 1998: Infertility in mice induced by a recombinant *Ectromelia* virus expressing mouse zona pellucida glycoprotein 3. *Biology of Reproduction 58*: 152–159.

Jiang, B.M.; Estes, M.K.; Barone, C.; Barniak, V.; O'Neal, C.M.; Ottaiano, A.; Madore, H.P.; Conner, M.E. 1999: Heterotypic protection from rotavirus infection in mice vaccinated with virus-like particles. *Vaccine 17*: 1005–1013.

Jolly, S. E. 1993: Biological control of possums. *New Zealand Journal of Zoology 20*: 335–339.

Jolly, S. E. 1994: Biotechnological control. Possum control for the 21st century. *Forest and Bird No. 272*: 26–31.

Jolly, S. E.; Scobie, S.; Spurr, E. B.; McAllum, C.; Cowan, P. E. 1998: Behavioural effects of reproductive inhibition in brushtail possums. *In:* Biological control of possums. *The Royal Society of New Zealand Miscellaneous Series 45*: 125–127.

Lintermans, P.; Degreve, H. 1995: Live bacterial vectors for mucosal immunization. *Advanced Drug Delivery Reviews 18*: 73–79.

Kreeger, T.J. (*ed.*) 1997: Contraception in wildlife management. *Technical Bulletin No. 1853, Animal and Plant Health Inspection Service.* Washington DC, USDA.

Loague, P. 1993: Pest control and animal welfare. *New Zealand Journal of Zoology 20*: 253–255.

Lou, Y-H.; Bagavant, H.; Ang, J.; McElveen, M. F.; Thai, H.; Tung, K. S. K. 1996: Influence of autoimmune ovarian disease pathogenesis on ZP3 contraceptive vaccine design. *Journal of Reproduction and Fertility Supplement 50*: 159–163.

Marsupial CRC 1998: Annual report of the Cooperative Research Centre for Conservation and Management of Marsupials. Sydney, Marsupial CRC, Macquarie University.

Mason, H. S.; Arntzen, C. J. 1995: Transgenic plants as vaccine production systems. *Trends in Biotechnology 13*: 389–392.

Meers, J.; Perrott, M.; Rice, M.; Wilks, C. 1998: The detection and isolation of viruses from possums in New Zealand. *In:* Biological control of possums. *The Royal Society of New Zealand Miscellaneous Series 45*: 25–28.

Melton, R. G.; Sherwood, R. F. 1996: Antibody-enzyme conjugates for cancer therapy. *Journal of the National Cancer Institute 88*: 153–165.

Miller, L. A. 1997: Delivery of immunocontraceptive vaccines for wildlife management. *In:* Kreeger, T. J. *ed.* Contraception in wildlife management. *Technical Bulletin No. 1853, Animal and Plant Health Inspection Service.* Washington DC, USDA. Pp. 49–58.

Morgan, D.; Eason, C. 1996: Improving conventional approaches to possum control. Annual Report from the National Science Strategy Committee on Possum/Bovine Tuberculosis Control. (Wellington, Ministry of Research, Science and Technology). Pp. 16–20.

National Science Strategy Committee (NSSC) for Possum and Bovine Tuberculosis Control. 1997: Annual Report. Wellington, New Zealand, NSSC. 79 p.

Obendorf, D.; Spratt, D.; Beveridge, I.; Presidente, P.; Coman, B. 1998: Parasites and diseases in Australian brushtail possums, *Trichosurus* spp.. *In:* Biological control of possums. *The Royal Society of New Zealand Miscellaneous Series 45*: 6–9.

O'Hara. P. J. 1997: Decision on the application to approve the importation of rabbit calicivirus as a biological control agent for feral rabbits. Wellington, Ministry of Agriculture.

O'Keefe, J.; Wickstrom, M. 1998: Viruses of the brushtail possum in New Zealand. *In:* Biological control of possums. *The Royal Society of New Zealand Miscellaneous Series 45*: 23–24.

Oogjes, G. 1997: Ethical aspects and dilemmas of fertility control of unwanted wildlife: an animal welfarist's perspective. *Reproduction, Fertility and Development 9*: 163–167.

Parkes, J.; Baker, A. N.; Ericksen, K. 1997: Possum control by the Department of Conservation: background, issues, and results from 1993 to 1995. Wellington, Department of Conservation. 40 p.

Parliamentary Commissioner for the Environment 1994: Possum Management in New Zealand. Wellington, New Zealand, Office of the Parliamentary Commissioner for the Environment. 196 p.

Ramsey, D. S. L. 1998: Population responses of wild possums to fertility control. *In:* Biological control of possums. *The Royal Society of New Zealand Miscellaneous Series 45*: 111–114.

Ramsay, A. J.; Ramshaw, I. A. 1997: Cytokine enhancement of immune responses important for immunocontraception. *Reproduction, Fertility and Development 9*: 91–98.

Roberts, M. G.; Dobson, A. P. 1995: The population dynamics of communities of parasitic helminths. *Mathematical Biosciences 126*: 191–214.

Robinson, A. J.; Jackson, R.; Kerr, P.; Merchant, J.; Parer, I.; Pech, R. 1997: Progress towards using recombinant myxoma virus as a vector for fertility control in rabbits. *Reproduction, Fertility and Development 9*: 77–83.

Sanson, R. 1994: EpiMAN — A decision support system for managing a foot and mouth disease epidemic. *Surveillance 21(1)*: 22–24.

Selwood, L.; Frankenberg, S.; Casey, N. 1998: An overview of the importance of the egg coats for embryonic survival in the common brushtail possum: normal development and targets for contraception. *In:* Biological control of possums. *The Royal Society of New Zealand Miscellaneous Series 45*: 89–95.

Shellam, G. R. 1994: The potential of murine cytomegalovirus as a viral vector for immunocontraception. *Reproduction, Fertility and Development 6*: 129–138.

Singer, P. 1997: Neither human nor natural: ethics and feral animals. *Reproduction, Fertility and Development 9*: 157–162.

Smith, G. 1996: Vaccine delivery systems: Genetically engineered viruses as candidate vaccines. *Vaccine 14*: 681–683.

Smith, G. L.; Walmsley, A.; Polkinghorne, I. 1997: Plant-derived immunocontraceptive vaccines. *Reproduction, Fertility and Development 9*: 85–89.

Stewart, H.; Jenkin, B.; McCarthy R.; Flinn, D. 1994: Public versus user perceptions of risks and benefits of plantation forestry in Victoria, Australia. *In:* Seawright, A. A.; Eason, C. T. *ed* Proceedings of the science workshop on 1080. *The Royal Society of New Zealand Miscellaneous Series 28*: 10–19.

Szostak, M. P.; Hensel, A.; Eko, F.O.; Klein, R.; Auer, T.; Mader, H.; Haslberger, A.; Bunka, S.; Wanner, G.; Lubitz, W. 1996: Bacterial ghosts: non-living candidate vaccines. *Journal of Biotechnology 44*: 161–170.

Taylor, A.; Cooper, D.; Fricke, B.; Cowan, P. 1998: Population genetic structure and mating system of brushtail possums in New Zealand — implications for biocontrol. *In:* Biological control of possums. *The Royal Society of New Zealand Miscellaneous Series 45*: 122–124.

Tyndale-Biscoe, C. H. 1994: Virus-vectored immunocontraception of feral mammals. *Reproduction, Fertility and Development 6*: 9–16.

Tyndale-Biscoe, C. H. 1997: Immunosterilization for wild rabbits: the options. *In:* Kreeger, T.J. *ed.* Contraception in wildlife management. *Technical Bulletin No. 1853, Animal and Plant Health Inspection Service.* Washington DC, USDA. Pp. 223–234.

Waage, J. K.; Mills, N. J. 1992: Biological control. *In:* Crawley, M. J. *ed.* Natural enemies. Oxford, Blackwell Scientific.

Williams, C. K. 1997: Development and use of virus-vectored immunocontraception. *Reproduction, Fertility and Development 9*: 169–178.

Williams, J. M. 1994: Food and fibre markets and societal trends: implications for pest management. *In:* Seawright, A. A.; Eason, C. T. *ed.* Proceedings of the science workshop on 1080. *Royal Society of New Zealand Miscellaneous Series 28*: 20–32.

CHAPTER TWENTY-FIVE

Development of Decision Support Systems for Possum Management

David Choquenot and John Parkes

What is a decision support system?

In its most general sense, a *decision support system* (DSS) is any tool that helps its users make better decisions. A DSS may include single or multiple elements. Information can be "contained" in groups of interested people, books, manuals, tables, videos, and photographs as well as in computer software (Stuth & Stafford-Smith 1993). Almost any integrated source of information that by itself or in combination with other sources helps people make better decisions can be represented as a DSS.

Where a DSS attempts to support decision-making processes that take many variables into account; consider large quantities of data; deal with some level of uncertainty; or attempt to integrate combinations of biological, physical, and economic factors; the complexity of decision making can rapidly become overwhelming. Under these conditions, the analytical speed of computers becomes a necessary adjunct to the information managers have available, to aid their decision making. For example, decision making in natural resource systems often requires managers to consider both the biological and economic value of a range of natural attributes. These attributes are affected by a multitude of biophysical factors and ecological processes. Further complexity is added in that many of these processes will have a significant degree of uncertainty associated with them. In this situation, a computer-based DSS will often be the only way that managers can take full account of the consequences of their decisions and the contingencies that affect these consequences.

In order to support decision making, a DSS must ultimately help a manager decide between various options for action. To that end, the support offered by a DSS can be thought of as proceeding through three sequential phases: (a) *informing* the manager, (b) allowing the manager to *compare* between alternative actions, and (c) helping the manager *identify* the "best" of these alternatives. Not every DSS will proceed to the third or even second phase and will simply present basic information that helps a manager reach a decision in a structured fashion. Good examples of this are expert systems that retain information in a structure that can be actively interrogated by a manager. However, it was recognised early in the development of more integrative DSS that, where a management system was complex, a way of comparing the outcomes of alternative actions or considering the effect of various contingencies was an essential part of the decision-making process (Thieraf 1982; Bennett 1983). Similarly, where a range of alternative actions are available to a manager, some structured way of identifying the options that result in the best net gains from management was desirable (Stuth & Stafford-Smith 1993). Because comparison of alternative management actions and identification of best options require much of the same information, these two processes are often linked in a DSS through models that simulate the effect of alternative management actions.

Using DSS to formulate and compare possum control strategies

Possums in New Zealand are controlled to reduce their impact on native and commercial forests, reduce their predation on native birds and invertebrates, and decrease the incidence of Tb in domestic livestock. Various DSS have been developed to help managers make decisions about when, where, and how to control possums. These include "EpiMAN (TB)", which integrates a number of computer-based tools that predict where possum control will be most effective for Tb management (MacKenzie 1997), and "Optimise", which is also a computer-based application, but focuses on estimating the relative costs of different options for possum control (McGlinchy 1996). The emphasis of these and other existing DSS is on enhancing the efficiency of possum management either by improving how control activities are targeted or identifying the most cost-effective alternatives for control.

Choquenot *et al.* (1998) describe a simple prototype DSS based on a computer simulation model that integrates these approaches. It uses available data on possum population dynamics and the costs associated with aerial poisoning with 1080 and ground hunting to predict and compare the cost-effectiveness of alternative strategies for possum control (Fig. 25.1). A control strategy is some combination of possum control techniques implemented on a fixed schedule or in relation to changes in possum density.

As an example, Figure 25.2 contrasts predicted variation in annual control costs and possum density for two control strategies over 100 years. The strategies contrasted are (a) a 10-year cycle of aerial baiting and (b) a one-off aerial-baiting campaign followed up with ground hunting every 3 years if possum density exceeds 0.1 possums/ha. The average annual costs associated with the first strategy (NZ$2.22/year) are clearly less than those associated with strategy (b) (NZ$4.89/year). However, the

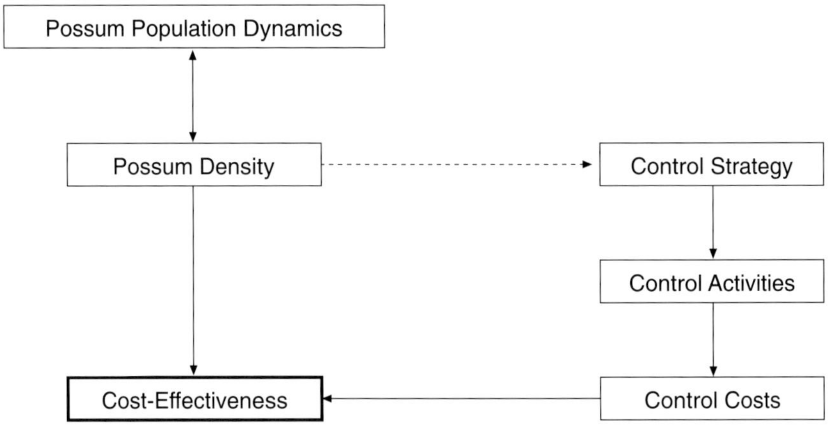

Fig. 25.1
The structure of a prototype DSS for possum management that integrates elements predicting possum population dynamics and the cost-effectiveness of different forms of possum control.

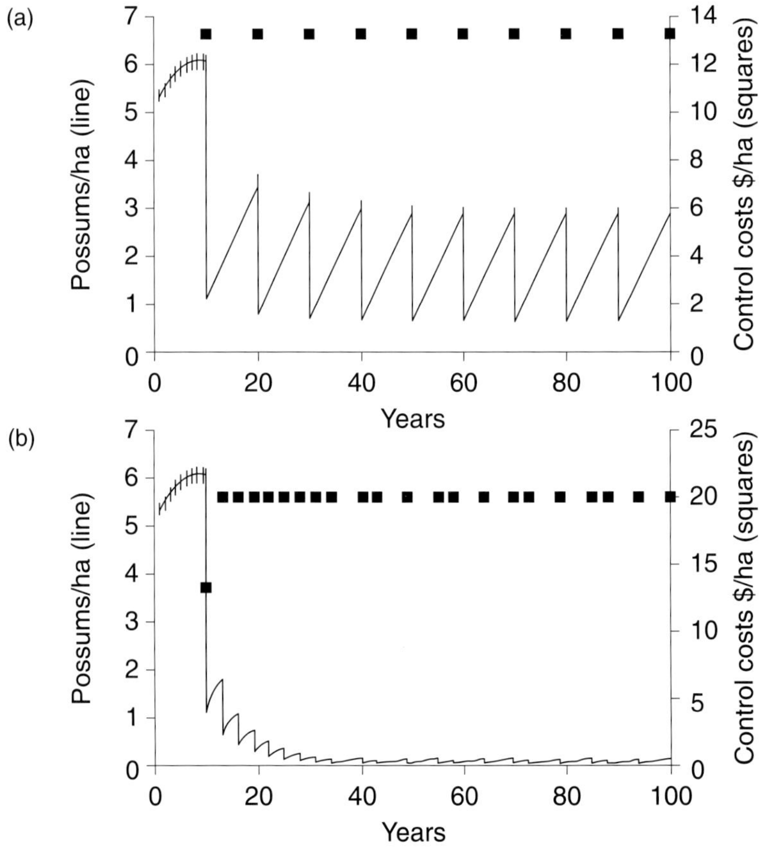

Fig. 25.2 (a) & (b)
Variation in predicted annual control costs and possum density for two control strategies over 100 years. The strategies are (a) a 10-year cycle of aerial baiting and (b) a one-off aerial-baiting campaign followed up with ground hunting every 3 years if possum density exceeds 0.1 possum/ha. Predictions come from a prototype possum management DSS (Choquenot *et al.* 1998).

second strategy results in much more intensive possum control, which dramatically decreases the average number of possums surviving between control operations. A manager could use these differences to determine which strategy best met the ultimate objective of specific possum control initiatives. For example, if possum control was being undertaken primarily to protect native forests, the average reduction in possum density achieved by the first strategy (c. 60%) may be sufficient to protect forest canopies. If so, the extra cost of the second strategy would be hard to justify. In contrast, if the objective of possum control was to reduce the incidence of Tb in cattle herds adjoining the control area, aerial baiting on a 10-year cycle may not hold possums at sufficiently low densities to break the disease cycle. Under these conditions, the extra cost incurred by the more intensive ground hunting strategy may well be necessary if the control objectives are to be met.

Using a DSS to identify "best" strategies for possum control

Of course the strategies illustrated in Figure 25.2 are only two of an almost infinite range of potential combinations of control techniques, treatment intervals, and/or threshold possum densities at which control could be undertaken. Rather than simply comparing two arbitrarily defined scenarios, a better solution is to use the DSS to identify the best strategy for achieving specific control objectives. The best strategy may be that which reduces possum density to some specified level most cheaply or most quickly, or that which maximises the overall reduction in possum density achieved for a given budget.

For example, as previously mentioned, an important objective of possum control will often be to reduce the incidence of Tb in cattle herds adjoining a control area. If the probability of Tb cycling in possum populations held below densities of 1 possum/ha was very low, a manager would be interested in identifying the most cost-effective strategy of maintaining possum density below this target level. The DSS described above can be used to estimate how long different maintenance control intervals take to *permanently* suppress possum densities below 1 possum/ha, and whether or not this outcome is dependent on the control technique used (Fig. 25.3). In this example, aerial baiting appears to achieve permanent suppression of possum density more rapidly than does ground hunting (e.g., aerial

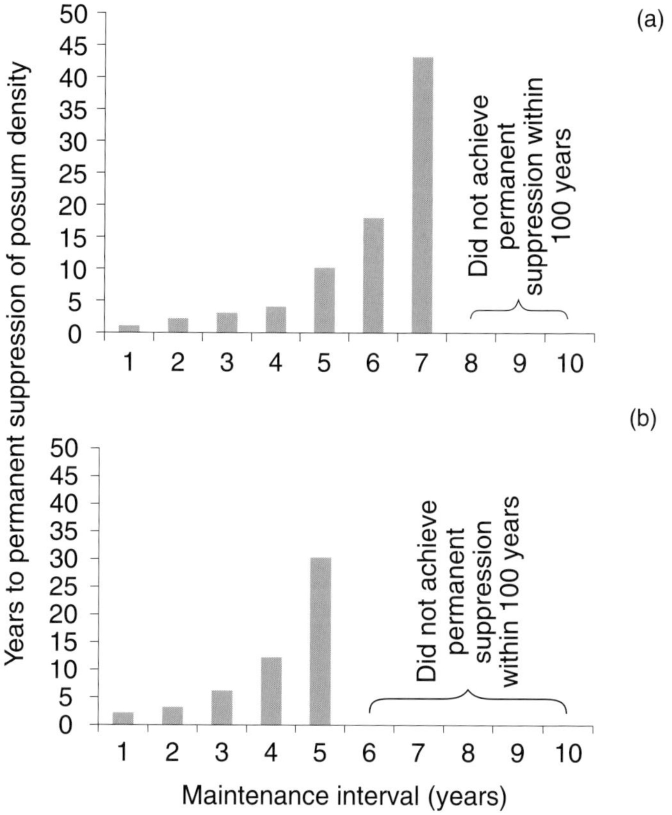

Fig. 25.3
The predicted time that different maintenance intervals employing two possum control techniques take to permanently suppress possum density below 1 possum/ha. The possum control techniques modelled are (a) aerial baiting, and (b) ground hunting. Predictions come from a prototype possum management DSS (Choquenot *et al.* 1998).

baiting requires 10 years to achieve permanent suppression of possum density using a 5-year maintenance interval, while ground hunting requires 30 years at the same interval). Aerial baiting also achieves permanent suppression at longer maintenance intervals (i.e., the maximum maintenance interval that achieved permanent suppression of possum density within 100 years was a 7-year interval for aerial baiting and a 5-year interval for ground hunting).

A manager primarily concerned with the ongoing cost of maintenance possum control in the Tb area would prefer the longer maintenance intervals and lower consequent costs associated with aerial baiting. For example, the average annual maintenance cost using a 7-year interval for aerial baiting is NZ$1.91/ha compared with NZ$4.26/ha for a 3-year interval. In contrast, a manager who required a rapid reduction in possum density below the threshold for disease persistence would prefer the 3-year maintenance interval because it achieved this reduction in 3 years compared with 43 years for the 7-year maintenance interval. To identify optimal possum control strategies in this way, a DSS must have a flexible structure in which various combinations of control strategy and techniques can be examined. To further improve its capacity to identify optimal control strategies, a DSS may use looped structures that allow the effect of combinations of management decisions to be automatically reiterated and their relative efficiency compared.

The future: impacts, uncertainty and spatial complexity

All of the examples given above assume that the possum control objectives that a DSS helps to achieve can be represented solely in terms of possum density. While this may be true for impacts where there is a threshold possum density above which impacts are unacceptable (e.g., the disease thresholds used to control possums for Tb management), the relationship between possum density and the extent of impacts on biological resources will generally be more complex. For example, a simple hypothetical model of change in forest canopy condition can be added to the prototype DSS described above, by assuming that future canopy condition depends on both its current condition and prevailing possum density (i.e., change in canopy condition is assumed to approximate logistic vegetation growth and be affected by possums in proportion to their prevailing density). The resultant model can be used to simulate change in possum density and canopy condition under different possum control regimes.

Figure 25.4 models these changes for a 10-year cycle of aerial baiting assuming both possum density and canopy condition start at around half their maximum levels. While the control strategy quickly locks variation in possum density into a relatively regular pattern, the reciprocal effect on canopy condition is neither simple nor necessarily consistent with that regular pattern. Figure 25.5 shows the reciprocal pattern of variation in possum density and canopy condition for the simulation shown in Figure 25.4. Clearly under these conditions a target possum density will be of limited use if the objective of possum management is to achieve some consistent response in terms of canopy condition. The trophic interactions that obscure any consistent relationship between possum density and canopy condition in this example are likely to be at least as complex for other likely impacts of possum consumption. For example, the existence of simple relationships between possum density and rates of change in affected bird species is unlikely given that eggs probably constitute an infrequent and minor part of the possum diet.

These complex relationships between the density of possums and changes in the resources they affect have important implications for DSS. In the absence of information on the form of these relationships, a DSS will only ever be able to help managers make decisions about the best ways of managing possums to prescribed densities. Given that managers are increasingly charged with managing the impact of possums, rather than possum density per se, a DSS that does not take the broader impacts of possums into account will be of limited management use. However, given the complexity of the functional processes that will generally underpin how resources affected by possums respond to possum management, the research needed to move a DSS to this next level of development is daunting. Further complicating our capacity to disentangle these functional processes is the growing realisation that they are subject to (a) high levels of uncertainty reflecting the influence of unpredictable factors such as the weather and capricious land management decisions, and (b) complex spatial dynamics that influence rates of change in possum density and their propensity to consume some resources. One potential solution to this complexity is to use large-scale experiments to directly measure the response of resources to management interventions, reducing the emphasis on some of the biological detail that drives these responses. This approach to experimental management is called *adaptive management*, and has been given considerable prominence in a range of resource management contexts over the past 20 years (Walters 1986).

DSS and adaptive management

Adaptive management is a term that appears to have wide usage but often little currency. At its most basic level, adaptive management simply means that by monitoring the effectiveness of changes to a management process, improvements in the process can be made. However, over the past 20 years, this simple principle has been extended to include many of the details and intricacies of more formal experimental design, including the need for replication and the use of non-treated (non-managed) controls (Walters 1986).

An increasing use of mathematical models as an integral part of adaptive management provides an obvious linkage to the process of developing and refining DSS for resource management systems. In an adaptive management programme, any alternative views that managers have about how their actions affect the way an ecosystem functions, are formulated as a series of "competing" models. These models are used to make predictions about how an ecosystem will respond to a change in the way it is managed. By comparing these predictions with the results of monitoring on sites where different

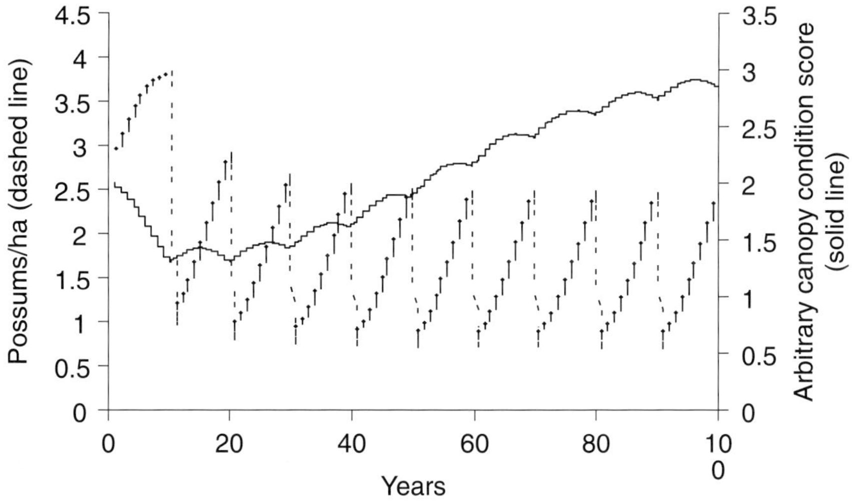

Fig. 25.4
Variation in predicted possum density and an arbitrary canopy condition score (0 is canopy collapsed, 4 is an unaffected canopy) over 100 years, with possums controlled using aerial baiting on a 10-year cycle. Predictions come from a prototype possum management DSS (Choquenot *et al.* 1998).

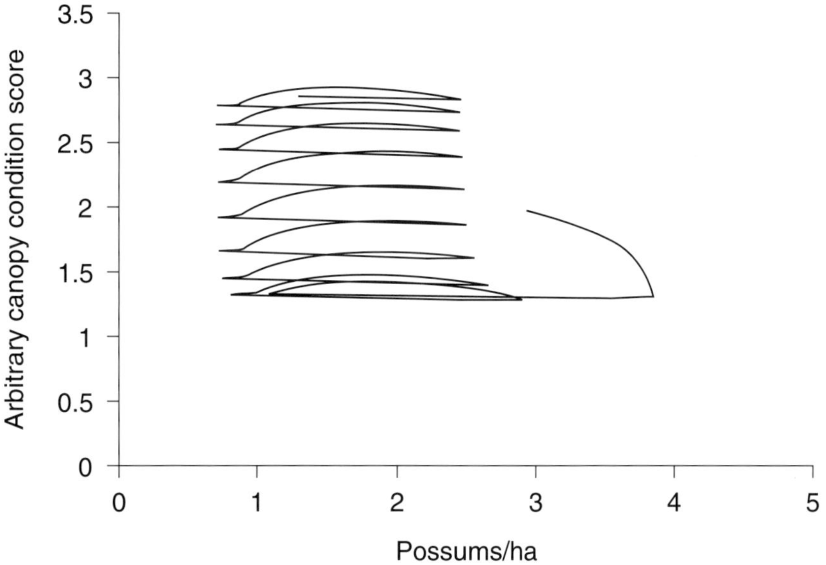

Fig. 25.5
Variation in predicted possum density and an arbitrary canopy condition score (0 is canopy collapsed, 4 is an unaffected canopy) as a function of possum density, with possums controlled using aerial baiting on a 10-year cycle. Predictions come from a prototype possum management DSS (Choquenot *et al.* 1998).

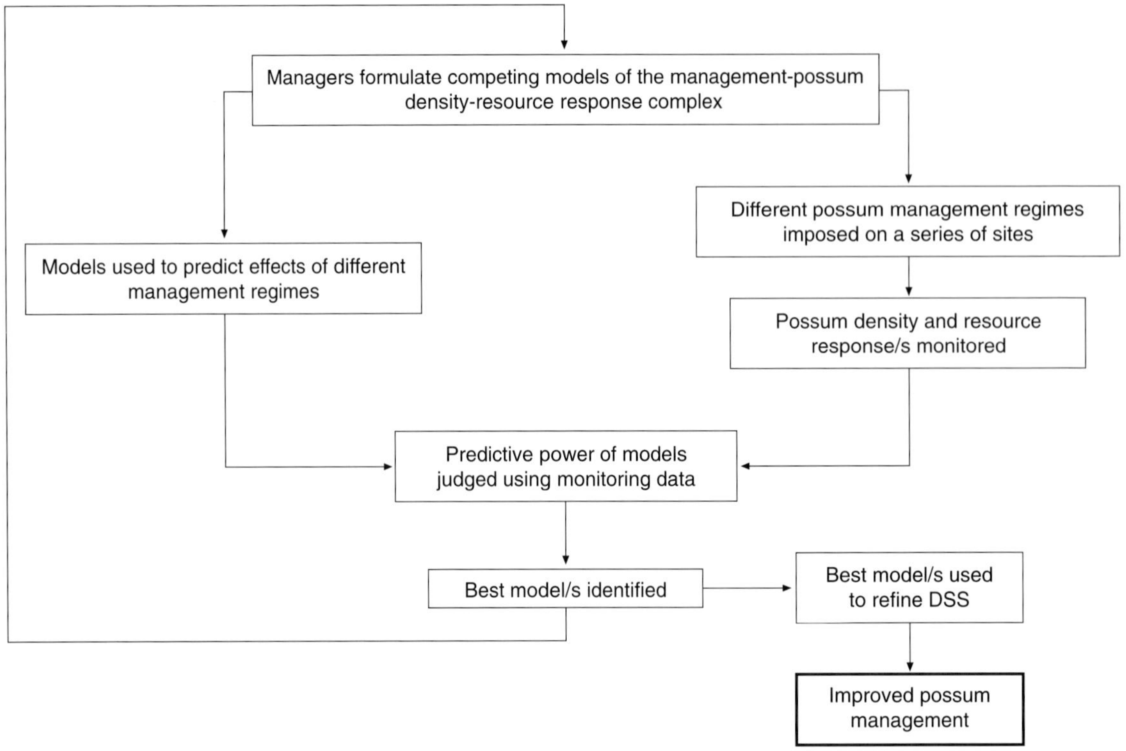
Managers formulate competing models of the management-possum density-resource response complex
Different possum management regimes imposed on a series of sites
Models used to predict effects of different management regimes
Possum density and resource response/s monitored
Predictive power of models judged using monitoring data
Best model/s identified
Best model/s used to refine DSS
Improved possum management

Fig. 25.6
A diagram of an adaptive management programme for possum management and how the programme leads to improvements in DSS used to help managers make possum management decisions. The programme leads to continuous refinement of understanding about how management influences possum density and responses in affected resources by linking modelling and monitoring as integral parts of the management process. Using the models that develop out of the adaptive management programme to DSS ensures that these refinements are captured in a way that makes them immediately available to possum managers not directly involved in the programme.

management regimes are implemented, the most accurate model can be identified. When viewed in this way, the competing models used in adaptive management programmes become the equivalent of the hypotheses that are tested in more formal scientific experiments.

Opinions vary about how possums and vegetation interact, about how forests respond to different possum control strategies, and about how the risk of domestic animal herds contracting Tb varies with possum density. Data currently available cannot unequivocally demonstrate which opinion is right or wrong. The lack of a consensus concerning how resources respond to various forms of possum control is perhaps best illustrated by the range of maintenance intervals used in possum control operations undertaken to protect conservation resources. Possums in some "mainland island" reserves are controlled annually or even continuously, while in other areas control is undertaken every 5 or 7 years. As discussed elsewhere in this volume, factors such as control technique, the potential for bait shyness, likely rates of reinvasion, and the susceptibility of the affected resource will complicate this issue still further. Research into the response of possum populations and the condition of affected resources to different possum control strategies has the potential to improve the effectiveness of possum control. If this work is conducted in a formal adaptive management framework, employing adequate replication of management "treatments" in order to distinguish between competing models of the managed possum-resource complex, it will also provide important guidance to where the results of more focused studies will most directly improve possum management.

An adaptive management programme that feeds directly into a DSS outcome is shown diagrammatically in Figure 25.6. Using "what-if" scenarios based on simple DSS such as that described above, managers identify their best-bet strategies for optimal possum management. The logic underlying these consensus decisions is formalised into a series of competing models that are then used to predict how possum populations and affected resources (i.e., forests, endangered species population abundance, or Tb prevalence) will respond under a range of different possum management regimes. These regimes are then imposed across a series of sites at an appropriate spatial and temporal scale, and the possum population and resources on those sites are intensively monitored. By contrasting model predictions and these monitoring data, the most accurate models are identified and used to formulate a new set of competing models and the adaptive management process continues. Progressively more accurate models of how the management-possum density-resource response complex behaves are an obvious outcome of the adaptive management process. By using these models as the basis for an evolving DSS, the adaptive management programme can be extrapolated to any number of possum management operations, and the DSS becomes a constantly evolving management tool.

Summary

- A decision support system (DSS) is any tool that helps managers make better decisions.
- Where decisions involve multiple variables, large quantities of data, different sorts of factors and outcomes, or some level of uncertainty, a DSS will generally have to rely on the analytical speed a computer-based system offers.
- Although several DSS already exist to help managers make some of the decisions involved in possum control, there are currently no comprehensive and robust systems that allow managers to simulate and compare alternative control strategies.
- In developing such systems, the need to predict changes in the resources possums affect rather than possum density per se must be recognised.
- Adaptive management is a potentially powerful approach to achieving the understanding of interactions between possum management, possum density, and resource condition necessary for the development of more sophisticated DSS.
- Because adaptive management programmes require the use of predictive models, they are highly conducive to the development of DSS.

References

Bennett, J. W. 1983: Building decision support systems. Mass., USA, Addison-Wesley.

Choquenot, D.; McGlinchy, A.; Parkes, J. 1998: Spatial decision support systems for pest control. — a progress report. Landcare Research Contract Report LC9798/132 (unpublished) 8 p.

MacKenzie, J. 1997: A decision support system to assist in control of possums for TB purposes. *In:* Seminar proceedings on possum and mustelid control research, National Possum Control Agencies, Wellington, New Zealand. Pp. 129–132.

McGlinchy, A. 1996: Decision support tools for possum management. *In:* Improving conventional control of possums. *The Royal Society of New Zealand Miscellaneous Series 35*: 35–38.

Stuth, J. W.; Stafford-Smith, M. 1993: Decision support systems for grazing lands: An overview. *In:* Stuth, J. W.; Lyons, B. G. *ed.* Decision support systems for the management of grazing lands: emerging issues. Paris, France, UNESCO. Pp. 1–35.

Thieraf, R. J. 1982: Decision support systems for effective planning and control. Englewood Cliffs, New Jersey, USA, Prentice Hall.

Walters, C. J. 1986: Adaptive management of renewable resources. New York, Macmillan.

Editor's Conclusion — The beginning of the end or the end of the beginning

In the 4 years it has taken to produce this book there has been a significant shift not only in the way the possum problem is perceived and managed but also in the commercial value of the possum. Some of the changes appear to have been prompted by the prevailing corporate culture and economic imperative to show a benefit for funds spent. Others have been led by scientific progress or prompted by the legislative requirements of the Resource Management Act (1991) and the Biosecurity Act (1993) that public consultation be an integral part of pest management. Whatever these changes have been, most have not been examined in previous chapters because in many instances they occurred after the pertinent chapter was written or were outside the brief given to authors. This concluding chapter reflects on comments prompted by previous chapters and uses the benefit of hindsight to draw some conclusions on the five areas critical to possum management.

Possum impacts, control and the need to show a benefit

Possums have been in New Zealand for almost 150 years and their status as pests in native forests has been the subject of debate for most of this century. Lamentably, their impact in native forests was not measured in any systematic way until the 1940s and 1950s, and only in recent years has it been measured quantitatively (Chapter 10). The impact of possums on forestry and agriculture has rarely been measured in anything more than qualitative terms (Chapter 9), which means it is still very difficult to assess whether some control operations are worth undertaking.

Large-scale poisoning of possums in forested areas (>1000 ha) commenced in 1954 when managers began using aircraft to drop carrot baits loaded with 1080 (Chapter 13). This practice has continued to the present day, but with a special focus on refining the technique. Little attention was paid to measuring the benefits of killing possums in forest and farm areas and the benefits are still poorly documented.

The reasons for this appear to have been manyfold. The benefits of reducing or removing possums from forested areas may take some time to become evident, as for example some plants grow slowly. Furthermore the benefits of possum control may be manifested in many ways such as increased flowering, fruiting, and seedling recruitment of forest plants (Chapter 21). They may also depend on the floristic nature of the vegetation, the presence of other mammals such as deer and goats, and site-specific characteristics of soil, climate, and topography. Historically speaking it was also assumed that high possum kills were bound to produce benefits, yet in hindsight we now know this is not always the case.

All this means is that we should have been measuring the benefits of controlling possums in forested areas right from when possum control operations first commenced, and assessing the success of operations based on the expected benefits. The doubt expressed in Chapter 22 as to whether we may ever know the direct benefits of controlling possums to native wildlife because of the confounding factor that poisoning operations often also kill rats, is pessimistic. Current video surveillance technology now offers a very easy method to monitor one or several nests and partition egg, chick, and adult losses due to rat or possum predation.

The unique and perverse nature of bovine tuberculosis (Tb) in New Zealand has also made it difficult to show clear benefits of poisoning possums when seeking to reduce the incidence of Tb in livestock. The reduction of possum numbers in and around herds once infected with bovine Tb has enabled many farms to become and remain free of Tb (Chapter 20). Data presented (in Fig. 20.3) are compelling and the 1999 results show a further decline so that herd reactor rates are now 56% below their peak in 1994. Similar examples are needed as unfortunately Hohotaka (see Fig. 20.2) is the only documented example with an experimental control, and most scientists will not be willing to generalise from one example. No doubt more evidence will accumulate with time.

Changes in techniques, technologies, and management of possum control

As already mentioned, in the last 4 years there has been quite a change in the techniques and technologies used to control possums. Until the mid-1990s most large-scale control operations were undertaken using poisoned 1080 carrots or cereal pellets dropped from aircraft. This has now changed so that hunters using traps, cyanide paste, and bait stations are routinely contracted to control possums in areas up to 5000 ha in size. This has been made possible by the introduction of new possum bait formulations such as Campaign® and Feratox®, and changes in the way pest control operations are administered.

In recent decades possum control operations that were once managed and undertaken by government organisations, such as the landowner-organised Pest Destruction Boards, and the New Zealand Forest Service, are now managed by regional authorities on behalf of the Animal Health Board, and by the Department of Conservation (DOC). Both DOC and regional authorities regularly contract the services of pest control operators, particularly for ground control operations. Tenders for ground control contracts are very competitive, and it appears this competition has accelerated the number of innovations in the way poisons are used. The introduction of operational monitoring of hunter performance (which measures the impact of management on possum numbers, see Chapter 12) has underpinned this process and even changed the reason why hunters are paid for hunting possums. In the past most possum hunters harvested possums as a resource because the possums themselves were valuable. Today hunters now mostly kill possums because possums are a pest, and it is their absence that is valued.

The motive for hunting possums, however, is set to change again because the value of the possum as a resource has recently increased. As pointed out in Chapter 23 possum skin prices fluctuate over time but the skin price, and consequently the motivation for hunting possums as a resource, has generally declined since 1980. This downward trend in the value of possums as a resource has been arrested by textile manufacturers that are now seeking possum fur fibre (fur removed from the skin) to blend with wool for the production of very soft, warm, high-value garments. There is now a shortage of possum fibre, currently valued at NZ$45/kg and hunters and inventors have quickly responded to this need by developing machines such as the Possum Plucker, for removing fibre from possums carcasses in the field. The demand for possum fibre will reintensify the conflict between the status of the possum as a resource and possums as a pest.

Possums as a resource and possums as a pest

With the development of the commercial demand for possum fur, fibre, and probably meat (the latter to be sold into Asian markets), it is likely that there will be pressure from some people in the community to continue harvesting possums as a resource. The requirements of managing possums as a resource conflict clearly with the requirements of managing them as a pest. As mentioned in Chapter 1 it appears that during the 1950s the spread of possums was assisted by hunters keen to take advantage of the bounty scheme operating at the time. Today, possums are present in 97% of New Zealand so their introduction for harvesting purposes into the remaining 3% of the country is less likely than in the past, but our need to prevent this happening must be even greater because there are so few areas remaining without possums. An additional risk with the increase in the value of possums as a resource is the possibility that possums will be moved from one area to another to supplement a local population for harvesting at a later date. Releases such as this could well result in the transmission of possum-borne diseases such as Tb, giardia, and other parasites and diseases listed in Chapter 7, as well as further damage to local natural resources. The likely conflict between those that see possums as a pest and those that see possums as a resource will need to be considered as the possum harvesting industry restrengthens and the promise of possum biocontrol agents and Tb vaccines become a reality.

Biological control — Can it live up to expectations?

The recent experience with rabbits and rabbit haemorrhagic disease (RHD) has shown how difficult it can be to gain consensus and approval for the use of a biocontrol agent for mammals. At present it seems likely that immunocontraception will be the first form of biological control used to manage possums in New Zealand as the introduction of any disease for pest control will be the subject of extensive testing on non-target species, and public consultation (Chapter 24). A public survey of how best to control possums (see Chapter 17) clearly showed that fertility control is the preferred form of possum biological control. It appears that any immunocontraceptive developed in the next 5–10 years will be dispensed in a bait because of the difficulty in finding or developing a delivery agent such as a possum-specific sexually transmitted disease. Perhaps the greatest challenge faced by those wishing to use biological techniques for possum control will be the need to ensure target specificity and approval by the national and international community. The use of a biocontrol agent so close to Australia where there are 24 endemic species of possum is of obvious concern. Another public concern is likely to be the innovative and unprecedented nature of such a development. Other than the use of diseases such as myxomatosis and RHD with rabbits and an experimental trial using salmonella for rat control, there are few examples of vertebrate biological control that will assist the public to feel confident about the use of the techniques. This difficulty is recognised by those developing the technology and by the Parliamentary Commissioner for the Environment who are now working together to explore the public acceptability of using biological control for possums.

Adaptive management

While the theory of adaptive management has been around for more than 20 years, an appreciation of what it has to offer agencies responsible for possum control, such as the Animal Health Board and the Department of Conservation, is only relatively recent. The second part of Chapter 25 mentions the potential for using adaptive management in association with decision support systems for possum control and, although the topic is not covered in detail, its importance should not be underestimated. The adaptive management process is a research tool that requires outcome predictions to be made and the replication of treatments across several sites. It helps managers build a database that can be interrogated and modelled at some later date, and helps store and build a corporate memory, which could be lost during corporate and departmental restructures. It certainly is not the silver bullet for the possum problem but the technique does offer great potential to those who have the responsibility of managing possums in the future.

Conclusion

It has been almost 150 years since possums were introduced to New Zealand and we can now be sure we are closer to the beginning of the end, rather than the end of the beginning, as far as their history in New Zealand is concerned. As the new millennium begins the only barrier that could conceivably prevent possums being driven to extinction in New Zealand, within the next 50 years, is a lack of imagination and the political will and skill to achieve such an end. This snapshot of our current knowledge of the possum undoubtedly provides a solid foundation for those wishing to take up the challenge and see it through to its final conclusion: the eradication of possums from New Zealand.

Appendix 1
Possum-related Legislation

Key points relevant to possum control

Health and Safety in Employment Act 1992 and Health and Safety in Employment Regulations 1995

- The object is to prevent illness or injury to :
  - employees carrying out possum control work
  - other people as a result of possum control activities.
- The Act imposes duties on principals, employers, the self-employed, employees, and persons in charge of work places, to avoid harm.
- The Act is administered by the Occupational Safety and Health service of the Department of Labour.

Noxious Substance Regulations 1954

- The Noxious Substance Regulations apply to some poisons used in pest control: refer to its attached schedule of noxious substances.
- The Regulations are administered by the Ministry of Health, Medical Officers of Health, and public health services.

Pesticides Act 1979

- This legislation provides for the registration of pesticides and regulates their use.
- The Act is enforced by the Pesticides Board and MAF Regulatory Authority.

Pesticides (Vertebrate Pest Control) Regulations 1983

- These regulations govern the use of controlled pesticides for possum control.
- Controlled pesticides include 1080, cyanide, arsenic, and strychnine: see its attached schedule.
- The Regulations are administered by the Pesticides Board, MAF Regulatory Authority, and Medical Officers of Health.

Toxic Substances Act 1979 and Toxic Substances Regulations 1983

- The Toxic Substances Act includes controls on hazardous substances.
- This legislation is administered by the Ministry of Health, Medical Officers of Health, and Health Protection Officers.

Opossum Regulations 1953

- These regulations control the taking, liberating, and harbouring of possums.
- These regulations also prohibit liberation and the keeping of live possums.
- These regulations are administered by the Department of Conservation.

Local Government Act 1974

- The Local Government Act creates regional, local, and unitary councils and contains their constitution.
- Pest control work by regional councils must be carried out within this framework.
- Regional and unitary councils may be:
  - Planners : Regional councils can adopt and implement regional pest management strategies; they may be given planning functions by national pest management strategies.
  - Operators : Regional councils can implement regional pest management strategies; they may undertake pest control work to implement national strategies.
  - Regulators : Regional councils administer the Resource Management Act and regional plans; the Act and regional plans may control discharges to land and/or water.

Public Bodies Contracts Act 1959

- This statute sets out how regional councils must make pest control contracts.
- Regional councils :
  - must make contracts worth more than $1,000 either in writing or by regional council resolution;
  - may make contracts worth up to $1,000 orally.

Wild Animal Control Act 1977

- A possum is a wild animal.
- This Act regulates the hunting, capture, and liberation of possums.
- This Act is administered by the Department of Conservation.

Conservation Act 1987

- The Director-General of Conservation has power to control possums in the conservation estate.
- This Act sets the framework for the administration of the conservation estate by the Minister and Department of Conservation.
- Management must be undertaken in accordance with the Act, ministerial policies, conservation management strategies, and conservation management plans.
- Conservation management strategies and plans may affect pest control.
- Any conservation area, or part of it, may be closed, for public safety or in accordance with a conservation management strategy or plan.

Wildlife Act 1953

- Possums are not wildlife for the purposes of the Act.
- Under the Act, wildlife sanctuaries, wildlife refuges, and wildlife management reserves can be established.
- Management of wildlife sanctuaries, refuges, and management reserves is undertaken in accordance with general policies issued by the Minister, conservation management strategies, and conservation management plans.
- Possums in wildlife sanctuaries, wildlife refuges, and wildlife management reserves are not protected.
- Control and eradication of possums, however, must be undertaken in accordance with Ministerial policies, and any conservation management strategies or plans.
- The Act is administered by the Department of Conservation.
- The Department can undertake pest control work. Possum programmes may also be undertaken by other persons with the consent of the Department, and the permission of any occupier.

Health Act 1956

- The purpose of this Act is to protect the public health.
- It is administered by the Ministry of Health, Medical Officers of Health (appointed by the Director-General of Health), Health Protection Officers (appointed by the Director-General of Health), and territorial local authorities (and environmental health officers appointed by them).

Operators need to:

- Avoid creating health hazards.
- Keep work premises and work sites clean and free of health hazards, both for workers and for the public.
- Avoid contamination of water supplies.
- Remove carcasses from positions where they are offensive or hazardous to health.

Animal Protection Act 1960 (repealed)

- The purpose of the Act was to protect animals from cruelty.
- This protection, with the exception of trapping, did not extend to possum control.
- The Act was administered by the Ministry of Agriculture and Forestry.

Animal Welfare Act 1999

- This Act replaces the Animal Protection Act 1960.
- The Act emphasises duty of care for animals – meeting their physical, health, and behavioural needs – rather than focusing on cruelty.
- The Act enables the National Animal Welfare Advisory Committee to prohibit traps that have unacceptable impacts on animal welfare.
- The Act sets checking times of restraining traps (leg-hold, box/cage traps) to 12 hours after sunrise.

- Kill traps do not need to be checked daily.
- Pest control methods are exempt from the Act.
- The Act encourages the development of voluntary codes of practice for pest control.

National Parks Act 1980

- The object is to preserve national parks in their natural state.
- The eradication of introduced animals, such as possums, promotes this goal.
- Biological controls may be used in national parks as a pest control method, with Ministerial consent.
- The Act is administered by the Minister and Department of Conservation.

New Zealand Walkways Act 1990

- The purpose of this Act is to establish public walking tracks.
- Poisoning and shooting on or in the vicinity of walkways is prohibited, unless permitted by the occupier, or someone authorised by the occupier.
- The Act is administered by the Department of Conservation.
- The Minister of Conservation may appoint controlling authorities for walkways.
- The controlling authority may be a department of state, a local authority, or other statutory body.

Reserves Act 1977

- This Act sets the legal framework for management of public reserves.
- There are several kinds of reserve: recreation, historic, scenic, nature, scientific, government purpose, and local purpose.
- The Act is administered by the Department of Conservation, territorial local authorities, and other organisations appointed as administering bodies of reserves.
- Administration must be in accordance with any ministerial policy for reserves, conservation management strategies, and conservation management plans.
- Ministerial consent is required for possum control work in scenic, historic, nature, and scientific reserves.
- The consent of the relevant administering body is required for pest control work in recreation, government purpose, and local purpose reserves.
- In general, but not always, recreation and local purpose reserves are administered by territorial local authorities.
- Biological control organisms may be used if authorised by the Minister.

Arms Act 1993

- This Act regulates the possession and use of firearms.
- It is administered by the police.

Operators need to:

- Hold a current firearms licence, if using firearms.
- Keep firearms locked and immobilised when not in use.
- Not leave in an unattended vehicle.
- Notify a police arms office of any change in address.
- Report loss, theft, or destruction of firearms to the police.
- Report injuries caused by firearms to police.
- Supply name, address, and date of birth to police on request.
- Remain alcohol and drug free while in charge of a firearm.
- Repair firearms if required by police.

Civil Aviation Act 1990

- This Act sets up the framework for the management of civil aviation to promote safety.
- General obligations are set out in the Act.
- The detailed operational requirements are contained in the Civil Aviation Regulations 1953 (reprint SR 1980/88) and the Civil Aviation Rules.
- The Act, Regulations, and Rules apply to all aircraft operations, operators, and pilots.
- Detailed provisions for aerial pest control are found in the Rules.
- The Act is administered by the Civil Aviation Authority (CAA).
- Operators: an operator is a person who flies or uses an aircraft or causes or permits the aircraft to fly whether on board or not.

Fire Service Act 1975

- Under this Act the Fire Service has power to deal with hazardous substance incidents.

Land Transport Act 1998 and Land Transport (Driver Licensing) Rules 1999 Land Transport Rule : Dangerous Goods 1999

- This legislation regulates the transportation of explosives and hazardous substances.
- Explosives and toxins used in possum control work are governed by the Act and Rules.
- The degree of control depends on the nature and quantity and use of dangerous goods. The quantities are: explosives 50 kg, small arms cartridges 250 kg, toxic substances 50 kg (solids), or 50 litres (liquids).
- If these quantities are exceeded, drivers must have a 'dangerous goods' endorsement on their drivers' licences.
- There are regulatory standards set in the rules for packaging, labelling and marking, accompanying documentation, segregation, transport procedures, emergency procedures and training.
- The legislation is administered by the Land Transport Safety Authority and the Police.

Source: *User's guide to legislation.* National Possum Control Agencies unpublished document.

Hazardous Substances and New Organisms Act 1996

- The purpose of this Act is to protect the environment, and the health and safety of people and communities, by preventing or managing the adverse effects of hazardous substances and new organisms.
- The Act establishes the Environmental Risk Management Authority (ERMA) to assess and decide on applications to introduce hazardous substances or new organisms into New Zealand.
- Import or manufacture of hazardous goods, and import, development, field testing, or release of new organsisms, are prohibited otherwise than in accordance with approval issued under this Act.
- The hazardous substances provisions of the Act will come into effect during 2000. Existing legislation relating to Dangerous Goods, Toxic Substances, and Pesticides will remain in force until then.

Biosecurity Act 1993

- This Act covers the exclusion, eradication, and effective management of pests and unwanted organisms.
- It provides for the implementation of national and regional pest management strategies.

Resource Management Act 1991

- The single, overarching purpose of this Act is to promote the sustainable management of natural and physical resources.
- It imposes restrictions on land use and the discharge of contaminants into or onto water, land, and air.
- It requires that steps be taken to avoid, remedy, or mitigate any adverse effects of activities on the environment.

Source: Pam Pye, Legal Coordinator, Landcare Research, Lincoln.

Index

Symbols

G

H

I

K

L

R

S

T

V

W